Transmission Lines and Lumped Circuits

Academic Press Series in
ELECTROMAGNETISM

EDITED BY
Isaak Mayergoyz, University of Maryland, College Park, Maryland

BOOKS PUBLISHED IN THE SERIES
Georgio Bertotti, *Hysterisis in Magnetism: For Physicists, Material Scientists, and Engineers*
Scipione Bobbio, *Electrodynamics of Materials: Forces, Stresses, and Energies in Solids and Fluids*
Alain Bossavit, *Computational Electromagnetism: Variational Formulations, Complementarity, Edge Elements*
M.V.K. Chari and S.J. Salon, *Numerical Methods in Electromagnetics*
Göran Engdahl, *Handbook of Giant Magnetostrictive Materials*
Vadim Kuperman, *Magnetic Resonance Imaging: Physical Principles and Applications*
John C. Mallinson, *Magneto-Resistive Heads: Fundamentals and Applications*
Isaak Mayergoyz, *Nonlinear Diffusion of Electromagnetic Fields*
Shan X. Wang and Alexander M. Taratorin, *Magnetic Information Storage Technology*

RELATED BOOKS
John C. Mallinson, *The Foundations of Magnetic Recording, Second Edition*
Reinaldo Perez, *Handbook of Electromagnetic Compatibility*

ACADEMIC PRESS SERIES IN ELECTROMAGNETISM

Electromagnetism is a classical area of physics and engineering that still plays a very important role in the development of new technology. Electromagnetism often serves as a link between electrical engineers, material scientists, and applied physicists. This series presents volumes on those aspects of applied and theoretical electromagnetism that are becoming increasingly important in modern and rapidly developing technology. Its objective is to meet the needs of researchers, students, and practicing engineering.

This is a volume in

ELECTROMAGNETISM

ISAAK MAYERGOYZ, SERIES EDITOR
UNIVERSITY OF MARYLAND
COLLEGE PARK, MARYLAND

Transmission Lines and Lumped Circuits

GIOVANNI MIANO

Dipartimento di Ingegneria Elettrica
Università degli Studi di Napoli Federico II

ANTONIO MAFFUCCI

Dipartimento di Ingegneria Elettrica
Università degli Studi di Napoli Federico II

ACADEMIC PRESS

A Harcourt Science and Technology Company

San Diego San Francisco New York Boston
London Sydney Tokyo

This book is printed on acid-free paper. ♾

Cover concept courtesy of Franco Lancio

Academic Press
A Harcourt Science and Technology Company
525 B Street, Suite 1900, San Diego, California 92101-4495, USA
http://www.academicpress.com

Academic Press
Harcourt Place, 32 Jamestown Road, London NW1 7BY, UK

Library of Congress Catalog Card Number: 00-108487

International Standard Book Number: 0-12-189710-9

PRINTED IN THE UNITED STATES OF AMERICA
00 01 02 03 04 05 MB 9 8 7 6 5 4 3 2 1

To Professor Scipione Bobbio

Contents

Foreword

This volume in the Academic Press Electromagnetism series presents an in-depth, modern, comprehensive and self-contained treatment of transmission line theory.

Analysis of transmission lines is a classical topic in electromagnetism. Usually, transmission lines are studied for simple loads that can be modeled as impedances. The comprehensive treatment of transmission lines with complex time-varying and nonlinear loads does not exist in monographic literature. Nevertheless, this treatment is becoming increasingly important in the area of high-density interconnects of VLSI circuits where unintentional delays, cross talk voltages, and signal losses are commonplace and of serious concern to circuit designers. This book, written by Professor Giovanni Miano and Dr. Antonio Maffucci, fills this critical void in technical literature.

This book is concerned with the time-domain analysis of linear and time-invariant transmission lines connected to nonlinear and time-varying lumped circuits. The basic idea that runs through the entire book is time-domain multiport representations of transmission lines. This approach has many advantages that are clearly and convincingly described in the book. Another unique feature of the book is its broad scope. It covers with equal depth the topics related to multiconductor, ideal and lossy, spatially homogeneous and inhomogeneous transmission lines as well as phenomena of bifurcation and chaos in transmission lines with nonlinear resitive loads. To make the book self-contained and easily readable, the authors present in detail the various mathematical facts, concepts and techniques. The book is written with a strong emphasis on conceptual depth and mathematical rigor which are trademarks of Neapolitan school of electrical engineering.

I believe that this book will be a valuable reference for students, researchers, practicing electrical and electronic engineers, applied mathematicians and physicists. This book will enrich its readers with many ideas and insights, important facts, and concepts.

Isaak Mayergoyz
Series Editor

Preface

This book deals with the analysis of networks composed of transmission lines and lumped circuits. It is intended for senior and graduate students in electrical and electronic engineering. It will also be a useful reference for industrial professionals and researchers concerned with computer-aided circuit analysis and design. As far as the transmission line model is concerned, it is assumed that the reader is acquainted with the subject at the level of undergraduate courses in electrical engineering (e.g., Collins, 1992; Paul, 1994; Franceschetti, 1997).

Introductory courses on transmission line theory deal with the basic concepts of traveling and standing waves, and analyze, in the frequency domain, transmission lines connected to independent sources and impedances. They introduce several basic notions, such as those of traveling wave, standing wave, characteristic impedance, power flux, reflection coefficients, voltage standing-wave ratio, and impedance transformation, tackle the problem of impedance matching, and teach how to use the Smith chart. However, frequency domain analysis techniques are not useful for high-speed electronic circuits and distribution systems of electrical energy, which consist of many transmission lines connected to many nonlinear and time-varying lumped circuits.

This book concerns the time domain analysis of electrical networks composed of linear time invariant transmission lines and lumped circuits that, in general, can be nonlinear and/or time varying. The theory of wave propagation in linear time invariant transmission lines, two-conductor or multiconductor, without losses or with losses, with parameters dependent or not upon frequency, uniform or nonuniform, is presented in a way that is new, completely general, and yet concise. The terminal behavior of these lines is characterized in the time domain through convolution relations with delays. A characterization dealing exclusively with the voltages and currents at the line ends is a prerequisite to tackling the study of networks composed of transmission lines and lumped circuits by way of all those techniques of analysis and computation that are typical of lumped circuit theory. The most widely used circuit simulator, SPICE, simulates transmission lines by using this approach.

A unique feature of this book is the extension of some concepts of lumped circuit theory, such as those of associated discrete circuit (Chua and Lin, 1975) and associated resistive circuit (Hasler and Neirynck, 1986) to networks composed of lumped and distributed elements. The qualitative study of equations relevant to networks composed of transmission lines and lumped circuits is carried out. In particular, it is shown that transmission lines connecting nonlinear locally active resistors may exhibit fascinating nonlinear phenomena such as bifurcations and chaos.

This book has profited by the many suggestions and comments made by our colleagues over the years. In particular, we appreciate the support for the development of this text from the Department of Electrical Engineering of the University of Naples, Federico II.

The authors are indebted to Isaak D. Mayergoyz for his encouragement throughout this project. In particular, Giovanni Miano expresses his sincere gratitude to him for his hospitality during the initial stage of writing. The writing of this book has benefited enormously from the scientific atmosphere at the Department of Electrical Engineering of the University of Maryland.

We express our special thanks to Gregory T. Franklin, Marsha Filion, and Angela Dooley of Academic Press for their refined courtesy and continued assistance and to Franco Lancio who designed the cover of this book.

During the last months this book has taken up time usually devoted to our wives, Gabriella and Michela. The book was written thanks to their patience and understanding.

Introduction

Due to the rapid increase in signal speed and advances in electronic circuit technology, *interconnections* between electronic devices may behave as *transmission lines*. Unintentional delays, crosstalk voltages, reflections, signal losses, and voltage overshoots on terminal devices affect the correct operation of high-speed electronic circuits (e.g., Paul, 1992). "To meet the challenges of high-speed digital processing, today's multilayer printed circuit boards must: *a*) reduce propagation delay between devices; *b*) manage transmission line reflections and crosstalk (signal integrity); *c*) reduce signal losses; *d*) allow for high density interconnections" (Montrose, 1998). Crosstalk, delays and multireflections are also important phenomena in on-chip interconnections for high-performance microprocessors (e.g., Deutsch *et al.*, 1995, 1997; Motorola Inc., 1989). Therefore, for design and verification of these circuits, accurate and efficient simulation techniques are needed. As most electronic devices are time-varying and nonlinear, the analysis of these systems and hence of the transmission lines themselves, must be performed in the time domain.

The analysis of transmission lines in the time domain is also important for power systems to predict the transient behavior of long power lines either excited by external electromagnetic fields, for example, emission of high-power radars, nuclear electromagnetic pulses, lightning strokes, or under disruptions such as short circuits at some places (e.g., Paul, 1992).

This book concerns the time domain analysis of electrical networks composed of transmission lines and lumped circuits. The time domain analysis of lumped circuits is a fundamental and well-understood subject in electrical engineering. Nevertheless, the time domain analysis of networks composed of transmission lines and lumped circuits, and of the transmission lines themselves, are not so well understood.

The primary purposes of this book are threefold.

First, the theory of wave propagation in transmission lines, whether two-conductor or multiconductor, lossless or with losses, with parameters depending or not on the frequency, uniform or nonuniform, is presented in a way that is new, completely general, and yet concise.

The second objective is to give an original and general method to characterize the terminal behavior of transmission lines in both the frequency and time domains. A characterization of the lines dealing exclusively with the voltages and currents at their ends is a prerequisite to tackling the study of networks composed of transmission lines and lumped circuits through all those techniques of analysis typical of lumped circuit theory.

The last objective, but not the least, is the qualitative study of the equations relevant to networks composed of transmission lines and lumped circuits. A unique feature of this book is the extension of some of the concepts of lumped circuit theory, such as those of *associated discrete circuit* (Chua and Lin, 1975) and *associated resistive circuit* (Hasler and Neirynck, 1986), to networks composed of lumped and distributed elements.

Whether the transmission line model accurately describes actual interconnections or not is a basic question that is beyond the scope of this text. The reader is assumed to be acquainted with it. Here we only recall that, under the assumption of a *quasi-transverse electromagnetic* (quasi-TEM) *mode of propagation*, interconnections may be modeled as transmission lines (e.g., Lindell and Gu, 1987; Collin, 1992; Paul, 1994). Whether this assumption is satisfied or not for actual interconnections depends on the frequency spectrum of the signals propagating along them, on their cross-sectional dimensions, and on the electromagnetic properties of the conductors and the medium embedding them. In most cases of interest a simple criterion is the following. The distance between the conductors must be much smaller than the lowest characteristic wavelength of the signals propagating along them.

The first chapter of this book is devoted to the transmission line model. First, we shall recall some of the basic aspects of the quasi-TEM approximation and the equations of the transmission lines. Then, we shall examine closely some general properties of the transmission line equations that are of considerable importance and that we shall be using widely in this book.

A SURVEY OF THE PROBLEM

Let us consider a generic electrical network consisting of transmission lines and lumped circuits. The behavior of the overall network is the result of the reciprocal effects of two requirements. The first is that each part of the network should behave in compatibility with its specific nature, and the second is that such behavior should be in turn compatible with all the other parts of the network.

The behavior of the transmission lines is described by *transmission line equations*. The *characteristic equations* of the single lumped circuit elements and *Kirchhoff's laws* regulate the behavior of the lumped circuits. The interactions between these and the transmission lines, and between the transmission lines themselves, are described by the continuity conditions for the voltages and currents at the "boundaries" between the transmission lines and the lumped circuit elements, and between the transmission lines themselves.

Those transmission lines of practical interest have losses, parameters depending on the frequency, and may be spatially nonuniform. In many cases the physical parameters of the line are uncertain and a description of statistical type is required.

Lumped circuits in general are very complex. They consist of dynamic elements (e.g., inductors, capacitors, and transformers), resistive elements that may be nonlinear and time varying (e.g., diodes, transistors, operational amplifiers, logic gates, and inverters), and integrated circuits.

Although the single parts of these networks are in themselves complex, the main difficulty lies in the fact that it is necessary to resolve coupled problems of a profoundly different nature. The transmission line equations are linear and time-invariant partial differential equations of the hyperbolic type, while the lumped circuit equations are algebraic-ordinary differential equations, which in general are time varying and nonlinear.

The equations of these networks have to be solved with given initial conditions for the distributions of voltages and currents along the transmission lines, the charges of the capacitors, and the fluxes of the inductors.

The solution of the transmission line equations requires, beside the initial conditions, knowledge of the voltages or currents at the line ends. Therefore, for each line we have to solve an initial value problem with boundary conditions where, however, the values of the voltages and currents at the line ends are themselves unknowns.

When the lumped circuits are linear and time invariant, the problem is on the whole linear and time invariant, and its solution does not present particular difficulties. For example, all the network equations can be solved simultaneously by using the Fourier transform.

Instead, when the lumped circuits are time varying and/or nonlinear, solving the overall network in the frequency domain serves no purpose. Therefore, the problem has to be studied directly in the time domain and the difficulties may be considerable. This makes the choice of the solution method critical.

The most obvious way to solve a problem of this kind is the following. First, we determine analytically the general solution of the line equations in the time domain, which involves arbitrary functions. Then we impose the initial conditions, the continuity conditions for the voltages and currents at the line ends, and the lumped circuit equations in order to determine these arbitrary functions and, hence, the voltage and current distributions along the lines, together with the voltages and currents of the lumped parts of the network.

Unfortunately, this procedure is not generally applicable because only for uniform transmission lines without losses and with parameters independent of frequency is it possible to determine analytically the general solution of the line equation in the time domain.

When the line parameters are independent of frequency, the transmission line equations may be solved numerically by making a finite difference approximation to the partial derivatives (e.g., Djordjevic *et al.*, 1987; Paul, 1994). Approximated methods based on finite elements also may be used to solve the transmission line equations (e.g., Silvester and Ferrari, 1990). The time domain equations for a line with parameters that depend on the frequency are convolution-partial differential equations and thus in these cases, we have to approximate partial derivatives as well as convolution integrals numerically (e.g., Paul, 1994).

The numerical procedures based on finite difference and finite element approximations of the transmission line equations can be easily interfaced to the time-stepping procedures for solving the lumped circuits to which the lines are connected. However, they require a lot of computer memory and computer time to execute because the current and voltage distributions along the lines are involved.

In the spatial discretization of any line the number of cells should be such that the transit time of the signal considered over the generic cell is much shorter than its smallest rise (or fall) characteristic time in order to reduce parasitic oscillations and noncausal effects. In consequence, too many cells greatly increase running time.

In implementation of time discretization, the time step should be much smaller than the smallest transit time along the cells to guarantee numerical stability and control parasitic oscillations. In consequence, the computer running time becomes prohibitively large if accurate responses are expected for long transmission lines, namely, transmission lines whose transit times are much greater than the smallest rise (or fall) characteristic times of the signals transmitted along the lines.

The discretization with respect to the space coordinate is equivalent to the approximation of the transmission lines obtained by cascades of lumped element cells if finite backward, forward or central differences are used to approximate the spatial derivatives.

THE IDEA OF THE BOOK

There is another way to resolve the problem that is at the same time general, simple, effective, and elegant. It is based on the assumption that the interactions between the transmission lines and the lumped circuits, and between the transmission lines themselves, depend only on the terminal behavior of the lines. In consequence, each transmission line can be represented as a multiport in the time domain, irrespective of what it is connected to: two-conductor lines as two-ports and $(n+1)$-conductor lines as $2n$-ports. As the lines we are considering are linear and time invariant, the time domain characteristic relations of the multiports representing them can be expressed through convolution equations.

In 1967, Branin proposed an equivalent two-port, based on linear resistors and controlled voltage sources, to describe the terminal behavior of ideal two-conductor lines (Branin, 1967). The governing laws of the controlled sources are linear algebraic difference relations with one delay. In 1973, Marx extended that representation to ideal multiconductor lines. In 1986, Djordjevic, *et al.*, characterized the terminal behaviour of a lossy multiconductor transmission line in the time domain by the convolution technique (Djordjevic, Sarkar, and Harrington, 1986).

Before looking at why it is possible to represent a transmission line as a multiport and how it is done, let us see why it is convenient to do so.

What makes this approach particularly interesting is the fact that the analysis of the parts that are always linear and time-invariant, (the transmission lines), is separated from the analysis of the parts that are generally nonlinear and time-varying, (the lumped circuits). There are additional but no less important advantages. First of all, we separate the analysis of the parts that are described by partially differential equations of hyperbolic type, from the analysis of the parts that are described by algebraic-ordinary differential equations. In this way the problem is reduced to the study of a network in which the transmission lines are modeled in the same way as the lumped elements: multiports representing the transmission lines differ from multiports representing the lumped elements only in their characteristic relations.

Once the characteristic relations of the multiports representing the transmission lines have been determined, the problem is reduced to the analysis of a network whose equations consist of the characteristic equations of the single elements, lumped or distributed, and of the Kirchhoff's equations describing the interaction between them.

Interest in a reduced model of this type, where the unknowns are only the voltages and currents at the ends of the lines and those of the lumped circuits, also lies in the fact that, in very many applications, we are interested only in the dynamics of these variables. However, when the voltages and currents at the ends of the lines have been evaluated, we need only solve the transmission line equations with given boundary and initial conditions to determine the current and voltage distribution along the whole lines.

Let us now see why this is possible. For the sake of simplicity we refer to a network consisting of a two-conductor line connecting two lumped circuits, as sketched in Fig. 1, the left end of the line being connected to a generic lumped circuit C_1 and the right end to a generic lumped circuit C_2.

The foundations of the transmission line model are the following (see next chapter):

(*i*) the electric current $i(x, t)$ and the electric voltage $v(x, t)$ at any position along the line and at any time are well defined;

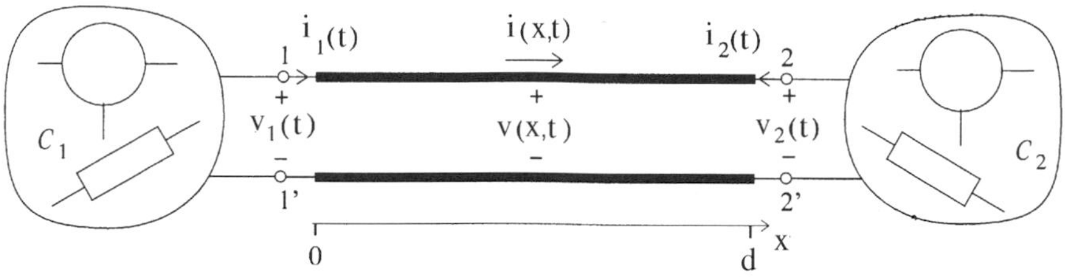

Figure 1. Lumped circuits connected by a two-conductor transmission line.

(*ii*) the transmission line interacts with the rest of the network only through the terminals.

In consequence of the first property, each end of the transmission line can be represented as a *port* of a single *two-port* — each port is characterized by one current and one voltage. Let $v_1(t)$ indicate the voltage among the terminals 1-1′ of the line and $i_1(t)$ the current flowing through the terminal 1,

$$i_1(t) = i(x = 0,\ t), \tag{1}$$

$$v_1(t) = v(x = 0,\ t), \tag{2}$$

and $v_2(t)$ indicate the voltage among the terminals 2-2′ and $i_2(t)$ the current flowing through the terminal 2,

$$i_2(t) = -i(x = d,\ t), \tag{3}$$

$$v_2(t) = v(x = d,\ t). \tag{4}$$

The reference directions are chosen according to the normal convention for the two-ports corresponding to the two ends of the line.

In consequence of the foregoing properties, the line itself imposes a well-defined relation between the terminal voltages $v_1(t)$, $v_2(t)$, and the terminal currents $i_1(t)$, $i_2(t)$. In fact, as we shall show in the next chapter, the solution of transmission line equations is unique when, besides the initial conditions, the voltage or the current at each line end is imposed. Instead, no solution would be found if one tried to impose more than two variables, for example, the voltages at both ends as well as the current at one end. Obviously these relations depend only on the transmission line equations and not on the elements to which the line is connected. In other words, we are saying

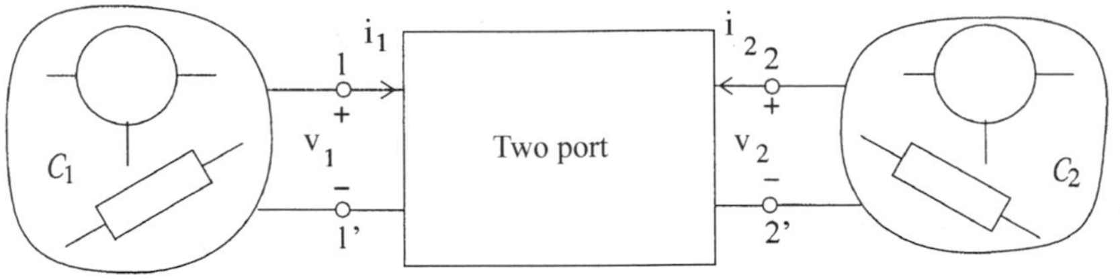

Figure 2. Equivalent representation of the circuit shown in Fig. 1.

that the transmission line at the terminals behaves as a two-port, and hence the interaction of the transmission line with the rest of the circuit can be described through the equivalent circuit represented in Fig. 2.

Thus, it is clear that in general the dynamics of the electric variables at the terminals of a $(n + 1)$-conductor transmission line can be described by representing the line as a $2n$-port, irrespective of the type of the line itself (with losses, parameters depending on the frequency and parameters nonuniform in space).

Let us now see how to determine the characteristic relations of the multiport representing the line.

The representation of a line as a multiport in the time domain is a basic step that can present difficulties. The degree of difficulty obviously depends on the nature of the line. However, when dealing with linear and time-invariant transmission lines, it is always possible to use those techniques of analysis and representation appropriate to linear and time-invariant systems, namely the Laplace (or Fourier) transform and the convolution theorem. As we shall see, this is a key point in our approach. By using the Laplace transform, it will easily be possible to characterize the terminal behavior of a generic multiconductor line, with losses, with parameters depending on the frequency, and with spatially nonuniform parameters. Once a characterization is obtained in the Laplace domain, the corresponding time domain characterization is achieved by applying the convolution theorem.

For the sake of simplicity, we refer here to a uniform two-conductor transmission line. Let us consider the characteristic relation obtained by choosing the currents i_1 and i_2 as *control* variables, and the voltages v_1 and v_2 as *dependent* variables. This is the so-called *current-controlled* representation. Representations in which the control variables are the voltages v_1 and v_2 (*voltage-controlled* representations), or the current at one end and the voltage at the other end (*hybrid* representations) are also possible. As all of these representations are completely equivalent, in this text we shall deal with the problem of the current-controlled representation in detail. For other representations, which can be easily obtained from this, we shall give only the expressions.

Transmission lines are systems with an *internal state* and thus two different types of descriptions of their terminal behavior are possible: *external descriptions*, or *input-output descriptions*, and *internal descriptions*, or *input-state-output descriptions*.

In the input-output descriptions only the terminal variables are involved. For instance, in the current-controlled input-output description, the terminal behavior of the line is described through the convolution relations (Djordjevic, Sarkar, and Harrington, 1986, 1987),

$$v_1(t) = (z_d * i_1)(t) + (z_t * i_2)(t) + e_1(t), \tag{5}$$

$$v_2(t) = (z_t * i_1)(t) + (z_d * i_2)(t) + e_2(t); \tag{6}$$

and in this book $(z * i)(t)$ indicates the convolution product. The impulse response $z_d(t)$ is the inverse Laplace transform of the driving-point impedance at port $x = 0$ (or at $x = d$) when the port at $x = d$ (respectively, at $x = 0$) is kept open circuited, and $z_t(t)$ is the inverse Laplace transform of the *transfer impedance* when the port at $x = 0$ (or at $x = d$) is kept open circuited. The independent voltage sources $e_1(t)$ and $e_2(t)$ take into account the effects of the initial conditions and distributed independent sources along the line. Analog relations may be written for the voltage-controlled and hybrid input-output representations.

Representations in which the control variables are the currents or the voltages at the same end (*transmission* representations) are also possible. However, these representations in the time domain are useless, as we shall see in detail in Chapter 4, §4.7.4.

Unlike the input-output descriptions, several *input-state-output* descriptions are possible, depending on the choice of the state variables. Due to the linearity, the propagation along the transmission line can always be represented through the superposition of a forward voltage wave propagating toward the right and a backward voltage wave propagating toward the left. For instance, doubling the amplitude of the backward voltage wave at the left line end, $w_1(t)$, and doubling the amplitude of the forward voltage wave at the right line end, $w_2(t)$, allows the state of the line to be completely specified.

As we shall show in this book, the input-state-output description obtained by choosing $w_1(t)$ and $w_2(t)$ as state variables,

$$v_1(t) = (z_c * i_1)(t) + w_1(t), \tag{7}$$

$$v_2(t) = (z_c * i_2)(t) + w_2(t), \tag{8}$$

has a comprehensible physical meaning and leads to a mathematical model that is at the same time elegant and extremely simple. It can be effectively solved by recursive algorithms. The impulse response $z_c(t)$ is the inverse Laplace transform of the characteristic impedance of the line. It would describe the behavior of the transmission line at one end if the line at the other end were *perfectly matched* (it is equal to the current-based impulse response of a semiinfinite transmission line too). Instead, $w_1(t)$ and $w_2(t)$ take into account the effects due to the initial conditions, the distributed independent sources along the line, and the actual circuits connected to the line ends. The equations governing the state dynamics — the so-called *state equations* — are

$$w_1(t) = [p * (2v_2 - w_2)](t), \tag{9}$$

$$w_2(t) = [p * (2v_1 - w_1)](t). \tag{10}$$

The impulse response $p(t)$ is the inverse Laplace transform of the *global propagation operator* $P(s)$ that links the amplitude of the forward voltage wave at the end $x = d$ to the amplitude that the same wave assumes at the end $x = 0$, where it is excited. For uniform lines, $P(s)$ also links the amplitude of the backward voltage wave at the end $x = 0$ to the amplitude that the same wave assumes at the end $x = d$, where it is excited. The operator $P(s)$ contains the factor e^{-sT} describing the delay introduced by the finite propagation velocity of quasi-TEM mode propagating along the guiding structure represented by the transmission line. Therefore, $w_1(t)$ depends only on the backward voltage waves excited at the right line end in the time interval $(0, t - T)$, and $w_2(t)$ depends only on the forward voltage waves excited at the left line end in the time interval $(0, t - T)$. A description of this type was independently proposed by Lin and Ku (1992) and by Gordon *et al.* (1992) for lossy multiconductor transmission lines with frequency-dependent parameters. Recently, the authors also have extended this type of representation to nonuniform transmission lines (Maffucci and Miano, 1999d; 2000b).

As we shall show in this book, the uniqueness of the input-state-output description is that the impulse responses $z_c(t)$ and $p(t)$ have a duration much shorter than that of the impulse responses $z_d(t)$ and $z_t(t)$ characterizing the input-output description. Therefore, the transmission line characterization based on the input-state-output description has two main advantages compared with that based on the input-output description:

(*i*) the amount of frequency domain data necessary to obtain the impulse responses $z_c(t)$ and $p(t)$ is modest; and
(*ii*) the computational cost of the convolution relations based on the input-state-output description is low.

It is clear that the input-state-output description is more suitable for time domain transient analysis, whereas the input-output description is preferable when dealing with sinusoidal, quasi-sinusoidal and periodic steady-state operating conditions in the frequency domain.

The two-port representing a uniform transmission line always can be modeled by a Thévenin equivalent circuit of the type shown in Fig. 3, whatever the nature of the transmission line is. The behavior of each end of the transmission line is described through a linear time-invariant one-port connected in series with a linear-controlled voltage source — $z_c(t)$ is the current-based impulse response of the linear one-port, and the state equations are the governing laws of the controlled voltage sources.

Due to the delay T introduced by the line, the value of w_1 at time t depends solely on the values that the voltages v_2 and the state variable w_2 assume in the interval $(0, t-T)$. This fact, which is a direct consequence of the propagation phenomenon, allows the controlled voltage source w_1 to be treated as an independent source if the problem is solved by means of an iterative procedure. The same considerations hold for the other controlled voltage source.

To determine the impulse responses characterizing the transmission line as two-port, we have to solve the line equations with the proper initial conditions by considering the currents i_1 and i_2 at the line ends as if they were known. The degree of difficulty in solving a problem of this kind obviously will depend on the nature of the line.

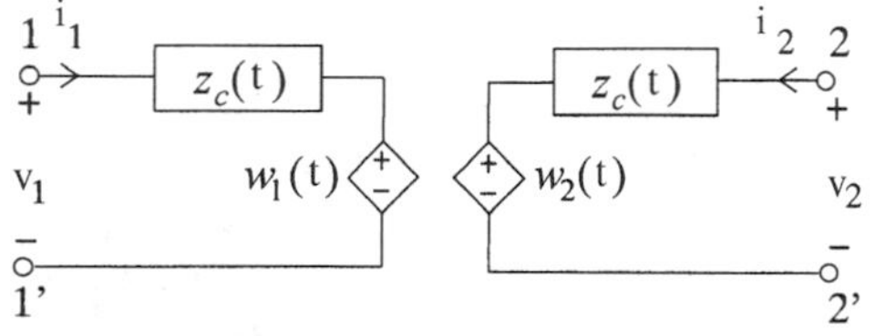

Figure 3. Equivalent circuit of Thévenin type of a uniform transmission line.

In Chapter 2 we shall find the characteristic relations of ideal two-conductor lines directly in the time domain by using the d'Alembert form to represent the general solution of the line equations. We shall extend these results to ideal multiconductor transmission lines in Chapter 3, by introducing the concept of propagation mode.

When there are losses and the parameters depend on the frequency, the general solution of the line equations cannot be expressed in closed form in the time domain. By using the Laplace transform, uniform lossy two-conductor lines and uniform two-conductor lines with frequency dependent parameters will be characterized as two-ports in Chapters 4 and 5, respectively. Chapter 6 deals with uniform multiconductor lines with losses and parameters depending on the frequency.

In Chapter 7 we shall show a simple idea that may be useful for extending these characterizations to the more general case of nonuniform transmission lines.

In Chapter 8 we outline a notable property of transmission line equations. Since the equations of transmission lines with frequency-independent parameters are partial differential equations of the *hyperbolic type*, they can be written in a particularly simple and expressive form, which is called *characteristic form*, where only *total derivatives* appear.

Transmission line equations in characteristic form are interesting for at least two reasons. First, as we shall see in Chapter 8, there is the fact that it is possible to solve lossy and nonuniform line equations in characteristic form by using simple iterative methods. Another aspect, of similar importance, is that from transmission line equations in characteristic form it is possible to understand how irregularities present in the initial and/or boundary conditions are propagated along the line. We shall also be verifying the possibility of extending the characteristic form to transmission lines with frequency-dependent parameters.

Chapter 9 will deal with the equations of networks composed of lumped nonlinear circuits interconnected by transmission lines, and obtained by representing the lines through equivalent circuits of the type shown in Fig. 3. In particular, we deal first with the conditions that ensure the existence and uniqueness of the solution of such equations. Once the existence and uniqueness have been established, the problem of the numerical evaluation of the solution is tackled. In this chapter the concepts of *associated resistive circuit* and *associated discrete circuit* also will be introduced.

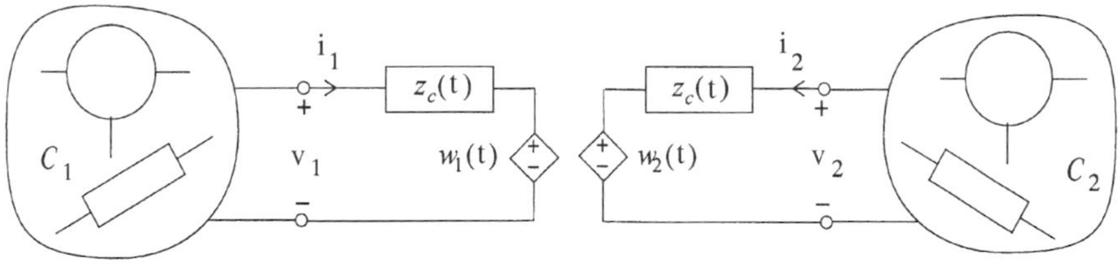

Figure 4. Equivalent circuit of the network shown in Fig. 1.

Interest in a reduced model of the type shown in Fig. 3 lies not only in the fact that in very many applications we are interested only in the dynamics of the terminal variables of the lines, but above all in the greater simplicity of the model itself. Consider, for example, the circuit in Fig. 4. In the generic time interval $[iT, (i+1)T]$ with $i \geqslant 1$, the behavior of w_1 depends only on the time history of v_2 and w_2 in the time interval $(0, iT)$, while the behavior of w_2 depends only on the time history of v_1 and w_1 in the same time interval. Consequently, if the solution for $t \leqslant iT$ is known, both w_1 and w_2 are known for $iT \leqslant t \leqslant (i+1)T$. For $0 \leqslant t \leqslant T$, w_1 and w_2 depend solely on the initial conditions and distributed independent sources. This property allows us to analyze networks consisting of transmission lines and lumped circuits with common circuit simulators. The most widely used circuit simulator SPICE (Simulation Program with Integrated Circuit Emphesys) uses equivalent representations of the type shown in Fig. 3 to simulate ideal transmission lines. In Chapter 9 we shall consider these matters in detail.

Chapter 10 deals with the qualitative study of the asymptotic behavior and stability of the solutions of networks composed of transmission lines and nonlinear resistors. In particular, we shall show that, in a transmission line connecting locally active resistors, fascinating nonlinear phenomena such as *bifurcations* and *chaos* may arise.

CHAPTER 1

Transmission Line Equations and Properties

In the first part of this chapter, we shall remember some basic concepts of the transmission line model, and then we shall recall the equations governing the different types of transmission lines.

The reader is assumed to be acquainted with the basic elements of the transmission line model. Thus, it is obvious that the short summary presented here is intended merely to familiarize the reader with those concepts and equations that are frequently used throughout the book. The reader is referred to the many excellent books and reviews existing in the literature for a complete and comprehensive treatment of the subject (e.g., Heaviside, 1893; Marcuvitz and Schwinger, 1951; Schelkunoff, 1955; Djordjevic, Sarkar, and Harrington, 1987; Collin, 1992; Paul, 1994; Franceschetti, 1997).

The rest of this chapter is concerned with some general properties of the transmission line equations mutuated from the theory of Maxwell equations (e.g., Stratton, 1941; Franceschetti, 1997), which are of considerable importance and widely used in this book. There is a profound link between the transmission line equations and Maxwell equations. That such a link exists is shown also by the fact that the structure of the line equations strongly resembles that of Maxwell equations. Consequently, it is not by chance, for example, that an energy theorem analogous to the Poynting theorem is valid for line equations.

We shall first enunciate a Poynting theorem for transmission lines with frequency-independent parameters, extend the theorem to the frequency domain, and then deal with the more general case of transmission lines with frequency-dependent parameters. This theorem is important not only for its physical significance but also because it makes it possible to tackle other important questions such as the problem of the uniqueness of the solution and the reciprocity properties.

By using the Poynting theorem it will be possible to establish conditions that, once satisfied, ensure that the solution of the transmission line equations is determined and unique. The problem of the uniqueness of the solution is fundamental to the characterization of two-conductor lines as two-ports and, more widely, of multiconductor lines as multiports.

Some properties of two-ports and, more generally, of multiports that represent the terminal behaviors of transmission lines are a direct consequence of another fundamental property of line equations, common to many linear systems: the *property of reciprocity*. By using Poynting theorem in the Laplace domain we shall demonstrate various forms of the property of reciprocity for transmission lines.

1.1. TRANSMISSION LINE MODEL

The electrical transmission of signals and energy by electrical conductors is one of the most important contributions of engineering technology to modern civilization.

Among its visible manifestations, the most impressive are the high-voltage transmission lines on tall steel towers that cross our countries in all directions. Equally obvious and important, if neither impressive nor attractive, are the flat pack or ribbon cables used to connect electronic systems, the coupled microstrips used to build microwave filters, metal strips on printed circuit boards to carry digital data or control signals, and integrated circuit interconnections (e.g., Tripathi and Sturdivant, 1997).

It is evident that we have to deal with complex systems where the electronic devices behave as *lumped circuital elements*, the interconnections behave as *multiconnected guiding structures*, and the interactions between the interconnections and the electronic devices do not take place only through terminals.

To ensure the correct operation of high-speed electronic circuits, in printed circuit boards we must reduce propagation delay between devices, manage transmission line reflections and crosstalk, and reduce signal losses. For these purposes, the models of the interconnections must be able to describe the delays due to the finite propagation velocity of the electromagnetic field, the multireflections due to the terminations, the attenuation and the distortion of the signals due to losses and dispersion in time, and the crosstalk between adjacent conductors (e.g., Dai, 1992; Paul, 1992; Montrose, 1998). Crosstalk, delays, and multireflections are also important phenomena in on-chip interconnections for high-performance microprocessors (e.g., Deutsch *et al.*, 1990, 1995, 1997).

To predict the transient behavior of long power lines either excited by external electromagnetic fields, e.g., emission of high-power radars, nuclear electromagnetic pulses, lightning strokes, or under disruptions, such as a short circuit at some place, it is crucial to take into account the delay introduced by the finite propagation velocity of the electromagnetic field and the coupling between close conductors (e.g., Paul, 1992).

To take into account these phenomena, is a complete description of the dynamics of the electromagnetic field generated around the guiding structures needed?

If the cross-sectional dimensions of the guiding structure, for example, the conductor separations, are much smaller than the lowest characteristic wavelength of the signals propagating along them, these phenomena may be described accurately, and hence foreseen, through the *transmission line model* (e.g., Lindell and Gu, 1987; Collin, 1992; Paul, 1994). A *distributed circuit model* of the interconnections, even if approximated, matches the circuit character of the overall electrical system we are considering.

The transmission line model is based on the assumptions that:

(i) the configuration of the electromagnetic field surrounding the guiding structures, whether they are made by two conductors or by many conductors, is of the *quasi*-TEM *type* with respect to the guiding structure axis; and
(ii) the total current flowing through each transverse section is equal to zero[1].

[1]In actual guiding structures the total current flowing through any cross section is not equal to zero (e.g., Paul, 1994). Let us refer, for the sake of simplicity, to the two-conductor case. The so-called *differential mode* currents are equal in magnitude at the same cross section and are oppositely directed. These correspond to the currents that are predicted by the transmission line model. The other currents are the so-called *common mode* currents, which are equal in magnitude at the same cross section but have the same directions. The common mode currents may be significant when the two wires behave as an antenna and then radiate in the transverse direction. The transmission line model cannot describe the common mode current.

A *transverse electromagnetic* (TEM) *field configuration* is one in which the electric and magnetic fields in the space surrounding the conductors are transverse to the conductor axis, which we shall choose to be the **x** axis of a rectangular coordinate system.

The basic properties of TEM fields are:

(i) The line integral of the electric field **E** along a generic closed line belonging to the transverse plane $x = \text{const}$ is always equal to zero; and
(ii) the line integral of the magnetic field **H** along a generic closed line γ belonging to the transverse plane $x = \text{const}$ is equal to the electric current linked with γ.

The TEM fields are the fundamental modes of propagation of *ideal* multiconnected guiding structures (e.g., Collin, 1992; Franceschetti, 1997), that is, guiding structures with uniform transverse section along the conductor axis, made by perfect conductors, and embedded in a perfect dielectric.

In actual multiconnected guiding structures the electromagnetic field is never exactly of the TEM type.

In ideal shielded guiding structures, high-order non-TEM modes with discrete spectra can propagate as well as the TEM fundamental modes. In unshielded guiding structures there are also non-TEM propagating modes with continuous spectra.

Actual guiding structures are most frequently embedded in a transversally nonhomogeneous medium, and thus TEM modes cannot exist in them. However, even if the medium were homogeneous, due to the losses, the guiding structure could not support purely TEM modes. Furthermore, the field structure is complicated by the influence of the nonuniformities present along the axis of the guiding structures (bends, crossover, etc.).

However, when the cross-sectional dimensions of the guiding structure are less than the smallest characteristic wavelength of the electromagnetic field propagating along it, the transverse components of the electromagnetic field give the "significant contribution" to the overall field and to the resulting terminal voltages and currents (e.g., Lindell and Gu, 1987). In other words, we have that:

(a) Even if the magnetic induction field **B** has a longitudinal component different from zero, the following relation holds:

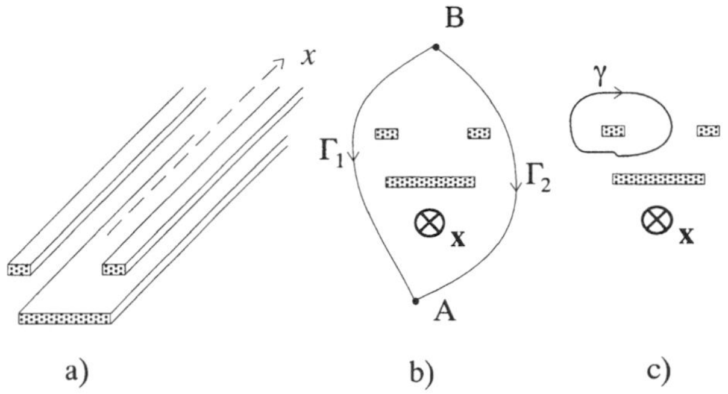

Figure 1.1. Longitudinal view and cross section of a three-wire guiding structure.

$$\left|\oint_{\Gamma} \mathbf{E}\cdot\mathbf{t}\,dl = -\frac{d}{dt}\iint_{S_\Gamma} B_x\,dS\right| \ll \left|\int_{\Gamma_1} \mathbf{E}\cdot\mathbf{t}\,dl\right| \approx \left|\int_{\Gamma_2} \mathbf{E}\cdot\mathbf{t}\,dl\right|, \qquad (1.1)$$

where Γ_1 and Γ_2 are two generic "admissible"[2] lines belonging to the transverse plane x = const and linking two generic points A and B, Γ is the closed line composed of Γ_1 and Γ_2 and S_Γ is the surface in the plane x = const that has the closed line Γ as contour, see Fig. 1.1b;

(b) Although the longitudinal component of the electric induction field **D** is different from zero, the following relation holds:

$$\left|\oint_{\gamma} \mathbf{H}\cdot\mathbf{t}\,dl - \iint_{S_\gamma} J_x\,dS = \frac{d}{dt}\iint_{S_\gamma} D_x\,dS\right| \ll \left|\oint_{\gamma} \mathbf{H}\cdot\mathbf{t}\,dl\right| \approx \left|\iint_{S_\gamma} J_x\,dS\right|, \qquad (1.2)$$

where S_γ is the surface in the plane x = const that has the closed line γ as contour, see Fig. 1.1c. Thus from Eqs. (1.1) and (1.2) we obtain

$$\oint_{\Gamma} \mathbf{E}\cdot\mathbf{t}\,dl \cong 0, \qquad (1.3)$$

and

$$\oint_{\gamma} \mathbf{H}\cdot\mathbf{t}\,dl \cong i_\gamma(x,t). \qquad (1.4)$$

When conditions (1.3) and (1.4) are satisfied, the configuration of the electromagnetic field is said to be of *quasi-TEM type*.

[2]An *admissible line* is a line that has a length of the same order of magnitude as the largest characteristic transverse dimension of the guiding structure.

The operation of a multiconnected guiding structure is conditioned by the topology of the circuit into which it is inserted. Let us first consider, for the sake of simplicity, a two-conductor guiding structure. The simplest, and at the same time the most significant example, is that in which each end of the guiding structure is connected to a one-port. In this case the current that enters a given terminal is equal to the current that exits from the other terminal at the same end. In consequence, if the guiding structure interacts with the rest of the network only through the terminals, then the total current flowing through any transverse section must be zero.

This simple example illustrates a very general result that also holds for multiconductor lines. If the guiding structure interacts with the rest of the network only through the ends, it is by the way it is connected to the network that the total current flowing through any of its cross sections is equal to zero — for instance, when one-ports link the terminals at each line end in pairs.

Under the assumption of quasi-TEM field distribution and zero total current the *distributed circuit model* holds: namely, throughout the guiding structures the electric current $i = i(x;t)$ through any conductor and the voltage $v = v(x;t)$ across any pair of conductors, at any abscissa x and at any time t, are well-defined. Any guiding structure that satisfies these conditions is called *transmission line*.

The equations governing the dynamics of the currents flowing through the conductors and the voltages between the conductors are the so-called *transmission line equations*. Under the assumption of quasi-TEM field distribution and zero total current they can be derived from the integral form of Maxwell equations. In the next two sections we shall briefly recall the equations of different types of transmission line models and the properties of their parameters. The reader who wants to examine closely how transmission line equations are related to the Maxwell equations is referred to specialized texts (e.g., Collin, 1992; Paul, 1994; Franceschetti, 1997). However, in Chapter 5, where we shall deal with transmission lines with frequency-dependent parameters, we shall briefly outline the main steps of the reasoning through which the transmission line equations may be obtained from the Maxwell equations.

In conclusion, the importance of the transmission line models resides in their ability to describe a large variety of guiding structures in the "low frequency regime", offering a powerful unified approach to their study. Transmission lines model guiding structures such as cables, wires, power lines, printed circuit board traces, buses for carrying digital data or control signals in modern electronic circuits, VLSI interconnections, coupled microstrips, and microwave circuits.

Even if transmission line models describe only approximately the electromagnetic behavior of these systems they are particularly important for the engineering application in view of its inherent simplicity, physical intuition, and scalar description of the problem.

1.2. TWO-CONDUCTOR TRANSMISSION LINE EQUATIONS

Let us consider a transmission line with two conductors, Fig. 1.2. Let the x-axis be oriented along the line axis, with $x = 0$ corresponding to the left end of the line and $x = d$ corresponding to the right end. Let $v = v(x, t)$ represent the voltage between the two conductors and $i = i(x, t)$ the current flowing through the upper conductor, at the abscissa x (with $0 \leqslant x \leqslant d$) and time t. The references for the current and voltage directions are those shown in Fig. 1.2. According to the assumptions, the current distribution along the other wire is equal to $-i(x, t)$.

1.2.1. Ideal Transmission Lines

Ideal transmission lines model ideal guiding structures, that is, interconnections without losses, uniform in space and with parameters independent of frequency. The equations for the voltage and the current distributions are (e.g., Heaviside, 1893; Franceschetti, 1997),

$$-\frac{\partial v}{\partial x} = L\frac{\partial i}{\partial t}, \tag{1.5}$$

$$-\frac{\partial i}{\partial x} = C\frac{\partial v}{\partial t}, \tag{1.6}$$

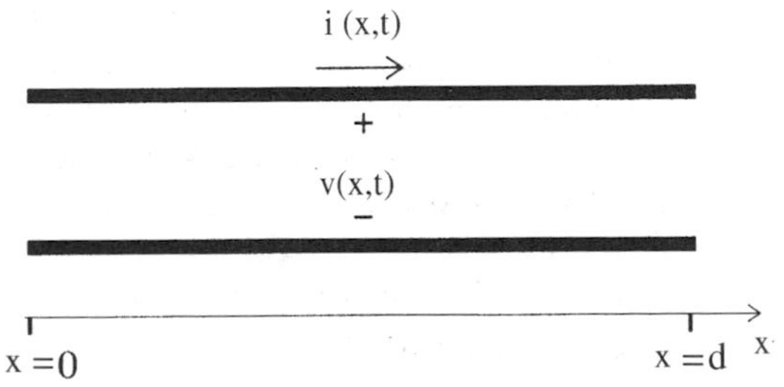

Figure 1.2. Sketch of a two-conductor transmission line.

where L and C are, respectively, the *inductance and capacitance per-unit-length* of the guiding structure.

The parameter L represents the magnetic flux linkage per-unit-length when a constant unitary current is flowing in one conductor and returning via the other. The parameter C represents the electric charge per-unit-length accumulated on the conductor labeled with the sign + when a constant unitary voltage is applied between the two conductors (on the other conductor the per-unit-length electric charge is equal to $-C$). Both L and C are positive definite (passive two-conductor guiding structures). In fact, the quantities $Li^2/2$ and $Cv^2/2$ represent, respectively, the per-unit-length energy associated with the magnetic and electric field.

1.2.2. Lossy Transmission Lines

The transmission line equations for lossy lines with constant parameters are in the time domain (e.g., Heaviside, 1893; Paul, 1994),

$$-\frac{\partial v}{\partial x} = L\frac{\partial i}{\partial t} + Ri, \tag{1.7}$$

$$-\frac{\partial i}{\partial x} = C\frac{\partial v}{\partial t} + Gv, \tag{1.8}$$

where L and C are always the inductance and capacitance per unit length of the transmission line, respectively, and R and G are the *per-unit-length longitudinal resistance and transverse conductance* of the line, respectively.

The parameter R represents the resulting per-unit-length voltage drop due to the conductor ohmic losses when a constant unitary current is flowing in one conductor and returning via the other. The parameter G represents the per-unit-length transverse current due to the embedding medium losses when a constant unitary voltage is applied between the two conductors. Both R and G are positive definite (passive and dissipative two-conductor guiding structures). In fact, the quantity $Ri^2 + Gv^2$ represents the per-unit-length power dissipated along the guiding structure.

The following important properties hold for transmission lines surrounded by homogeneous media (e.g., Paul, 1994),

$$LC = \mu\varepsilon, \tag{1.9}$$

$$GL = \sigma\mu, \tag{1.10}$$

where ε is the dielectric constant, μ is the magnetic permeability, and σ is the electrical conductivity of the embedding medium.

1.2.3. Nonuniform Transmission Lines

The per-unit-length parameters depend on the cross-sectional dimensions of the guiding structure represented by the transmission line. If these dimensions vary along the line axis, then the per-unit-length parameters will be functions of the position variable x, $L = L(x)$, $C = C(x)$, $R = R(x)$, and $G = G(x)$ (e.g., Franceschetti, 1997). The principal causes of the spatial nonuniformity are the variations in the wire cross sections and/or in the distance between the wires. This type of nonuniformity occurs frequently in printed circuit boards and on-chip interconnections. The nonuniformity also may be due to the nonhomogeneity of the surrounding medium.

Equations (1.9) and (1.10) also hold for nonuniform lines, if the embedding medium is homogeneous.

1.2.4. Transmission Lines with Distributed Sources

Equations (1.5) and (1.6) (or Eqs. (1.7) and (1.8)) describe two-conductor transmission lines excited at the ends by lumped sources. Transmission lines also may be excited by known incident electromagnetic fields generated by nearby radiating structures. The effects of these incident fields may be approximately taken into account by including suitable *distributed sources* along the line (e.g., Paul, 1994),

$$-\frac{\partial v}{\partial x} = L\frac{\partial i}{\partial t} + Ri - e_s, \tag{1.11}$$

$$-\frac{\partial i}{\partial x} = C\frac{\partial v}{\partial t} + Gv - j_s. \tag{1.12}$$

The distributed sources $e_s(x, t)$ and $j_s(x, t)$ depend on the incident electromagnetic field and on the structure of the guiding system modeled by the transmission line.

1.2.5. Transmission Lines with Frequency-Dependent Parameters

It is worth noting that Eqs. (1.11) and (1.12) are only valid for lines with parameters that do not depend on frequency. If this is not the case, the line has to be described by more complicated equations.

In the frequency domain, Eqs. (1.11) and (1.12) transform as follows:

$$-\frac{dV}{dx} = Z(i\omega)I - E_s, \tag{1.13}$$

$$-\frac{dI}{dx} = Y(i\omega)V - J_s, \tag{1.14}$$

where $V(x; i\omega)$, $I(x; i\omega)$, $E_s(x; i\omega)$, and $J_s(x; i\omega)$ are the Fourier transforms of $v(x, t)$, $i(x, t)$, $e_s(x, t)$, and $j_s(x, t)$, respectively, and

$$Z(i\omega) = i\omega L + R, \tag{1.15}$$

$$Y(i\omega) = i\omega C + G. \tag{1.16}$$

For the sake of simplicity, we have assumed that the line is initially at rest.

Most signals in high-speed digital circuits have short rise and fall times, which result in a very broad frequency spectra. Therefore, the modeling of the guiding structure involved in these circuits requires an accurate description of their broadband characteristics. In general, due to the losses inside the conductors and dielectrics, the transverse nonhomogeneity, and time dispersion of the embedding medium, the per-unit-length parameters depend on the frequency. For these lines, Eqs. (1.13) and (1.14) also hold, but the *per-unit-length longitudinal impedance* Z and the *per-unit-length transverse admittance* Y are more complicated functions of the frequency.

The actual expressions of the functions $Z(i\omega)$ and $Y(i\omega)$ depend on the physical phenomenon causing the dispersion in time. In Chapter 5 we shall outline briefly the expression of Z and Y for transmission lines of practical interest, such as lines with frequency-dependent dielectric losses, lines with transverse nonhomogenous dielectrics, lines with normal and anomalous skin effect, superconducting transmission lines, and power lines above a finite conductivity ground.

As we shall see in Chapter 5, the per-unit-length admittance $Y(i\omega)$ for all these lines has an asymptotic expression of the type

$$Y(i\omega) = (i\omega C_\infty + G_\infty) + Y_r(i\omega), \tag{1.17}$$

where the parameters C_∞ and G_∞ are independent of frequency. The terms G_∞ and ωC_∞ are, respectively, the leading terms of the real and imaginary parts of Y for $\omega \to \infty$, whereas the remainder $Y_r(i\omega)$ has the following asymptotic behavior[3]:

$$Y_r(i\omega) = O(\omega^{-1}) \quad \text{as } \omega \to \infty. \tag{1.18}$$

As we shall see in Chapter 5, the per-unit-length impedance $Z(i\omega)$ for transmission lines with *classical skin effect* has an asymptotic expression of the type

$$Z(i\omega) = (i\omega L_\infty + K\sqrt{i\omega} + R_\infty) + Z_r(i\omega), \tag{1.19}$$

where the parameters L_∞, K, and R_∞ are independent of frequency. The terms $K\sqrt{\omega/2}$ and ωL_∞ are, respectively, the leading terms of the real and imaginary parts of Z for $\omega \to \infty$, whereas

$$Z_r(i\omega) = O\left(\frac{1}{\sqrt{i\omega}}\right) \quad \text{as } \omega \to \infty. \tag{1.20}$$

The asymptotic expression of $Z(i\omega)$ for transmission lines with *anomalous skin effect* (see Chapter 5) may be described through equations similar to Eqs. (1.19) and (1.20) if we replace $\sqrt{i\omega}$ with $(i\omega)^{2/3}$. Instead, in all the cases in which the skin effect is negligible the asymptotic expression of the per-unit-length impedance $Z(i\omega)$ is similar to that of per-unit-length admittance $Y(i\omega)$.

Let us consider the real and imaginary parts of $Z(i\omega)$ and $Y(i\omega)$:

$$\mathscr{R}(\omega) = Re\{Z\}, \ \chi(\omega) = Im\{Z\}, \ \mathscr{G}(\omega) = Re\{Y\}, \ \mathscr{B}(\omega) = Im\{Y\}. \tag{1.21}$$

For lossy lines we must necessarily have

$$\mathscr{R}(\omega) > 0 \quad \text{and} \quad \mathscr{G}(\omega) > 0 \tag{1.22}$$

for any ω. These inequalities are straightforward. In fact, under sinusoidal steady-state operation,

$$v(x, t) = Re\{\bar{V}(x)e^{i\omega t}\}, \tag{1.23}$$

$$i(x, t) = Re\{\bar{I}(x)e^{i\omega t}\}, \tag{1.24}$$

the per-unit-length time average electric power dissipated along the line, due to the conductor and dielectric losses, is given by

$$S(x) = \tfrac{1}{2}\mathscr{R}(\omega)|\bar{I}(x)|^2 + \tfrac{1}{2}\mathscr{G}(\omega)|\bar{V}(x)|^2, \tag{1.25}$$

[3]The statement "$g(x) = O(x^m)$ as $x \to \infty$" indicates that the limit $\lim_{x\to\infty} g(x)/x^m = \kappa$, exists, it is bounded and it is different from zero; m and κ are constants.

where $\bar{V}(x)$ and $\bar{I}(x)$ are, respectively, the representative phasors of the voltage and the current distributions.

For lines with parameters independent of frequency, the quantities $\chi(\omega)|\bar{I}(x)|^2/2$ and $\mathscr{B}(\omega)|\bar{V}(x)|^2/2$ represents 2ω times the time average values of the per unit length energy associated with the magnetic field and electric field, respectively, and hence, both $\chi(\omega)$ and $\mathscr{B}(\omega)$ are positive definite. For transmission lines with frequency-dependent parameters these terms may have no particular physical significance.

The time domain equations for line with parameters depending on frequency are convolution-partial differential equations of the type,

$$-\frac{\partial v}{\partial x} = L_{\infty}\frac{\partial i}{\partial t} + R_{\infty}i + (\hat{z}_r * i)(t) - e_s, \tag{1.26}$$

$$-\frac{\partial i}{\partial x} = C_{\infty}\frac{\partial v}{\partial t} + G_{\infty}v + (y_r * v)(t) - j_s, \tag{1.27}$$

where $y_r(t)$ is the inverse transforms of $Y_r(i\omega)$ and $\hat{z}_r(t)$ is the inverse transforms of $\hat{Z}_r(i\omega)=K(i\omega)^{\alpha}+Z_r(i\omega)$: $\hat{Z}_r(i\omega)$ behaves as $(i\omega)^{\alpha}$ for $\omega \to \infty$, where $\alpha = 1/2$ or $2/3$ for lines with skin effect and $\alpha < 0$ for all other cases.

1.3. MULTICONDUCTOR TRANSMISSION LINE EQUATIONS

Equations for two-conductor transmission lines can be generalized to multiconductor transmission lines. Let us consider a transmission line having a total of $(n + 1)$ conductors, Fig. 1.3. We assume that the conductor indicated with "0" is the reference conductor for the voltages. Let $v_k = v_k(x, t)$ represent the voltage between the kth conductor and the reference one, and $i_k = i_k(x, t)$ represent the current flowing through the kth conductor, with $k = 1, \ldots, n$. The electric current flowing in the reference conductor is equal to $-\Sigma_{k=1}^{n} i_k(x, t)$.

1.3.1. Ideal Multiconductor Transmission Lines

The line equations for ideal multiconductor transmission lines are (e.g., Schelkunoff, 1955; Paul, 1994; Franceschetti, 1997),

$$-\frac{\partial \mathbf{v}}{\partial x} = \mathrm{L}\frac{\partial \mathbf{i}}{\partial t}, \tag{1.28}$$

$$-\frac{\partial \mathbf{i}}{\partial x} = \mathrm{C}\frac{\partial \mathbf{v}}{\partial t}. \tag{1.29}$$

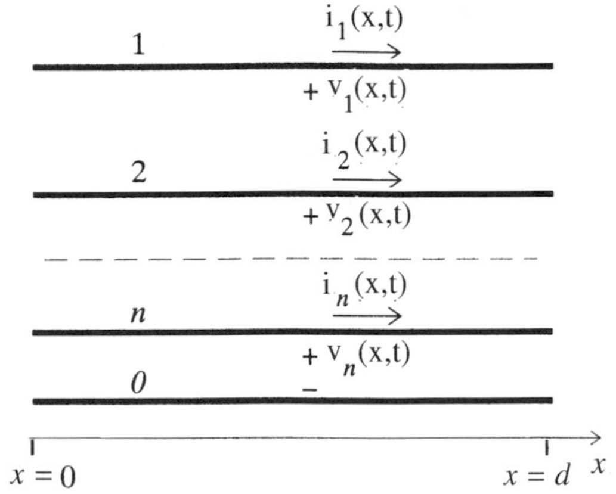

Figure 1.3. Sketch of a multiconductor transmission line.

We are using a matrix notation to have a compact form. The voltage and the current column vectors are defined as

$$\mathbf{v}(x,t) = |v_1(x,t)\ v_2(x,t) \cdots v_n(x,t)|^{\mathrm{T}}, \tag{1.30}$$

$$\mathbf{i}(x,t) = |i_1(x,t)\ i_2(x,t) \cdots i_n(x,t)|^{\mathrm{T}}, \tag{1.31}$$

and the per-unit-length $n \times n$ matrices L and C are given by

$$\mathrm{L} = \begin{bmatrix} l_{11} & l_{12} & \cdots & l_{1n} \\ l_{21} & l_{22} & \cdots & l_{2n} \\ \cdots & \cdots & \ddots & \cdots \\ l_{n1} & l_{n2} & \cdots & l_{nn} \end{bmatrix}, \quad \mathrm{C} = \begin{bmatrix} c_{11} & c_{12} & \cdots & c_{1n} \\ c_{21} & c_{22} & \cdots & c_{2n} \\ \cdots & \cdots & \ddots & \cdots \\ c_{n1} & c_{n2} & \cdots & c_{nn} \end{bmatrix}; \tag{1.32}$$

l_{ii} is the per-unit-length static self-inductance of the loop defined by the ith conductor and the reference conductor; l_{ij} is the static mutual-inductance between the ith conductor and the jth conductor; c_{ii} is the per-unit-length static self-capacitance of the ith conductor with respect to all the other conductors held at the reference potential; c_{ij} is the per-unit-length static mutual-capacitance between the ith conductor and the jth conductor.

The per-unit-length parameter matrices L and C have important properties. The primary ones are that they are symmetric,

$$\mathrm{L} = \mathrm{L}^{\mathrm{T}}, \quad \mathrm{C} = \mathrm{C}^{\mathrm{T}}, \tag{1.33}$$

and positive definite,

$$\mathbf{I}^{\mathrm{T}}\mathrm{L}\mathbf{I} > 0, \quad \text{for any } \mathbf{I} \neq \mathbf{0}, \tag{1.34}$$

$$\mathbf{V}^{\mathrm{T}}\mathrm{C}\mathbf{V} > 0, \quad \text{for any } \mathbf{V} \neq \mathbf{0}. \tag{1.35}$$

The symmetry of C is a direct consequence of the reciprocity theorem for the electrostatic field, while the symmetry of L is a direct consequence of the reciprocity theorem for the stationary magnetic field (e.g., Fano, Chu, and Adler, 1960; Paul, 1994). The only hypothesis necessary for the validity of the reciprocity theorem is that the medium must be linear and reciprocal.

Equations (1.34) and (1.35) are consequence of the passivity of the line. The distributions along the line of the per-unit-length energies associated to the electric and magnetic fields are, respectively,

$$w_e = \tfrac{1}{2}\mathbf{v}^{\mathrm{T}}\mathrm{C}\mathbf{v}, \tag{1.36}$$

$$w_m = \tfrac{1}{2}\mathbf{i}^{\mathrm{T}}\mathrm{L}\mathbf{i}. \tag{1.37}$$

At any time t and abscissa x, $w_e(x,t)$ must be *positive* for all choices of voltages, and $w_m(x,t)$ must be *positive* for all choices of currents; furthermore, $w_e = 0$ iff $\mathbf{v}(x,t) = \mathbf{0}$, and $w_m = 0$ iff $\mathbf{i}(x,t) = \mathbf{0}$ (we exclude degenerate situations that do not have a physical significance). In consequence, matrices L and C satisfy Eqs. (1.34) and (1.35) (e.g., Fano, Chu, and Adler, 1960; Paul, 1994).

For multiconductor transmission lines immersed in a homogeneous medium, the per-unit-length parameter matrices are related by (e.g., Paul, 1994)

$$\mathrm{LC} = \mathrm{CL} = \frac{1}{c^2}\mathrm{I}, \tag{1.38}$$

where I is the identity matrix and $c = 1/\sqrt{\varepsilon\mu}$.

1.3.2. Lossy Multiconductor Transmission Lines

Similarly to the two-conductor line case, for lossy multiconductor lines the equations are

$$-\frac{\partial \mathbf{v}}{\partial x} = \mathrm{L}\frac{\partial \mathbf{i}}{\partial t} + \mathrm{R}\mathbf{i}, \tag{1.39}$$

$$-\frac{\partial \mathbf{i}}{\partial x} = \mathrm{C}\frac{\partial \mathbf{v}}{\partial t} + \mathrm{G}\mathbf{v}. \tag{1.40}$$

The $n \times n$ matrices L and C are the per-unit-length static inductance and capacitance matrices characterizing the guiding structure that is supposed to be lossless. The $n \times n$ matrices R and G are given by

$$\mathrm{R} = \begin{bmatrix} r_1 + r_0 & r_0 & \cdots & r_0 \\ r_0 & r_2 + r_0 & \cdots & r_0 \\ \cdots & \cdots & \ddots & \cdots \\ r_0 & r_0 & \cdots & r_n + r_0 \end{bmatrix}, \quad \mathrm{G} = \begin{bmatrix} g_{11} & g_{12} & \cdots & g_{1n} \\ g_{21} & g_{22} & \cdots & g_{2n} \\ \cdots & \cdots & \ddots & \cdots \\ g_{n1} & g_{n2} & \cdots & g_{nn} \end{bmatrix}, \tag{1.41}$$

where r_i is the per-unit-length static resistance of the ith conductor, g_{ii} and g_{ij} are, respectively, the per-unit-length static self- and mutual conductances taking into account the dc dielectric losses and the surface leakage.

The general properties of L and C have been described in the previous paragraph. They are symmetric and positive definite. Here we shall look briefly at the general properties of the per-unit-length matrices R and G, the primary ones of which are that they are symmetric,

$$\mathrm{R} = \mathrm{R}^{\mathrm{T}}, \tag{1.42}$$

$$\mathrm{G} = \mathrm{G}^{\mathrm{T}}, \tag{1.43}$$

and positive definite,

$$\mathbf{i}^{\mathrm{T}}\mathrm{R}\mathbf{i} > 0 \quad \text{for any } \mathbf{i} \neq \mathbf{0}, \tag{1.44}$$

$$\mathbf{v}^{\mathrm{T}}\mathrm{G}\mathbf{v} > 0 \quad \text{for any } \mathbf{v} \neq \mathbf{0}. \tag{1.45}$$

The symmetry of G is a direct consequence of the reciprocity theorem for the stationary current field (e.g., Fano, Chu, and Adler, 1960), while the symmetry of R is a direct consequence of its simple expression (see Eq. (1.41)).

The distribution along the line of the per-unit-length electric power dissipated in the imperfect conductors $s_{||}$ and embedding materials $s_{\perp}$ is given by

$$s_{||}(x,t) = \mathbf{i}^{\mathrm{T}}\mathrm{R}\mathbf{i}, \tag{1.46}$$

$$s_{\perp}(x,t) = \mathbf{v}^{\mathrm{T}}\mathrm{G}\mathbf{v}. \tag{1.47}$$

Therefore R and G have to be positive definite.

Finally, for multiconductor transmission lines embedded in a homogeneous medium, matrices L and G are related by (e.g., Paul, 1994)

$$\mathrm{LG} = \mathrm{GL} = \tau_0 \mathrm{I}, \tag{1.48}$$

where $\tau_0 = \mu\sigma$ and I is the identity matrix.

Similarly to the two-conductor line case, when the cross-section dimensions of the guiding structure vary along the line axis, the matrices depend on the position variable x. However, (1.38) and (1.48) still hold when the conductors are embedded in a homogeneous medium.

1.3.3. Multiconductor Transmission Lines with Distributed Sources

As with the two-conductor line case, the effects of known incident electromagnetic fields radiated by nearby radiating structures may be described through suitable distributed sources (of vectorial type) defined along the multiconductor lines. The line equations are (e.g., Paul, 1994):

$$-\frac{\partial \mathbf{v}}{\partial x} = \mathrm{L}\frac{\partial \mathbf{i}}{\partial t} + \mathrm{R}\mathbf{i} - \mathbf{e}_s, \tag{1.49}$$

$$-\frac{\partial \mathbf{i}}{\partial x} = \mathrm{C}\frac{\partial \mathbf{v}}{\partial t} + \mathrm{G}\mathbf{v} - \mathbf{j}_s. \tag{1.50}$$

1.3.4. Multiconductor Transmission Lines with Frequency-Dependent Parameters

Equations (1.49) and (1.50) are no longer valid when the line parameters depend on frequency. As in the two-conductor case, the behavior of these lines can be easily described in the frequency domain by

$$-\frac{d\mathbf{V}}{dx} = \mathrm{Z}(i\omega)\mathbf{I} - \mathbf{E}_s, \tag{1.51}$$

$$-\frac{d\mathbf{I}}{dx} = \mathrm{Y}(i\omega)\mathbf{V} - \mathbf{J}_s, \tag{1.52}$$

where $\mathbf{V}(x;i\omega)$, $\mathbf{I}(x;i\omega)$, $\mathbf{E}_s(x;i\omega)$, and $\mathbf{J}_s(x;i\omega)$ are the Fourier transforms of $\mathbf{v}(x,t)$, $\mathbf{i}(x,t)$, $\mathbf{e}_s(x;t)$, and $\mathbf{j}_s(x;t)$, respectively. The actual expressions for the *per-unit-length longitudinal matrix impedance* Z and the *per-unit-length transverse matrix admittance* Y depend on the actual physical constitution of the guiding structure represented by the transmission line model (see Chapter 6). Similarly to the two-conductor line case, each entry of Z may be represented through Eq. (1.19) and each entry of Y may be represented through Eq. (1.17).

The symmetry property is a general property that also holds for matrices Z and Y:

$$\mathrm{Z} = \mathrm{Z}^{\mathrm{T}}, \tag{1.53}$$

$$\mathrm{Y} = \mathrm{Y}^{\mathrm{T}}. \tag{1.54}$$

For lossy multiconductor lines we have inequalities similar to those of Eq. (1.22). Let us consider again a sinusoidal steady-state operation. The voltage and the current can be represented by

$$\mathbf{v}(x,t) = Re\{\bar{\mathbf{V}}(x)e^{i\omega t}\}, \tag{1.55}$$

$$\mathbf{i}(x,t) = Re\{\bar{\mathbf{I}}(x)e^{i\omega t}\}, \tag{1.56}$$

where the complex vector $\bar{\mathbf{V}}(x)$ is the representative phasor of the vector distribution of the sinusoidal voltages, and $\bar{\mathbf{I}}(x)$ is the representative phasor of the vector distribution of the sinusoidal currents. Placing

$$\mathscr{R}(\omega) = Re\{\mathrm{Z}\},\ \chi(\omega) = Im\{\mathrm{Z}\},\ \mathscr{G}(\omega) = Re\{\mathrm{Y}\},\ \mathscr{B}(\omega) = Im\{\mathrm{Y}\}, \tag{1.57}$$

as Z and Y are symmetric, we have

$$S_{||}(x) = \tfrac{1}{2}Re\{\bar{\mathbf{I}}^{*\mathrm{T}}\mathrm{Z}(i\omega)\bar{\mathbf{I}}\} = \tfrac{1}{2}\bar{\mathbf{I}}^{*\mathrm{T}}\mathscr{R}(\omega)\bar{\mathbf{I}}, \tag{1.58}$$

$$S_{\perp}(x) = \tfrac{1}{2}Re\{\bar{\mathbf{V}}^{*\mathrm{T}}\mathrm{Y}(i\omega)\bar{\mathbf{V}}\} = \tfrac{1}{2}\bar{\mathbf{V}}^{*\mathrm{T}}\mathscr{G}(\omega)\bar{\mathbf{V}}, \tag{1.59}$$

$$\tfrac{1}{2}Im\{\bar{\mathbf{I}}^{*\mathrm{T}}\mathrm{Z}(i\omega)\bar{\mathbf{I}}\} = \tfrac{1}{2}\bar{\mathbf{I}}^{*\mathrm{T}}\chi(\omega)\bar{\mathbf{I}}, \tag{1.60}$$

$$\tfrac{1}{2}Im\{\bar{\mathbf{V}}^{*\mathrm{T}}\mathrm{Y}(i\omega)\bar{\mathbf{V}}\} = \tfrac{1}{2}\bar{\mathbf{V}}^{*\mathrm{T}}\mathrm{B}(\omega)\bar{\mathbf{V}}; \tag{1.61}$$

with $\bar{\mathbf{A}}^*$ we indicate the conjugate of the complex vector $\bar{\mathbf{A}}$; $S_{||}$ and $S_{\perp}$ represent, respectively, the per unit length time-average electrical power dissipated into the conductors and embedding medium. Therefore, for lossy multiconductor lines there must be

$$\bar{\mathbf{I}}^{*\mathrm{T}}\mathscr{R}(\omega)\bar{\mathbf{I}} > 0, \quad \bar{\mathbf{V}}^{*\mathrm{T}}\mathscr{G}(\omega)\bar{\mathbf{V}} > 0 \tag{1.62}$$

for any ω and for any $\bar{\mathbf{I}} \neq \bar{\mathbf{0}}$, and $\bar{\mathbf{V}} \neq \bar{\mathbf{0}}$.

For multiconductor lines with frequency independent parameters, $\bar{\mathbf{I}}^{*\mathrm{T}}\chi(\omega)\bar{\mathbf{I}}/2$ and $\bar{\mathbf{V}}^{*\mathrm{T}}\mathscr{B}(\omega)\bar{\mathbf{V}}/2$ correspond, respectively, to 2ω times the per-unit-length time-average energy associated with the magnetic and the electric fields, so in these cases matrices $\chi(\omega)$ and $\mathscr{B}(\omega)$ are positive definite.

1.4. POYNTING'S THEOREM FOR LINES WITH FREQUENCY-INDEPENDENT PARAMETERS

One of the best known results related to Maxwell equations is the Poynting theorem (e.g., Stratton, 1941; Franceschetti, 1997). We shall now show that a similar result is obtained for transmission lines.

1.4.1. Two-Conductor Transmission Lines

Let us multiply Eq. (1.11) by i and Eq. (1.12) by v and sum them termwise. In so doing we obtain

$$\frac{\partial p}{\partial x} + \frac{\partial w_{em}}{\partial t} + s_d = s_s, \tag{1.63}$$

where

$$w_{em} = \frac{Li^2}{2} + \frac{Cv^2}{2}, \tag{1.64}$$

$$s_d = Ri^2 + Gv^2, \tag{1.65}$$

$$s_s = e_s i + v j_s, \tag{1.66}$$

and

$$p = iv. \tag{1.67}$$

In the quasi-TEM model the value w_{em} represents the energy per unit length associated to the electromagnetic field along the guiding structure described by the transmission line. The value s_d represents the electrical power dissipated per unit length and s_s represents the per-unit-length electrical power generated by the distributed sources. The quantity $p(x, t)$ represents the flux of the Poynting vector through the transversal section of the line at abscissa x and at time t in forward direction.

Integrating Eq. (1.63) termwise in the spatial interval $(0, d)$, we obtain

$$\frac{d}{dt}\int_0^d w_{em}(x,t)dx + \int_0^d s_d(x,t)\,dx + [p(d,t) - p(0,t)] = \int_0^d s_s(x,t)dx. \tag{1.68}$$

This is the *Poynting theorem* for two-conductor transmission lines, according to which the power generated by the distributed sources along the line is equal to the sum of three terms:

(i) the variations in the time of a term corresponding to the sum of the energies associated to the electric and magnetic fields, and stored along the line;
(ii) the power dissipated along the line due to the Joule effect in the conductors and the losses in the dielectric; and
(iii) the flux of the Poynting vector coming out from the ends of the line, $p(d,t) - p(0,t)$.

The significance of Eq. (1.68), and, therefore, of flux $p(d,t) - p(0,t)$, from an energy point of view can be understood by referring to the first principle of thermodynamics: $p(d,t) - p(0,t)$ corresponds to the per-unit-time electrical energy outcoming from the ends of the line.

If we introduce the electrical power absorbed by the line at the end $x = 0$

$$p_1(t) = p(0,t), \tag{1.69}$$

and the electrical power absorbed by the line at the end $x = d$

$$p_2(t) = -p(d,t), \tag{1.70}$$

then Eq. (1.68) becomes

$$\frac{d}{dt}\int_0^d w_{em}(x,t)dx + \int_0^d s_d(x,t)dx = \int_0^d s_s(x,t)dx + [p_1(t) + p_2(t)]. \tag{1.71}$$

1.4.2. Multiconductor Transmission Lines

Now let us extend the Poynting theorem to multiconductor lines with parameters independent of frequency, described by Eqs. (1.49) and (1.50). As the matrices L and C are symmetric we have the identity

$$\mathbf{i}^{\mathrm{T}}\mathrm{L}\frac{\partial \mathbf{i}}{\partial t} + \mathbf{v}^{\mathrm{T}}\mathrm{C}\frac{\partial \mathbf{v}}{\partial t} = \frac{\partial w_{em}}{\partial t}, \tag{1.72}$$

where

$$w_{em} = \tfrac{1}{2}\mathbf{i}^{\mathrm{T}}\mathrm{L}\mathbf{i} + \tfrac{1}{2}\mathbf{v}^{\mathrm{T}}\mathrm{C}\mathbf{v} \tag{1.73}$$

is the energy per unit length associated with the electromagnetic field.

Let us left multiply both members of Eq. (1.49) by $\mathbf{i}^{\mathrm{T}}$ and both members of Eq. (1.50) by $\mathbf{v}^{\mathrm{T}}$, and sum them member by member. Using identity (1.72) and integrating on the spatial interval $(0, d)$ we once again obtain Eq. (1.68), where

$$s_d = \mathbf{i}^{\mathrm{T}}\mathrm{R}\mathbf{i} + \mathbf{v}^{\mathrm{T}}\mathrm{G}\mathbf{v}, \tag{1.74}$$

$$s_s = \mathbf{i}^{\mathrm{T}}\mathbf{e}_s + \mathbf{j}_s^{\mathrm{T}}\mathbf{v}, \tag{1.75}$$

and

$$p = \mathbf{i}^{\mathrm{T}}\mathbf{v}. \tag{1.76}$$

If we indicate the electrical power absorbed by the multiconductor line at the end $x = 0$ with

$$p_1(t) = \mathbf{i}^{\mathrm{T}}(0, t)\mathbf{v}(0, t), \tag{1.77}$$

and the electrical power absorbed by the line at the end $x = d$ with

$$p_2(t) = -\mathbf{i}^{\mathrm{T}}(d, t)\mathbf{v}(d, t), \tag{1.78}$$

Eq. (1.71) still holds.

Remark

It is to be observed that if the matrices L and C were not symmetric, relation equation (1.72) would no longer hold and it would be pointless to speak of energy associated with the electromagnetic field. It is evident that this would be a physical absurdity and so matrices L and C must necessarily be symmetric. ◇

1.5. UNIQUENESS OF THE SOLUTION OF TRANSMISSION LINE EQUATIONS

It is certainly very important to succeed in establishing conditions that ensure, once satisfied, that the solution of the transmission line equations is determined and unique.

1.5.1. Two-Conductor Transmission Lines

Now we shall demonstrate that the solution of system equations (1.11) and (1.12) is determined and unique if the voltage and current distributions along the line at the initial instant (*initial conditions*) and an electric variable at each line end (*boundary conditions*) are assigned.

Let $v(x,t)$, $i(x,t)$ be a solution of Eqs. (1.11) and (1.12) that satisfies the assigned initial and boundary conditions, and let $v'(x,t)$, $i'(x,t)$ be another solution produced by the same distributed sources that satisfies the same initial and boundary conditions.

As Eqs. (1.11) and (1.12) are linear, voltage $\Delta v = v - v'$ and current $\Delta i = i - i'$ are solutions of these equations rendered homogeneous,

$$-\frac{\partial \Delta v}{\partial x} = L\frac{\partial \Delta i}{\partial t} + R\Delta i, \tag{1.79}$$

$$-\frac{\partial \Delta i}{\partial x} = C\frac{\partial \Delta v}{\partial t} + G\Delta v. \tag{1.80}$$

The Poynting theorem applied to these equations gives the relation

$$\frac{d\Delta W}{dt} + \Delta p_d(t) = \Delta p(t), \tag{1.81}$$

where

$$\Delta W(t) = \int_0^d \left(\frac{1}{2}L\Delta i^2 + \frac{1}{2}C\Delta v^2\right)dx, \tag{1.82}$$

$$\Delta p_d(t) = \int_0^d (R\Delta i^2 + G\Delta v^2)dx, \tag{1.83}$$

$$\Delta p(t) = [\Delta i(0,t)\Delta v(0,t) - \Delta i(d,t)\Delta v(d,t)]. \tag{1.84}$$

As the two solutions must satisfy the same boundary conditions, we must have $\Delta v(0,t)\Delta i(0,t) = 0$ and $\Delta v(d,t)\Delta i(d,t) = 0$, that is, $\Delta p(t) = 0$ and, therefore,

$$\frac{d\Delta W}{dt} = -\Delta p_d(t). \tag{1.85}$$

As the lines we are considering are passive, R and G are positive, and hence

$$\Delta p_d(t) = \int_0^d (R\Delta i^2 + G\Delta v^2)dx \geqslant 0. \tag{1.86}$$

Therefore, for any t we must have

$$\frac{d\Delta W}{dt} \leqslant 0, \tag{1.87}$$

and so $\Delta W(t)$ can never increase. Moreover, as L and C are positive for passive lines, $\Delta W(t)$ can never assume negative values. Since at instant $t = 0$, $\Delta v = v - v' = 0$, and $\Delta i = i - i' = 0$ for $0 \leqslant x \leqslant d$, then $\Delta W(t = 0) = 0$. As $\Delta W(t)$ cannot become negative and cannot increase, we must necessarily have $\Delta W(t) = 0$ for $t > 0$, and so $\Delta v(x, t) = 0$, $\Delta i(x, t) = 0$ for $0 \leqslant x \leqslant d$ and $t > 0$.

The conclusion is that to specify the solution of transmission line equations (1.11) and (1.12) entirely, it is sufficient to assign the voltage and current along the line at initial instant $t = 0$ (*initial conditions*), the distributed sources for $t \geqslant 0$, and the voltage or current values at the ends of the line (*boundary conditions*). It is also possible to assign the current at one end, and the voltage at the other.

Remark

An important consequence of this result is the possibility of representing the behavior of a two-conductor transmission line at the ends by means of a two-port. Consider, for example, the network obtained by connecting the ends of the line to two independent current sources. From the uniqueness theorem already illustrated here, this network admits a unique solution for the assigned initial distribution of the current and voltage along the line. In consequence, the transmission line imposes a well-determined and unique relation between the currents at the two ends and the voltages at the two ends. The equations that expresses this link do not depend on what is connected to the line. They depend only on the physical nature of the line and nothing else. ◇

1.5.2. Multiconductor Transmission Lines

To determine the conditions that ensure the uniqueness of the solution of Eqs. (1.49) and (1.50) it is possible to proceed as for a two-conductor line. Let us place $\Delta \mathbf{v} = \mathbf{v} - \mathbf{v}'$ and $\Delta \mathbf{i} = \mathbf{i} - \mathbf{i}'$, where $\mathbf{v}$, $\mathbf{i}$ and $\mathbf{v}'$, $\mathbf{i}'$ are two possible solutions of Eqs. (1.49) and (1.50) that satisfy the same initial and boundary conditions. For the linearity property $\Delta \mathbf{v}''$, $\Delta \mathbf{i}''$ is the solution of the same equations rendered homogeneous. Applying the Poynting theorem to Eqs. (1.49) and (1.50) after they have been made homogeneous, we once more obtain Eq. (1.81), where now

$$\Delta W(t) = \int_0^d \left(\frac{1}{2} \Delta \mathbf{i}^{\mathrm{T}} \mathrm{L} \Delta \mathbf{i} + \frac{1}{2} \Delta \mathbf{v}^{\mathrm{T}} \mathrm{C} \Delta \mathbf{v} \right) dx, \tag{1.88}$$

$$\Delta p_d(t) = \int_0^d \left(\Delta \mathbf{i}^{\mathrm{T}} \mathrm{R} \Delta \mathbf{i} + \Delta \mathbf{v}^{\mathrm{T}} \mathrm{G} \Delta \mathbf{v} \right) dx, \tag{1.89}$$

$$\Delta p(t) = [\Delta \mathbf{i}^{\mathrm{T}}(0, t) \Delta \mathbf{v}(0, t) - \Delta \mathbf{i}^{\mathrm{T}}(d, t) \Delta \mathbf{v}(d, t)]. \tag{1.90}$$

As $\Delta \mathbf{v}$ and/or $\Delta \mathbf{i}$ are zero at the ends of the line, $\Delta p(t) = 0$, and once more we obtain Eq. (1.85). Because of the passivity of the line, matrices L, C, R, and G are positive definite, and so $\Delta W(t) \geqslant 0$ and $d\Delta W/dt \leqslant 0$. The conclusion is the same as that reached for the two-conductor case: To specify the solution entirely it is sufficient to assign the initial conditions of the voltages and currents along the line, and the voltage and current value in correspondence with each conductor at both terminations.

An important consequence of this result is the possibility of representing the behavior of a multiconductor transmission line at the ends by means of a multiport. The reasoning is the same as that we have outlined for the two-conductor line.

1.6. POYNTING'S THEOREM FOR LINES IN THE FREQUENCY DOMAIN

It is also possible to formulate the Poynting theorem for sinusoidal steady-state operations. This theorem is applicable to the more general case of transmission lines with frequency-dependent parameters.

1.6.1. Two-Conductor Transmission Lines

Let us consider two-conductor transmission lines in sinusoidal steady state. The voltage and current distribution may be represented through Eqs. (1.23) and (1.24), respectively. The equations for phasors $\bar{V}(x)$ and $\bar{I}(x)$ are

$$\frac{d\bar{V}}{dx} = -Z\bar{I} + \bar{E}_s, \tag{1.91}$$

$$\frac{d\bar{I}}{dx} = -Y\bar{V} + \bar{J}_s. \tag{1.92}$$

They are the same as Eqs. (1.13) and (1.14); $\bar{E}_s(x)$ and $\bar{J}_s(x)$ are the representative phasors of the distributed sources.

Multiplying the first equation by $\bar{I}^*/2$, and the complex conjugate of Eq. (1.92) by $\bar{V}/2$, and then summing them termwise we obtain

$$\frac{dP}{dx} = -\frac{1}{2}Z|\bar{I}|^2 - \frac{1}{2}Y^*|\bar{V}|^2 + \frac{1}{2}\bar{E}_s\bar{I}^* + \frac{1}{2}\bar{V}\bar{J}_s \tag{1.93}$$

where

$$P = \tfrac{1}{2}\bar{V}\bar{I}^*. \tag{1.94}$$

Integrating along the line, we obtain

$$P_1 + P_2 = \frac{1}{2}\int_0^d Z|\bar{I}|^2\,dx + \frac{1}{2}\int_0^d Y^*|\bar{V}|^2\,dx - \frac{1}{2}\int_0^d (\bar{E}_s\bar{I}^* + \bar{V}\bar{J}_s)dx, \tag{1.95}$$

where

$$P_1 = P(0), \tag{1.96}$$

$$P_2 = -P(d). \tag{1.97}$$

The real parts of P_1 and P_2 are, respectively, the average value in the period $2\pi/\omega$ of the electric power absorbed by the line at the end $x = 0$ and the end $x = d$, while the imaginary part is the reactive power. Separating the real part from the imaginary part in Eq. (1.95), we have

$$P_a = \frac{1}{2}\int_0^d \mathscr{R}|\bar{I}|^2\,dx + \frac{1}{2}\int_0^d \mathscr{G}|\bar{V}|^2\,dx - Re\left\{\frac{1}{2}\int_0^d (\bar{E}_s\bar{I}^* + \bar{V}\bar{J}_s{}^*)dx\right\}, \tag{1.98}$$

$$Q = \frac{1}{2}\int_0^d \chi|\bar{I}|^2\,dx - \frac{1}{2}\int_0^d \mathscr{B}|\bar{V}|^2\,dx - Im\left\{\frac{1}{2}\int_0^d (\bar{E}_s\bar{I}^* + \bar{V}\bar{J}_s^*)dx\right\}, \tag{1.99}$$

where

$$P_a = Re\{P_1 + P_2\} \tag{1.100}$$

is the average electrical power absorbed by the line and

$$Q = Im\{P_1 + P_2\} \tag{1.101}$$

is the reactive power absorbed by the line.

1.6.2. Multiconductor Transmission Lines

Let us consider multiconductor transmission lines in sinusoidal steady state. The voltage and current distribution may be represented through Eqs. (1.55) and (1.56), respectively. The equations for the representative phasors $\bar{\mathbf{V}}(x)$ and $\bar{\mathbf{I}}(x)$ are

$$\frac{d\bar{\mathbf{V}}}{dx} = -\mathrm{Z}\bar{\mathbf{I}} + \bar{\mathbf{E}}_s, \tag{1.102}$$

$$\frac{d\bar{\mathbf{I}}}{dx} = -\mathrm{Y}\bar{\mathbf{V}} + \bar{\mathbf{J}}_s, \tag{1.103}$$

which are the same as Eqs. (1.51) and (1.52); $\bar{\mathbf{E}}_s(x)$, $\bar{\mathbf{J}}_s(x)$ are the representative phasors of the sinusoidal distributed sources.

Multiplying to the left the first equation by $\bar{\mathbf{I}}^{*\mathrm{T}}/2$ and the complex conjugate of Eq. (1.103) by $\bar{\mathbf{V}}^{\mathrm{T}}/2$, and then summing them termwise we obtain

$$\frac{dP}{dx} = -\frac{1}{2}\bar{\mathbf{I}}^{*\mathrm{T}}\mathrm{Z}\bar{\mathbf{I}} - \frac{1}{2}\bar{\mathbf{V}}^{\mathrm{T}}\mathrm{Y}^*\bar{\mathbf{V}}^* + \frac{1}{2}\bar{\mathbf{I}}^{*\mathrm{T}}\bar{\mathbf{E}}_s + \frac{1}{2}\bar{\mathbf{V}}^{T}\bar{\mathbf{J}}_s^*, \tag{1.104}$$

where $P(x) = \bar{\mathbf{V}}^{\mathrm{T}}\bar{\mathbf{I}}^*/2$. At this point, integrating along the line, we obtain a form analogous to Eq. (1.95).

1.7. UNIQUENESS OF THE SOLUTION OF TRANSMISSION LINE EQUATIONS WITH FREQUENCY-DEPENDENT PARAMETERS

In Section 1.5 we have shown the uniqueness of the solution for transmission lines with frequency independent parameters by using the Poynting theorem in the time domain. Now by using the Poynting theorem in the frequency domain, we shall demonstrate that the solution of a generic transmission line with losses is unique if, at each end, the values of the voltages or the values of the currents are assigned.

1.7.1. Two-Conductor Transmission Lines

We shall now demonstrate that the solution of Eqs. (1.91) and (1.92) is determined and unique if the voltage or the current is specified at each end of the line, so long as $Z(i\omega)$ and $Y(i\omega)$ have a real part greater than zero, even if arbitrarily small.

Let $\bar{V}(x)$, $\bar{I}(x)$ be a solution of Eqs. (1.91) and (1.92) and let $\bar{V}'(x)$, $\bar{I}'(x)$ be another solution that satisfies the same boundary conditions. As the equations are linear, voltage $\Delta\bar{V} = \bar{V} - \bar{V}'$ and current

$\Delta\bar{I} = \bar{I} - \bar{I}'$ are solutions of Eqs. (1.91) and (1.92) rendered homogeneous,

$$\frac{d\Delta\bar{V}}{dx} = -Z\Delta\bar{I}, \tag{1.105}$$

$$\frac{d\Delta\bar{I}}{dx} = -Y\Delta\bar{V}. \tag{1.106}$$

These equations must be solved by imposing that at each end of the line either the voltage or the current is zero.

The Poynting theorem applied to the homogeneous system equations (1.105) and (1.106) gives equation

$$\Delta P(0) - \Delta P(d) = \frac{1}{2}\int_0^d Z|\Delta\bar{I}|^2\,dx + \frac{1}{2}\int_0^d Y^*|\Delta\bar{V}|^2\,dx, \tag{1.107}$$

where $\Delta P = \Delta\bar{V}\Delta\bar{I}^*/2$. As the two solutions must satisfy the same boundary conditions, $\Delta P(x)$ is zero at both ends of the line. Placing $\Delta P(0) - \Delta P(d) = 0$ in Eq. (1.107) and considering the real part, we obtain

$$\int_0^d \mathscr{R}|\Delta\bar{I}|^2\,dx + \int_0^d \mathscr{G}|\Delta\bar{V}|^2\,dx = 0. \tag{1.108}$$

As $\mathscr{R}(\omega) > 0$ and $\mathscr{G}(\omega) > 0$, we must necessarily have

$$\Delta\bar{V}(x) = 0, \quad \Delta\bar{I}(x) = 0, \tag{1.109}$$

which shows the uniqueness of the solution in the frequency domain, and hence in the time domain.

Remark

If the transmission line were ideal, $\mathscr{R}(\omega) = 0$ and $\mathscr{G}(\omega) = 0$, Eq. (1.108) should be an identity and Eq. (1.109) could not be true. From Eq. (1.107), by placing $\Delta P(0) - \Delta P(d) = 0$ and considering the imaginary part, we obtain

$$\int_0^d \chi(\omega)|\Delta\bar{I}(x)|^2\,dx = \int_0^d \mathscr{B}(\omega)|\Delta\bar{V}(x)|^2\,dx. \tag{1.110}$$

This equation, too, does not exclude that there can be solutions other than zero. In fact, in ideal transmission lines, there exist sinusoidal solutions that satisfy null boundary conditions. They are the natural modes of oscillation of the line when it is connected to short circuits and/or open circuits, as we shall see in Chapter 2.

1.7.2. Multiconductor Transmission Lines

To determine the conditions that ensure the uniqueness of the solution of Eqs. (1.102) and (1.103) it is possible to proceed as for the two-conductor transmission line. Let us place $\Delta\bar{\mathbf{V}} = \bar{\mathbf{V}} - \bar{\mathbf{V}}'$ and $\Delta\bar{\mathbf{I}} = \bar{\mathbf{I}} - \bar{\mathbf{I}}'$, where $\bar{\mathbf{V}}$, $\bar{\mathbf{I}}$ and $\bar{\mathbf{V}}'$, $\bar{\mathbf{I}}'$ are two possible solutions of Eqs. (1.102) and (1.103) that satisfy the same boundary conditions. Owing to the linearity, $\Delta\bar{\mathbf{V}}$, $\Delta\bar{\mathbf{I}}$ is the solution of the same equations rendered homogeneous. If $\Delta\bar{\mathbf{V}}$ or $\Delta\bar{\mathbf{I}}$ are zero at each end of the line, applying the Poynting theorem to Eqs. (1.102) and (1.103), after they have been made homogeneous, we obtain

$$\int_0^d \Delta\bar{\mathbf{I}}^{*\mathrm{T}}(x)\mathscr{R}(\omega)\Delta\bar{\mathbf{I}}(x)dx + \int_0^d \Delta\bar{\mathbf{V}}^{*\mathrm{T}}(x)\mathscr{G}(\omega)\Delta\bar{\mathbf{V}}(x)dx = 0, \quad (1.111)$$

$$\int_0^d \Delta\bar{\mathbf{I}}^{*\mathrm{T}}(x)\chi(\omega)\Delta\bar{\mathbf{I}}(x)dx - \int_0^d \Delta\bar{\mathbf{V}}^{*\mathrm{T}}(x)\mathscr{B}(\omega)\Delta\bar{\mathbf{V}}(x)dx = 0. \quad (1.112)$$

As matrices $\mathscr{R}(\omega)$ and $\mathscr{G}(\omega)$ are positive definite, from Eq. (1.111) it follows $\Delta\bar{\mathbf{V}} = \mathbf{0}$ and $\Delta\bar{\mathbf{I}} = \mathbf{0}$, which demonstrates the uniqueness of the solution. For ideal multiconductor transmission lines the same considerations hold as for two-conductor lines.

1.8. TRANSMISSION LINE EQUATIONS IN THE LAPLACE DOMAIN

In this book we shall always use the method of the Laplace transform to study nonideal transmission lines. In the Laplace domain the line equations are

$$-\frac{dV}{dx} = Z(s)I - E_s, \quad (1.113)$$

$$-\frac{dI}{dx} = Y(s)V - J_s, \quad (1.114)$$

for two-conductor transmission lines, and

$$-\frac{d\mathbf{V}}{dx} = \mathrm{Z}(s)\mathbf{I} - \mathbf{E}_s, \quad (1.115)$$

$$-\frac{d\mathbf{I}}{dx} = \mathrm{Y}(s)\mathbf{V} - \mathbf{J}_s, \quad (1.116)$$

for multiconductor lines. These equations are obtained, respectively, by analytical prolongation of the equation systems (1.13) and (1.14) and (1.51) and (1.52) in the complex plane. In this way, $V(x;s)$, $I(x;s)$, $\mathbf{V}(x;s)$, and $\mathbf{I}(x;s)$ represent, respectively, the Laplace transform of $v(x,t)$, $i(x,t)$, $\mathbf{v}(x,t)$, and $\mathbf{i}(x,t)$ in the convergence region of the solution (see Appendix B). If the line is not ideal, the abscissa of convergence is always less than zero. The effects of the initial conditions may be taken into account through independent distributed sources.

1.9. RECIPROCITY THEOREMS FOR TWO-CONDUCTOR TRANSMISSION LINES

In this section we briefly illustrate three different forms of reciprocity in the Laplace domain for transmission lines with losses, parameters depending on the frequency, and without distributed sources. The line can also be nonuniform.

1.9.1. The First Form of the Reciprocity Theorem

Consider two equal lines $\mathscr{T}\mathscr{L}'$ and $\mathscr{T}\mathscr{L}''$, the first connected to a voltage source at $x=0$, and to a short circuit at $x=d$, the other connected to a short circuit at $x=0$, and to a voltage source at $x=d$, Fig. 1.4. Let us indicate all the electric variables of $\mathscr{T}\mathscr{L}'$ with an apex, and all those relative to $\mathscr{T}\mathscr{L}''$ with two apices.

Circuit $\mathscr{T}\mathscr{L}'$ is described by equations

$$-\frac{dV'}{dx} = ZI', \tag{1.117}$$

$$-\frac{dI'}{dx} = YV', \tag{1.118}$$

with the boundary conditions

$$V'(0;s) = V_1'(s), \tag{1.119}$$

$$V'(d;s) = 0, \tag{1.120}$$

where V_1' is the voltage imposed by the voltage source at $x=0$, and circuit $\mathscr{T}\mathscr{L}''$ is described by equations

$$-\frac{dV''}{dx} = ZI'', \tag{1.121}$$

$$-\frac{dI''}{dx} = YV'', \tag{1.122}$$

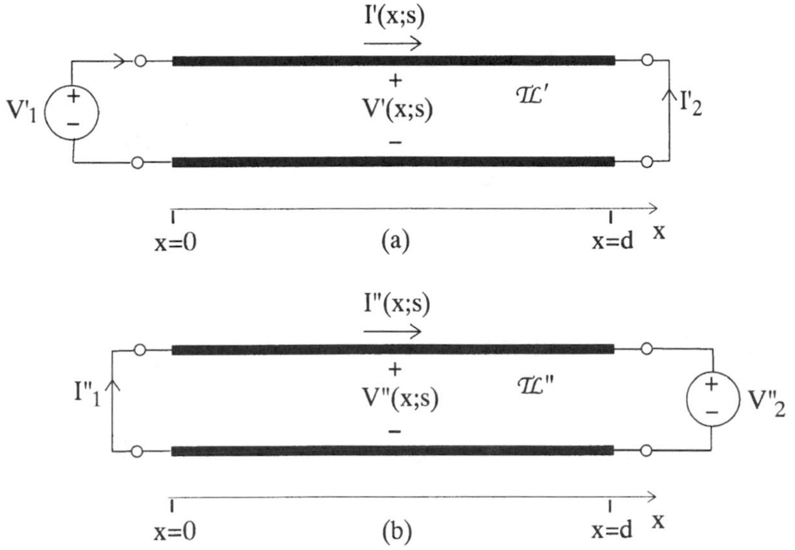

Figure 1.4. Two reciprocal circuits.

with boundary conditions

$$V''(0;s) = 0, \tag{1.123}$$

$$V''(d;s) = V_2''(s), \tag{1.124}$$

where V_2'' is the voltage imposed by the voltage source at $x = d$. Multiplying both members of Eq. (1.117) by I'' and those of Eq. (1.122) by V' and then summing them termwise, we obtain

$$I''\frac{dV'}{dx} + V'\frac{dI''}{dx} = -ZI'I'' - YV'V''. \tag{1.125}$$

Multiplying both members of Eq. (1.118) by V'' and those of Eq. (1.121) by I' and then summing them, termwise, we obtain

$$I'\frac{dV''}{dx} + V''\frac{dI'}{dx} = -ZI'I'' - YV'V''. \tag{1.126}$$

Combining Eqs. (1.125) and (1.126) we arrive immediately at

$$\frac{d(V'I'')}{dx} = \frac{d(V''I')}{dx}. \tag{1.127}$$

Integrating Eq. (1.127) along the line we obtain

$$V_2'I_2'' + V_1'I_1'' = V_2''I_2' + V_1''I_1', \tag{1.128}$$

where $V_2' = V'(d,s)$, $V_1'' = V''(0,s)$, $I_1' = I'(0,s)$, $I_2' = -I'(d,s)$, $I_1'' = I''(0,s)$, and $I_2'' = -I''(d,s)$ (the references for the directions are in agreement with the normal convention at both the line ends).

Now imposing boundary condition (1.119), (1.120), (1.123), and (1.124), we finally have the first form of the reciprocity theorem,

$$V_1'I_1'' = V_2''I_2'. \tag{1.129}$$

Equation (1.129) in the time domain becomes

$$(v_1' * i_1'')(t) = (v_2'' * i_2')(t). \tag{1.130}$$

Two other forms of the reciprocity theorem exist. We shall limit ourselves to introducing them, as the demonstration is obtained by proceeding as before.

1.9.2. The Second Form of the Reciprocity Theorem

Consider two equal lines, the first connected to a current source at $x = 0$ and to an open circuit at $x = d$, and the second connected to an open circuit at $x = 0$ and to a current source at $x = d$. The second form of the reciprocity theorem says that

$$I_1'V_1'' = I_2''V_2'. \tag{1.131}$$

1.9.3. The Third Form of the Reciprocity Theorem

Consider two equal lines, the first connected to a current source at $x = 0$ and to a short circuit at $x = d$, and the second connected to an open circuit at $x = 0$ and to a voltage source at $x = d$. The third form of the reciprocity theorem says that

$$I_1'V_1'' = -I_2'V_2''. \tag{1.132}$$

1.10. RECIPROCITY THEOREMS FOR MULTICONDUCTOR TRANSMISSION LINES

In this section we shall extend the three forms of the reciprocity theorem to multiconductor lines of finite length without distributed sources. Moreover, we shall present an interesting property of reciprocity for multiconductor semi-infinite lines with no distributed sources.

Consider again two equal transmission lines, $\mathcal{TL}'$ and $\mathcal{TL}''$. Let us indicate with one apex the electric variables of the line $\mathcal{TL}'$ and with two apices those relevant to the line $\mathcal{TL}''$. The electric variables of the two lines are the solutions of the equations

$$\frac{d\mathbf{V}'}{dx} = -\mathbf{Z}\mathbf{I}', \tag{1.133}$$

$$\frac{d\mathbf{I}'}{dx} = -\mathbf{Y}\mathbf{V}', \tag{1.134}$$

and

$$\frac{d\mathbf{V}''}{dx} = -\mathbf{Z}\mathbf{I}'', \tag{1.135}$$

$$\frac{d\mathbf{I}''}{dx} = -\mathbf{Y}\mathbf{V}''. \tag{1.136}$$

Left multiplying both terms of Eq. (1.133) by $\mathbf{I}''^{\mathrm{T}}$ and both terms of Eq. (1.136) by $\mathbf{V}'^{\mathrm{T}}$, and then summing them termwise, we obtain

$$\mathbf{I}''^{\mathrm{T}}\frac{d\mathbf{V}'}{dx} + \mathbf{V}'^{\mathrm{T}}\frac{d\mathbf{I}''}{dx} = -\mathbf{I}''^{\mathrm{T}}\mathbf{Z}\mathbf{I}' - \mathbf{V}'^{\mathrm{T}}\mathbf{Y}\mathbf{V}''. \tag{1.137}$$

Operating in a similar way with the other two equations we obtain

$$\mathbf{I}'^{\mathrm{T}}\frac{d\mathbf{V}''}{dx} + \mathbf{V}''^{\mathrm{T}}\frac{d\mathbf{I}'}{dx} = -\mathbf{I}'^{\mathrm{T}}\mathbf{Z}\mathbf{I}'' - \mathbf{V}''^{\mathrm{T}}\mathbf{Y}\mathbf{V}'. \tag{1.138}$$

Using the symmetry of matrices Z and Y, from Eqs. (1.137) and (1.138) we obtain

$$\frac{d}{dx}(\mathbf{I}''^{\mathrm{T}}\mathbf{V}') = \frac{d}{dx}(\mathbf{I}'^{\mathrm{T}}\mathbf{V}''). \tag{1.139}$$

1.10.1. The First Form of the Reciprocity Theorem

Let us suppose that $\mathcal{TL}'$ and $\mathcal{TL}''$ are of finite length d, see Fig. 1.5. It is assumed, for the line $\mathcal{TL}'$

$$\mathbf{V}'(0;s) = \mathbf{V}_1'(s), \tag{1.140}$$

$$\mathbf{V}'(d;s) = \mathbf{0}, \tag{1.141}$$

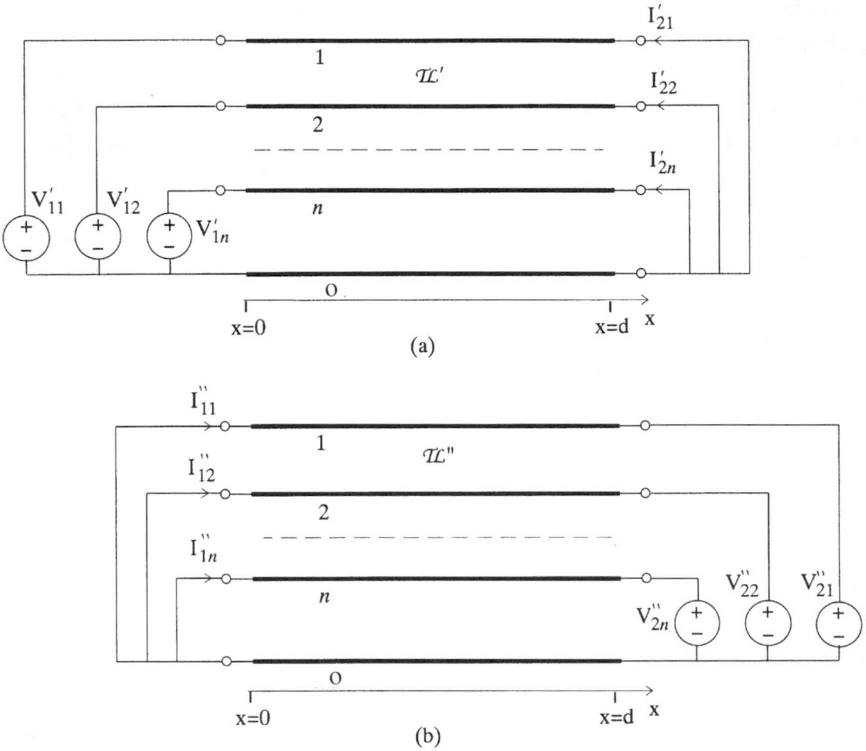

Figure 1.5. Two reciprocal circuits.

and, for the line $\mathcal{TL}''$

$$\mathbf{V}''(0;s) = \mathbf{0}, \tag{1.142}$$

$$\mathbf{V}''(d;s) = \mathbf{V}_2''(s). \tag{1.143}$$

Integrating Eq. (1.139) along the line and using the boundary condition equations (1.140)–(1.143), we obtain the first form of the reciprocity theorem

$$\mathbf{I}_1''^{\mathrm{T}}\mathbf{V}_1' = \mathbf{I}_2'^{\mathrm{T}}\mathbf{V}_2'', \tag{1.144}$$

where

$$\mathbf{I}_1''(s) = \mathbf{I}''(0;s), \tag{1.145}$$

$$\mathbf{I}_2'(s) = -\mathbf{I}'(d;s). \tag{1.146}$$

1.10.2. The Second Form of the Reciprocity Theorem

Consider again two lines of finite length $\mathscr{T}\mathscr{L}'$ and $\mathscr{T}\mathscr{L}''$, and assume for $\mathscr{T}\mathscr{L}'$

$$\mathbf{I}'(0;s) = \mathbf{I}'_1(s), \tag{1.147}$$

$$\mathbf{I}'(d;s) = \mathbf{0}, \tag{1.148}$$

and for $\mathscr{T}\mathscr{L}''$

$$\mathbf{I}''(0;s) = \mathbf{0}, \tag{1.149}$$

$$\mathbf{I}''(d;s) = -\mathbf{I}''_2(s). \tag{1.150}$$

In this case the reciprocity theorem assumes the form

$$\mathbf{I}'^{\mathrm{T}}_1\mathbf{V}''_1 = \mathbf{I}''^{\mathrm{T}}_2\mathbf{V}'_2. \tag{1.151}$$

1.10.3. The Third Form of the Reciprocity Theorem

Consider again two lines of finite length, $\mathscr{T}\mathscr{L}'$ and $\mathscr{T}\mathscr{L}''$, and assume for $\mathscr{T}\mathscr{L}'$

$$\mathbf{I}'(0;s) = \mathbf{I}'_1(s), \tag{1.152}$$

$$\mathbf{V}'(d;s) = \mathbf{0}, \tag{1.153}$$

and for $\mathscr{T}\mathscr{L}''$

$$\mathbf{I}''(0;s) = \mathbf{0}, \tag{1.154}$$

$$\mathbf{V}''(d;s) = \mathbf{V}''_2(s). \tag{1.155}$$

In this case the reciprocity theorem takes the form

$$\mathbf{I}'^{\mathrm{T}}_1\mathbf{V}''_1 = -\mathbf{I}'^{\mathrm{T}}_2\mathbf{V}''_2. \tag{1.156}$$

1.10.4. Reciprocity Theorem for a Semi-infinite Transmission Line

Consider, now, two semi-infinite lines, $0 \leqslant x < \infty$, and assume that both the voltage and the current are regular at infinity, that is,

$$\lim_{x\to+\infty} \mathbf{V}'(x;s) = \mathbf{0}, \quad \lim_{x\to+\infty} \mathbf{I}'(x;s) = \mathbf{0}, \tag{1.157}$$

$$\lim_{x\to+\infty} \mathbf{V}''(x;s) = \mathbf{0}, \quad \lim_{x\to+\infty} \mathbf{I}''(x;s) = \mathbf{0}. \tag{1.158}$$

From Eq. (1.139), using Eqs. (1.157) and (1.158), we obtain the reciprocity property

$$\mathbf{I}'^{\mathrm{T}}(x;s)\mathbf{V}''(x;s) = \mathbf{I}''^{\mathrm{T}}(x;s)\mathbf{V}'(x;s) \quad \text{for } 0 \leqslant x < \infty. \tag{1.159}$$

CHAPTER 2

Ideal Two-Conductor Transmission Lines Connected to Lumped Circuits

In this chapter we develop in detail the principal theme of this book by referring to ideal two-conductor transmisson lines. Let us consider for the sake of simplicity a network consisting of an ideal two-conductor line connecting two lumped circuits as sketched in Fig. 2.1. The two lumped circuits could also interact between them through other coupling mechanisms, for instance, controlled sources of the circuit C_1 that are governed by some electrical variables of the circuit C_2, and transformers.

The two-port representing the ideal two-conductor transmission line is directly characterized in the time domain. Both an input-output description and input-state-output description are deeply investigated. Simple equivalent circuits of Thévenin and Norton type, obtained from the input-state-output description, are examined. Each port of the Thévenin (Norton) equivalent circuit consists of a linear resistor in series (parallel) with a controlled voltage (current) source. The laws governing the controlled sources are retarded linear algebraic equations with one delay. The most widely used circuit simulator, SPICE, uses equivalent circuits of this type to simulate ideal transmission lines.

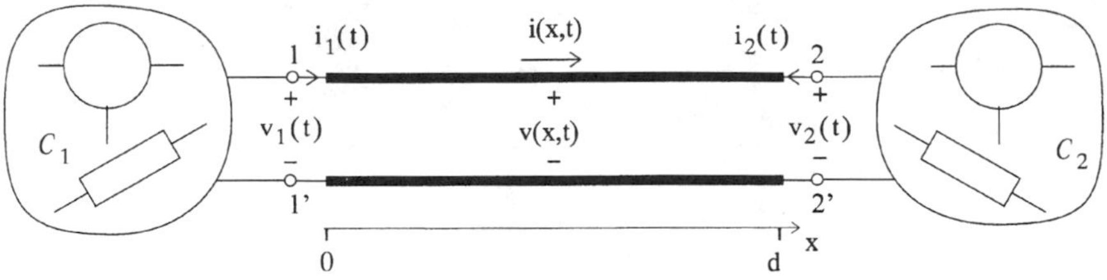

Figure 2.1. Lumped circuits connected by a two-conductor transmission line.

2.1. d'ALEMBERT SOLUTION OF TWO-CONDUCTOR TRANSMISSION LINE EQUATIONS

The equations for ideal two-conductor transmission lines are (see §1.2.1),

$$-\frac{\partial v}{\partial x} = L\frac{\partial i}{\partial t}, \tag{2.1}$$

$$-\frac{\partial i}{\partial x} = C\frac{\partial v}{\partial t}, \tag{2.2}$$

where L and C are positive constants. For the moment we do not consider distributed independent sources along the line, but at the end of this chapter we shall show how to take their effects into account.

The determination of the general solution of Eqs. (2.1) and (2.2) may be made by operating directly in the time domain. Starting from these equations, by derivation and substitution, we obtain the following system of uncoupled equations:

$$\frac{\partial^2 v}{\partial x^2} - LC\frac{\partial^2 v}{\partial t^2} = 0, \tag{2.3}$$

$$\frac{\partial^2 i}{\partial x^2} - LC\frac{\partial^2 i}{\partial t^2} = 0. \tag{2.4}$$

Each of these has the form of a *wave equation* (e.g., Smirnov, 1964a). By placing

$$c = \frac{1}{\sqrt{LC}}, \tag{2.5}$$

the general solution of Eqs. (2.3) and (2.4) in d'Alembert form is

$$v(x,t) = v^{+}(t - x/c + \alpha^{+}) + v^{-}(t + x/c + \alpha^{-}), \tag{2.6}$$

$$i(x,t) = i^{+}(t - x/c + \alpha^{+}) + i^{-}(t + x/c + \alpha^{-}), \tag{2.7}$$

where α^{+} and α^{-} are arbitrary constants. The functions v^{+}, v^{-}, i^{+}, and i^{-} are arbitrary. Generally they can be continuous or indeed generalized functions in the sense of the theory of distributions — for example, unit step functions, Dirac functions, etc. (e.g., Courant and Hilbert, 1989).

The term $v^{+}(t - x/c + \alpha^{+})$ represents a *traveling voltage wave*, which propagates in the positive direction of the x axis, with constant velocity c, without being distorted. It is the so-called *forward voltage wave*. Similarly, $v^{-}(t + x/c + \alpha^{-})$ is a *traveling voltage wave* that propagates in the direction of negative x. It is the so-called *backward voltage wave*. Analogous considerations hold for i^{+} and i^{-}, which are called, respectively, *forward current wave* and *backward current wave*.

Obviously, the set of all the possible solutions of Eqs. (2.3) and (2.4) is much ampler than the set of the solutions of the original system equations (2.1) and (2.2). As Eqs. (2.6) and (2.7) are solutions of the wave equations, it is sufficient to ensure that they satisfy any of the two equations of the original system. Substituting Eqs. (2.6) and (2.7) in Eq. (2.1), we obtain

$$i^{+} = \frac{v^{+}}{R_c}, \tag{2.8}$$

$$i^{-} = -\frac{v^{-}}{R_c}, \tag{2.9}$$

where

$$R_c = \sqrt{L/C} \tag{2.10}$$

is the *characteristic "resistance"* of the line. Finally, substituting Eqs. (2.8) and (2.9) in Eq. (2.7) we obtain

$$i(x,t) = \frac{1}{R_c}[v^{+}(t - x/c + \alpha^{+}) - v^{-}(t + x/c + \alpha^{-})]. \tag{2.11}$$

In this way, the general solution of the line equations is represented in terms of forward and backward voltage waves only. Clearly, we can also represent it through the superposition of forward and backward current waves. We need only to use Eqs. (2.8) and (2.9).

Remark

The equations of an ideal line also admit solutions of the type $v = V$, $i = I$, where V, I are two arbitrary uncorrelated constants. We may disregard such solutions because they, in general, are not compatible with the initial and boundary conditions. ◇

Let us consider now a line of finite length d. Contrary to what is done in the literature, in this chapter and in the next one, we shall choose constants α^+ and α^- so that $v^+(t)$ represents the amplitude of the forward voltage wave at the right end of the line, $x = d$, and at time instant t, while $v^-(t)$ represents the amplitude of the backward voltage wave at the left-hand side, $x = 0$, and at time instant t. Consequently, we must choose

$$\alpha^+ = T, \quad \alpha^- = 0, \tag{2.12}$$

where the *one-way transit time*

$$T = \frac{d}{c}, \tag{2.13}$$

is the time necessary for a wave leaving one end of the line to reach the other end. With this choice, both v^+ and v^- are defined in the interval $(0, +\infty)$. The physical significance of this choice is very simple: an observer at end $x = 0$ can see the entire backward wave and an observer at $x = d$ can see the entire forward wave. Clearly, such a choice is significant only when the line has a finite length. When we shall consider lines of infinite length we shall always choose $\alpha^+ = \alpha^- = 0$.

In conclusion, the general solution for the voltage and current distributions along an ideal two-conductor transmission line always can be represented through the superposition of two *waves*, which do not interact mutually along the line — a forward and a backward wave. The amplitudes of these waves are determined by imposing the initial distribution of the voltage and current along the line, that is, the *initial conditions*

$$i(x, t = 0) = i_0(x) \quad \text{for } 0 \leqslant x \leqslant d, \tag{2.14}$$

$$v(x, t = 0) = v_0(x) \quad \text{for } 0 \leqslant x \leqslant d, \tag{2.15}$$

and by indicating what happens at the line ends, that is, the *boundary conditions*.

In Chapter 1, §1.5.1, we have shown that there is only one solution of the two-conductor transmission line equations compatible with assigned initial conditions and boundary values at each line end for the voltage or current. However, in general, the values of the voltage and current at line ends are not known, but are themselves unknowns of the problem; they depend on the actual network in which the line lies. The forward wave and the backward one interact mutually only through the circuits connected to the line. Now we shall deal with this problem.

Throughout we shall always assume that the values at the line ends of the initial current and voltage distributions are compatible with the operation of the circuits to which the line is connected, at the initial time instant.

2.2. SOME ELEMENTARY NETWORKS

Even if this text assumes that the reader is familiar with the behavior of an ideal two-conductor transmission line and the fundamental concepts concerning the propagation along them, in order to introduce the reader gradually to the difficulties inherent in the problems dealt with in this book, as a next step we shall consider some elementary networks. In this way we also have the opportunity to recall briefly some of the fundamental concepts relevant to transmission line theory.

In particular, in this section we shall show how forward and backward voltage waves are determined once the initial conditions have been assigned and the boundary conditions specified for semi-infinite transmission lines connected to simple lumped elements. In particular we shall gradually outline the difficulties that arise when we have to impose the boundary conditions, because the voltage and current values at the line ends, unlike their initial distributions along the line, are themselves unknowns of the problem.

2.2.1. An Infinite Line

Let us first consider a line that extends spatially from $-\infty$ to $+\infty$. The particularity of such a problem, as we shall now see, lies in the fact that the solution for any x and t (finite) depends only on the initial conditions and not on what is eventually connected to the two ends of the line at infinity.

For the general solution of a line of infinite length let us consider

$$v(x,t) = v^+(t - x/c) + v^-(t + x/c), \tag{2.16}$$

$$i(x,t) = \frac{1}{R_c}[v^+(t - x/c) - v^-(t + x/c)]. \tag{2.17}$$

In this case the voltage and the current are defined in the space interval $(-\infty, +\infty)$, and so $v^{+}(\tau)$ and $v^{-}(\tau)$ are defined in the interval $(-\infty, +\infty)$.

As we now show, v^{+} and v^{-} are univocally determined from the initial conditions (2.14) and (2.15). Requiring that (2.16) and (2.17) satisfy Eqs. (2.14) and (2.15) for $-\infty < x < +\infty$, we immediately obtain the forward and backward voltage waves,

$$v^{+}(\tau) = \frac{1}{2}[v_0(-c\tau) + R_c i_0(-c\tau)] \quad \text{for } -\infty < \tau < +\infty, \tag{2.18}$$

$$v^{-}(\tau) = \frac{1}{2}[v_0(c\tau) - R_c i_0(c\tau)] \quad \text{for } -\infty < \tau < +\infty. \tag{2.19}$$

In this case, the regularity properties of the solution depend solely on the regularity of the initial distribution of the voltage and current along the line.

A generic initial condition produces both forward and backward waves, whereas for particular initial conditions only forward waves or backward ones are excited. When $v_0(x) = R_c i_0(x)$, there is only a forward wave, while when $v_0(x) = -R_c i_0(x)$, there is only a backward one. In both these cases the absolute value of the ratio between the voltage and the current at any x and t is equal to the characteristic resistance of the line.

2.2.2. A Semi-infinite Line Connected to an Ideal Current Source

Earlier we considered a line in free evolution, which extends spatially from $-\infty$ to $+\infty$. Its solution at finite t and x is not affected by what it is connected to at its ends. To highlight the influence of the ends it is necessary to place oneself at a finite distance from them.

For this reason let us consider a line that extends spatially from $x = 0$ to $+\infty$ (semi-infinite line), connected to an independent current source at $x = 0$ acting for $t \geqslant 0$, as shown in Fig. 2.2. The boundary condition at $x = 0$ is a further constraint to be imposed on the general solution (2.16) and (2.17). This is the simplest example of a problem with boundary conditions. In this case, function v^{+} is defined in the time interval $(-\infty, +\infty)$, while v^{-} is defined in $(0, \infty)$.

The initial conditions determine the forward voltage wave v^{+} only for $-\infty < \tau \leqslant 0$ and the backward voltage wave v^{-} for $0 \leqslant \tau < +\infty$. For the sake of simplicity, here we assume that initially there is only the backward voltage wave along the line, that is, $v_0(x) = -R_c i_0(x)$,

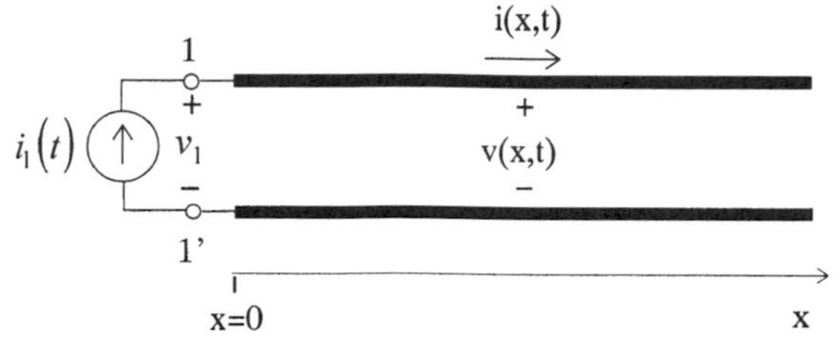

Figure 2.2. Semi-infinite line connected to an independent current source.

hence

$$v^+(\tau) = 0 \quad \text{for } -\infty < \tau < 0, \tag{2.20}$$

$$v^-(\tau) = v_0(c\tau) \quad \text{for } 0 \leqslant \tau < \infty. \tag{2.21}$$

Instead, the forward voltage wave for $t > 0$ depends on the boundary condition at $x = 0$,

$$i(x = 0, t) = i_1(t) \quad \text{for } t \geqslant 0; \tag{2.22}$$

$i_1(t)$ is the current imposed by the independent source. So that the initial current distribution along the line is compatible with this boundary condition, we require that $R_c i_1(t = 0) = -v_0(x = 0)$. This compatibility condition will imply that v^+ is continuous at $t = 0$.

By imposing that the general solution of the line equations satisfies Eq. (2.22), we obtain

$$v^+(t) = R_c i_1(t) + v^-(t) \quad \text{for } t \geqslant 0, \tag{2.23}$$

where v^- is known because it is completely determined by the initial conditions. Thus the voltage and current distributions along the line are

$$v(x, t) = [R_c i_1(t - x/c) + v^-(t - x/c)]u(t - x/c) + v^-(t + x/c), \tag{2.24}$$

$$i(x, t) = \frac{1}{R_c}\{[R_c i_1(t - x/c) + v^-(t - x/c)]u(t - x/c) - v^-(t + x/c)\}, \tag{2.25}$$

where $u(t)$ is the Heaviside unit step function.

We observe that the line voltage at end $x = 0$, $v_1(t) = v(x = 0, t)$, is related to the current $i_1(t)$ through the expression

$$v_1(t) = R_c i_1(t) + 2v^-(t). \tag{2.26}$$

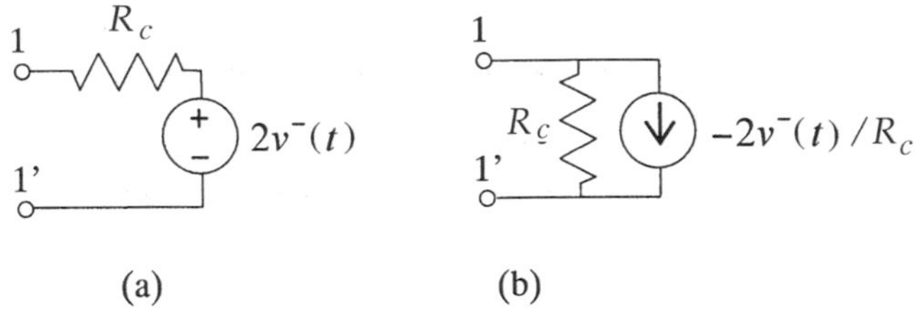

Figure 2.3. Equivalent circuit of Thévenin (a) and Norton (b) type of a semi-infinite two-conductor transmission line.

Note that the relation (2.26) would not change if the forward wave v^+ were different from zero because the line is infinite in the positive x-direction.

This result is very interesting. The behavior of the semi-infinite line at end $x = 0$ can be represented by a very simple lumped circuit, consisting of a linear resistor of resistance R_c connected in series with an independent voltage source imposing the voltage $2v^-$, see Fig. 2.3a. This is an equivalent representation of the semi-infinite line of Thévenin type. Figure 2.3b illustrates the corresponding equivalent circuit of Norton type: the resistor of resistance R_c is connected in parallel with an independent current source that imposes the current $-2v^-/R_c$.

If the line were initially at rest its behavior would be equivalent to that of a linear resistor of resistance R_c.

2.2.3. A Semi-infinite Line Connected to a Linear Resistor; Reflection Coefficient

When a semi-infinite line is connected to an independent current or voltage source, it is simple to impose the corresponding boundary condition because the value of the current or voltage at the line end is known.

In practice, the action of the element to which the line is connected can be more complex than that made by a single independent source — for example, if one thinks of a line connected to a resistor, or to a capacitor, or indeed to a very complex lumped circuit. In all of these cases the boundary conditions are equations themselves, and the voltage and current at the line end are themselves unknowns.

Let us consider a semi-infinite line connected to a linear resistor at $x = 0$, see Fig. 2.4a. As in the previous case, it is assumed that

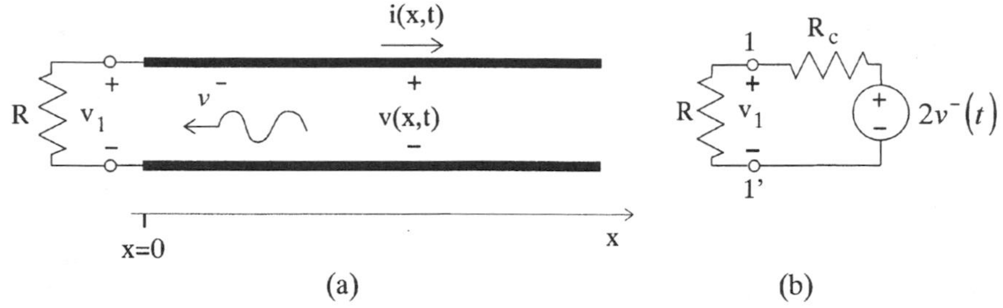

Figure 2.4. Semi-infinite line connected to a linear resistor (a); a lumped equivalent circuit (b).

initially there is only the backward voltage wave along the line. For $t > 0$, a forward voltage wave is excited because of the reflection due to the presence of the resistor at $x = 0$.

To determine the reflected wave at the termination, we must impose that the general solution (2.16) and (2.17) satisfy the boundary equation

$$v(x = 0, t) = -Ri(x = 0, t) \quad \text{for } t \geqslant 0; \tag{2.27}$$

the resistance R is defined according to the normal convention, so for passive resistors it is positive.

For $R \neq R_c$, the assigned initial condition is compatible with boundary condition (2.27) only if $v_0(x = 0) = 0$, while for $R = R_c$ $v_0(x = 0)$ may assume any value. This compatibility condition will always imply that v^+ is continuous at $t = 0$.

Imposing boundary condition (2.27), we obtain the simple relation

$$v^+(t) = \Gamma v^-(t) \quad \text{for } 0 \leqslant t < +\infty, \tag{2.28}$$

where

$$\Gamma = \frac{R - R_c}{R + R_c} \tag{2.29}$$

is the *voltage reflection coefficient* of the resistive termination.

We can solve this problem in a simpler way. The line behavior at the end $x = 0$ can be represented through the Thévenin equivalent one-port shown in Fig. 2.3a. By replacing the line with this equivalent circuit, we obtain the circuit shown in Fig. 2.4b. For the voltage v_1 we immediately obtain

$$v_1(t) = \frac{2R}{R + R_c} v^-(t) \quad \text{for } 0 \leqslant t < +\infty. \tag{2.30}$$

Since for $t \geqslant 0$

$$v^{+}(t) = v_{1}(t) - v^{-}(t), \tag{2.31}$$

from relation (2.30) we immediately obtain Eq. (2.28).

Remark

When $R = R_c$ the reflection coefficient is equal to zero and the forward wave amplitude is zero, that is, there is no wave reflected by the resistor at $x = 0$. In this case, at $x = 0$, the line behaves as if it were extended to $-\infty$, and for this reason it is called *matched* at $x = 0$.

When $R \neq R_c$ the reflection coefficient is no longer zero and a forward wave is excited. If the line is terminated with a short circuit, $R = 0$, the reflection coefficient is equal to -1, while when the line is terminated with an open circuit, $R = \infty$, the reflection coefficient is equal to $+1$.

When the line is terminated with a short circuit the forward voltage wave is the negative specular image of the backward wave with respect to $x = 0$. When the line is terminated with an open circuit the forward voltage wave is the positive specular image of the backward wave, with respect to $x = 0$. For strictly passive resistors, that is, R positive and bounded, the absolute value of the reflection coefficient is <1. For active linear resistors, the absolute value of the reflection coefficient is >1 because $R < 0$. ◇

2.2.4. A Semi-infinite Line Connected to a Linear Capacitor

Let us analyze now what happens when the line is connected to a linear dynamic one-port, for example, a capacitor, see Fig. 2.5. In cases like this, the boundary condition to be imposed is a linear ordinary differential equation.

Let us again assume the initial condition (2.20) and (2.21). To determine v^{+} for $0 \leqslant t < +\infty$ we need to impose the boundary equation

$$i_1 = -C\frac{dv_1}{dt}, \tag{2.32}$$

where

$$i_1(t) = i(x = 0; t), \quad v_1(t) = v(x = 0; t). \tag{2.33}$$

Capacitance C is defined according to the normal convention, thus it

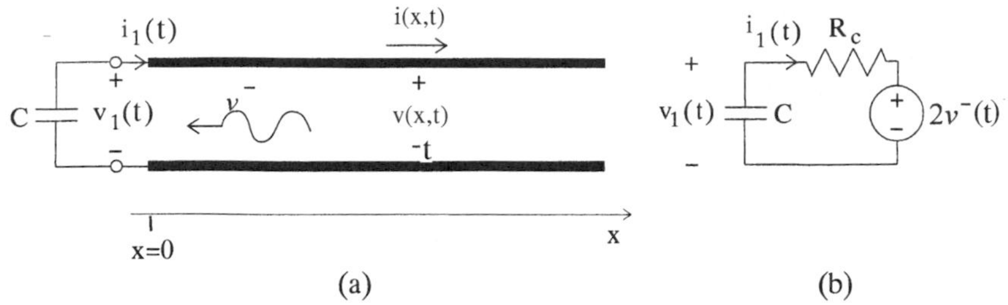

Figure 2.5. Semi-infinite line connected to a linear capacitor (a); a lumped equivalent circuit (b).

is positive if the capacitor is passive. For the compatibility condition the initial value of the voltage of the capacitor $v_1(t=0)$ must be equal to $v_0(x=0)$.

We can solve this problem imposing that the general solution (2.16) and (2.17) of the line equations satisfies Eq. (2.32). In this way we obtain directly the equation for the unknown v^+, which will be a linear ordinary differential equation.

The problem may be resolved in a different way, that is, by solving directly the equivalent lumped circuit shown in Fig. 2.5b, where the line behavior at $x = 0$ is again represented by the equivalent Thévenin one-port. From the circuit of Fig. 2.5b we immediately obtain, for $t \geqslant 0$,

$$C\frac{dv_1}{dt} + \frac{v_1}{R_c} = \frac{2v^-}{R_c}. \tag{2.34}$$

This ordinary differential equation has to be solved with the initial condition $v_1(t=0) = v_0(x=0)$. The right-hand term of Eq. (2.34) is known because for $t \geqslant 0$ the function v^- is completely determined by the initial conditions of the line. Once v_1 is known, the function v^+ is determined through Eq. (2.31).

Remark

The solution of Eq. (2.34) is more complex than those we have previously discussed because we must now resolve a nonhomogeneous ordinary differential equation. As the equation is linear and time invariant, this problem can be easily solved by using the Laplace transform (see Appendix B).

For the sake of simplicity, let us assume that $v_1(t = 0) = v_0(x = 0) = 0$. By applying the Laplace transform to differential equation (2.34), we obtain

$$V_1(s) = \frac{2Z(s)}{R_c + Z(s)} V^-(s), \tag{2.35}$$

where $V_1(s)$ and $V^-(s)$ are, respectively, the Laplace transforms of $v_1(t)$ and $v^-(t)$, and $Z = 1/sC$ is the impedance operator corresponding to the capacitor. Now by using Eq. (2.31), from Eq. (2.35) we obtain

$$V^+(s) = \Gamma(s) V^-(s), \tag{2.36}$$

where the voltage reflection "operator" is defined as

$$\Gamma(s) = \frac{Z(s) - R_c}{Z(s) + R_c}. \tag{2.37}$$

Therefore, in the Laplace domain, the action of the capacitor can be represented by the reflection coefficient $\Gamma(s)$ given by Eq. (2.37), which depends on the complex variable s. In the considered case $\Gamma(s)$ has a pole at $s = -1/R_c C$.

Using the convolution theorem we obtain in the time domain,

$$v^+(t) = (\gamma * v^-)(t), \tag{2.38}$$

where $\gamma(t)$ is the Laplace inverse transform of $\Gamma(s)$, given by

$$\gamma(t) = -\delta(t) + \frac{2}{\tau_0} e^{-t/\tau_0} u(t), \tag{2.39}$$

and $\tau_0 = R_c C$.

In periodic and quasi-periodic operating conditions it is more suitable to evaluate first $V^+(s)$ along the imaginary axis and then to calculate $v^+(t)$ by making the inverse Fourier transform of $V^+(i\omega)$.

In general, the action of a dynamic one-port, initially at rest, can always be represented by a convolution relation of the type (2.38); the impulse response $\gamma(t)$ is the wave reflected by the one-port when an impulsive backward wave of unitary amplitude reaches it. ◇

2.2.5. A Semi-infinite Line Connected to a Nonlinear Resistor

Let us now consider the case in which the semi-infinite line is connected to a nonlinear resistor that we assume, for example, to be voltage controlled,

$$i_1 = -g(v_1); \tag{2.40}$$

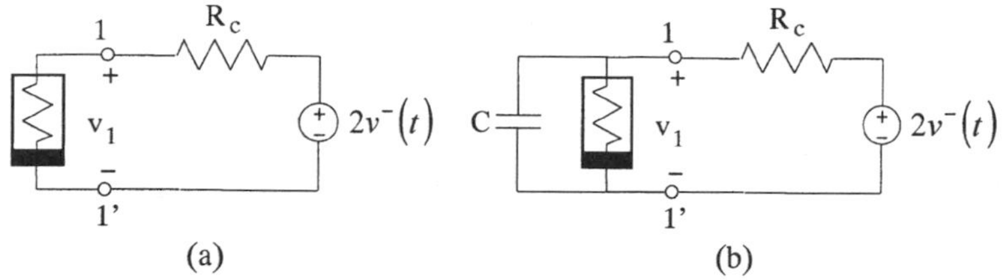

Figure 2.6. Equivalent circuit of a semi-infinite transmission line connected to a nonlinear resistor (a), and to a nonlinear resistor in parallel with a linear capacitor (b).

$g = g(v_1)$ represents the characteristic of the resistor defined according to the normal convention. It is again assumed that Eqs. (2.20) and (2.21) express the initial conditions. For the compatibility of the initial distributions of the voltage and current at the line end $x = 0$ with the boundary equation (2.40) we must have $v_0(x = 0) - R_c g[v_0(x = 0)] = 0$.

Figure 2.6a shows the equivalent circuit obtained by representing the line behavior at the left end through the Thévenin equivalent one-port of Fig. 2.3a. The equation for v_1 is immediately obtained,

$$g(v_1) + \frac{v_1}{R_c} = \frac{2v^-}{R_c}. \tag{2.41}$$

Therefore, for $t > 0$ the forward wave amplitude is the solution of the nonlinear algebraic equation (we have used Eq. (2.31))

$$R_c g[v^+(t) + v^-(t)] + v^+(t) - v^-(t) = 0. \tag{2.42}$$

Here v^- is always a known quantity.

This equation implicitly defines function $v^+ = F(v^-)$, which expresses the reflected wave amplitude v^+ as a function of the incident wave amplitude v^-. For the time being we shall not consider all the problems that are connected with resolving the nonlinear equation (2.41) or the equivalent Eq. (2.42), which, let us say at once, can be resolved exactly only in few cases.

Let us briefly point out that, in general, Eq. (2.41), and hence Eq. (2.42), may not have a solution or may have more than one solution for a given v^-. This depends solely on the characteristic curve of the nonlinear resistor to which the line is connected. This may be seen graphically in the plane (v_1, i_1), see Fig. 2.7.

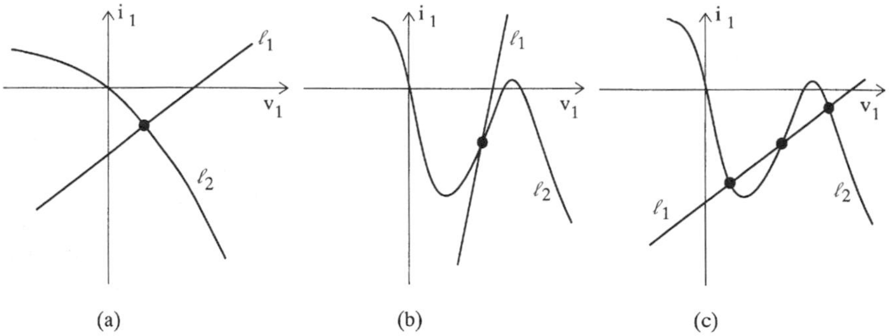

Figure 2.7. Graphical solution of Eq. (2.41) for different characteristic curves of the nonlinear resistor.

The solutions of Eq. (2.41) are represented by the intersections of the straight line ℓ_1 described by equation

$$i_1 = \frac{v_1 - 2v^-}{R_c}, \tag{2.43}$$

with the characteristic curve ℓ_2 of the nonlinear resistor described by Eq. (2.40). Figure 2.7 shows some possible situations.

If the characteristic curve ℓ_2 is strictly monotone, there is only one intersection between ℓ_1 and ℓ_2. In this case the function F is certainly single valued, see Fig. 2.8a.

Figure 2.7b and 2.7c shows what can happen when the characteristic curve ℓ_2 is not strictly monotone. It is clear that, if

$$\frac{dg}{dv_1} > -\frac{1}{R_c}, \tag{2.44}$$

then the slope of ℓ_2 is always greater than the local slope of ℓ_2, and so there is only one intersection, Fig. 2.7b. Also in this case, the function F is single valued. If, however, Eq. (2.44) is not satisfied, then the slope of ℓ_1 can be less than that of ℓ_2. In this case there are three intersections between ℓ_1 and ℓ_2, Fig. 2.7c, and function F has three branches, see Fig. 2.8b. Clearly, this situation is not acceptable from the physical point of view. As we shall see in Chapter 9, when there are situations of this type, it means that the model we are solving does

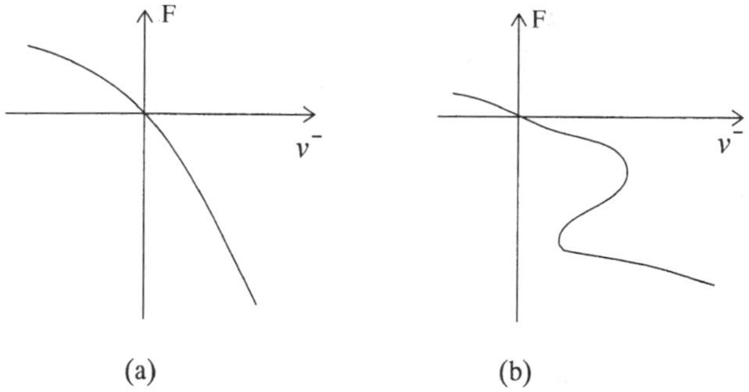

Figure 2.8. Nonlinear functions F.

not adequately represent the corresponding physical system, and hence a more accurate model is required.

2.2.6. A Semi-infinite Line Connected to a Nonlinear Resistor in Parallel with a Linear Capacitor

Let us introduce a linear capacitor in parallel with the nonlinear resistor into the network that we have just analyzed. The equivalent circuit is illustrated in Fig. 2.6b.

From direct inspection of the circuit in Fig. 2.6b we find

$$C\frac{dv_1}{dt} + \frac{v_1}{R_c} + g(v_1) = \frac{2v^-}{R_c}. \tag{2.45}$$

This equation must be solved together with the initial condition for v_1.

Because of the presence of the capacitor, the relation between v_1 and v^- is no longer of the instantaneous type, but is of the functional type, with memory. This is a further difficulty, added to that due to the presence of the nonlinearity. Once v_1 is known, the function v^+ is determined through Eq. (2.31).

In this case, unlike that considered in the previous paragraph, the solution is always unique irrespective of the particular shape of the characteristic curve of the voltage-controlled resistor (if the capacitance is different from zero). We shall analyze this question in detail in Chapter 9.

Clearly, then, difficulties will gradually proliferate as we consider more and more complex circuital elements.

2.3. NATURAL FREQUENCIES OF A FINITE LENGTH TRANSMISSION LINE CONNECTED TO SHORT CIRCUITS

The dynamics of the voltages and currents in transmission lines of finite length are affected by both the ends. One of the most important effects is the appearance of evolutionary natural modes of oscillating type. Here we briefly recall this feature.

Consider, for example, a transmission line of length d connected to a short circuit at both ends, see Fig. 2.9a, and assume that along it there is initially only the backward voltage wave v^-, whose initial profile is that shown in Fig. 2.9b.

Here we use again Eqs. (2.16) and (2.17) to express the general solution of the line equations. Therefore v^+ is defined in the interval $(-T, +\infty)$ and v^- in $(0, +\infty)$. Imposing the boundary condition $v(x = 0, t) = 0$ at the line left end, we obtain

$$v^+(t) = -v^-(t) \quad \text{for } t \geqslant 0. \tag{2.46}$$

Therefore, the voltage $v(x, t)$ for $(t - x/c) \geqslant 0$ is given by

$$v(x, t) = -v^-(t - x/c) + v^-(t + x/c). \tag{2.47}$$

The boundary condition $v(x = d, t) = 0$ at the line right end also imposes

$$v^-(t - T) = v^-(t + T) \quad \text{for } t \geqslant T. \tag{2.48}$$

This implies that v^- must be periodic with period $2T$ for $t > T$. This result is obviously independent of the distribution of the initial conditions of the line.

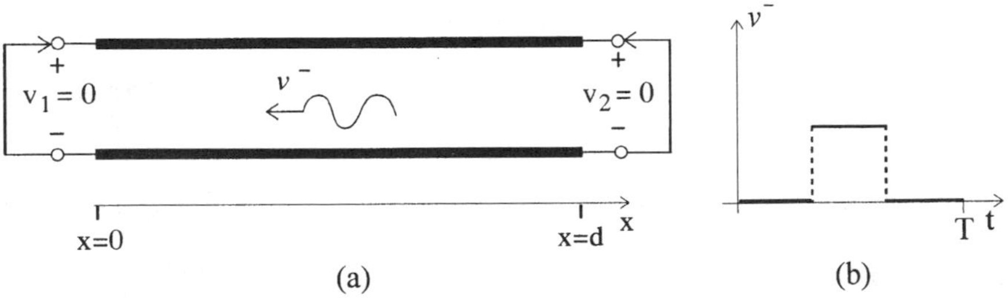

Figure 2.9. Transmission line connected to two short circuits.

The fact that the problem we are considering admits only periodic solutions is a direct consequence of an important property — all the natural modes of evolution of this circuit are of harmonic type and their characteristic frequencies are an integer multiple of the fundamental characteristic frequency

$$\omega_0 = \frac{\pi}{T} = \frac{\pi c}{d}. \tag{2.49}$$

This property can easily be demonstrated. Among all the possible solutions, consider the one obtained by subtracting two traveling sinusoidal waves, one forward and the other backward, both with the same frequency ω and the same amplitude A,

$$\begin{aligned} v(x,t) &= A\sin[\omega(t + x/c)] - A\sin[\omega(t - x/c)] \\ &= 2A\cos(\omega t)\sin(\omega x/c). \end{aligned} \tag{2.50}$$

The superposition of two traveling sinusoidal waves generates a *standing sinusoidal wave*.

This solution of the wave equations naturally satisfies the boundary condition at $x = 0$, irrespective of the value of ω. This agrees with the fact that when a backward wave reaches the short circuit a forward wave is excited, again of a sinusoidal type, with the same frequency but opposite in amplitude. As the stationary voltage wave must also satisfy the other boundary condition, that is, $v(x = d, t) = 0$, frequency ω cannot be arbitrary; only some discrete values are compatible with the stationary voltage wave configuration and the boundary conditions, that is, only those that satisfy the condition:

$$\omega = n\omega_0 \quad n = 1, 2, 3, \ldots, \tag{2.51}$$

where ω_0 is given by Eq. (2.49). Frequency ω_0 is the smallest compatible with the boundary conditions and is called the *fundamental frequency* of the circuit; all the others are integer multiples of it.

One must demonstrate immediately that the same result is reached if, instead of two short circuits at the line ends, one considers two open circuits. However, if we have a line connected to a short circuit at one end and to an open one at the other, Eq. (2.51) still holds, but the fundamental frequency is double that obtained when the line is connected to two short circuits.

As we shall show later, oscillating natural modes exist even when the line is connected to other types of one-ports. In fact it can be said that the presence of these modes is substantially the result of two contemporary actions, the multireflections due to the elements connected to the line ends, and the delay introduced by the finite length of the line. The presence of dissipation in the elements connected to the line implies that these oscillations will have amplitudes decaying in time. Linear dynamic elements such as inductors and capacitors modify the oscillation frequencies, as we shall show later in this chapter.

2.4. TWO-CONDUCTOR TRANSMISSION LINES AS TWO-PORTS

We have already seen in the study of semi-infinite lines that the resolution of the problem can be very much simplified if we first determine the voltage and current at the line end $x = 0$ by representing the line behavior through the Thévenin or Norton equivalent circuits shown in Fig. 2.3. Then, once the voltage or the current at the end $x = 0$ is known, we can calculate both the voltage and current distribution all along the line.

The same approach can be adopted to study transmission lines of finite length connecting generic lumped circuits. The interaction of two-conductor transmission lines with lumped circuits can be described by representing the lines as two-ports, see Fig. 2.10. The advantages of this method are widely discussed in the introduction to this book, so we shall not return to them.

The voltages and currents at line ends are related to the voltage and current distributions along the line through the relations

$$v_1(t) = v(x = 0, t), \quad i_1(t) = i(x = 0, t) \tag{2.52}$$

and

$$v_2(t) = v(x = d, t), \quad i_2(t) = -i(x = d, t). \tag{2.53}$$

The reference directions are always chosen according to the normal convention for the two ports corresponding to the two ends of the line.

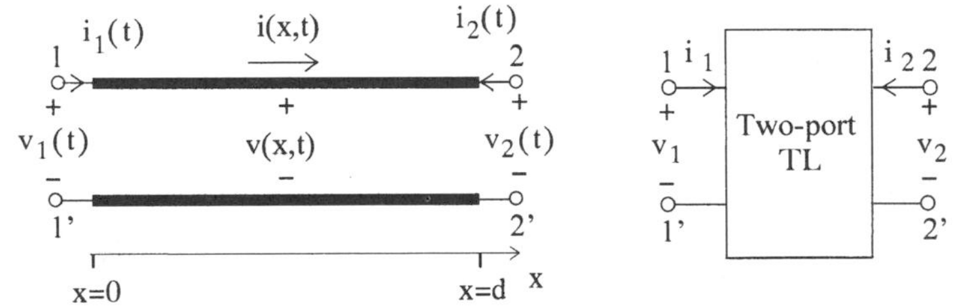

Figure 2.10. Two-conductor transmission line as a two-port element.

To determine the relations between the variables v_1, v_2, i_1, and i_2 imposed by the transmission line we shall do as follows:

- we first impose the initial conditions along the line; and
- then we particularize the solution of the line equations at the line ends.

For the general solution of the line equations we shall use

$$v(x,t) = v^+(t - x/c + T) + v^-(t + x/c), \tag{2.54}$$

$$i(x,t) = \frac{1}{R_c}[v^+(t - x/c + T) - v^-(t + x/c)], \tag{2.55}$$

where T is given by Eq. (2.13). We remember that with this choice both $v^+(t)$ and $v^-(t)$ are defined in the interval $(0, +\infty)$; they represent, respectively, the forward voltage wave amplitude at end $x = d$ and the backward voltage wave amplitude at end $x = 0$.

2.4.1. State Variables of the Line

The voltage and current distributions along the line are completely identified by the functions $v^+(t)$ and $v^-(t)$, and vice versa. Thus $v^+(t)$ and $v^-(t)$ completely specify the *state* of the line. We shall consider them as *state variables* of the line.

It is easy to verify that the per-unit-length energy w_{em} associated with the electromagnetic field along the guiding structure, see §1.4.1, may be expressed as

$$w_{em}(x,t) = \tfrac{1}{2}Li^2 + \tfrac{1}{2}Cv^2 = C[v^+(t - x/c + T)]^2 + C[v^-(t + x/c)]^2. \tag{2.56}$$

Furthermore, the electrical power absorbed by an ideal line may be expressed as

$$i_1 v_1 + i_2 v_2 = \frac{[v^+(t+T)]^2 + [v^-(t+T)]^2}{R_c} - \frac{[v^+(t)]^2 + [v^-(t)]^2}{R_c}. \tag{2.57}$$

2.4.2. Transmission Line Behavior at the Ends

The initial conditions fix the state of the line v^+ and v^- in the time interval $(0, T)$. By placing

$$v_0^+(t) = \tfrac{1}{2}\{v_0[c(T-t)] + R_c i_0[c(T-t)]\} \quad 0 \leqslant t \leqslant T, \tag{2.58}$$

$$v_0^-(t) = \tfrac{1}{2}\{v_0(ct) - R_c i_0(ct)\} \quad 0 \leqslant t \leqslant T, \tag{2.59}$$

and imposing that expressions (2.54) and (2.55) satisfy the initial conditions (2.14) and (2.15), we obtain

$$v^+(t) = v_0^+(t) \quad \text{and} \quad v^-(t) = v_0^-(t) \quad \text{for } 0 \leqslant t \leqslant T. \tag{2.60}$$

The state of the line for $t > T$ depends on the values of the voltage and current at the line ends.

Specifying expressions (2.54) and (2.55) at end $x = 0$, we obtain relations

$$v_1(t) = v^+(t + T) + v^-(t), \tag{2.61}$$

$$R_c i_1(t) = v^+(t + T) - v^-(t), \tag{2.62}$$

whereas specifying them at the end $x = d$, we obtain relations

$$v_2(t) = v^+(t) + v^-(t + T), \tag{2.63}$$

$$-R_c i_2(t) = v^+(t) - v^-(t + T). \tag{2.64}$$

Therefore, for $t > T$ the amplitude of the outgoing forward (backward) wave at $x = 0$ ($x = d$) is equal to the difference between the voltage and amplitude of incoming backward (forward) wave at the same end and at the same time.

Subtracting Eqs. (2.61) and (2.62) termwise, and summing Eqs. (2.63) and (2.64) termwise, we have, respectively, for $t > 0$:

$$v_1(t) - R_c i_1(t) = 2v^-(t), \tag{2.65}$$

$$v_2(t) - R_c i_2(t) = 2v^+(t). \tag{2.66}$$

If the state of the line were completely known at any t, these equations would completely determine the terminal behavior of the line.

Different formulations of the equations governing the state dynamics for $t > T$ are possible. From Eqs. (2.61) and (2.63) we immediately obtain

$$v^+(t) = v_1(t - T) - v^-(t - T) \quad \text{for } t > T, \tag{2.67}$$

$$v^-(t) = v_2(t - T) - v^+(t - T) \quad \text{for } t > T, \tag{2.68}$$

whereas from Eqs. (2.62) and (2.64) we immediately obtain

$$v^+(t) = R_c i_1(t - T) + v^-(t - T) \quad \text{for } t > T, \tag{2.69}$$

$$v^-(t) = R_c i_2(t - T) + v^+(t - T) \quad \text{for } t > T. \tag{2.70}$$

Instead, summing Eqs. (2.61) and (2.62) termwise and subtracting Eqs. (2.63) and (2.64) termwise, we have, respectively,

$$v^{+}(t) = \tfrac{1}{2}[v_1(t-T) + R_c i_1(t-T)] \quad \text{for } t > T, \tag{2.71}$$

$$v^{-}(t) = \tfrac{1}{2}[v_2(t-T) + R_c i_2(t-T)] \quad \text{for } t > T. \tag{2.72}$$

Equations (2.67) and (2.68) (or Eqs. (2.69) and (2.70)) describe in *implicit form* the relation between the state of the line and the electrical variables at the line ends, whereas Eqs. (2.71) and (2.72) provide the same relation, but in an *explicit form.*

Remark: Matched Line

From Eqs. (2.71) and (2.72) we immediately deduce a very important property of ideal two-conductor transmission lines. Equation (2.71) states that v^{+} would be equal to zero for any $t > T$ if the line were connected at the left end to a resistor with resistance R_c — *matched line at the left end.* The same considerations hold for the backward wave if the line is *matched at the right end.* ◇

Equations (2.65) and (2.66), joined to the equations describing the state dynamics, give in implicit form the relations between the voltages v_1 and v_2 and the currents i_1 and i_2. Nevertheless, the role of Eqs. (2.65) and (2.66) is different from that of the equations describing the state dynamics.

Relations (2.65) and (2.66) are two fundamental results of considerable importance. They are linear and algebraic. Let us consider, for example, Eq. (2.65). In substance it says that the voltage at any time t at end $x = 0$ is equal to the sum of two terms as in the case of the semi-infinite line analyzed in §2.2. The term $R_c i_1(t)$ would be the voltage at left line end at time t if the backward voltage wave were equal to zero, as, for example, happens in a semi-infinite line initially at rest, or in a finite line matched at the other end. In this case the line at the left end behaves like a resistor with resistance R_c. Instead, the term $2v^{-}$ is the voltage that would appear at time t if the line were connected to end $x = 0$ with an open circuit. The factor agrees with the following consideration — when the current at the left end is zero, the amplitude of the forward wave is equal to that of the backward one, and then the voltage is two times the amplitude of v^{-}.

If the backward voltage wave v^{-} were known for each $t > 0$, the behavior of the line at the left end would be entirely determined by Eq. (2.65) as in the semi-infinite line case. Instead, for transmission lines of finite length, the backward voltage wave is known only for

$0 \leqslant t \leqslant T$ (from the initial conditions). For $t > T$ it depends on both what is connected to the right and left ends of the line, due to the reflections, and so it is an unknown of the problem.

Similar considerations can be given to Eq. (2.66).

What is the physical meaning of the equations describing the state dynamics? Let us consider, for example, the state variable v^+. Equation (2.67) (or (2.69)) describes the dynamics of v^+ in an implicit form, whereas Eq. (2.71) describes the dynamics of v^+ in an explicit form. The expression $v_1(t - T) - v^-(t - T)$ in Eq. (2.67), the expression $R_c i_1(t - T) - v^-(t - T)$ in Eq. (2.69) and the expression $[v_1(t - T) + R_c i_1(t - T)]/2$ in Eq. (2.71) are all equal to the forward wave amplitude at the line end $x = 0$ and at time instant $t - T$, whereas $v^+(t)$ is the amplitude of the forward wave at the time instant t and at the end $x = d$. Therefore, Eqs. (2.67), (2.69), and (2.71) all state the same property, the amplitude of the forward voltage wave at time t and at the right end of the line is equal to the amplitude of the same wave at the left end and at the previous time instant $t - T$ (for any $t > T$). This is one of the fundamental properties of an ideal two-conductor transmission line. Similar considerations hold for the equations of the other state variable v^-. Thus, for $t > T$ the values of state variables at the time instant t only depend on the values of the voltages and/or currents at the line ends, and of the state variables themselves at the time instant $t - T$.

It is clearly evident that Eqs. (2.67) and (2.68) are completely equivalent to Eqs. (2.69) and (2.70), or to Eqs. (2.71) and (2.72) from the computational point of view as well. For this reason in the following we shall use Eqs. (2.67) and (2.68) to describe the state dynamics.

Two different descriptions of the characteristic relations of the two-port representing the line are possible.

The set of equations (2.65)–(2.68) describes the terminal properties of the line, as well as the internal state. We call them the *internal* or *input-state-output description* of the line. Equations (2.67) and (2.68) govern the dynamics of the state, whereas Eqs. (2.65) and (2.66) describe the terminal properties. Therefore we call Eqs. (2.67) and (2.68) *state equations*, and Eqs. (2.65) and (2.66) *output equations*.

A description in which only the terminal variables are involved is possible. In fact by substituting the expression of v^+ given by Eq. (2.71) into Eq. (2.66), and the expression of v^- given by Eq. (2.72) into Eq. (2.65), we eliminate the state variables. This formulation of the characteristic relations of the two-port representing the line is called the *external* or *input-output description*.

2.5. THE INPUT-OUTPUT DESCRIPTION

Substituting the expression of v^+ given by Eq. (2.71) into Eq. (2.66), and the expression of v^- given by Eq. (2.72) into Eq. (2.65), we obtain two linearly independent equations in terms of v_1, v_2, i_1, and i_2,

$$v_1(t) - R_c i_1(t) = v_2(t - T) + R_c i_2(t - T) \quad \text{for } t \geqslant T, \tag{2.73}$$

$$v_2(t) - R_c i_2(t) = v_1(t - T) + R_c i_1(t - T) \quad \text{for } t \geqslant T, \tag{2.74}$$

In this way we have eliminated the internal state of the line, and only the terminal variables are involved. To solve the system (2.73) and (2.74) we need to know v_1, v_2, i_1, and i_2 for $0 \leqslant t \leqslant T$. By imposing the initial conditions (2.60) for v^+ and v^-, from Eqs. (2.65) and (2.66) we obtain

$$v_1(t) - R_c i_1(t) = 2v_0^-(t) \quad \text{for } 0 \leqslant t \leqslant T, \tag{2.75}$$

$$v_2(t) - R_c i_2(t) = 2v_0^+(t) \quad \text{for } 0 \leqslant t \leqslant T. \tag{2.76}$$

These equations do not determine univocally v_1, v_2, i_1, and i_2 for $0 \leqslant t \leqslant T$. They can be determined only when the circuits connected to the line are specified.

Equations (2.73) and (2.74) describe the relations between the terminal variables v_1, v_2, i_1, and i_2 in an implicit form. As we learned in circuit theory there are six possible explicit representations of two of the four variables v_1, v_2, i_1, and i_2 in terms of the remaining two (e.g., Chua, Desoer, and Kuh, 1987).

The *current controlled representation* expresses v_1 and v_2 as functions of i_1 and i_2, the *voltage controlled representation* expresses i_1 and i_2 as functions of v_1 and v_2, the *hybrid representations* express v_1 and i_2 as functions of i_1 and v_2 and the converse, and the *transmission representations* express v_1 and i_1 as functions of v_2 and i_2, and the converse. As all of these representations are completely equivalent, in this text we shall then deal with the problem of the current controlled representation. For other representations, which can easily be obtained from this, we shall give only the expressions.

To find these representations we have to solve the system of difference equations with one delay (2.73) and (2.74). As these equations are linear and time invariant, the solution to this problem is immediate and neat if we use the method of Laplace transform.

We postpone the complete solution of this problem and the determination of all possible input-output representations to Chapter 4, where we shall deal with the more general case of lossy transmission lines. Here we report only the current controlled representation in order to outline a drawback that is peculiar to all input-output descriptions.

As we shall show in Chapter 4, in the time domain the *current controlled* representation is given by

$$v_1(t) = (z_d * i_1)(t) + (z_t * i_2)(t) + e_1(t), \tag{2.77}$$

$$v_2(t) = (z_t * i_1)(t) + (z_d * i_2)(t) + e_2(t), \tag{2.78}$$

where the impulse response

$$z_d(t) = R_c\delta(t) + 2R_c \sum_{i=1}^{\infty} \delta(t - 2iT), \tag{2.79}$$

is the inverse Laplace transform of the driving-point impedance at port $x = 0$ (or at $x = d$) when port at $x = d$ (respectively, at $x = 0$) is kept open circuited, and

$$z_t(t) = 2R_c \sum_{i=0}^{\infty} \delta[t - (2i + 1)T], \tag{2.80}$$

is the inverse Laplace transform of the *transfer impedance* when the port at $x = 0$ (or at $x = d$) is kept open circuited; $\delta(t)$ indicates the Dirac function. Equations (2.77) and (2.78) are the same as Eqs. (5) and (6) outlined in the introduction of this book. The independent voltage sources $e_1(t)$ and $e_2(t)$ take into account the effects of the initial conditions.

Relations (2.77) and (2.78) express the voltages $v_1(t)$ and $v_2(t)$ as functions of the initial state of the line and of the history of the "input variables" i_1 and i_2 on the whole time interval $(0, t)$. In consequence of the multireflections due the definitions of $z_d(t)$ and $z_t(t)$, the number of Dirac functions they contain increases with the time. Therefore, numerical simulators of transmission lines based on a description of this type are time consuming if evolutions much longer than T have to be analyzed. This problem, as we shall see in the next section, may be overcome by using the input-state-output description based on the forward and backward voltage waves.

2.6. THE INPUT-STATE-OUTPUT DESCRIPTION AND THE EQUIVALENT CIRCUITS OF THÉVENIN AND NORTON TYPE

It is useful to rewrite Eqs. (2.65) and (2.66), and state equations (2.67) and (2.68) in terms of the new state variables w_1 and w_2 defined as

$$w_1 \equiv 2v^-, \tag{2.81}$$

$$w_2 \equiv 2v^+. \tag{2.82}$$

Then equations representing the line behavior at the ends can be rewritten as

$$v_1(t) - R_c i_1(t) = w_1(t) \quad \text{for } t > 0, \tag{2.83}$$

$$v_2(t) - R_c i_2(t) = w_2(t) \quad \text{for } t > 0. \tag{2.84}$$

These equations are the same as Eqs. (7) and (8) outlined in the introduction to this book. Thus, for ideal two-conductor lines the impulse response $z_c(t)$ is given by

$$z_c(t) = R_c \delta(t). \tag{2.85}$$

For $0 \leqslant t \leqslant T$, the state variables w_1 and w_2 depend only on the initial conditions of the line and are given by

$$w_1(t) = 2v_0^-(t) \quad \text{for } 0 \leqslant t \leqslant T, \tag{2.86}$$

$$w_2(t) = 2v_0^+(t) \quad \text{for } 0 \leqslant t \leqslant T. \tag{2.87}$$

For $t > T$, $w_1(t)$ and $w_2(t)$ are related, respectively, to the values of the voltages v_2 and v_1 at the time instant $t - T$, and to the values of the state variables w_2 and w_1 at the time instant $t - T$ through equations

$$w_1(t) = [2v_2(t-T) - w_2(t-T)] \quad \text{for } t > T, \tag{2.88}$$

$$w_2(t) = [2v_1(t-T) - w_1(t-T)] \quad \text{for } t > T. \tag{2.89}$$

These equations are the same as Eqs. (9) and (10) outlined in the introduction to this book. Thus, for ideal two-conductor lines the impulse response $p(t)$ is given by

$$p(t) = \delta(t - T). \tag{2.90}$$

Equations (2.88) and (2.89) are linear algebraic difference relations with one delay, which have to be solved with the initial conditions (2.86) and (2.87).

Remark

The uniqueness of the input-state-output description of an ideal two-conductor line is that the impulse responses $z_c(t)$ and $p(t)$ contain one and only one Dirac pulse, whereas the impulse responses $z_d(t)$ and $z_t(t)$ given by Eqs. (2.79) and (2.80), characterizing the current controlled input-output description, contain an infinite number of Dirac pulses. The input-state-output description, even though of the implicit type, has two great advantages over the input-output description discussed in the previous paragraph: it is undoubtedly much simpler in form, and the rule of updating the state is expressed by simple equations of recurrent type. It is clear that the input-state-output

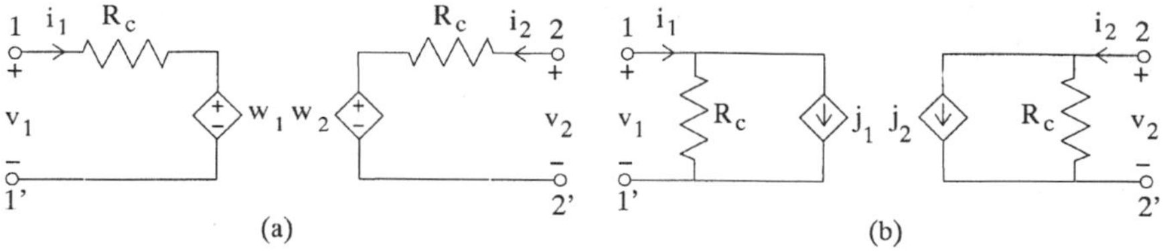

Figure 2.11. Time domain equivalent circuit of Thévenin type (a), and Norton type (b), as implemented in SPICE.

description is more suitable for transient analysis, whereas the input-output description is preferable to deal with sinusoidal and periodic steady-state operating conditions in the frequency domain. We postpone to Chapter 4, §4.7, a thorough analysis of this question. ◇

The Eqs. (2.83) and (2.84) suggest the equivalent circuit of Thévenin type shown in Fig. 2.11a. Each port of the transmission line behaves as a linear resistor of resistance R_c connected in series with a controlled voltage source. The state equations (2.88) and (2.89) govern the controlled voltage sources w_1 and w_2. An equivalent circuit of this type, as far as we know, was proposed for the first time by Branin (Branin, 1967). He used the method of the characteristic curves (see Chapter 8) to determine it.

The equivalent circuits that we have determined use resistors and controlled sources, namely, "lumped" elements. Unlike the classical controlled sources of the circuit theory, in this case the control laws are represented by linear algebraic difference relations with one delay. This fact allows the controlled sources that represent the line to be treated as independent sources if the problem is resolved by means of iterative procedures.

Let us consider, for example, the circuit in Fig. 2.12. It is the equivalent circuit of the network of Fig. 2.1 obtained by replacing the line with the equivalent circuit of Thévenin type of Fig. 2.11a. In the generic time interval $[iT, (i+1)T]$ with $i \geqslant 1$, voltages $w_1(t)$ and $w_2(t)$ depend, respectively, only on $v_2(t-T)$, $w_2(t-T)$ and $v_1(t-T)$, $w_1(t-T)$. Consequently, if the solution for $t \leqslant iT$ were known, both $w_1(t)$ and $w_2(t)$ would be known for $iT \leqslant t \leqslant (i+1)T$; for $0 \leqslant t \leqslant T$, $w_1(t)$ and $w_2(t)$ are given by the initial conditions. This property suggests analyzing networks consisting of transmission lines and lumped circuits iteratively; at each step of the iteration we have to solve an equivalent circuit composed only of lumped elements. The

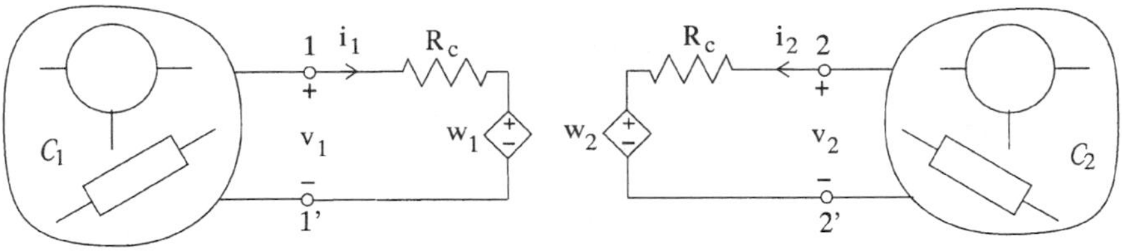

Figure 2.12. Equivalent circuit of the network of Fig. 2.1.

most widely used numerical circuit simulator SPICE uses equivalent circuits of the type shown in Fig. 2.11 to simulate ideal two-conductor transmission lines. In Chapter 9 we shall consider these matters in detail.

Remark

Figure 2.11b shows an equivalent circuit of Norton type. The controlled current sources $j_1(t)$ and $j_2(t)$ are related to the controlled voltage sources through the relations

$$j_1 = -\frac{w_1}{R_c}, \tag{2.91}$$

$$j_2 = -\frac{w_2}{R_2}. \tag{2.92}$$

The governing laws of these sources may be expressed as

$$j_1(t) = [-2i_2(t-T) + j_2(t-T)] \quad \text{for } t > T, \tag{2.93}$$

$$j_2(t) = [-2i_1(t-T) + j_1(t-T)] \quad \text{for } t > T. \tag{2.94}$$

They have to be solved with the initial conditions

$$j_1(t) = -2v_0^-(t)/R_c \quad \text{for } 0 \leqslant t \leqslant T, \tag{2.95}$$

$$j_2(t) = -2v_0^+(t)/R_c \quad \text{for } 0 \leqslant t \leqslant T. \tag{2.96}$$

2.7. LINES CONNECTED TO LINEAR LUMPED CIRCUITS

In this section we shall use the equivalent circuits of the two-conductor line that we have determined in the previous section to analyze some notable networks that can be solved analytically. We

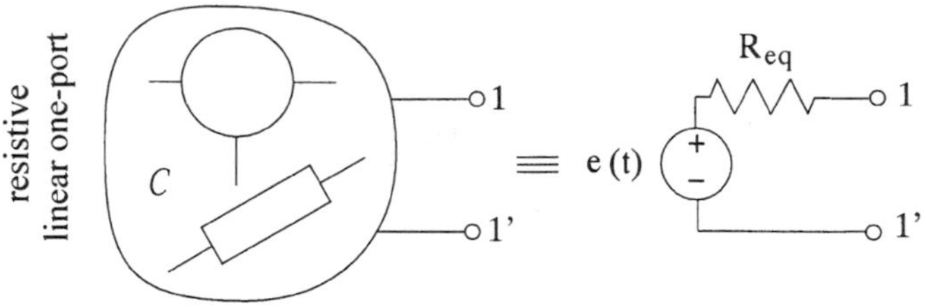

Figure 2.13. Thévenin equivalent one-port of a linear resistive one-port.

first analyze, in depth, a two-conductor transmission line connected at each end to a one-port consisting of independent sources and linear time-invariant resistive elements (resistors, linear controlled sources, ideal transformers, ideal linear operational amplifiers, and gyrators). Then we shall make some remarks on the most general case of a transmission line connected to linear time-invariant dynamic lumped circuits. For the sake of simplicity, we shall assume the line to be initially at rest.

Any one-port C (except for an independent current source), consisting of linear resistive elements and independent sources, can be substituted by its Thévenin equivalent one-port (e.g., Chua, Desoer, and Kuh, 1987), see Fig. 2.13. By using this equivalent representation, the study of a two-conductor line connecting two linear resistive circuits can always be reduced to the study of a transmission line connected at each end to a linear resistor in series with an ideal voltage source, such as the one illustrated in Fig. 2.14.

Determining the electrical variables at the line ends is immediate if the line is represented through its equivalent circuit of Thévenin

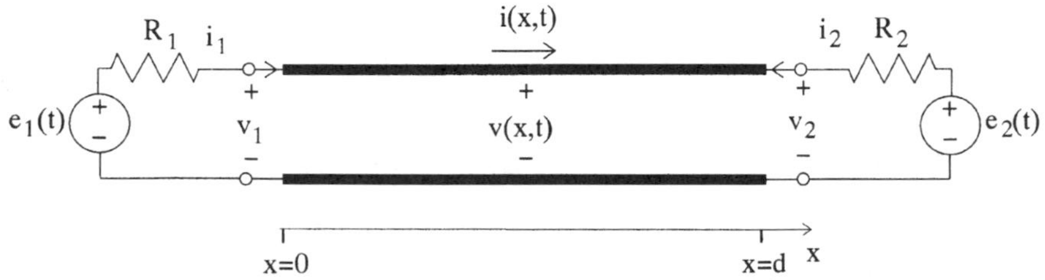

Figure 2.14. Transmission line connected to two linear resistive one-ports.

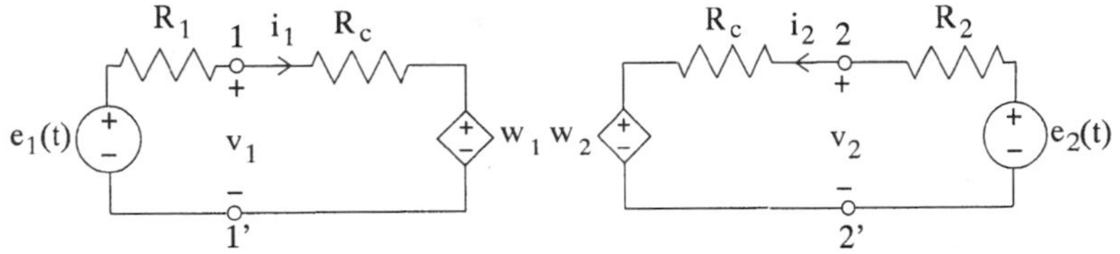

Figure 2.15. Equivalent circuit of network shown in Fig. 2.14.

type, Fig. 2.11a. The equivalent network obtained in this way is shown in Fig. 2.15. The unknowns of this circuit are the electrical variables at the transmission line ends and the voltages w_1 and w_2. It is immediate that

$$v_1(t) = \frac{R_1}{R_1 + R_c} w_1(t) + \frac{R_c}{R_1 + R_c} e_1(t), \tag{2.97}$$

$$v_2(t) = \frac{R_2}{R_2 + R_c} w_2(t) + \frac{R_c}{R_2 + R_c} e_2(t). \tag{2.98}$$

These equations together with the control laws (2.88) and (2.89) of the sources w_1 and w_2 determine the solution of the problem.

2.7.1. State Equations in Normal Form

Substituting expressions (2.97) and (2.98) into Eqs. (2.88) and (2.89), we obtain for $w_1(t)$ and $w_2(t)$

$$w_1(t) = \Gamma_2 w_2(t - T) + \frac{2R_c}{R_2 + R_c} e_2(t - T) \quad \text{for } t > T, \tag{2.99}$$

$$w_2(t) = \Gamma_1 w_1(t - T) + \frac{2R_c}{R_1 + R_c} e_1(t - T) \quad \text{for } t > T, \tag{2.100}$$

where Γ_1 and Γ_2 are the voltage reflection coefficients that we would find at the corresponding ends if the independent voltage sources were switched off,

$$\Gamma_1 = \frac{R_1 - R_c}{R_1 + R_c}, \tag{2.101}$$

$$\Gamma_2 = \frac{R_2 - R_c}{R_2 + R_c}. \tag{2.102}$$

Equations (2.99) and (2.100) are the state equations in normal form of the circuit of Fig. 2.15: the state of the circuit at the time instant t is expressed as a function of the state of the circuit and the independent sources at the time instant $t - T$.

When $R_1 = R_c$ we have $\Gamma_1 = 0$ and the line is matched to the left end; in this case the forward voltage wave depends only on the voltage of the independent source present at the right end. There is an analogous case for $R_2 = R_c$.

Once $w_1(t)$ and $w_2(t)$ are known, the two voltages at the line ends, and therefore the currents, are soon calculated, as they are linked to each other by algebraic Eqs. (2.97) and (2.98).

The simplicity with which we have reached Eqs. (2.99) and (2.100) and the neatness of their form are surprising. These equations are valid even when the terminal circuits are time varying. In this case R_1 and R_2 and therefore Γ_1 and Γ_2 are time dependent. In the following we only deal with the time-invariant situations.

2.7.2. Natural Frequencies of the Network

The natural modes of evolution of a line connected to two linear resistors are the solutions of the set of homogeneous equations

$$w_1(t) = \Gamma_2 w_2(t - T), \tag{2.103}$$

$$w_2(t) = \Gamma_1 w_1(t - T). \tag{2.104}$$

They have solutions of the type

$$w_1(t) = A_1 e^{\lambda t}, \tag{2.105}$$

$$w_2(t) = A_2 e^{\lambda t}, \tag{2.106}$$

where A_1 and A_2 are arbitrary constants and the *natural frequency* λ is the solution of the characteristic equation

$$1 - \Gamma_1 \Gamma_2 e^{-2\lambda T} = 0. \tag{2.107}$$

Equation (2.107) has an infinite number of roots, and they are the solution of the equation

$$e^{2\lambda T} = \Gamma_1 \Gamma_2 e^{2k\pi i} \quad k = 0, \pm 1, \pm 2, \ldots, \tag{2.108}$$

from which we obtain

$$\lambda_k = \frac{1}{T}\ln(\sqrt{|\Gamma_1 \Gamma_2|}) + i\frac{\pi}{T}\begin{cases} k & \text{for } \Gamma_1\Gamma_2 > 0, \\ k + \frac{1}{2} & \text{for } \Gamma_1\Gamma_2 < 0. \end{cases} \tag{2.109}$$

The natural frequencies have all the same real part, that is,

$$\alpha = \frac{1}{T}\ln(\sqrt{|\Gamma_1\Gamma_2|}). \tag{2.110}$$

It is less than zero for $|\Gamma_1\Gamma_2| < 1$, that is, for strictly passive resistors. In these cases the amplitudes of all the natural modes decay exponentially in time with the time constant $\tau_0 = |1/\alpha|$. It is to be observed that the natural frequencies could have a positive real part if at least one resistor were active, that is, if at least one of the two resistors had a negative resistance (in this case it could be $|\Gamma_1\Gamma_2| > 1$). When the line is connected at both ends to short circuits and/or open circuits, the real part of λ_k is equal to zero (see §2.3).

The imaginary part of the natural frequencies depends only on the sign of $\Gamma_1\Gamma_2$. If both the terminations are short circuits or open circuits, we have $\Gamma_1\Gamma_2 = 1$, and the functions $w_1(t)$ and $w_2(t)$ are periodic of period $2T$. When one termination is a short circuit and the other is an open circuit, we have $\Gamma_1\Gamma_2 = -1$ and the solutions are periodic with period $4T$ (see §2.3). When both resistors are strictly passive, that is, $R_1 > 0$, $R_2 > 0$, we have $|\Gamma_1\Gamma_2| < 1$ and the solutions still oscillate with the period $2T$ if $\Gamma_1\Gamma_2/|\Gamma_1\Gamma_2| = 1$ and with period $4T$ if $\Gamma_1\Gamma_2/|\Gamma_1\Gamma_2| = -1$, but amplitude decreases in time with the time constant τ_0. If at least one of the two resistors is active, $|\Gamma_1\Gamma_2|$ can be greater than one. In this case the amplitude of the oscillations grows in time with the time constant τ_0.

2.7.3. Solution in the Laplace Domain

For time-invariant circuits, Eqs. (2.99) and (2.100) can easily be solved with the Laplace transform. Here, for the sake of simplicity, we consider only the case in which $w_1(t) = w_2(t) = 0$ for $0 \leqslant t \leqslant T$, that is, the transmission line is initially at rest.

Transforming both sides of Eqs. (2.99) and (2.100) we obtain

$$W_1(s) - \Gamma_2 P(s) W_2(s) = P(s)\frac{2R_c}{R_2 + R_c}E_2(s), \tag{2.111}$$

$$-\Gamma_1 P(s) W_1(s) + W_2(s) = P(s)\frac{2R_c}{R_1 + R_c}E_1(s), \tag{2.112}$$

where $W_1(s)$, $W_2(s)$, $E_1(s)$, and $E_2(s)$ are, respectively, the Laplace transforms of $w_1(t)$, $w_2(t)$, $e_1(t)$, and $e_2(t)$, and

$$P(s) = \exp(-sT) \tag{2.113}$$

is the Laplace domain operator corresponding to an ideal delay. Equations (2.111) and (2.112) are the state equations in normal form in the Laplace domain. The solution of the system of algebraic Eqs. (2.111) and (2.112) is

$$W_1(s) = \frac{P}{1-\Gamma_1\Gamma_2 P^2}\left(\Gamma_2 P\frac{2R_c}{R_c+R_1}E_1 + \frac{2R_c}{R_c+R_2}E_2\right), \tag{2.114}$$

$$W_2(s) = \frac{P}{1-\Gamma_1\Gamma_2 P^2}\left(\frac{2R_c}{R_c+R_1}E_1 + \Gamma_1 P\frac{2R_c}{R_c+R_2}E_2\right). \tag{2.115}$$

Observe that the poles of the function $H(s) = 1/[1 - \Gamma_1\Gamma_2 P^2(s)]$ coincide with the natural frequencies (2.109).

Equations (2.114) and (2.115) are immediately transformed in the time domain if we represent $(1 - \Gamma_1\Gamma_2 P^2)^{-1}$ through the geometric series

$$\frac{1}{1-\Gamma_1\Gamma_2 P^2} = \sum_{i=0}^{\infty} (\Gamma_1\Gamma_2 P^2)^i, \tag{2.116}$$

in the region $Re\{s\} > \alpha$ (where α is given by Eq. (2.110)), because there $H(s)$ is analytic and $|\Gamma_1\Gamma_2 e^{-2sT}| < 1$. Thus, there certainly exists a real nonnegative number σ_c, such that, for $Re\{s\} > \sigma_c$, Eqs. (2.114) and (2.115) can be expressed as

$$\begin{aligned} W_1(s) &= \frac{2R_c\Gamma_2}{R_c+R_1}E_1(s)\sum_{i=0}^{\infty}(\Gamma_1\Gamma_2)^i P^{2i+2}(s) \\ &\quad + \frac{2R_c}{R_c+R_2}E_2(s)\sum_{i=0}^{\infty}(\Gamma_1\Gamma_2)^i P^{2i+1}(s), \end{aligned} \tag{2.117}$$

$$\begin{aligned} W_2(s) &= \frac{2R_c}{R_c+R_1}E_1(s)\sum_{i=0}^{\infty}(\Gamma_1\Gamma_2)^i P^{2i+1}(s) \\ &\quad + \frac{2R_c\Gamma_1}{R_c+R_2}E_2(s)\sum_{i=0}^{\infty}(\Gamma_1\Gamma_2)^i P^{2i+2}(s). \end{aligned} \tag{2.118}$$

If σ_0 is the convergence abscissa of independent sources E_1 and E_2, then we have $\sigma_c = \max(\sigma_0, \alpha)$.

Admitting that it is licit to inversely transform by series, and using the property

$$L^{-1}\{F(s)P^m(s)\} = \int_{0^-}^{t^+} \delta(t - mT - \tau)f(\tau)\,d\tau = f(t - mT)u(t - mT), \tag{2.119}$$

from Eqs. (2.117) and (2.118) we immediately deduce

$$\begin{aligned} w_1(t) = & \frac{2R_c\Gamma_2}{R_c+R_1}\sum_{i=0}^{N(t-2T)}(\Gamma_1\Gamma_2)^i e_1[t-2(i+1)T] \\ & + \frac{2R_c}{R_c+R_2}\sum_{i=0}^{N(t-T)}(\Gamma_1\Gamma_2)^i e_2[t-(2i+1)T], \end{aligned} \tag{2.120}$$

$$\begin{aligned} w_2(t) = & \frac{2R_c}{R_c+R_1}\sum_{i=0}^{N(t-T)}(\Gamma_1\Gamma_2)^i e_1[t-(2i+1)T] \\ & + \frac{2R_c\Gamma_1}{R_c+R_2}\sum_{i=0}^{N(t-2T)}(\Gamma_1\Gamma_2)^i e_2[t-2(i+1)T] \end{aligned} \tag{2.121}$$

where

$$N(t) = \mathrm{int}(t/2T). \tag{2.122}$$

The symbol $\mathrm{int}(y)$ indicates the function that gives the entire part of a real number when it is greater than zero, and is equal to zero for $y \leqslant 0$.

The sums in Eqs. (2.120) and (2.121) can be expressed analytically only for particular wave forms of the sources. When the voltages imposed by the sources are both constant in time, $e_1(t) = E_1$ and $e_2(t) = E_2$ (for $t > 0$), we obtain

$$\begin{aligned} w_1(t) = & \frac{2R_c\Gamma_2}{R_c+R_1}E_1\frac{1-(\Gamma_1\Gamma_2)^{N(t-2T)+1}}{1-\Gamma_1\Gamma_2} \\ & + \frac{2R_c}{R_c+R_2}E_2\frac{1-(\Gamma_1\Gamma_2)^{N(t-T)+1}}{1-\Gamma_1\Gamma_2}, \end{aligned} \tag{2.123}$$

$$\begin{aligned} w_2(t) = & \frac{2R_c}{R_c+R_1}E_1\frac{1-(\Gamma_1\Gamma_2)^{N(t-T)+1}}{1-\Gamma_1\Gamma_2} \\ & + \frac{2R_c\Gamma_1}{R_c+R_2}E_2\frac{1-(\Gamma_1\Gamma_2)^{N(t-2T)+1}}{1-\Gamma_1\Gamma_2}. \end{aligned} \tag{2.124}$$

If both the resistors are strictly passive we have $\lim_{m\to\infty}[1-(\Gamma_1\Gamma_2)^m]=1$, and hence, for $t \to +\infty$ the solution tends to a stationary steady state. Finally, if the voltages of the two sources are periodic in time with period $2T$ (in this case the sources are in resonance with the line) for w_1 and w_2 we obtain expressions similar to Eqs. (2.123) and (2.124). The voltage waves $w_1(t)$ and $w_2(t)$ increase as $N(t-T)+1$ for $\Gamma_1\Gamma_2 = 1$.

Remark

The problem resolved previously is much more significant than it might appear. Equations (2.117) and (2.118) may represent the solution, as we shall now see, of much more complex problems than those treated hitherto.

Consider a two-conductor line that connects two one-ports consisting not only of time-invariant linear resistive elements and independent sources, but also of linear and time-invariant capacitors, inductors, and transformers. The solution of a network of this type is immediate if we operate in the Laplace domain. For the sake of simplicity, let us again assume that the network is initially at rest.

In the Laplace domain both lumped terminal circuits can be represented by their Thévenin equivalent one-ports (e.g., Chua, Desoer, and Kuh, 1987), see Fig. 2.16. Figure 2.17 shows the equivalent circuit in the Laplace domain of the overall network we are considering. The control laws of the two controlled sources are (see Eqs. (2.88) and (2.89))

$$W_1(s) = P(s)[2V_2(s) - W_2(s)], \tag{2.125}$$

$$W_2(s) = P(s)[2V_1(s) - W_1(s)]. \tag{2.126}$$

Resolving the circuit shown in Fig. 2.17, and using Eqs. (2.125) and (2.126) we once again obtain the state equations in the normal form (2.114) and (2.115), so long as the impedances $Z_1(s)$ and $Z_2(s)$ are considered instead of R_1 and R_2.

In this case, too, there certainly exists a finite positive real number σ_c, so that, for $\mathrm{Re}\{s\} > \sigma_c$, the expression

$$H(s) = [1 - \Gamma_1(s)\Gamma_2(s)P^2(s)]^{-1}$$

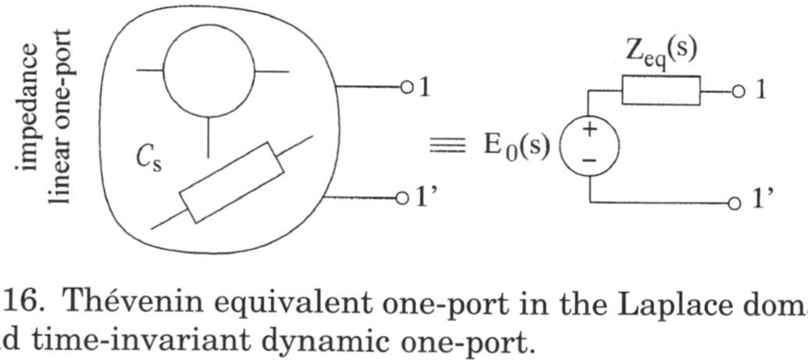

Figure 2.16. Thévenin equivalent one-port in the Laplace domain of a linear and time-invariant dynamic one-port.

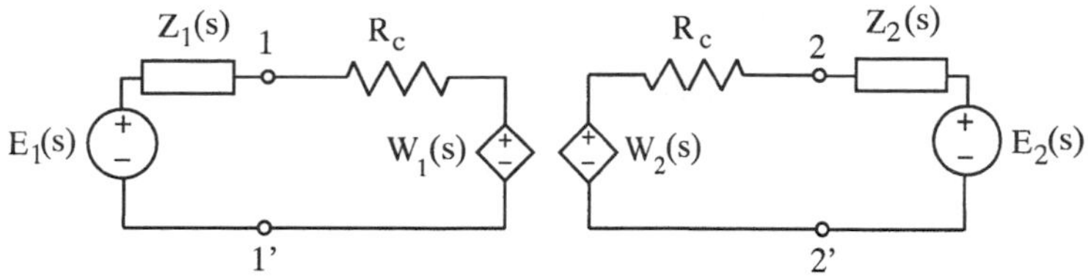

Figure 2.17. Equivalent circuit in the Laplace domain of a line connecting two time-invariant linear one-ports containing dynamic elements.

can be represented by the geometric series, and hence, we once again obtain expressions for W_1 and W_2 formally equal to those given by Eqs. (2.117) and (2.118).

As impedances Z_1, Z_2 and reflection coefficients Γ_1, Γ_2 are functions of s, the inverse Laplace transforms of W_1 and W_2 cannot, in general, be expressed in closed form. They can be evaluated only approximately, for example, by using numeral algorithms based on fast Fourier transforms (FFT).

As we shall see in Chapter 9, in these situations, with few exceptions, the reduction of the circuit equations to the state equations in normal form is not an efficient method if the state equations have to be solved numerically, whether we have to determine transients or steady-state solutions.

Let us give a simple example of the difficulties encountered in trying to solve analytically a problem of this type. Consider, for instance, the case where $Z_1(s) = 0$ and $Z_2(s) = 1/sC$, and let us determine the natural frequencies. They are the roots of the characteristic equation

$$1 + \Gamma_2(s)P^2(s) = 0 \tag{2.127}$$

and the poles of the reflection coefficient $\Gamma_2(s)$, that is, the roots of the equation

$$R_c + Z_2(s) = 0. \tag{2.128}$$

In this case the reflection coefficient $\Gamma_2(s)$ is given by

$$\Gamma_2(s) = \frac{1 - \tau_0 s}{1 + \tau_0 s}, \tag{2.129}$$

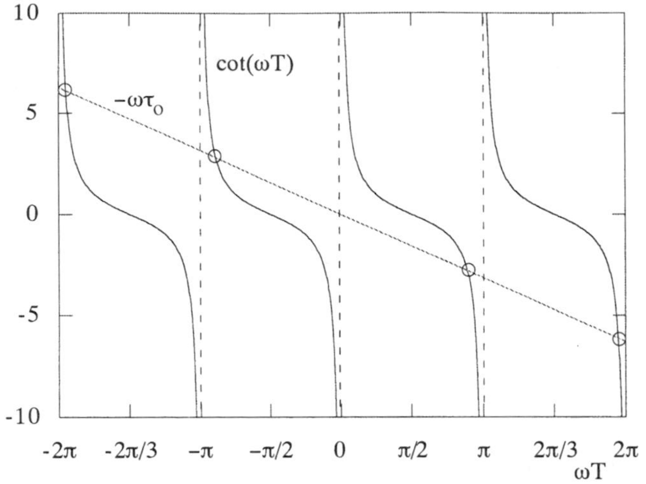

Figure 2.18. Graphical solution of Eq. (2.130).

where $\tau_0 = R_c C$. Therefore, one natural frequency of the system is real and all the others are purely imaginary. The real natural frequency is the pole $-1/\tau_0$, of Γ_2 and all the purely imaginary frequencies are the roots of the equation

$$\omega\tau_0 + \cot(\omega T) = 0. \tag{2.130}$$

This is a transcendent equation that cannot be resolved analytically. Its graphic solution is reported in Fig. 2.18. ◇

2.8. A GLIMPSE AT A TRANSMISSION LINE CONNECTED TO A NONLINEAR ONE-PORT: STATE EQUATIONS IN NORMAL FORM

So far we have considered only linear and time-invariant lumped elements connected to the line. When the lumped elements are nonlinear and/or time varying, the Laplace transform is no longer useful to determine the state equations in normal form and the problem must be resolved by operating directly in the time domain.

Now let us highlight the difficulties that arise when the property of linearity is lacking, a property that is the basis of the method described in the preceding section. Let us consider first a line that connects two one-ports of resistive type, one of which is nonlinear. Then we shall see how the problem is further complicated if we insert a linear capacitor in parallel with the nonlinear resistor.

2.8.1. A Line Connected to a Nonlinear Resistor

As an example we shall refer to the circuit in Fig. 2.19. The line is connected to a short circuit at $x = 0$ and to a voltage-controlled nonlinear resistor at $x = d$,

$$i_2 = -g(v_2). \tag{2.131}$$

Now we want to determine the state equations in normal form. In this case it is immediate that

$$w_2(t) = -w_1(t - T), \tag{2.132}$$

because the line is connected to a short circuit at $x = 0$. To determine the other state equation we first express the voltages v_2 as functions of w_2, and then substitute the expressions so obtained in Eq. (2.88), which governs the controlled voltage source w_1. Voltage v_2 is the solution of the nonlinear algebraic equation

$$R_c g[v_2(t)] + v_2(t) = w_2(t), \tag{2.133}$$

as in the case dealt with in §2.2.5. This equation implicitly defines an instantaneous nonlinear relation between voltage v_2 and the state w_2 that we indicate with

$$v_2(t) = F[w_2(t)]. \tag{2.134}$$

The function F is single valued only if Eq. (2.133) admits a unique solution for every value of the "independent" variable w_2. We have already dealt with this question in §2.2.5. However, we shall come back to it in Chapter 9, where we shall study this question in detail.

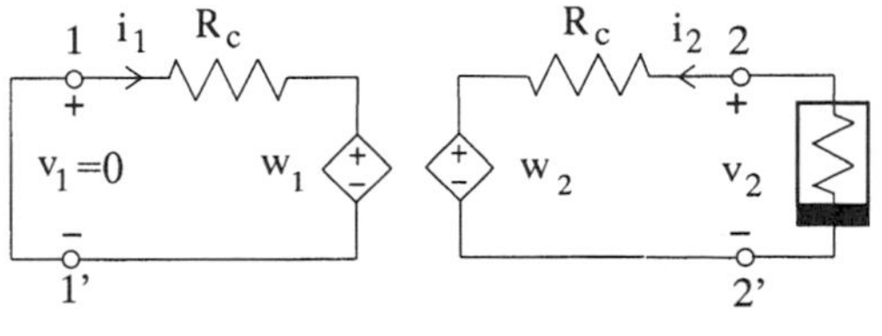

Figure 2.19. Equivalent circuit of a transmission line connected to a short circuit at one end and to a nonlinear resistor at the other end.

Let us put ourselves in the situation where F is single valued. By substituting expressions (2.134) into Eq. (2.88) we obtain

$$w_1(t) = -w_2(t-T) + 2F(w_2(t-T)) \quad \text{for } t > T. \tag{2.135}$$

Equations (2.132) and (2.135) are the state equations in normal form of the circuit of Fig. 2.19. They are a system of algebraic difference equations with delay T. They must be solved with the initial conditions (2.86) and (2.87).

The system of equations (2.132) and (2.135) can be reduced to the scalar equation with delay $2T$

$$w_1(t) = w_1(t-2T) + 2F(-w_1(t-2T)) \quad \text{for } t > 2T. \tag{2.136}$$

Equation (2.136) allows the solution to be determined recursively. Initial conditions and Eq. (2.135) fix w_1 for $0 \leqslant t \leqslant 2T$. By using Eq. (2.136) we determine w_1 for $2T \leqslant t \leqslant 4T$. Once w_1 is known for $2T \leqslant t \leqslant 4T$, we determine w_1 for $4T \leqslant t \leqslant 6T$, and so forth. In Chapter 10 we shall study in detail the properties and the behavior of the solution of recurrent equations of the type (2.136).

2.8.2. A Line Connected to a Nonlinear Resistor in Parallel with a Linear Capacitor

Let us introduce a linear capacitor in parallel with the nonlinear resistor into the network that we have just analyzed. The equivalent circuit is illustrated in Fig. 2.20.

One state equation is again given by Eq. (2.132). To determine the other state equation we proceed as in the following. From direct inspection of the circuit in Fig. 2.20 we find

$$C\frac{dv_2}{dt} + \frac{v_2}{R_c} + g(v_2) = \frac{w_2(t)}{R_c}. \tag{2.137}$$

Because of the presence of the capacitor, the relation between v_2 and w_2 is no longer of the instantaneous type, but is of the functional type, with memory, $v_2(t) = \mathscr{F}\{w_2\}(t)$, as in the case dealt with in §2.2.6. There is a further difficulty, as in this case w_2 is an unknown of the problem. It is related to w_1 through Eq. (2.132) and w_1 depends on v_2 through the state equation (2.88). In this case the system of state equations in normal form is for $t > T$

$$\begin{aligned} w_1(t) &= 2\mathscr{F}\{w_2\}(t-T) - w_2(t-T), \\ w_2(t) &= -w_1(t-T). \end{aligned} \tag{2.138}$$

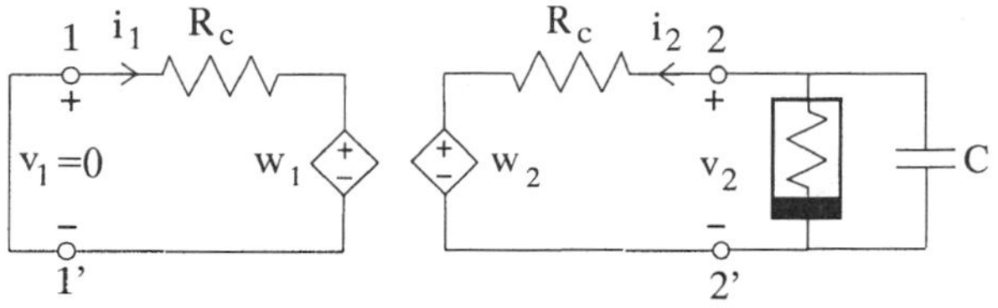

Figure 2.20. Equivalent circuit of a line linked at one end to a short circuit and at the other end to a parallel connection of a nonlinear resistor and a linear capacitor.

The first equation of this system is a nonlinear functional difference equation with one delay.

This problem also can be resolved by using a recurrent procedure, but in this case at each step of the iteration we have to resolve the nonlinear ordinary differential equation (2.137). The initial conditions assign w_1 and w_2 for $0 \leqslant t \leqslant T$ and v_2 at time $t = 0$. Resolving Eq. (2.137) for $0 \leqslant t \leqslant T$, we determine v_2 for $0 \leqslant t \leqslant T$, and then using Eqs. (2.88) and (2.132) we shall determine w_1 and w_2 for $T \leqslant t \leqslant 2T$. Once they have been evaluated, we repeat the procedure and determine v_2 for $T \leqslant t \leqslant 2T$ and w_1, w_2 for $2T \leqslant t \leqslant 3T$. Proceeding in this way we solve the circuit.

In this case it is also possible to reduce the problem to the solution of one scalar equation. By using Eqs. (2.88) and (2.132) from Eq. (2.137) we obtain for $t > 2T$

$$\frac{C}{2}\frac{d}{dt}[w_1(t) - w_1(t-2T)] + \frac{1}{2R_c}[w_1(t) - w_1(t-2T)] + g\left\{\frac{1}{2}[w_1(t) - w_1(t-2T)]\right\} + \frac{w_1(t-2T)}{R_c} = 0 \tag{2.139}$$

In the classification of functional differential equations, Eq. (2.139) is called a *neutral type differential equation with one delay* because it also contains the time derivative of the delayed unknown function (e.g., Hale and Verduyn Lunel, 1993). The solution of a problem of this type is very hard.

It is clear that as the complexities of the system gradually increase so do the difficulties in resolving the problem. Thus it is obvious that we shall be forced to turn to solution methods of a numerical type. Then we must ask ourselves which formulation of the problem will best fit numerical methods. As we mentioned at the end of §2.6, and as we shall see in detail in Chapter 9, by representing a line through the equivalent circuits shown in Fig. 2.11, we can resolve any network consisting of ideal transmission lines and lumped circuits by applying those numerical techniques proper of the computer-aided analysis of electronic circuits (e.g., Chua and Lin, 1975).

2.9. IDEAL TWO-CONDUCTOR TRANSMISSION LINES WITH DISTRIBUTED SOURCES

Until now we have referred to ideal two-conductor transmission lines without distributed sources. Here we show how the two-port model obtained in §2.6 can be extended to lines with distributed sources. The equations for these lines are (see Chapter 1)

$$\begin{aligned} \frac{\partial v}{\partial x} + L\frac{\partial i}{\partial t} &= e_s, \\ \frac{\partial i}{\partial x} + C\frac{\partial v}{\partial t} &= j_s. \end{aligned} \tag{2.140}$$

The general solution of these equations may be expressed in the form

$$v(x,t) = v^+(t - x/c + T) + v^-(t + x/c) + v_p(x,t), \tag{2.141}$$

$$i(x,t) = \frac{1}{R_c}[v^+(t - x/c + T) - v^-(t + x/c)] + i_p(x,t), \tag{2.142}$$

where $v_p(x,t)$, $i_p(x,t)$ is a particular solution of the system (2.140).

2.9.1. A Particular Solution of the Line Equations with Distributed Sources

A particular solution of the system of equations (2.140) is that it satisfies null initial conditions and the boundary conditions of the line as if the line were perfectly matched. This ensures particular solutions of short duration. To determine this solution let us use Green's function method.

Applying the property of sampling of the Dirac function, we rewrite the system (2.140) as

$$\frac{\partial v}{\partial x} + L\frac{\partial i}{\partial t} = \int_{0^-}^{t^+} \delta(t-\tau)e_s(x,\tau)\,d\tau, \tag{2.143}$$

$$\frac{\partial i}{\partial x} + C\frac{\partial v}{\partial t} = \int_{0^-}^{t+} \delta(t-\tau) j_s(x,\tau)\, d\tau. \tag{2.144}$$

Let us indicate with $h_v^{(e)}(x,t;\tau)$, $h_i^{(e)}(x,t;\tau)$ the solution of the "auxiliary problem" defined by the equations

$$\frac{\partial h_v^{(e)}}{\partial x} + L\frac{\partial h_i^{(e)}}{\partial t} = \delta(t-\tau) e_s(x,\tau), \tag{2.145}$$

$$\frac{\partial h_i^{(e)}}{\partial x} + C\frac{\partial h_v^{(e)}}{\partial t} = 0, \tag{2.146}$$

the initial conditions

$$h_v^{(e)}(x, t=0; \tau) = 0 \quad \text{for } 0 \leqslant x \leqslant d, \tag{2.147}$$

$$h_i^{(e)}(x, t=0; \tau) = 0 \quad \text{for } 0 \leqslant x \leqslant d, \tag{2.148}$$

and the boundary conditions

$$h_v^{(e)}(x=0, t; \tau) = -R_c h_i^{(e)}(x=0, t; \tau) \quad \text{for } t > 0, \tag{2.149}$$

$$h_v^{(e)}(x=d, t; \tau) = R_c h_i^{(e)}(x=d, t; \tau) \quad \text{for } t > 0. \tag{2.150}$$

Moreover, let $h_v^{(j)}(x,t;\tau)$, $h_i^{(j)}(x,t;\tau)$ be the solution of the equations

$$\frac{\partial h_v^{(j)}}{\partial x} + L\frac{\partial h_i^{(j)}}{\partial t} = 0, \tag{2.151}$$

$$\frac{\partial h_i^{(j)}}{\partial x} + C\frac{\partial h_v^{(j)}}{\partial t} = \delta(t-\tau) j_s(x,\tau), \tag{2.152}$$

with the same initial and boundary conditions for $h_v^{(e)}$ and $h_i^{(e)}$. Using the superposition property, a particular solution of Eqs. (2.140) can be expressed in the form

$$v_p(x,t) = \int_0^t [h_v^{(e)}(x,t;\tau) + h_v^{(j)}(x,t;\tau)]\, d\tau, \tag{2.153}$$

$$i_p(x,t) = \int_0^t [h_i^{(e)}(x,t;\tau) + h_i^{(j)}(x,t;\tau)]\, d\tau. \tag{2.154}$$

This particular solution is, by construction, equal to zero at the initial time and satisfies the perfect matching conditions at the line ends.

Let us begin to resolve the first auxiliary problem. We soon observe that for $0 \leqslant t \leqslant \tau^-$ the solution is identically null and, for $t \geqslant \tau^+$, Eqs. (2.145) and (2.146) become homogeneous; $\tau^- = \tau - \varepsilon$ and

$\tau^+ = \tau + \varepsilon$ where ε is a positive and arbitrarily small number. Thus if we know the state of the line at $t = \tau^+$, we can solve the problem by using d'Alembert's solution, starting from that time instant.

Let us assume that function $e_s(x, t)$ is continuous and derivable with respect to x. We integrate both sides of Eqs. (2.145) and (2.146) in the time from $t = \tau - \varepsilon$ to $t = \tau + \varepsilon$. Using initial conditions and considering the limit $\varepsilon \to 0$, we determine the state of the line at time instant $t = \tau^+$,

$$h_i^{(e)}(x, t = \tau^+; \tau) = \frac{1}{L} e_s(x, \tau), \tag{2.155}$$

$$h_v^{(e)}(x, t = \tau^+; \tau) = 0. \tag{2.156}$$

Having assumed that $e_s(x, t)$ is continuous and derivable in respect to x, the terms $\partial h_v^{(e)}/\partial x$ and $\partial h_i^{(e)}/\partial x$ are limited and therefore do not contribute.

Since the solutions we are looking for have to be equal to zero for $t < \tau$, we consider d'Alembert solutions of the form

$$h_v^{(e)}(x, t; \tau) = \{v^{+(e)}[x - c(t - \tau); \tau] + v^{-(e)}[x + c(t - \tau); \tau]\} u(t - \tau), \tag{2.157}$$

$$h_i^{(e)}(x, t; \tau) = \frac{1}{R_c} \{v^{+(e)}[x - c(t - \tau); \tau] - v^{-(e)}[x + c(t - \tau); \tau]\} u(t - \tau) \tag{2.158}$$

The unknowns $v^{+(e)}(\xi; \tau)$ and $v^{-(e)}(\xi; \tau)$ are defined for $-\infty < \xi < +\infty$. Imposing conditions (2.155) and (2.156) and boundary conditions (2.149) and (2.150) we obtain

$$-v^{-(e)}(\xi; \tau) = v^{+(e)}(\xi; \tau) = \begin{cases} \dfrac{R_c}{2L} e_s(\xi, \tau) & \text{for } 0 \leqslant \xi \leqslant d, \\ 0 & \text{otherwise,} \end{cases} \tag{2.159}$$

Proceeding in the same way, we determine the solution of the other auxiliary problem. Assuming that the function $j_s(x, t)$ is continuous and derivable in respect to x one has

$$h_i^{(j)}(x, t = \tau^+; \tau) = 0, \tag{2.160}$$

$$h_v^{(j)}(x, t = \tau^+; \tau) = \frac{1}{C} j_s(x, \tau). \tag{2.161}$$

Thus the solution to the second auxiliary problem can be expressed as

$$h_v^{(j)}(x,t;\tau) = \{v^{+(j)}[x - c(t-\tau);\tau] + v^{-(j)}[x + c(t-\tau);\tau]\}u(t-\tau), \tag{2.162}$$

$$h_i^{(j)}(x,t;\tau) = \frac{1}{R_c}\{v^{+(j)}[x - c(t-\tau);\tau] - v^{-(j)}[x + c(t-\tau);\tau]\}u(t-\tau) \tag{2.163}$$

where the functions $v^{+(j)}$ and $v^{-(j)}$ are given by

$$v^{+(j)}(\xi;\tau) = v^{-(j)}(\xi;\tau) = \begin{cases} \dfrac{1}{2C} j_s(\xi,\tau) & \text{for } 0 \leqslant \xi \leqslant d, \\ 0 & \text{otherwise,} \end{cases} \tag{2.164}$$

2.9.2. Characterization as Two-Ports

Specifying expressions (2.141) and (2.142) at the end $x = 0$ and subtracting them termwise, we obtain the equation

$$v_1(t) - R_c i_1(t) = 2v^-(t) + e_1(t), \tag{2.165}$$

where

$$e_1(t) = v_p(x=0,t) - R_c i_p(x=0,t). \tag{2.166}$$

Specifying expressions (2.141) and (2.142) at end $x = d$ and summing them termwise, we obtain the equation

$$v_2(t) - R_c i_2(t) = 2v^+(t) + e_2(t), \tag{2.167}$$

where

$$e_2(t) = v_p(x=d,t) + R_c i_p(x=d,t). \tag{2.168}$$

Thus it is clear that a line with distributed sources can be represented with the equivalent two-port of Thévenin type shown in Fig. 2.11a, as long as we insert two independent voltage sources that supply voltages e_1 and e_2 connected in series with the controlled voltage sources w_1 and w_2, respectively.

The equations governing the state dynamics are

$$v^+(t) = v_1(t-T) - v^-(t-T) - v_p(x=0,t-T), \tag{2.169}$$

$$v^-(t) = v_2(t-T) - v^+(t-T) - v_p(x=d,t-T). \tag{2.170}$$

CHAPTER 3

Ideal Multiconductor Transmission Lines

In Chapter 2 we have shown how ideal two-conductor transmission lines can be characterized as two-ports in the time domain, and then how their behavior can be represented through simple equivalent lumped circuits of Thévenin and Norton type. Here, we shall extend these results to ideal multiconductor transmission lines by using the matrix theory. First, we shall diagonalize the ideal multiconductor line equations, introduce the concept of propagation mode, and extend the d'Alembert solution to these equations. Then, using this solution we shall determine the characteristic relation of the $2n$-port representing a generic ideal multiconductor line, and introduce some equivalent circuits of Thévenin and Norton type.

3.1. d'ALEMBERT SOLUTION FOR IDEAL MULTICONDUCTOR TRANSMISSION LINES

Consider an ideal multiconductor transmission line consisting of $n + 1$ conductors (see Fig. 1.3). The equations describing its behavior are (see §1.3.1)

$$-\frac{\partial \mathbf{v}}{\partial x} = \mathrm{L}\frac{\partial \mathbf{i}}{\partial t}, \tag{3.1}$$

$$-\frac{\partial \mathbf{i}}{\partial x} = \mathrm{C}\frac{\partial \mathbf{v}}{\partial t}. \tag{3.2}$$

where the $n \times n$ matrices L and C are symmetric and positive definite.

For the moment we have not considered distributed independent sources along the line, but at the end of this chapter we shall show how to take their effects into account.

By derivation and substitution of Eqs. (3.1) and (3.2), we obtain the two uncoupled equation systems

$$\frac{\partial^2 \mathbf{v}}{\partial x^2} - \mathrm{LC}\frac{\partial^2 \mathbf{v}}{\partial t^2} = \mathbf{0}, \tag{3.3}$$

$$\frac{\partial^2 \mathbf{i}}{\partial x^2} - \mathrm{CL}\frac{\partial^2 \mathbf{i}}{\partial t^2} = \mathbf{0}. \tag{3.4}$$

Each of these systems bears a resemblance to the scalar wave equation, and in fact, both systems (3.3) and (3.4) consist of n coupled scalar partial differential equations of the second order.

The set of all the possible solutions of the equation systems (3.3) and (3.4) is obviously much wider than that of the possible solutions of the starting equations (3.1) and (3.2). Consequently, once the general solutions of systems of equations (3.3) and (3.4) have been determined, we have to impose that both solutions satisfy one of the two equation systems (3.1) and (3.2).

Be careful! Even though the matrices L and C are symmetric, their product is not generally a symmetric matrix, thus the commutative property does not hold, and consequently LC $\neq$ CL. Therefore, the two equation systems (3.3) and (3.4) are characterized by different parameters.

The great difficulty in dealing with multiconductor lines derives from the fact that it is necessary to solve systems of coupled partial differential equations of the second order. This difficulty is overcome by using *similarity transformations* (e.g., Korn and Korn, 1961; Horn and Johnson, 1985), which diagonalize the matrices LC and CL. In this way the equation systems (3.3) and (3.4) are transformed into two new systems, each consisting of n uncoupled scalar equations of the same type as Eqs. (2.3) and (2.4).

Let us consider the change of variables

$$\mathbf{v}(x,t) = \mathrm{T_v}\mathbf{u}(x,t), \tag{3.5}$$

$$\mathbf{i}(x,t) = \mathrm{T_i}\mathbf{f}(x,t), \tag{3.6}$$

where $\mathrm{T_v}$ and $\mathrm{T_i}$ are two nonsingular matrices to be determined so that the set of equations for $\mathbf{u}(x,t)$ and the set of equations for $\mathbf{f}(x,t)$ consist of n uncoupled scalar partial differential equations. Substituting

Eq. (3.5) into equation systems (3.3), and (3.6) into equation system (3.4) we obtain

$$\frac{\partial^2 \mathbf{u}}{\partial x^2} - (\mathrm{T}_\mathrm{v}^{-1}\mathrm{LCT}_\mathrm{v})\frac{\partial^2 \mathbf{u}}{\partial t^2} = \mathbf{0}, \tag{3.7}$$

$$\frac{\partial^2 \mathbf{f}}{\partial x^2} - (\mathrm{T}_\mathrm{i}^{-1}\mathrm{CLT}_\mathrm{i})\frac{\partial^2 \mathbf{f}}{\partial t^2} = \mathbf{0}. \tag{3.8}$$

Is there a similarity transformation T_v that diagonalizes the matrix LC, and a similarity transformation that diagonalizes the matrix CL? The answer is "yes," and this is only because of the properties of the matrices L and C. Let us first examine the diagonalization problem of the matrix LC, and then extend the results to the matrix CL.

3.1.1. Properties and Diagonalization of the Matrices LC and CL

Let β indicate the generic eigenvalue of the matrix LC and $\mathbf{e}$ the corresponding right-hand eigenvector. They are the solution to the problem

$$(\mathrm{LC})\mathbf{e} = \beta\mathbf{e}. \tag{3.9}$$

The matrix LC, as we shall show later, possesses n linearly independent eigenvectors $\mathbf{e}_1, \mathbf{e}_2, \ldots, \mathbf{e}_n$, even when some or all of its eigenvalues coincide. This is one of the notable properties of the matrix LC and is a direct consequence of the fact that the matrices L and C are symmetric and strictly positive definite. Thus the matrix

$$\mathrm{T}_\mathrm{v} = |\mathbf{e}_1 \vdots \mathbf{e}_2 \vdots \cdots \vdots \mathbf{e}_n| \tag{3.10}$$

is invertible and it is the similarity transformation that diagonalizes LC,

$$\mathrm{T}_\mathrm{v}^{-1}\mathrm{LCT}_\mathrm{v} = \mathrm{B}; \tag{3.11}$$

B is the diagonal matrix

$$\mathrm{B} = \mathrm{diag}(\beta_1, \beta_2, \ldots, \beta_n), \tag{3.12}$$

and $\beta_1, \beta_2, \ldots, \beta_n$ are the eigenvalues corresponding to the eigenvectors $\mathbf{e}_1, \mathbf{e}_2, \ldots, \mathbf{e}_n$.

Let us now demonstrate that the matrix LC has n linearly independent eigenvectors.

It is well known that, if the eigenvalues of a matrix are all distinct, the corresponding eigenvectors are all linearly independent (e.g., Korn and Korn, 1961; Horn and Johnson, 1985). In reality, this is only a sufficient condition for the eigenvectors to be independent, but not a necessary one.

Let us recall that the algebraic multiplicity m_j of an eigenvalue λ_j of a matrix A is the multiplicity of the root λ_j of the secular equation $\det(\mathrm{A} - \lambda \mathrm{I}) = 0$. The geometric multiplicity m'_j of the eigenvalue λ_j is the number of linearly independent eigenvectors of A corresponding to λ_j. Clearly, the geometric multiplicity of an eigenvalue is less, or at most equal, to its algebraic multiplicity (e.g., Korn and Korn, 1961; Horn and Johnson, 1985).

As the matrices L and C are symmetric and strictly positive definite, the geometric multiplicity of each eigenvalue of LC is always equal to its algebraic multiplicity. This property is a direct consequence of one of the fundamental properties of a linear algebraic problem not often mentioned in the literature. However, it is an extremely interesting problem known as the *generalized eigenvalue problem* (e.g., Korn and Korn, 1961).

The matrix L is positive definite, and thus is invertible. Moreover, as L is symmetric, L^{-1} will also be symmetric as well as positive definite. As a result, the eigenvalue problem (3.9) can be rewritten in the equivalent form

$$\mathrm{C}\mathbf{e} = \beta \mathrm{L}^{-1}\mathbf{e}. \tag{3.13}$$

This is an example of a generalized eigenvalue problem. As C is symmetric and L^{-1} is symmetric and positive definite, the following properties hold (a demonstration is given in Appendix A):

(a) the eigenvalues $\beta_1, \beta_2, \ldots, \beta_n$ of LC are real and greater than zero;
(b) the geometric multiplicity m'_j of eigenvalue β_j is always equal to its algebraic multiplicity m_j; and
(c) the eigenvectors $\mathbf{e}_1, \mathbf{e}_2, \ldots, \mathbf{e}_n$ of LC are orthogonal with respect to the scalar product

$$\langle \mathbf{a}, \mathbf{b} \rangle_{\mathrm{L}} = \mathbf{a}^T \mathrm{L}^{-1}\mathbf{b}. \tag{3.14}$$

Normalizing the eigenvectors so that $\langle \mathbf{e}_j, \mathbf{e}_j \rangle_{\mathrm{L}} = 1$, we also have

$$\langle \mathbf{e}_i, \mathbf{e}_j \rangle_{\mathrm{L}} = \delta_{ij}, \tag{3.15}$$

where δ_{ij} is the Kronecker symbol.

Because all the eigenvalues are real, we may arrange them so that $\beta_1 \leqslant \beta_2 \leqslant \cdots \leqslant \beta_n$.

Remark

From the notable identity

$$\mathbf{a} = \sum_{h=1}^{n} \mathbf{e}_h \langle \mathbf{e}_h, \mathbf{a} \rangle, \tag{3.16}$$

where **a** is a generic vector, we have

$$\mathbf{a} = \sum_{h=1}^{n} \mathbf{e}_h (\mathbf{e}_h^{\mathrm{T}} \mathrm{L}^{-1}) \mathbf{a}. \tag{3.17}$$

As the matrix L is symmetric, from Eq. (3.17) it follows that

$$\sum_{h=1}^{n} \mathbf{e}_h (\mathrm{L}^{-1} \mathbf{e}_h)^{\mathrm{T}} = \mathrm{I}. \tag{3.18}$$

Recalling that $\mathbf{e}_1, \mathbf{e}_2, \ldots, \mathbf{e}_n$ are the columns of the similarity transformation matrix T_{v}, from Eq. (3.18), we have

$$\mathrm{T}_{\mathrm{v}} (\mathrm{L}^{-1} \mathrm{T}_{\mathrm{v}})^{\mathrm{T}} = \mathrm{I}, \tag{3.19}$$

and thus

$$\mathrm{T}_{\mathrm{v}} \mathrm{T}_{\mathrm{v}}^{\mathrm{T}} = \mathrm{L}. \tag{3.20}$$

If we had $\mathrm{L} = \mathrm{I}$, the transformation matrix T_{v} would be unitary, which is in agreement with the fact that the matrix C is symmetric. The matrix $\mathrm{T}_{\mathrm{v}}^{-1}$ can be expressed through the matrix $\mathrm{T}_{\mathrm{v}}^{\mathrm{T}}$,

$$\mathrm{T}_{\mathrm{v}}^{-1} = (\mathrm{L}^{-1} \mathrm{T}_{\mathrm{v}})^{\mathrm{T}}. \tag{3.21}$$

In particular, we have

$$\mathbf{s}_h = \mathrm{L}^{-1} \mathbf{e}_h, \tag{3.22}$$

where $\mathbf{s}_h^{\mathrm{T}}$ is the hth row of the matrix $\mathrm{T}_{\mathrm{v}}^{-1}$.

It is evident that the scalar product $\langle \mathbf{e}_h, \mathbf{a} \rangle_{\mathrm{L}}$ may be expressed as

$$\langle \mathbf{e}_h, \mathbf{a} \rangle_{\mathrm{L}} = \mathbf{s}_h^{\mathrm{T}} \mathbf{a}, \tag{3.23}$$

and hence the notable property holds

$$\begin{bmatrix} \langle \mathbf{e}_1, \mathbf{a} \rangle_{\mathrm{L}} = \mathbf{s}_1^{\mathrm{T}} \mathbf{a} \\ \langle \mathbf{e}_2, \mathbf{a} \rangle_{\mathrm{L}} = \mathbf{s}_2^{\mathrm{T}} \mathbf{a} \\ \cdots\cdots\cdots\cdots\cdots\cdots \\ \langle \mathbf{e}_n, \mathbf{a} \rangle_{\mathrm{L}} = \mathbf{s}_n^{\mathrm{T}} \mathbf{a} \end{bmatrix} = \mathrm{T}_{\mathrm{v}}^{-1} \mathbf{a}. \quad \diamond \tag{3.24}$$

Now let us consider the matrix CL. Let $\gamma_1, \gamma_2, \ldots, \gamma_n$ indicate the eigenvalues of the matrix CL, and $\mathbf{g}_1, \mathbf{g}_2, \ldots, \mathbf{g}_n$ indicate the corresponding right-hand eigenvectors. They are the solution to the problem

$$(\mathrm{CL})\mathbf{g} = \gamma\mathbf{g}. \tag{3.25}$$

Let us immediately demonstrate that the eigenvalues of CL coincide with those of LC. The eigenvalues of CL are the solutions of the secular equation

$$\det(\mathrm{CL} - \gamma\mathrm{I}) = 0. \tag{3.26}$$

Using identity $\mathrm{L}^{-1}\mathrm{L} = \mathrm{I}$, from Eq. (3.26) we obtain

$$\det[(\mathrm{C} - \gamma\mathrm{L}^{-1})\mathrm{L}] = \det(\mathrm{C} - \gamma\mathrm{L}^{-1})\det(\mathrm{L}) = 0. \tag{3.27}$$

As L is invertible ($\det(\mathrm{L}) \neq 0$), the roots of the secular equation (3.26) are (and only are) the n roots of the algebraic equation

$$\det(\mathrm{C} - \gamma\mathrm{L}^{-1}) = 0. \tag{3.28}$$

This is the secular equation associated to the generalized problem (3.13) and consequently the eigenvalues of CL coincide with those of LC. Thus,

$$(\mathrm{CL})\mathbf{g} = \beta\mathbf{g}. \tag{3.29}$$

For the matrix CL, all the properties that we have reported for the matrix LC also hold. In particular, we have

$$\mathrm{T}_\mathrm{i}^{-1}\mathrm{CLT}_\mathrm{i} = \mathrm{B}, \tag{3.30}$$

where the similarity transformation matrix T_i is given by

$$\mathrm{T}_\mathrm{i} = |\mathbf{g}_1 \vdots \mathbf{g}_2 \vdots \cdots \vdots \mathbf{g}_n|, \tag{3.31}$$

and the diagonal matrix B is given by Eq. (3.12). Moreover, the eigenvectors $\mathbf{g}_1, \mathbf{g}_2, \ldots, \mathbf{g}_n$ are orthogonal with respect to the scalar product $\langle \mathbf{a}, \mathbf{b} \rangle_\mathrm{C}$ defined as in Eq. (3.14). Normalizing the eigenvectors so that $\langle \mathbf{g}_j, \mathbf{g}_j \rangle_\mathrm{C} = 1$, we obtain

$$\langle \mathbf{g}_i, \mathbf{g}_j \rangle_\mathrm{C} = \delta_{ij}. \tag{3.32}$$

Furthermore, we have

$$\mathrm{T}_\mathrm{i}\mathrm{T}_\mathrm{i}^\mathrm{T} = \mathrm{C}. \tag{3.33}$$

3.1.2. Characteristic Resistance and Conductance Matrices

Generally, the eigenvectors of CL and those of LC, although different, are mutually related by means of very simple relations.

Let us consider the vectors $\mathbf{q}_h$ defined (for $h = 1, 2, \ldots, n$) as

$$\mathbf{q}_h = a_h \mathrm{L}^{-1} \mathbf{e}_h, \tag{3.34}$$

where a_h is a constant, which for the moment we shall assume to be arbitrary. From Eq. (3.13) we obtain

$$\mathrm{CL}\mathbf{q}_h = \beta_h \mathbf{q}_h, \tag{3.35}$$

and thus $\mathbf{q}_h$ is one of the eigenvectors of CL corresponding to the eigenvalue β_h. Thus a complete set of eigenvectors of CL is given by

$$\mathbf{g}_h = a_h \mathrm{L}^{-1} \mathbf{e}_h \quad \text{for } h = 1, 2, \ldots, n. \tag{3.36}$$

Imposing that $\langle \mathbf{g}_i, \mathbf{g}_i \rangle_C = 1$, we obtain for the constants a_k

$$a_h = \sqrt{\beta_h} \quad \text{for } h = 1, 2, \ldots, n. \tag{3.37}$$

Reasoning in the same way, we can determine a complete set of eigenvectors of LC starting from the eigenvectors of CL. From Eq. (3.25) we obtain (for $h = 1, 2, \ldots, n$)

$$\mathbf{e}_h = b_h \mathrm{C}^{-1} \mathbf{g}_h, \tag{3.38}$$

where b_h is an arbitrary constant. Imposing that $\langle \mathbf{e}_i, \mathbf{e}_i \rangle_L = 1$ we obtain

$$b_h = \sqrt{\beta_h} \quad \text{for } h = 1, 2, \ldots, n. \tag{3.39}$$

Placing

$$\mathrm{A} = \mathrm{diag}(\sqrt{\beta_1}, \sqrt{\beta_2}, \ldots, \sqrt{\beta_n}), \tag{3.40}$$

from Eqs. (3.36) to (3.39) we have, respectively,

$$\mathrm{T_i} = \mathrm{L}^{-1} \mathrm{T_v} \mathrm{A}, \tag{3.41}$$

$$\mathrm{T_v} = \mathrm{C}^{-1} \mathrm{T_i} \mathrm{A}. \tag{3.42}$$

Thus, once the similarity matrix $\mathrm{T_v}$ is known, the similarity matrix $\mathrm{T_i}$ can be directly evaluated from Eq. (3.41) (or Eq. (3.42)), and vice versa. Therefore, the matrix A satisfies the equation

$$\mathrm{A}^2 = \mathrm{T_v}^{-1} \mathrm{LC} \mathrm{T_v} = \mathrm{T_i}^{-1} \mathrm{CL} \mathrm{T_i} = \mathrm{B}. \tag{3.43}$$

Using the square root operation of a matrix from Eq. (3.42) we have

$$A = \sqrt{B} = T_v^{-1}\sqrt{LC}\,T_v = T_i^{-1}\sqrt{CL}T_i. \tag{3.44}$$

The general definition of a function of a matrix is given in Appendix A. Substituting Eq. (3.44) into Eq. (3.41) or (3.42) we obtain the notable relation

$$T_i = G_c T_v \tag{3.45}$$

where

$$G_c = L^{-1}\sqrt{LC} = \sqrt{(CL)^{-1}}C \tag{3.46}$$

is the *characteristic "conductance" matrix* of the line. The inverse of the conductance matrix G_c is known as the *characteristic "resistance" matrix*,

$$R_c = G_c^{-1} = \sqrt{(LC)^{-1}}L = C^{-1}\sqrt{CL}. \tag{3.47}$$

We shall show in §3.9 that both the matrices R_c and G_c are always invertible, positive definite and symmetric.

Remark

From Eqs. (3.36) and (3.37) (or from Eq. (3.38) and (3.39)) and from the property (3.15) (or (3.32)) we obtain:

$$\mathbf{e}_j^T \mathbf{g}_i = \sqrt{\beta_i}\,\delta_{ji}. \tag{3.48}$$

Therefore, the generic eigenvector $\mathbf{e}_i$ is orthogonal to the eigenvector $\mathbf{g}_j$ for $i \neq j$. Thus, between the transformation matrices T_v and T_i there is also the notable relation

$$T_i^T T_v = \sqrt{B}. \tag{3.49}$$

The property of orthogonality between $\mathbf{e}_i$ and $\mathbf{g}_j$ can be demonstrated in another way (Marx, 1973). The matrices L and C are symmetric and so we have

$$(LC)^T = CL. \tag{3.50}$$

Thus the matrix CL is the *adjoint matrix* of the matrix LC, and Eq. (3.29) is the adjoint equation of Eq. (3.9) (e.g., Korn and Korn, 1961). The property (3.48) is a direct consequence of this. Indeed, if we consider the nth eigenvector of LC and the jth eigenvector of CL, Eqs. (3.9) and (3.29) give

$$\mathbf{g}_j^{\mathrm{T}}\mathrm{LC}\mathbf{e}_i - \mathbf{e}_i^{\mathrm{T}}\mathrm{CL}\mathbf{g}_j = (\beta_i - \beta_j)\mathbf{e}_i^{\mathrm{T}}\mathbf{g}_j. \tag{3.51}$$

Due to property (3.50), the left-hand term of Eq. (3.51) is zero, and hence $\mathbf{e}_i^{\mathrm{T}}\mathbf{g}_j = 0$ for $\beta_i \neq \beta_j$. However, when there are coincident eigenvalues, we have to resort to the demonstration given in the previous paragraph. ◇

3.1.3. Natural Modes of Propagation

Now we introduce the natural modes of propagation of ideal multiconductor transmission lines and extend the d'Alembert solution to this type of line. Substituting, respectively, Eqs. (3.11) and (3.30) into Eqs. (3.7) and (3.8) we obtain

$$\frac{\partial^2 \mathbf{u}}{\partial x^2} - \mathrm{B}\frac{\partial^2 \mathbf{u}}{\partial t^2} = \mathbf{0}, \tag{3.52}$$

$$\frac{\partial^2 \mathbf{f}}{\partial x^2} - \mathrm{B}\frac{\partial^2 \mathbf{f}}{\partial t^2} = \mathbf{0}. \tag{3.53}$$

The equations of the system (3.52) and those of the system (3.53) are all uncoupled, because the matrix B is diagonal. From Eq. (3.52) we have

$$\frac{\partial^2 u_h}{\partial x^2} - \beta_h\frac{\partial^2 u_h}{\partial t^2} = 0 \quad \text{for } h = 1, 2, \ldots, n, \tag{3.54}$$

and from Eq. (3.53) we have

$$\frac{\partial^2 f_h}{\partial x^2} - \beta_h\frac{\partial^2 f_h}{\partial t^2} = 0 \quad \text{for } h = 1, 2, \ldots, n, \tag{3.55}$$

where u_h and f_h are, respectively, the hth component of the vectors $\mathbf{u}$ and $\mathbf{f}$. We have brought the problem of determining the general solution of the vectorial wave equations (3.3) and (3.4) back to that of determining the general solution of the scalar wave equations (3.54) and (3.55).

Remark

From the relations (3.41) and (3.42) we immediately obtain the other notable relation

$$\mathrm{T}_{\mathrm{v}}^{-1}\mathrm{L}\mathrm{T}_{\mathrm{i}} = \mathrm{T}_{\mathrm{i}}^{-1}\mathrm{C}\mathrm{T}_{\mathrm{v}} = \mathrm{A}. \tag{3.56}$$

Thus, using the two linear transformations T_v and T_i, it is possible to diagonalize the matrices L and C directly and simultaneously. Applying these transformations to the set of equations (3.1) and (3.2), we obtain the set

$$-\frac{\partial u_h}{\partial x} = \sqrt{\beta_h}\frac{\partial f_h}{\partial t}, \tag{3.57}$$

$$-\frac{\partial f_h}{\partial x} = \sqrt{\beta_h}\frac{\partial u_h}{\partial t}, \tag{3.58}$$

for $1 \leqslant h \leqslant n$. From Eqs. (3.57) and (3.58) we immediately obtain Eqs. (3.54) and (3.55). Obviously, the general solution of Eqs. (3.54) and that of Eq. (3.55) must satisfy at least one of the two equations of the set (3.57) and (3.58). ◇

The general solutions of Eqs. (3.54) and (3.55) in the d'Alembert form are, respectively (see Chapter 2),

$$u_h(x,t) = u_h^+(t - x/c_h + \alpha_h^+) + u_h^-(t + x/c_h + \alpha_h^-), \tag{3.59}$$

$$f_h(x,t) = f_h^+(t - x/c_h + \alpha_h^+) + f_h^-(t + x/c_h + \alpha_h^-), \tag{3.60}$$

where

$$c_h = \frac{1}{\sqrt{\beta_h}} > 0, \tag{3.61}$$

u_h^+, f_h^+, u_h^-, and f_h^- are arbitrary functions, and α_h^+, α_h^- are arbitrary constants.

Imposing the condition that the solutions (3.59) and (3.60) satisfy at least one of the two equations of set (3.57) and (3.58), we have

$$f^+ = u^+, \tag{3.62}$$

$$f^- = -u^-. \tag{3.63}$$

Then, the general solution of Eqs. (3.1) and (3.2) can be represented as

$$\mathbf{v}(x,t) = \sum_{h=1}^{n} \mathbf{e}_h[u_h^+(t - x/c_h + \alpha_h^+) + u_h^-(t + x/c_h + \alpha_h^-)], \tag{3.64}$$

$$\mathbf{i}(x,t) = \mathrm{R}_c^{-1} \sum_{h=1}^{n} \mathbf{e}_h[u_h^+(t - x/c_h + \alpha_h^+) - u_h^-(t + x/c_h + \alpha_h^-)]. \tag{3.65}$$

The general solutions for the voltages and for the currents of a multiconductor transmission line with $(n+1)$ conductors are expressed as the sum of n terms. Each term represents a *natural quasi*-TEM *propagation mode* of the line, and c_h is the corresponding propagation velocity. Each mode consists of the superposition of a forward voltage wave u_h^+ and a backward voltage wave u_h^-. The forward waves u_h^+ and the backward waves u_h^- must be determined by imposing the initial and the boundary conditions.

Now it is clear why R_c is called the characteristic resistance matrix of the line. If in each propagation mode the backward wave is absent, we have $\mathbf{v}(x,t) = R_c\mathbf{i}(x,t)$; if the forward wave is absent, we have $\mathbf{v}(x,t) = -R_c\mathbf{i}(x,t)$.

3.2. INFINITE MULTICONDUCTOR TRANSMISSION LINES

Let us now consider an infinite ideal multiconductor transmission line and by imposing the initial conditions, show how the forward and the backward waves are determined for each propagation mode of the line.

As for the two-conductor case (see §2.2.1), for a multiconductor line of infinite length we choose $\alpha_h^+ = \alpha_h^- = 0$ for each mode. Thus, the general solution of the line equations (3.64) and (3.65) becomes

$$\mathbf{v}(x,t) = \sum_{h=1}^{n} \mathbf{e}_h[u_h^+(t - x/c_h) + u_h^-(t + x/c_h)], \tag{3.66}$$

$$\mathbf{i}(x,t) = R_c^{-1} \sum_{h=1}^{n} \mathbf{e}_h[u_h^+(t - x/c_h) - u_h^-(t + x/c_h)]. \tag{3.67}$$

In this case the functions u_h^+ and u_h^-, defined in the interval $(-\infty, +\infty)$ for every h, are univocally determined imposing that the solutions (3.66) and (3.67) satisfy the initial conditions

$$\mathbf{v}(x, t=0) = \mathbf{v}_0(x) \quad \text{for } -\infty < x < +\infty, \tag{3.68}$$

$$\mathbf{i}(x, t=0) = \mathbf{i}_0(x) \quad \text{for } -\infty < x < +\infty. \tag{3.69}$$

Left multiplying both members of Eqs. (3.68) and (3.69) by $\mathbf{e}_h^T L^{-1}$, and using the properties of orthogonality of $\mathbf{e}_1, \mathbf{e}_2, \ldots, \mathbf{e}_n$ with respect to the scalar product $\langle \mathbf{a}, \mathbf{b} \rangle_L$, we obtain for every h and for $-\infty < \tau < +\infty$

$$u_h^+(\tau) + u_h^-(\tau) = \langle \mathbf{e}_h, \mathbf{v}_0(-c_h\tau) \rangle_L, \tag{3.70}$$

$$u_h^+(\tau) - u_h^-(\tau) = \langle \mathbf{e}_h, R_c\mathbf{i}_0(c_h\tau) \rangle_L. \tag{3.71}$$

From these equations for every mode we immediately have

$$u_h^+(\tau) = \tfrac{1}{2}\langle \mathbf{e}_h, [\mathbf{v}_0(-c_h\tau) + \mathrm{R}_c\mathbf{i}_0(-c_h\tau)]\rangle_{\mathrm{L}}, \tag{3.72}$$

$$u_h^-(\tau) = \tfrac{1}{2}\langle \mathbf{e}_h, [\mathbf{v}_0(c_h\tau) - \mathrm{R}_c\mathbf{i}_0(c_h\tau)]\rangle_{\mathrm{L}}. \tag{3.73}$$

Which modes are generated, and the intensity with which they are generated, depends on the initial distribution of the voltages and currents along the line. A generic initial distribution of the currents and voltages generates both the forward and backward waves of each of the modes that the line can sustain. Clearly, it is always possible, at least in principle, to prepare the distribution of the initial values of the voltages and the currents so that a single mode is excited. Furthermore, when $\mathbf{v}_0(x) = \mathrm{R}_c\mathbf{i}_0(x)$, all the modes have only the forward wave, while when $\mathbf{v}_0(x) = -\mathrm{R}_c\mathbf{i}_0(x)$, all the modes have only the backward wave.

3.3. SEMI-INFINITE MULTICONDUCTOR TRANSMISSION LINES AND EQUIVALENT CIRCUITS

Now let us consider a semi-infinite transmission line $0 \leqslant x < +\infty$, and let us assume that the currents at the end $x = 0$ $(i_{11}, i_{12}, \ldots, i_{1n})$ are known (Fig. 3.1),

$$\mathbf{i}(x = 0, t) = \mathbf{i}_1(t) \quad \text{per } t \geqslant 0; \tag{3.74}$$

$\mathbf{i}_1$ is the vector representative of the currents at the end $x = 0$,

$$\mathbf{i}_1 = |i_{11} i_{12} \cdots i_{1n}|^{\mathrm{T}}. \tag{3.75}$$

For the sake of simplicity, let us assume that the initial distributions of the voltages and the currents along the line are both zero. In this case, the functions, u_h^-, which are defined in the interval $(0, \infty)$, are identically zero, and the functions u_h^+ that are defined in the interval $(-\infty, +\infty)$, are identically zero for $\tau < 0$, while for $\tau \geqslant 0$ they depend solely on the boundary conditions (3.74). As long as the initial conditions are compatible with the constraint imposed by the boundary conditions at the initial instant, we must have $\mathbf{i}_1(0) = \mathbf{0}$.

Imposing now that the currents satisfy the boundary condition (3.74), we obtain

$$\sum_{h=1}^{n} \mathbf{e}_h u_h^+(t) = \mathrm{R}_c\mathbf{i}_1(t) \quad \text{for } t \geqslant 0. \tag{3.76}$$

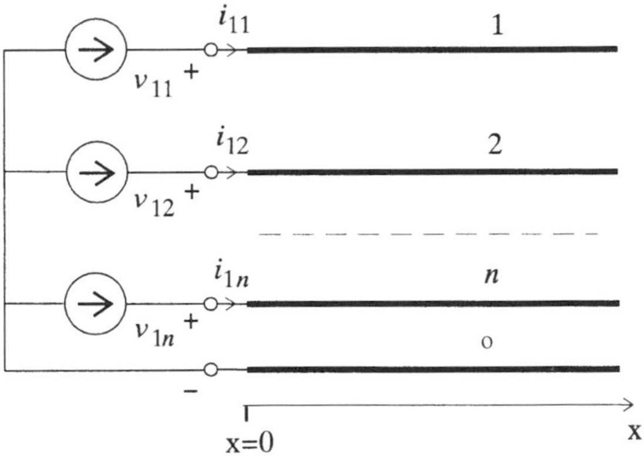

Figure 3.1. Semi-infinite line connected to independent current sources.

The mode that is generated, and the intensity with which it is generated, depend only on the actual values of the current sources.

Let us indicate with

$$\mathbf{v}_1 = |v_{11} v_{12} \cdots v_{1n}|^{\mathrm{T}} \tag{3.77}$$

the representative vector of the voltages at the end $x = 0$; we have chosen the normal convention for the current and voltage references (Fig. 3.1). Because

$$\mathbf{v}_1(t) = \mathbf{v}(x = 0, t) = \sum_{h=1}^{n} \mathbf{e}_h u_h^+(t), \tag{3.78}$$

we have

$$\mathbf{v}_1(t) = \mathrm{R}_c \mathbf{i}_1(t). \tag{3.79}$$

If there were also backward waves (which happens when the initial conditions are not zero), we would have to add a known term to the right-hand side of Eq. (3.79),

$$\mathbf{v}_1(t) = \mathrm{R}_c \mathbf{i}_1(t) + \mathbf{w}_1(t), \tag{3.80}$$

where

$$\mathbf{w}_1(t) = 2 \sum_{h=1}^{n} \mathbf{e}_h u_h^-(t). \tag{3.81}$$

From Eq. (3.79) it is evident that the behavior of the line at the end $x = 0$, when the line is initially at rest, is equivalent to that of a

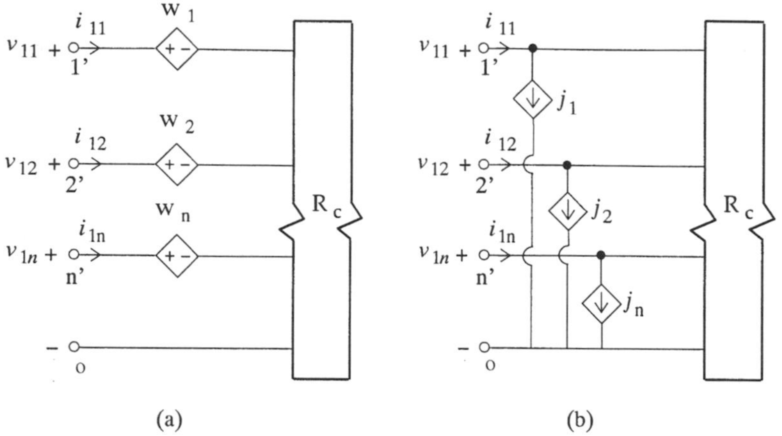

Figure 3.2. (a) Thévenin equivalent representation of a semi-infinite multiconductor transmission line; (b) Norton equivalent representation $\mathbf{j}_1 = -R_c^{-1}\mathbf{w}_1$.

linear n-port of the resistive type, with the resistance matrix equal to the characteristic resistance matrix of the line. If backward waves were present, we would have to connect suitable independent voltage sources in series to the terminals $1, 2, \ldots, n$ (Fig. 3.2a). This is the Thévenin equivalent representation of an ideal semi-infinite multiconductor line. The behavior of the line at the end $x = 0$ can also be described by means of the Norton equivalent representation of Fig. 3.2b.

3.4. IDEAL MULTICONDUCTOR TRANSMISSION LINES AS MULTIPORTS

The interaction of a multiconductor transmission line of finite length with lumped circuits can be described by representing the line as a $2n$-port. Now we extend all the results obtained for the two-conductor case given in §2.4 to ideal multiconductor lines.

At each end of the line, the pair composed by the hth terminal (with $h = 1, \ldots, n$), and the (0-th terminal, that is the reference conductor, forms a port (Fig. 3.3): v_{1h} and i_{1h} are, respectively, the voltage and the current of the hth port at the end $x = 0$; v_{2h} and i_{2h} are, respectively, the voltage and the current of the hth port at the end $x = d$. Let us indicate with $\mathbf{v}_1$ and $\mathbf{i}_1$ the vectors representing, respectively, the voltages and currents at the end $x = 0$, and with $\mathbf{v}_2$

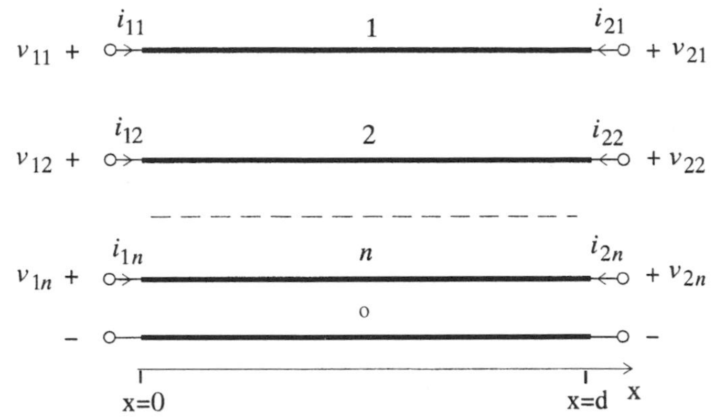

Figure 3.3. Sketch of a finite-length multiconductor transmission line.

and $\mathbf{i}_2$ the vectors representing, respectively, the voltages and currents at the end $x = d$.

The voltages and the currents at the line ends are related to the voltage and the current distributions along the line through the relations

$$\mathbf{i}_1(t) = \mathbf{i}(x = 0, t) \quad \text{for } t \geqslant 0, \tag{3.82}$$

$$\mathbf{v}_1(t) = \mathbf{v}(x = 0, t) \quad \text{for } t \geqslant 0, \tag{3.83}$$

and

$$\mathbf{i}_2(t) = -\mathbf{i}(x = d, t) \quad \text{for } t \geqslant 0, \tag{3.84}$$

$$\mathbf{v}_2(t) = \mathbf{v}(x = d, t) \quad \text{for } t \geqslant 0. \tag{3.85}$$

To determine the characteristic relations of the multiport representing an ideal multiconductor line we shall proceed in the same way as for the ideal two-conductor transmission lines dealt with in §2.4. First, we impose the condition that the general solution of the line equations satisfies the initial conditions along the line; then we particularize the solution at the line ends.

As for two-conductor lines, for multiconductor lines of finite length we choose the constants α_h^+ and α_h^- so that $u_h^+(t)$ represents the amplitude of the forward wave associated with the hth mode at the right end of line $x = d$, while $u_h^-(t)$ represents the amplitude of the backward wave associated with the hth mode at the left end $x = 0$. Consequently, it must be that

$$\alpha_h^+ = T_h, \quad \alpha_h^- = 0, \tag{3.86}$$

where

$$T_h = \frac{d}{c_h} = d\sqrt{\beta_h} \tag{3.87}$$

is the transit time of the hth mode. With this choice, $u_h^+(t)$ and $u_h^-(t)$ are defined in the interval $(0, +\infty)$ for every h. Then, the expressions for the general solution of the voltages and the currents become

$$\mathbf{v}(x,t) = \sum_{h=1}^{n} \mathbf{e}_h[u_h^+(t - x/c_h + T_h) + u_h^-(t + x/c_h)], \tag{3.88}$$

$$\mathbf{i}(x,t) = \mathrm{R}_c^{-1} \sum_{h=1}^{n} \mathbf{e}_h[u_h^+(t - x/c_h + T_h) - u_h^-(t + x/c_h)]. \tag{3.89}$$

These expressions can be rewritten as

$$\mathbf{v}(x,t) = \boldsymbol{v}^+(x,\mathrm{t}) + \boldsymbol{v}^-(x,t), \tag{3.90}$$

$$\mathbf{i}(x,t) = \mathrm{R}_c^{-1}[\boldsymbol{v}^+(x,t) - \boldsymbol{v}^-(x,t)], \tag{3.91}$$

where the variables

$$\boldsymbol{v}^+(x,t) = \sum_{h=1}^{n} \mathbf{e}_h u_h^+(t - x/c_h + T_h), \tag{3.92}$$

$$\boldsymbol{v}^-(x,t) = \sum_{h=1}^{n} \mathbf{e}_h u_h^-(t + x/c_h), \tag{3.93}$$

represent, respectively, the "actual voltage wave" traveling in the forward direction and the "actual voltage wave" traveling in the backward direction.

The voltage and current distributions along the line are completely identified by the set of functions $u_1^+(t)$, $u_2^+(t), \ldots, u_n^+(t)$ and $u_1^-(t)$, $u_2^-(t), \ldots, u_n^-(t)$, and vice-versa. Thus the set of the functions $u_1^+(t)$, $u_2^+(t), \ldots, u_n^+(t)$ and $u_1^-(t), u_2^-(t), \ldots, u_n^-(t)$ completely specify the internal state of the ideal multiconductor line.

3.4.1. Characterization of the Transmission Line Behavior at the Ends

The initial conditions

$$\mathbf{v}(x, t=0) = \mathbf{v}_0(x) \quad \text{for } 0 \leqslant x \leqslant d, \tag{3.94}$$

$$\mathbf{i}(x, t=0) = \mathbf{i}_0(x) \quad \text{for } 0 \leqslant x \leqslant d, \tag{3.95}$$

fix $u_h^+(t)$ and $u_h^-(t)$ in the time interval $(0, T_h)$ for every mode, whereas the boundary conditions determine $u_h^+(t)$ and $u_h^-(t)$ for $t > T_h$. By placing

$$u_{0h}^+(t) = \begin{cases} \frac{1}{2}\langle \mathbf{e}_h, \{\mathbf{v}_0[c_h(T_h - t)] + \mathrm{R_c}\mathbf{i}_0[c_h(T_h - t)]\}\rangle_{\mathrm{L}} & \text{for } 0 \leqslant t \leqslant T_h, \\ 0 & \text{for } t > T_h, \end{cases} \tag{3.96}$$

$$u_{0h}^-(t) = \begin{cases} \frac{1}{2}\langle \mathbf{e}_h, \{\mathbf{v}_0(c_h t) - \mathrm{R_c}\mathbf{i}_0(c_h t)\}\rangle_{\mathrm{L}} & \text{for } 0 \leqslant t \leqslant T_h, \\ 0 & \text{for } t > T_h. \end{cases} \tag{3.97}$$

and imposing the condition that expressions (3.88) and (3.89) satisfy the initial conditions, we obtain (see §3.2)

$$u_h^+(t) = u_{0h}^+(t) \quad \text{for } 0 \leqslant t \leqslant T_h, \tag{3.98}$$

$$u_h^-(t) = u_{0h}^-(t) \quad \text{for } 0 \leqslant t \leqslant T_h. \tag{3.99}$$

We recall that we have arranged the eigenvalues β_h of the matrix LC so as to have

$$T_1 \leqslant T_2 \leqslant \cdots \leqslant T_n, \tag{3.100}$$

that is, the first mode is the fastest and the nth is the slowest.

Specifying the expressions (3.90) and (3.91) at the end $x = 0$, we obtain

$$\mathbf{v}_1(t) = \boldsymbol{v}^+(0,t) + \boldsymbol{v}^-(0,t), \tag{3.101}$$

$$\mathrm{R_c}\mathbf{i}_1(t) = \boldsymbol{v}^+(0,t) - \boldsymbol{v}^-(0,t), \tag{3.102}$$

whereas specifying them at the end $x = d$, we obtain

$$\mathbf{v}_2(t) = \boldsymbol{v}^+(d,t) + \boldsymbol{v}^-(d,t), \tag{3.103}$$

$$\mathrm{R_c}\mathbf{i}_2(t) = \boldsymbol{v}^-(d,t) - \boldsymbol{v}^+(d,t), \tag{3.104}$$

where the expressions of the forward voltage $\boldsymbol{v}^+(x,t)$ and the backward voltage $\boldsymbol{v}^-(x,t)$ waves are given by Eqs. (3.92) and (3.93).

As in the case of the two-conductor line, it is possible to eliminate the forward wave $\mathbf{v}^+(0,t)$ from Eqs. (3.101) and (3.102) and the backward wave $\mathbf{v}^-(d,t)$ from Eqs. (3.103) and (3.104) to obtain

$$\mathbf{v}_1(t) - \mathrm{R_c}\mathbf{i}_1(t) = 2\mathrm{T_v}\mathbf{u}^-(t), \tag{3.105}$$

$$\mathbf{v}_2(t) - \mathrm{R_c}\mathbf{i}_2(t) = 2\mathrm{T_v}\mathbf{u}^+(t), \tag{3.106}$$

where $\mathbf{u}^+(t)$ and $\mathbf{u}^-(t)$ are the vectors representing the state functions $u_1^+(t), u_2^+(t), \ldots, u_n^+(t)$ and $u_1^-(t), u_2^-(t), \ldots, u_n^-(t)$.

Equations (3.105) and (3.106) are forms that are familiar from the theory of two-conductor transmission lines developed in §2.4. As for the two-conductor case, if the state of the line were completely known at every t, these equations would completely determine the terminal behavior of the line. In particular, if the backward waves represented by $\mathbf{u}^-(t)$ were known, then Eq. (3.105) and a knowledge of the network connected at $x = 0$ would suffice to determine $\mathbf{v}_1$ and $\mathbf{i}_1$. Similarly, if the forward waves represented by $\mathbf{u}^+(t)$ were known, then Eqs. (3.106) and a knowledge of the network connected at $x = d$ would suffice to determine $\mathbf{v}_2$ and $\mathbf{i}_2$. Actually, as for two-conductor transmission lines, the voltage waves $\mathbf{u}^+(t)$ and $\mathbf{u}^-(t)$ are themselves unknowns.

Different formulations of the equations governing the dynamics of the state are possible.

From Eqs. (3.101) and (3.103), for $t > 0$, we immediately obtain

$$\sum_{h=1}^{n} \mathbf{e}_h[u_h^+(t + T_h) + u_h^-(t)] = \mathbf{v}_1(t), \tag{3.107}$$

$$\sum_{h=1}^{n} \mathbf{e}_h[u_h^-(t + T_h) + u_h^+(t)] = \mathbf{v}_2(t). \tag{3.108}$$

Similar equations involving the terminal variables $\mathbf{i}_1$ and $\mathbf{i}_2$ instead of $\mathbf{v}_1$ and $\mathbf{v}_2$ may be obtained from Eqs. (3.102) and (3.104). As these equations are completely equivalent to Eqs. (3.107) and (3.108), hereafter we shall refer mainly to the latter.

Equations (3.107) and (3.108) express the forward and the backward voltage waves as functions of the values of the voltages at the line ends. As for two-conductor lines (see §2.4), the state equations are difference linear relations. It is soon observed, however, that for the multiconductor line there is a difficulty: Because the propagation modes have different velocities, each difference equation contains terms with different delay times, which, generally, are incommensurable. This difficulty is immediately overcome by again using the orthogonality property of the eigenvectors $\mathbf{e}_1, \mathbf{e}_2, \ldots, \mathbf{e}_n$ with respect to the scalar product $\langle \mathbf{a}, \mathbf{b} \rangle_{\mathrm{L}}$.

Left-multiplying each member of Eqs. (3.107) and (3.108) for the row vector $\mathbf{e}_i^{\mathrm{T}}\mathrm{L}^{-1}$, and using Eq. (3.15), for any h, we obtain

$$u_h^+(t) = -u_h^-(t - T_h) + \langle \mathbf{e}_h, \mathbf{v}_1(t - T_h) \rangle_{\mathrm{L}} \quad \text{for } t > T_h, \tag{3.109}$$

$$u_h^-(t) = -u_h^+(t - T_h) + \langle \mathbf{e}_h, \mathbf{v}_2(t - T_h) \rangle_{\mathrm{L}} \quad \text{for } t > T_h. \tag{3.110}$$

These equations have to be solved with the initial conditions given by Eqs. (3.98) and (3.99).

Instead, summing Eqs. (3.101) and (3.102) termwise, and summing Eqs. (3.103) and (3.104) similarly, for $t > 0$ we have

$$2 \sum_{h=1}^{n} \mathbf{e}_h u_h^+(t + T_h) = \mathbf{v}_1(t) + \mathrm{R}_c \mathbf{i}_1(t), \tag{3.111}$$

$$2 \sum_{h=1}^{n} \mathbf{e}_h u_h^-(t + T_h) = \mathbf{v}_2(t) + \mathrm{R}_c \mathbf{i}_2(t). \tag{3.112}$$

Left multiplying each member of Eqs. (3.111) and (3.112) by the row vector $\mathbf{e}_h^{\mathrm{T}}\mathrm{L}^{-1}$, using the property (3.15), for any h we obtain

$$u_h^+(t) = \tfrac{1}{2}\langle \mathbf{e}_h, [\mathbf{v}_1(t - T_h) + \mathrm{R}_c \mathbf{i}_1(t - T_h)]\rangle_{\mathrm{L}} \quad \text{for } t > T_h, \tag{3.113}$$

$$u_h^-(t) = \tfrac{1}{2}\langle \mathbf{e}_h, [\mathbf{v}_2(t - T_h) + \mathrm{R}_c \mathbf{i}_2(t - T_h)]\rangle_{\mathrm{L}} \quad \text{for } t > T_h. \tag{3.114}$$

Equations (3.109) and (3.110) describe in *implicit form* the relation between the state of the line and the electrical variables at the line ends, whereas Eqs. (3.113) and (3.114) provide the same relation, but in an *explicit form*.

The advantage of representing the solution through the natural propagation modes is in our being able to bring the analysis of the state dynamics back to the study of as many uncoupled problems as there are line propagation modes. Each problem has the same structure as that we have dealt with in the case of the two-conductor line, thus, all the considerations we have outlined in that case can be repeated. In particular, the functions $u_h^+(t)$ and $u_h^-(t)$ in the time interval $(0, T_h)$ are given by the initial conditions u_{0h}^+ and u_{0h}^-, whereas for $t > T_h$ they are expressed as functions of $u_h^+(t - T_h)$, $u_h^-(t - T_h)$ and the delayed terminal voltages through the relations (3.109) and (3.110). Equations (3.113) and (3.114) express $u_h^+(t)$ and $u_h^-(t)$ as functions of the delayed terminal voltages and currents.

Remark: Matched Ideal Multiconductor Line

From Eq. (3.113) it follows that $u_h^+(t)$ would be equal to zero for $t > T_h$ if the line were connected at the left end to a linear resistive multiport with resistance matrix R_c — *matched ideal multiconductor line at the left end*. The same consideration would hold for the backward waves if the line were matched at the right end. ◇

Equations (3.105) and (3.106), joined to the equations describing the state dynamics, give in implicit form the relations between the voltages $\mathbf{v}_1$, $\mathbf{v}_2$ and the currents $\mathbf{i}_1, \mathbf{i}_2$.

It is clearly evident that Eqs. (3.109) and (3.110) are completely equivalent to Eqs. (2.67) and (2.68) whereas Eqs. (3.113) and (3.114) are completely equivalent to Eqs. (2.71) and (2.72).

The set of equations (3.105), (3.106), (3.109), and (3.110) describe the terminal properties of the line, as well as the internal state. We call them either *internal* or *input-state-output description* of the ideal multiconductor line. Equations (3.109) and (3.110) govern the dynamics of the state, whereas Eqs. (3.105) and (3.106) describe the terminal properties.

A description in which only the terminal variables are involved is possible as in the two-conductor case (see §2.5). In fact, we may eliminate the state variables by substituting the expression of u_h^+ given by Eq. (3.113) into Eq. (3.106), and the expression of u_h^- given by Eq. (3.114) into Eq. (3.105). We postpone the study of this description and the determination of all possible input-output representations to Chapter 6, where we shall deal with the more general case of lossy multiconductor transmission lines with frequency-dependent parameters. Hereafter in this chapter we shall only deal with the input-state-output description. However, as in the two-conductor case, the input-state-output description is more suitable for time domain transient analysis, whereas to deal with sinusoidal and periodic steady-state operating conditions in the frequency domain the input-output description is preferable.

3.5. THE INPUT-STATE-OUTPUT DESCRIPTION AND THE EQUIVALENT CIRCUITS OF THÉVENIN AND NORTON TYPE

The matrix T_v is invertible, consequently, the two variables $\mathbf{w}_1$ and $\mathbf{w}_2$, defined as

$$\mathbf{w}_1(t) = 2T_v\mathbf{u}^-(t) = 2\sum_{h=1}^{n} \mathbf{e}_h \mathbf{u}_h^-(t), \quad \mathbf{w}_2(t) = 2T_v\mathbf{u}^+(t) = 2\sum_{h=1}^{n} \mathbf{e}_h u_h^+(t) \tag{3.115}$$

completely specify the state of the line too. By using the property (3.24) it is immediate that for any h

$$u_h^-(t) = \tfrac{1}{2}\langle \mathbf{e}_h, \mathbf{w}_1(t)\rangle_L \quad \text{and} \quad u_h^+(t) = \tfrac{1}{2}\langle \mathbf{e}_h, \mathbf{w}_2(t)\rangle_L. \tag{3.116}$$

Because in Eqs. (3.105) and (3.106) the state of the line appears through $\mathbf{w}_1$ and $\mathbf{w}_2$, we rewrite these equations as

$$\mathbf{v}_1(t) - \mathrm{R}_c\mathbf{i}_1(t) = \mathbf{w}_1(t), \tag{3.117}$$

$$\mathbf{v}_2(t) - \mathrm{R}_c\mathbf{i}_2(t) = \mathbf{w}_2(t), \tag{3.118}$$

and we shall consider $\mathbf{w}_1$ and $\mathbf{w}_2$ as *state variables*.

The equations for $\mathbf{w}_1$ and $\mathbf{w}_2$ are obtained by substituting the expressions of $u_h^-(t)$ given by Eq. (3.114) and the expressions of $u_h^+(t)$ given by Eq. (3.113) into the right-hand side of Eq. (3.115). In this manner we obtain

$$\mathbf{w}_1(t) = \sum_{h=1}^{n} \mathbf{e}_h\langle\mathbf{e}_h, [-\mathbf{w}_2(t - T_h) + 2\mathbf{v}_2(t - T_h)]\rangle_{\mathrm{L}} u(t - T_h) + \mathbf{w}_1^{(0)}(t) \tag{3.119}$$

$$\mathbf{w}_2(t) = \sum_{h=1}^{n} \mathbf{e}_h\langle\mathbf{e}_h, [-\mathbf{w}_1(t - T_h) + 2\mathbf{v}_1(t - T_h)]\rangle_{\mathrm{L}} u(t - T_h) + \mathbf{w}_2^{(0)}(t), \tag{3.120}$$

where

$$\mathbf{w}_1^{(0)}(t) = 2\mathrm{T}_v\mathbf{u}_0^-(t) \quad \text{and} \quad \mathbf{w}_2^{(0)}(t) = 2\mathrm{T}_v\mathbf{u}_0^+(t) \tag{3.121}$$

take into account the effects of the initial conditions, $\mathbf{u}_0^-(t)$ and $\mathbf{u}_0^+(t)$ are given by Eqs. (3.96) and (3.97), respectively, and $u(t)$ is the Heaviside unit step function. Note that the hth components of $\mathbf{w}_1^{(0)}(t)$ and $\mathbf{w}_2^{(0)}(t)$ are equal to zero for $t > T_h$. Equations (3.119) and (3.120) are a set of recurrence equations with multiple delays $T_1, T_2, \ldots, T_n$.

Equations (3.117) and (3.118) suggest that the terminal behavior of ideal multiconductor lines can also be described through an equivalent circuit of Thévenin type similar to that we have introduced in the two-conductor case (see §2.6). The equivalent circuit consists of two linear resistive multiports and controlled voltage sources (Marx, 1973) (Fig. 3.4a).

Let us indicate with $w_{1h}(t)$ and $w_{2h}(t)$ the hth component of the vectors $\mathbf{w}_1$ and $\mathbf{w}_2$, respectively. The resistance matrix of each multiport is equal to R_c. A controlled voltage source of value $w_{1h}(w_{2h})$ is connected in series at the h-terminal of the resistive multiport on the left (right). Note that the amplitudes of the controlled sources at the generic time $t > T_1$ are dependent only on the history of the voltages at the ends of the line in the interval $(0, t - T_1)$. This fact allows the controlled sources representing the line to be treated as independent sources if the problem is resolved by means of iterative procedures, as in the case of the two-conductor line.

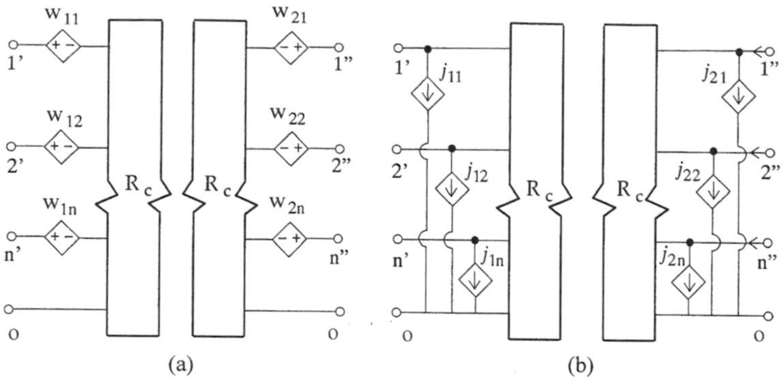

Figure 3.4. Time domain equivalent circuit of Thévenin (a) or Norton (b) type of an ideal multiconductor transmission line in terms of time-delayed controlled sources as implemented in SPICE.

Figure 3.4b illustrates an equivalent circuit of Norton type. A controlled current source is connected in parallel at each port,

$$\mathbf{j}_1(t) = -\mathrm{R}_c^{-1}\mathbf{w}_1(t), \quad \mathbf{j}_2(t) = -\mathrm{R}_c^{-1}\mathbf{w}_2(t); \tag{3.122}$$

j_{1h} and j_{2h} are, respectively, the hth components of the vectors $\mathbf{j}_1$ and $\mathbf{j}_2$.

Remark

The state equations (3.119) and (3.120) may be rewritten in a more compact form. Let us introduce the diagonal matrix

$$\Delta(t) = \mathrm{diag}(\delta(t - T_1), \delta(t - T_2), \ldots, \delta(t - T_n)). \tag{3.123}$$

By using the property (3.24), Eqs. (3.109) may be rewritten as ($\mathrm{T_v}$ does not depend on time)

$$\mathbf{u}^-(t) = \Delta(t) * [-\mathbf{u}^+(t) + \mathrm{T_v}^{-1}\mathbf{v}_2(t)] = [\Delta(t)\mathrm{T_v}^{-1}] * [-\mathrm{T_v}\mathbf{u}^+(t) + \mathbf{v}_2(t)]. \tag{3.124}$$

Here, the symbol $\mathrm{A}(t) * \mathrm{B}(t)$ indicates the usual matrix product, but the products between the entries are convolution products. From this equation we immediately obtain

$$2\mathrm{T_v}\mathbf{u}^-(t) = p(t) * [-2\mathrm{T_v}\mathbf{u}^+(t) + 2\mathbf{v}_2(t)], \tag{3.125}$$

where

$$p(t) = \mathrm{T_v}\Delta(t)\mathrm{T_v^{-1}} = \sum_{h=1}^{n} \mathbf{e}_h \mathbf{s}_h^{\mathrm{T}} \delta(t - T_h) = \sum_{h=1}^{n} \mathbf{e}_h (\mathrm{L}^{-1}\mathbf{e}_h)^{\mathrm{T}} \delta(t - T_h). \tag{3.126}$$

The same result holds for Eqs. (3.109). Therefore, the state equations may be rewritten as

$$\mathbf{w}_1(t) = p(t) * [2\mathbf{v}_2(t) - \mathbf{w}_2(t)] + \mathbf{w}_1^{(0)}(t), \tag{3.127}$$

$$\mathbf{w}_2(t) = p(t) * [2\mathbf{v}_1(t) - \mathbf{w}_1(t)] + \mathbf{w}_2^{(0)}(t), \tag{3.128}$$

The importance of this result is noteworthy. As we shall show in this book, whatever the type of the transmission line is (two-conductor or multiconductor, with losses or without, with parameters independent of frequency or depending on the frequency), the equations governing the state of the line may always be expressed in the form of Eqs. (3.127) and (3.128), provided that the line is uniform. The nature of the line determines only the actual expression of the matrix $p(t)$. In Chapter 7 we shall examine the case for nonuniform lines. ◇

3.6. MULTICONDUCTOR LINE WITH HOMOGENEOUS DIELECTRIC

For multiconductor transmission lines immersed in a homogeneous medium, the per-unit-length parameter matrices are related by Eq. (1.38). Consequently, all the natural modes have the same propagation velocity,

$$c_h = c_0 = \frac{1}{\sqrt{\varepsilon\mu}} \quad \text{for } h = 1, 2, \ldots, n, \tag{3.129}$$

and the transformation matrix $\mathrm{T_v}$ is any invertible matrix. Therefore, the general solution of the line equations in the d'Alembert form may be expressed as

$$\mathbf{v}(x,t) = \mathbf{v}^+(t - x/c_0 + T_0) + \mathbf{v}^-(t + x/c_0), \tag{3.130}$$

$$\mathbf{i}(x,t) = \mathrm{R_c^{-1}}[\mathbf{v}^+(t - x/c_0 + T_0) - \mathbf{v}^-(t + x/c_0)], \tag{3.131}$$

where

$$T_0 = \frac{d}{c_0}, \tag{3.132}$$

and, $\mathbf{v}^+$, $\mathbf{v}^-$ are arbitrary functions dependent on the initial and the boundary conditions. It is to be noted that all the components of the vector function $\mathbf{v}^+$ depend on the same argument $\xi = t - x/c_0 + T_0$ and all the components of the vector function $\mathbf{v}^-$ depend on the same argument $\zeta = t + x/c_0$. Both $\mathbf{v}^+$ and $\mathbf{v}^-$ are defined in the time interval $(0, \infty)$. Note that, in this case, the matrix operator $p(t)$ defined in Eq. (3.126) reduces to $\mathrm{I}\delta(t - T_0)$.

For these lines the characteristic resistance matrix is given by

$$\mathrm{R}_c = \mathrm{C}^{-1}/c_0 = c_0\mathrm{L}, \tag{3.133}$$

and the characteristic conductance matrix is given by

$$\mathrm{G}_c = c_0\mathrm{C} = \mathrm{L}^{-1}/c_0. \tag{3.134}$$

Let us now see how the characterization of this line as a multiport becomes simpler, as the problem reduces to a set of linear algebraic-difference equations with one delay T_0 as in the two-conductor line.

3.6.1. Characterization of the Transmission Line Behavior at the Ends

Imposing the initial conditions (3.94) and (3.95), one has

$$\mathbf{v}^+(t) = \mathbf{v}_0^+(t) \quad \text{for } 0 \leqslant t \leqslant T_0, \tag{3.135}$$

$$\mathbf{v}^-(t) = \mathbf{v}_0^-(t) \quad \text{for } 0 \leqslant t \leqslant T_0, \tag{3.136}$$

where

$$\mathbf{v}_0^+(t) = \tfrac{1}{2}\{\mathbf{v}_0[c_0(T_0 - t)] + \mathrm{R}_c\mathbf{i}_0[c_0(T_0 - t)]\} \quad \text{for } 0 \leqslant t \leqslant T_0, \tag{3.137}$$

$$\mathbf{v}_0^-(t) = \tfrac{1}{2}[\mathbf{v}_0(c_0 t) - \mathrm{R}_c\mathbf{i}_0(c_0 t)] \quad \text{for } 0 \leqslant t \leqslant T_0. \tag{3.138}$$

At the end $x = 0$, the voltage $\mathbf{v}_1$ and the current $\mathbf{i}_1$ at any time are given by

$$\mathbf{v}_1(t) = \mathbf{v}^+(t + T_0) + \mathbf{v}^-(t), \tag{3.139}$$

$$\mathrm{R}_c\mathbf{i}_1(t) = \mathbf{v}^+(t + T_0) - \mathbf{v}^-(t), \tag{3.140}$$

and at the end $x = d$, the voltage $\mathbf{v}_2$ and the current $\mathbf{i}_2$ are given by

$$\mathbf{v}_2(t) = \mathbf{v}^+(t) + \mathbf{v}^-(t + T_0), \tag{3.141}$$

$$-\mathrm{R}_c\mathbf{i}_2(t) = \mathbf{v}^+(t) - \mathbf{v}^-(t + T_0). \tag{3.142}$$

As in the case of the two-conductor line, it is possible to eliminate the forward wave $\mathbf{v}^+(t)$ from Eqs. (3.139) and (3.140), and the backward wave $\mathbf{v}^{-}(t)$ from Eqs. (3.141) and (3.142) to obtain

$$\mathbf{v}_1(t) - \mathrm{R}_c\mathbf{i}_1(t) = \mathbf{w}_1(t), \tag{3.143}$$

$$\mathbf{v}_2(t) - \mathrm{R}_c\mathbf{i}_2(t) = \mathbf{w}_2(t). \tag{3.144}$$

The equations governing the state dynamics are

$$\mathbf{w}_1(t) = -\mathbf{w}_2(t - T_0) + 2\mathbf{v}_2(t - T_0) \quad \text{for } t > T_0, \tag{3.145}$$

$$\mathbf{w}_2(t) = -\mathbf{w}_1(t - T_0) + 2\mathbf{v}_1(t - T_0) \quad \text{for } t > T_0. \tag{3.146}$$

This is a set of recurrence equations with one delay T_0 that recalls the state equations obtained for the ideal two-conductor line (see §2.6).

3.7. MULTICONDUCTOR TRANSMISSION LINE CONNECTED TO LINEAR RESISTIVE MULTIPORTS

Let us now apply the results described in the foregoing to the study of a simple time-invariant network. Let us consider an ideal multiconductor transmission line connected to lumped circuits at each end (Fig. 3.5). Only for the sake of simplicity shall we assume that the line is initially at rest. Each termination consists of independent sources and linear resistive elements.

By using the Thévenin equivalent representation, the behavior of C_1 and C_2 can be represented by the characteristic relations

$$\mathbf{v}_1 = -\mathrm{R}_1\mathbf{i}_1 + \mathbf{e}_1, \tag{3.147}$$

$$\mathbf{v}_2 = -\mathrm{R}_2\mathbf{i}_2 + \mathbf{e}_2. \tag{3.148}$$

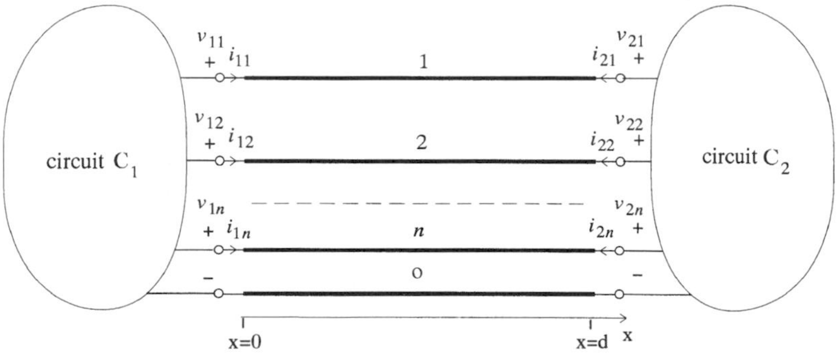

Figure 3.5. Multiconductor line connecting lumped circuits.

The set of equations that describe the behavior of this network, besides Eqs. (3.147) and (3.148), consists of Eqs. (3.117) and (3.118), which describe the behavior of the line at the ends, and of Eqs. (3.127) and (3.128) governing the dynamics of the state variables $\mathbf{w}_1$ and $\mathbf{w}_2$.

From Eqs. (3.117) and (3.147), for the voltages $\mathbf{v}_1$, we obtain

$$\mathbf{v}_1(t) = \mathrm{R}_1(\mathrm{R}_c + \mathrm{R}_1)^{-1}\mathbf{w}_1(t) + \mathrm{R}_c(\mathrm{R}_c + \mathrm{R}_1)^{-1}\mathbf{e}_1(t). \tag{3.149}$$

Similarly, from Eqs. (3.118) and (3.148), for $\mathbf{v}_2$, we obtain

$$\mathbf{v}_2(t) = \mathrm{R}_2(\mathrm{R}_c + \mathrm{R}_2)^{-1}\mathbf{w}_2(t) + \mathrm{R}_c(\mathrm{R}_c + \mathrm{R}_2)^{-1}\mathbf{e}_2(t). \tag{3.150}$$

Substituting these two expressions into Eqs. (3.127) and (3.128), respectively, we obtain the *state equations in normal form* of the circuit of Fig. 3.5,

$$\mathbf{w}_1(t) = p(t) * [\Gamma_2\mathbf{w}_2(t) + \mathbf{b}_1(t)], \tag{3.151}$$

$$\mathbf{w}_2(t) = p(t) * [\Gamma_1\mathbf{w}_1(t) + \mathbf{b}_2(t)]. \tag{3.152}$$

The matrices Γ_1 and Γ_2 are given by

$$\Gamma_1 = (\mathrm{R}_1 - \mathrm{R}_c)(\mathrm{R}_c + \mathrm{R}_1)^{-1}, \tag{3.153}$$

$$\Gamma_2 = (\mathrm{R}_2 - \mathrm{R}_c)(\mathrm{R}_c + \mathrm{R}_2)^{-1}, \tag{3.154}$$

and the known terms $\mathbf{b}_1$ and $\mathbf{b}_2$ have the following expressions:

$$\mathbf{b}_1(t) = 2\mathrm{R}_c(\mathrm{R}_c + \mathrm{R}_2)^{-1}\mathbf{e}_2(t), \tag{3.155}$$

$$\mathbf{b}_2(t) = 2\mathrm{R}_c(\mathrm{R}_c + \mathrm{R}_1)^{-1}\mathbf{e}_1(t). \tag{3.156}$$

The system of Eqs. (3.151) and (3.152) is a set of linear difference equations with more than one delay. Equations (3.153) and (3.154) are valid even when the terminal circuits are time varying. In this case, the matrices Γ_1 and Γ_2 are time dependent.

Remark

In §3.4.2 we have shown that an ideal multiconductor line is matched to an end when a linear resistive multiport with resistance matrix equal to the characteristic resistance matrix of the line R_c is connected at that end. The matrices Γ_1 and Γ_2 describe what happens at the ends of the line when it is not matched. These matrices are the generalization of the reflection coefficients that we have introduced for the two-conductor line in §2.2.3.

For $\mathrm{R}_1 \neq \mathrm{R}_c$ the termination at $x = 0$ introduces reflections and the matrix Γ_1 describes the amplitudes of the reflected forward waves for each mode. For $\mathrm{R}_2 \neq \mathrm{R}_c$ the termination at $x = d$ introduces reflections and the matrix Γ_2 describes the intensity of the reflected backward waves for each mode. For $R_1 = R_c$ and $R_2 = R_c$ we have, respectively, $\Gamma_1 = 0$ and $\Gamma_2 = 0$. For example, when $R_2 = R_c$, the backward waves are not excited at end $x = d$, that is, the terminal circuit does not introduce reflections and the terminal behavior of the transmission line is similar to that of a semi-infinite line. This is the natural extension of the matched line concept to multiconductor lines.

As the matrix R_c is symmetric (see §3.9), the eigenvalues of Γ_1 and Γ_2 are real if R_1 and R_2 are symmetric (the resistive elements of the lumped circuits are reciprocal). Furthermore, their absolute values are either less than unity or equal to unity if the lumped circuits are composed of passive elements, because the matrix R_c is positive definite (see §3.9). The demonstration of these properties is very simple. ◇

When the network is time invariant, to study the solutions of the state equations (3.151) and (3.152), it is useful to transform them into the Laplace domain,

$$\mathbf{W}_1(s) - P(s)\Gamma_2\mathbf{W}_2(s) = \mathbf{B}_1(s), \tag{3.157}$$

$$-P(s)\Gamma_1\mathbf{W}_1(s) + \mathbf{W}_2(s) = \mathbf{B}_2(s), \tag{3.158}$$

where

$$P(s) = T_v \operatorname{diag}(e^{-sT_1}, e^{-sT_2}, \ldots, e^{-sT_n})T_v^{-1} = \sum_{h=1}^{n} \mathbf{e}_h\mathbf{s}_h^T e^{-sT_h}, \tag{3.159}$$

and

$$\mathbf{B}_1(s) = 2R_c(R_c + R_2)^{-1}\mathbf{E}_2(s), \tag{3.160}$$

$$\mathbf{B}_2(s) = 2R_c(R_c + R_1)^{-1}\mathbf{E}_1(s). \tag{3.161}$$

As usual, we indicate the Laplace transform of $\mathbf{a}(t)$ with the corresponding capital letter $\mathbf{A}(s)$.

Combining Eqs. (3.157) and (3.158) we obtain

$$\mathbf{W}_1(s) = [I - P(s)\Gamma_2P(s)\Gamma_1]^{-1}[\mathbf{B}_1(s) + P(s)\Gamma_2\mathbf{B}_2(s)], \tag{3.162}$$

$$\mathbf{W}_2(s) = [I - P(s)\Gamma_1P(s)\Gamma_2]^{-1}[P(s)\Gamma_1\mathbf{B}_1(s) + \mathbf{B}_2(s)]. \tag{3.163}$$

Let A be a square matrix. The notable identity holds

$$(I - A)^{-1} = (I - A^n)^{-1}\sum_{i=0}^{n-1} A^i. \tag{3.164}$$

If the matrix A is diagonalizable through a similarity transformation and the absolute values of its eigenvalues are less than unity, then we obtain the geometric series

$$(\mathrm{I} - \mathrm{A})^{-1} = \sum_{i=0}^{\infty} \mathrm{A}^{i}. \tag{3.165}$$

Therefore, in the region of the complex plane where the absolute values of the eigenvalues of $\mathrm{P}\Gamma_2\mathrm{P}\Gamma_1$ (which coincide with the eigenvalues of $\mathrm{P}\Gamma_1\mathrm{P}\Gamma_2$) are less than unity, both the matrices $(\mathrm{I} - \mathrm{P}\Gamma_2\mathrm{P}\Gamma_1)^{-1}$ and $(\mathrm{I} - \mathrm{P}\Gamma_1\mathrm{P}\Gamma_2)^{-1}$ can be represented through the geometric series (3.165). Hence, we obtain expressions for $\mathbf{W}_1$ and $\mathbf{W}_2$ similar to those given by Eqs. (2.117) and (2.118) for the two-conductor case. Any term of each sum will be a linear combination of functions of the type $e^{-2msT_1}, e^{-2msT_2}, \ldots, e^{-2msT_n}$. For transmission lines embedded in a homogeneous dielectric we have

$$(\mathrm{I} - \mathrm{P}\Gamma_1\mathrm{P}\Gamma_2)^{-1} = (\mathrm{I} - \mathrm{P}\Gamma_2\mathrm{P}\Gamma_1)^{-1} = (\mathrm{I} - \Gamma_2\Gamma_1 e^{-2sT_0})^{-1}$$
$$= \sum_{m=0}^{\infty} (\Gamma_2\Gamma_1)^m e^{-2msT_0}. \tag{3.166}$$

Certainly there is a region of the complex plane where this series converges.

3.7.1. Natural Frequencies of the Network

The frequencies of the natural modes of the network are the roots of the equation

$$\det[\mathrm{I} - \mathrm{P}(s)\Gamma_1\mathrm{P}(s)\Gamma_2] = 0. \tag{3.167}$$

In general, it is not possible to determine the natural frequencies analytically given the complexity of Eq. (3.167).

This is possible, however, where a line is embedded in a homogeneous dielectric. In this case we have (see §3.6)

$$\mathrm{P}(s) = e^{-sT_0}\mathrm{I}, \tag{3.168}$$

and hence the natural frequencies of the circuit are the roots of the equation

$$\det(\mathrm{I} - e^{-2sT_0}\Gamma_1\Gamma_2) = 0. \tag{3.169}$$

Let us indicate with $\lambda_1, \lambda_2, \ldots, \lambda_n$ the eigenvalues of the matrix $\Gamma_1\Gamma_2$. They are all real if the resistive elements of the lumped parameter circuits are reciprocal. The natural frequencies are the solutions of the equations

$$e^{2s_{hk}T_0} = \lambda_h e^{2k\pi i} \quad \text{for } h = 1, 2, \ldots, n, \quad k = 0, \pm 1, \pm 2, \ldots, \tag{3.170}$$

and hence

$$s_{hk} = \frac{1}{T_0} ln(\sqrt{|\lambda_h|}) + i\frac{\pi}{T_0}\begin{cases} k & \text{for } \lambda_h > 0, \\ k + \frac{1}{2} & \text{for } \lambda_h < 0. \end{cases} \tag{3.171}$$

The absolute values of the eigenvalues λ_h are either less than or equal to unity if the lumped circuits are composed of passive elements.

Remark

The equations describing a network composed of an ideal multiconductor transmission line connecting two nonlinear resistive elements are a system of nonlinear algebraic difference equations with multiple delays. For multiconductor lines with homogeneous dielectric, they reduce to a system of nonlinear algebraic difference equations with one delay. The network equations become more complicated when the lumped part of the network also contains dynamics elements (e.g., capacitors, inductors, and transformers). Then, it is clear how, as the complexity of the transmission line model and of the lumped part of the network gradually increases, the difficulty in resolving the problem increases. Thus it is obvious that we shall be forced to turn to solution methods of numerical type. We shall deal with this in Chapter 9. ◇

3.8. A PARTICULAR SOLUTION OF THE IDEAL MULTICONDUCTOR TRANSMISSION LINE EQUATIONS WITH DISTRIBUTED SOURCES

Until now, we have referred to multiconductor lines without distributed sources. Now we shall see how to determine the general solution of the multiconductor line equations with distributed sources. We shall use the same method as that applied to the two-conductor case (see §2.9).

The equations for an ideal multiconductor transmission line with distributed sources are (see chapter 1)

$$\begin{aligned} \frac{\partial \mathbf{v}}{\partial x} + \mathrm{L}\frac{\partial \mathbf{i}}{\partial t} &= \mathbf{e}_s, \\ \frac{\partial \mathbf{i}}{\partial x} + \mathrm{C}\frac{\partial \mathbf{v}}{\partial t} &= \mathbf{j}_s, \end{aligned} \tag{3.172}$$

where $\mathbf{e}_s(x, t)$ and $\mathbf{j}_s(x, t)$ are assigned vector functions, defined for $0 \leqslant x \leqslant d$ and $t \geqslant 0$.

If $\mathbf{v}_p(x,t)$ and $\mathbf{i}_p(x,t)$ indicate any solution of the set (3.172), then the general solution of Eqs. (3.172) may be expressed as

$$\mathbf{v}(x,t) = \sum_{h=1}^{n} \mathbf{e}_h[u_h^+(t - x/c_h + T_h) + u_h^-(t + x/c_h)] + \mathbf{v}_p(x,t), \quad (3.173)$$

$$\mathbf{i}(x,t) = \mathrm{R}_c^{-1} \sum_{h=1}^{n} \mathbf{e}_h[u_h^+(t - x/c_h + T_h) - u_h^-(t + x/c_h)] + \mathbf{i}_p(x,t). \quad (3.174)$$

The multiport model obtained in §3.5 can easily be extended to multiconductor lines with distributed sources as for two-conductor lines (see §2.9.2) once a particular solution of Eqs. (3.172) has been determined.

As in the case of the two-conductor line, we shall choose as particular solution that which is zero at the instant, $t = 0$, and satisfies the boundary conditions (matched line):

$$\begin{aligned} \mathbf{v}_p(x=0,t) &= -\mathrm{R}_c\mathbf{i}_p(x=0,t), \\ \mathbf{v}_p(x=d,t) &= \mathrm{R}_c\mathbf{i}_p(x=d,t), \end{aligned} \quad (3.175)$$

for $t > 0$. To determine this solution we shall use the convolution method exactly as we have done for the two-conductor line in §2.9. Equations (3.172) can then be rewritten as

$$\frac{\partial \mathbf{v}}{\partial x} + \mathrm{L}\frac{\partial \mathbf{i}}{\partial t} = \int_{0^-}^{t^+} \delta(t-\tau)\mathbf{e}_s(x,\tau)d\tau, \quad (3.176)$$

$$\frac{\partial \mathbf{i}}{\partial x} + \mathrm{C}\frac{\partial \mathbf{v}}{\partial t} = \int_{0^-}^{t^+} \delta(t-\tau)\mathbf{j}_s(x,\tau)d\tau. \quad (3.177)$$

Let $\mathbf{h}_v^{(e)}(x,t;\tau)$ and $\mathbf{h}_i^{(e)}(x,t;\tau)$ indicate the solutions of the "auxiliary problem" defined by the equations

$$\frac{\partial \mathbf{h}_v^{(e)}}{\partial x} + \mathrm{L}\frac{\partial \mathbf{h}_i^{(e)}}{\partial t} = \delta(t-\tau)\mathbf{e}_s(x,\tau), \quad (3.178)$$

$$\frac{\partial \mathbf{h}_i^{(e)}}{\partial x} + \mathrm{C}\frac{\partial \mathbf{h}_v^{(e)}}{\partial t} = \mathbf{0}, \quad (3.179)$$

with the initial conditions

$$\mathbf{h}_v^{(e)}(x,t=0;\tau) = \mathbf{0}, \quad (3.180)$$

$$\mathbf{h}_i^{(e)}(x,t=0;\tau) = \mathbf{0}, \quad (3.181)$$

for $0 \leqslant x \leqslant d$, and the boundary conditions (3.175). Moreover, let $\mathbf{h}_v^{(j)}(x, t; \tau)$ and $\mathbf{h}_i^{(j)}(x, t; \tau)$ be the solution of the equations

$$\frac{\partial \mathbf{h}_v^{(j)}}{\partial x} + \mathrm{L}\frac{\partial \mathbf{h}_i^{(j)}}{\partial t} = \mathbf{0}, \tag{3.182}$$

$$\frac{\partial \mathbf{h}_i^{(j)}}{\partial x} + \mathrm{C}\frac{\partial \mathbf{h}_v^{(j)}}{\partial t} = \delta(t - \tau)\mathbf{j}_s(x, \tau), \tag{3.183}$$

with the initial conditions

$$\mathbf{h}_v^{(j)}(x, t = 0; \tau) = \mathbf{0}, \tag{3.184}$$

$$\mathbf{h}_i^{(j)}(x, t = 0; \tau) = \mathbf{0}, \tag{3.185}$$

for $0 \leqslant x \leqslant d$ and the boundary conditions (3.175).

Using the superposition property, a particular solution of Eqs. (3.172) can be expressed in the form

$$\mathbf{v}_p(x, t) = \int_0^t [\mathbf{h}_v^{(e)}(x, t; \tau) + \mathbf{h}_v^{(j)}(x, t; \tau)]d\tau, \tag{3.186}$$

$$\mathbf{i}_p(x, t) = \int_0^t [\mathbf{h}_i^{(e)}(x, t; \tau) + \mathbf{h}_i^{(j)}(x, t; \tau)]d\tau. \tag{3.187}$$

This particular solution is by construction equal to zero at the initial time and satisfies the perfect matching conditions at the line ends.

Let us assume that $\mathbf{e}_s(x, t)$ and $\mathbf{j}_s(x, t)$ are continuous and derivable with respect to x. Now let us proceed in the same way as for two-conductor lines (see §2.9). First, we integrate Eqs. (3.178) and (3.179) in the time interval $(\tau - \varepsilon, \tau + \varepsilon)$ where ε is an arbitrarily small positive number. At this point, using the initial conditions, and considering the limit $\varepsilon \to 0^+$, we obtain

$$\mathbf{h}_i^{(e)}(x, t = \tau^+; \tau) = \mathrm{L}^{-1}\mathbf{e}_s(x, \tau), \tag{3.188}$$

$$\mathbf{h}_v^{(e)}(x, t = \tau^+; \tau) = \mathbf{0}. \tag{3.189}$$

Then, using Eqs. (3.182) and (3.183), and considering the limit $\varepsilon \to 0^+$, we obtain

$$\mathbf{h}_i^{(j)}(x, t = \tau^+; \tau) = \mathbf{0}, \tag{3.190}$$

$$\mathbf{h}_v^{(j)}(x, t = \tau^+; \tau) = \mathrm{C}^{-1}\mathbf{j}_s(x, t). \tag{3.191}$$

The general solution of the equation systems (3.178) and (3.179) and of the equation systems (3.182) and (3.183) can be expressed through the d'Alembert solution for $t \geqslant \tau^+$. Imposing the initial conditions (3.188) and (3.189) and the boundary conditions (3.175), we obtain the solution of the first auxiliary problem

$$\mathbf{h}_v^{(e)}(x,t;\tau) = u(t-\tau)\sum_{h=1}^{n}\mathbf{e}_h\{u_h^{(e)}[x-c_h(t-\tau);\tau] - u_h^{(e)}[x+c_h(t-\tau);\tau]\}, \tag{3.192}$$

$$\mathbf{h}_i^{(e)}(x,t;\tau) = u(t-\tau)\mathrm{R}_c^{-1}\sum_{h=1}^{n}\mathbf{e}_h\{u_h^{(e)}[x-c_h(t-\tau);\tau] + u_h^{(e)}[x+c_h(t-\tau);\tau]\}, \tag{3.193}$$

where

$$u_h^{(e)}(\xi;\tau) = \begin{cases} \frac{1}{2}\langle\mathbf{e}_h, \mathrm{R}_c\mathrm{L}^{-1}\mathbf{e}_s(\xi,\tau)\rangle_{\mathrm{L}} & \text{for } 0 \leqslant \xi \leqslant d, \\ 0 & \text{otherwise.}\end{cases} \tag{3.194}$$

Proceeding in the same way, we determine the solution of the other auxiliary problem,

$$\mathbf{h}_v^{(j)}(x,t;\tau) = u(t-\tau)\sum_{h=1}^{n}\mathbf{e}_h\{u_h^{(j)}[x-c_h(t-\tau);\tau] + u_h^{(j)}[x+c_h(t-\tau);\tau]\}, \tag{3.195}$$

$$\mathbf{h}_i^{(j)}(x,t;\tau) = u(t-\tau)\mathrm{R}_c^{-1}\sum_{h=1}^{n}\mathbf{e}_h\{u_h^{(j)}[x-c_h(t-\tau);\tau] - u_h^{(j)}[x+c_h(t-\tau);\tau]\}, \tag{3.196}$$

where

$$u_h^{(j)}(\xi;\tau) = \begin{cases} \frac{1}{2}\langle\mathbf{e}_h, \mathrm{C}^{-1}\mathbf{j}_s(\xi,\tau)\rangle_{\mathrm{C}} & \text{for } 0 \leqslant \xi \leqslant d, \\ 0 & \text{otherwise.}\end{cases} \tag{3.197}$$

To characterize the line as a multiport, one proceeds as for the line without distributed sources. The multiport characterization obtained in §3.5 and §3.6 can be extended to lines with distributed sources proceeding in the same way as for the two-conductor case (see §2.9.2).We leave this task to the reader.

3.8.1. Transversally Homogeneous Lines

It is appropriate to deal with the case of the transversely homogeneous line separately. In this case the general solution of Eq. (3.172) is

$$\mathbf{v}(x,t) = \mathbf{v}^+(t - x/c_0 + T_0) + \mathbf{v}^-(t + x/c_0) + \mathbf{v}_p(x,t), \tag{3.198}$$

$$\mathbf{i}(x,t) = \mathrm{R}_c^{-1}[\mathbf{v}^+(t - x/c_0 + T_0) - \mathbf{v}^-(t + x/c_0)] + \mathbf{i}_p(x,t), \quad (3.199)$$

The particular solutions that satisfy zero initial conditions and the boundary conditions (3.175) are still of the type (3.186) and (3.187). Now the functions $\mathbf{h}_v^{(e)}(x,t;\tau)$ and $\mathbf{h}_i^{(e)}(x,t;\tau)$ are given by

$$\mathbf{h}_v^{(e)}(x,t;\tau) = u(t-\tau)\{\mathbf{v}^{(e)}[x - c_0(t-\tau);\tau] - \mathbf{v}^{(e)}[x + c_0(t-\tau);\tau]\}, \quad (3.200)$$

$$\mathbf{h}_i^{(e)}(x,t;\tau) = u(t-\tau)\mathrm{R}_c^{-1}\{\mathbf{v}^{(e)}[x - c_0(t-\tau);\tau] + \mathbf{v}^{(e)}[x + c_0(t-\tau);\tau]\}, \quad (3.201)$$

where

$$\mathbf{v}^{(e)}(\xi;\tau) = \begin{cases} \frac{1}{2}\mathrm{R}_c\mathrm{L}^{-1}\mathbf{e}_s(\xi,\tau) & \text{for } 0 \leqslant \xi \leqslant d, \\ \mathbf{0} & \text{otherwise.} \end{cases} \quad (3.202)$$

The functions $\mathbf{h}_v^{(j)}(x,t;\tau)$ and $\mathbf{h}_i^{(j)}(x,t;\tau)$ are given by

$$\mathbf{h}_v^{(j)}(x,t;\tau) = u(t-\tau)\{\mathbf{v}^{(j)}[x - c_0(t-\tau);\tau] - \mathbf{v}^{(j)}[x + c_0(t-\tau);\tau]\}, \quad (3.203)$$

$$\mathbf{h}_i^{(j)}(x,t;\tau) = u(t-\tau)\mathrm{R}_c^{-1}\{\mathbf{v}^{(j)}[x - c_0(t-\tau);\tau] - \mathbf{v}^{(j)}[x + c_0(t-\tau);\tau]\}, \quad (3.204)$$

where

$$\mathbf{v}^{(e)}(\xi;\tau) = \begin{cases} \frac{1}{2}\mathrm{C}^{-1}\mathbf{j}_s(\xi,\tau) & \text{for } 0 \leqslant \xi \leqslant d, \\ \mathbf{0} & \text{otherwise.} \end{cases} \quad (3.205)$$

3.9. PROPERTIES OF THE CHARACTERISTIC CONDUCTANCE MATRIX G_c AND RESISTANCE MATRIX R_c

Matrices G_c and R_c are symmetric and positive definite. We shall now demonstrate these properties for the matrix G_c. As $R_c = G_c^{-1}$, they also hold for the matrix R_c. The symmetry is a direct consequence of the reciprocity property shown in §1.10.4, whereas the passivity of the line implies positiveness.

Consider two equal semi-infinite transmission lines $\mathscr{TL}'$ and $\mathscr{TL}''$ initially at rest so that only forward waves are present along the two lines. In the time domain the solution of the line $\mathscr{T}\mathscr{L}'$ is

$$\mathbf{v}'(x,t) = \sum_{h=1}^{n} \mathbf{e}_h u_h'^{+}(t - x/c_h), \quad \mathbf{i}'(x,t) = \mathrm{G}_c \mathbf{v}'(x,t), \tag{3.206}$$

and that of the line $\mathscr{T}\mathscr{L}''$ is

$$\mathbf{v}''(x,t) = \sum_{h=1}^{n} \mathbf{e}_h u_h''^{+}(t - x/c_h), \quad \mathbf{i}''(x,t) = \mathrm{G}_c \mathbf{v}''(x,t). \tag{3.207}$$

The functions $u_h'^{+}(t)$ and $u_h''^{+}(t)$, defined for $-\infty < t < +\infty$, are identically zero for $t < 0$, while those for $t \geqslant 0$ depend only on the boundary conditions at $x = 0$.

Let $\mathbf{V}'(x;s)$, $\mathbf{I}'(x;s)$ and $\mathbf{V}''(x;s)$, and $\mathbf{I}''(x;s)$, respectively, indicate the Laplace transforms of the voltage and the current distributions along the lines $\mathscr{T}\mathscr{L}'$ and $\mathscr{T}\mathscr{L}''$. The solutions (3.206) and (3.207) in the Laplace domain become

$$\mathbf{V}'(x;s) = \sum_{h=1}^{n} \mathbf{e}_h U_h'^{+}(s) e^{-sx/c_h}, \quad \mathbf{I}'(x;s) = \mathrm{G}_c \mathbf{V}'(x;s), \tag{3.208}$$

$$\mathbf{V}''(x;s) = \sum_{h=1}^{n} \mathbf{e}_h U_h''^{+}(s) e^{-sx/c_h}, \quad \mathbf{I}''(x;s) = \mathrm{G}_c \mathbf{V}''(x;s), \tag{3.209}$$

where $U_h'^{+}(s)$ and $U_h''^{+}(s)$ are the Laplace transforms of the functions $u_h'^{+}(t)$ and $u_h''^{+}(t)$, respectively. Therefore, to the right of the imaginary axis the functions $\mathbf{V}'(x;s)$, $\mathbf{V}''(x;s)$, $\mathbf{I}'(x;s)$, and $\mathbf{I}''(x;s)$ tend to zero exponentially for $x \to +\infty$, and hence we can apply the reciprocity property shown in §1.10.4,

$$\mathbf{I}'^{\mathrm{T}}(x;s)\mathbf{V}''(x;s) = \mathbf{I}''^{\mathrm{T}}(x;s)\mathbf{V}'(x;s) \quad \text{for } 0 \leqslant x < \infty. \tag{3.210}$$

From Eq. (3.210) we have

$$\mathbf{I}''^{\mathrm{T}}(0;s)\mathbf{V}'(0;s) - \mathbf{I}'^{\mathrm{T}}(0;s)\mathbf{V}''(0;s) = 0, \tag{3.211}$$

thus

$$\mathbf{V}_1'^{\mathrm{T}}(\mathrm{G}_c - \mathrm{G}_c^{\mathrm{T}})\mathbf{V}_1'' = 0 \text{ for } 0 \leqslant x < \infty, \tag{3.212}$$

where $\mathbf{V}_1' = \mathbf{V}'(x=0;s)$ and $\mathbf{V}_1'' = \mathbf{V}''(x=0;s)$. As Eq. (3.212) must be satisfied for every $\mathbf{V}_1'$ and $\mathbf{V}_1''$, we must necessarily have

$$\mathrm{G}_c = \mathrm{G}_c^{\mathrm{T}}. \tag{3.213}$$

Note that, though the matrix CL is not symmetric, the matrix G_c is symmetric.

The property that G_c is positive definite is a direct consequence of the passivity of the line we are considering. Let us again consider a semi-infinite multiconductor line initially at rest and fed at $x = 0$ with n ideal step voltage sources

$$\mathbf{v}_1(t) = \mathbf{E}u(t); \tag{3.214}$$

$\mathbf{E}$ is a constant voltage vector. It is then evident that only forward waves exist, hence

$$\mathbf{i}_1(t) = G_c\mathbf{E}u(t). \tag{3.215}$$

Applying the Poynting theorem to an ideal multiconductor line (see §1.4.2), for $t \geqslant 0$, we have

$$\mathbf{v}_1^T(t)\mathbf{i}_1(t) = \frac{d}{dt}\left\{\int_0^\infty \left(\frac{1}{2}\mathbf{i}^T L\mathbf{i} + \frac{1}{2}\mathbf{v}^T C\mathbf{v}\right)dx\right\}. \tag{3.216}$$

By using Eqs. (3.214) and (3.215), Eq. (3.216) gives

$$(\mathbf{E}^T G_c\mathbf{E}) = \frac{d}{dt}\left\{\int_0^\infty \left(\frac{1}{2}\mathbf{i}^T L\mathbf{i} + \frac{1}{2}\mathbf{v}^T C\mathbf{v}\right)dx\right\}. \tag{3.217}$$

As the line is initially at rest, Eq. (3.217) for $t > 0$ gives

$$(\mathbf{E}^T G_c\mathbf{E})t = \int_0^\infty \left(\frac{1}{2}\mathbf{i}^T L\mathbf{i} + \frac{1}{2}\mathbf{v}^T C\mathbf{v}\right)dx. \tag{3.218}$$

Due to the passivity of the line, the matrix L and the matrix C are positive definite. Therefore, the right-hand term in Eq. (3.218), which represents the electromagnetic energy associated to the line, is always positive for $\mathbf{v}(x, t) \neq \mathbf{0}$ and $\mathbf{i}(x, t) \neq \mathbf{0}$; it is equal to zero when, and only when, $\mathbf{v}(x, t) = \mathbf{0}$ and $\mathbf{i}(x, t) = \mathbf{0}$ all along the line, that is, when, and only when, $\mathbf{E} = 0$. Thus, for any $\mathbf{E} \neq \mathbf{0}$ there must be

$$\mathbf{E}^T G_c\mathbf{E} > 0, \tag{3.219}$$

that is, G_c is positive definite.

Remark

By combining Eqs. (3.45) and (3.49) we obtain

$$G_c = T_i\sqrt{B^{-1}}T_i^T. \tag{3.220}$$

Note that the properties of symmetry and positiveness of G_c may also be derived from expression (3.220) (Marx, 1973). ◇

CHAPTER 4

Lossy Two-Conductor Transmission Lines

This chapter deals with lossy two-conductor transmission lines whose parameters are uniform in space and independent of frequency. The general solution of the lossy line equations can not be expressed in the d'Alembert form because of the dispersion due to the losses. However, as we are dealing with linear and time-invariant transmission lines, it is always possible to use those techniques of analysis and representation appropriate to linear and time-invariant problems, that is, the Laplace transform and the convolution theorem. This is a key point in our approach.

In the Laplace domain it is easy to solve the equations of lossy transmission lines because they become ordinary differential equations. Once the line has been characterized as a two-port in the Laplace domain, the representation in the time domain is immediately obtained by applying the convolution theorem.

This approach is most effective because it also allows us to characterize two-conductor transmission lines with parameters dependent on the frequency (Chapter 5), multiconductor transmission lines with losses and parameters dependent on the frequency (Chapter 6), and nonuniform transmission lines (Chapter 7).

The input-output description and the input-state-output description are closely investigated both in the Laplace domain and time domain. Simple equivalent circuits of Thévenin and Norton type based on the input-state-output description that extend those describing ideal two-conductor lines (see §2.6) also are closely studied. Furthermore, all the possible input-output descriptions will be accurately examined.

4.1. LOSSY TRANSMISSION LINES ARE DISPERSIVE

In the time domain the equations for lossy transmission lines are (see §1.2.2),

$$-\frac{\partial v}{\partial x} = L\frac{\partial i}{\partial t} + Ri, \tag{4.1}$$

$$-\frac{\partial i}{\partial x} = C\frac{\partial v}{\partial t} + Gv, \tag{4.2}$$

where L, C, R, and G are positive constant parameters.

The system of equations (4.1) and (4.2) may be transformed into a system of two uncoupled second-order partial differential equations as we have done for ideal transmission line equations. Starting from Eqs. (4.1) and (4.2), by derivation and substitution we obtain the following equation for the voltage distribution:

$$\frac{\partial^2 v}{\partial t^2} - c^2\frac{\partial^2 v}{\partial x^2} + (\alpha + \beta)\frac{\partial v}{\partial t} + \alpha\beta v = 0, \tag{4.3}$$

where

$$c^2 = \frac{1}{LC}, \quad \alpha = \frac{R}{L}, \quad \beta = \frac{G}{C}; \tag{4.4}$$

c would be the propagation velocity of the quasi-TEM mode if the line were lossless, α the inductive and β the capacitive damping factors. The current distribution satisfies a similar equation. If we introduce a new unknown function $u(x, t)$ (e.g., Smirnov, 1964a) such that

$$v(x, t) = e^{-\mu t}u(x, t), \tag{4.5}$$

and

$$\mu = \frac{1}{2}(\alpha + \beta) = \frac{1}{2}\left(\frac{R}{L} + \frac{G}{C}\right), \tag{4.6}$$

we obtain the simpler equation

$$\frac{\partial^2 u}{\partial t^2} - c^2\frac{\partial^2 u}{\partial x^2} = \nu^2 u, \tag{4.7}$$

where

$$\nu = \frac{1}{2}(\alpha - \beta) = \frac{1}{2}\left(\frac{R}{L} - \frac{G}{C}\right). \tag{4.8}$$

Note that it is always $0 \leqslant |\nu| \leqslant \mu$.

Let us look for a harmonic wave solution of Eq. (4.7) of the type

$$u(x, t) = U \cos(\omega t - \gamma x). \tag{4.9}$$

The dispersion relation is given by

$$\omega^2 + \nu^2 = c^2\gamma^2, \tag{4.10}$$

and the phase velocity is

$$c_{ph} = \frac{\omega}{\gamma} = c\frac{\omega}{\sqrt{\omega^2 + \nu^2}}. \tag{4.11}$$

Because the phase velocity c_{ph} depends upon the frequency, any signal is propagated with distortion. Therefore, a lossy two-conductor transmission line is dispersive for $\nu \neq 0$.

4.1.1. The Heaviside Condition

Equation (4.7) reduces to the dispersionless wave equation (see Eq. (2.3)) iff

$$\nu = 0, \text{ that is, } \alpha = \beta. \tag{4.12}$$

This is the so-called *Heaviside condition* that Oliver Heaviside (1887) discovered while studying the possibility of distortionless transmission of telegraph signals along cables. In this case the solution of Eq. (4.3) is an "undistorted" traveling wave:

$$v^{\pm}(x, t) = e^{-\mu t}u(t \mp x/c) \tag{4.13}$$

where u is an arbitrary function. The losses only cause a damping of the wave. This result has been important for telegraphy. It shows that, given appropriate values for the per-unit-length parameters of the line, signals can be transmitted in an undistorted form, even if damped in time. "*The distortionlessless state forms a simple and natural boundary between two diverse kind of propagation of a complicated nature, in each of which there is continuous distortion ...*" (Heaviside, 1893). The reader is referred to the critical and historical introduction by Ernst Weber to *Electromagnetic Theory* by Oliver Heaviside, (Heaviside, 1893) for further information on this question.

When $\nu \neq 0$, the general solution of Eq. (4.7) can not be expressed in the d'Alembert form because the phase velocity c_{ph} depends upon the frequency. For lines with frequency dependent losses, the attenuation will be different for various frequency components and there will be a further variation in the phase velocity (see next chapter).

4.2. SOLUTION OF THE LOSSY TRANSMISSION LINE EQUATIONS IN THE LAPLACE DOMAIN

There are several methods of solving the problem of lossy two-conductor transmission lines (e.g., Smirnov, 1964a; Doetsch, 1974). In this book we shall use the method based on the Laplace transform because it is the more general. In fact this method also allows us to deal with lossy multiconductor lines with parameters depending on the frequency and nonuniform transmission lines, as we shall see in the following three chapters.

Here, we shall determine the general solution of the line equations by using the Laplace transform. In the Laplace domain it is easy to solve the equations of lossy two-conductor transmission lines because they become a system of two ordinary differential equations. Once the solution is determined in the Laplace domain, the corresponding expression in the time domain is easily obtained by applying the convolution theorem. Thus, we shall only consider transmission lines initially at rest. Nonzero initial conditions can be dealt with by using equivalent distributed sources along the line. At the end of this chapter we shall tackle the problem of solving transmission lines with distributed sources.

Having assumed that the initial distributions of the voltage and current along the line are equal to zero, in the Laplace domain systems of equations (4.1) and (4.2) become

$$\frac{dV}{dx} = -Z(s)I, \tag{4.14}$$

$$\frac{dI}{dx} = -Y(s)V, \tag{4.15}$$

where $V(x;s)$ is the Laplace transform of the voltage distribution, $I(x;s)$ is the Laplace transform of the current distribution, and

$$Z(s) = R + sL, \tag{4.16}$$

$$Y(s) = G + sC. \tag{4.17}$$

The parameters Z and Y are, respectively, the per-unit-length longitudinal impedance and transverse admittance of the line in the Laplace domain.

Henceforth, the steps we have followed in the analysis of ideal transmission line equations can be pursued. Starting from Eqs. (4.14) and (4.15), by derivation and substitution we obtain the two uncoupled second-order differential equations

$$\frac{d^2V}{dx^2} - k^2(s)V = 0, \tag{4.18}$$

$$\frac{d^2I}{dx^2} - k^2(s)I = 0, \tag{4.19}$$

where we have introduced the function[1]

$$k(s) = \sqrt{Z(s)Y(s)}. \tag{4.20}$$

By substituting Eqs. (4.16) and (4.17) in Eq. (4.20) we obtain

$$k(s) = \sqrt{(R + sL)(G + sC)} = \frac{s}{c}\sqrt{(1 + \mu/s)^2 - (\nu/s)^2}, \tag{4.21}$$

where the parameters c, μ, and ν are as defined in the previous section. This function is multivalued, having two branches (e.g., Smirnov, 1964b). It has two branch points of the first order along the negative real axis, one at $s_R = -R/L$ and the other at $s_G = -G/C$. Both the points s_R and s_G are branch points of regular type; see Appendix B. When the Heaviside condition $\nu = 0$ is satisfied, the two branch points coincide and hence cancel each other out.

We can choose any one of the two branches of $k(s)$ to determine the general solution to our problem. Hereafter, we shall consider the branch that has positive real part for $Re\{s\} > 0$; thus in the lossless limit $k \to +s/c$. Consequently, we must operate in the domain C_{cut} obtained by cutting the complex plane along the segment belonging to the negative real axis whose ends are s_R and s_G. This branch of $k(s)$ has a positive imaginary part for $Im\{s\} > 0$.

The general solution of the line equations (4.18) and (4.19) may be written in the so-called "traveling wave" form

$$V(x, s) = V^+(s)e^{-k(s)(x-x^+)} + V^-(s)e^{k(s)(x-x^-)}, \tag{4.22}$$

$$I(x, s) = I^+(s)e^{-k(s)(x-x^+)} + I^-(s)e^{k(s)(x-x^-)}, \tag{4.23}$$

[1]The expression $-ik(s = i\omega)$ is the so-called *propagation constant*. For ideal transmission lines it is equal to ω/c.

where V^+, V^-, I^+, and I^- are arbitrary functions and x^+ and x^- are arbitrary constants. This form highlights forward and backward waves along the line, as we shall see later on. A representation of the general solution alternative to Eqs. (4.22) and (4.23) is provided by the superposition of the two particular solutions $\cos[k(s)x]$ and $\sin[k(s)x]$. This is the so-called "standing wave" form (e.g., Collin, 1992; Franceschetti, 1997). We shall not consider this representation because it is not suitable for our purposes. As we have already seen in the previous chapters and we shall see in the next ones, a time domain characterization of the terminal behavior of a line based on the forward and backward waves is more general, elegant and effective.

Obviously, the set of all possible solutions for Eqs. (4.18) and (4.19) is much ampler than the set of the solutions of the original systems of equations (4.14) and (4.15). As Eqs. (4.22) and (4.23) are solutions of Eqs. (4.18) and (4.19), respectively, it is necessary to ensure that they satisfy one of the two equations of the original system. For example, substituting Eqs. (4.22) and (4.23) in Eq. (4.14) we obtain

$$I(x;s) = \frac{1}{Z_c(s)}[V^+(s)e^{-k(s)(x-x^+)} - V^-(s)e^{k(s)(x-x^-)}], \tag{4.24}$$

where

$$Z_c(s) = \sqrt{\frac{Z(s)}{Y(s)}} \tag{4.25}$$

is the *characteristic impedance* of the line. Clearly, we can also represent the solution through the functions $I^+(s)$ and $I^-(s)$. It is easy to show that $I^+ = V^+/Z_c$ and $I^- = -V^-/Z_c$.

By substituting Eqs. (4.16) and (4.17) in Eq. (4.25) we obtain for $Z_c(s)$

$$Z_c(s) = \sqrt{\frac{R+sL}{G+sC}} = R_c\sqrt{\frac{1+(\mu+\nu)/s}{1+(\mu-\nu)/s}}, \tag{4.26}$$

where

$$R_c = \sqrt{\frac{L}{C}}. \tag{4.27}$$

The function $Z_c(s)$ also has two branch points of the first order — one at $s_R = -R/L$ and the other at $s_G = -G/C$; s_R is a branch point of regular type and s_G is a branch point of polar type. However, according to the choice we have made for the function $k(s)$, $Z_c(s)$ in the domain C_{cut} is single valued and its real part is always positive.

For ideal transmission lines $\mu = \nu = 0$, and Eqs. (4.21) and (4.26) reduce to $k = s/c$ and $Z_c = R_c$, respectively. The parameter R_c would be the characteristic resistance of the transmission line if it were without losses.

The arbitrary functions $V^+(s)$ and $V^-(s)$ have to be determined by indicating what happens at the line ends, that is, by imposing the *boundary conditions*. In Chapter 1 it is shown that there is only one solution of the transmission line equations (4.14) and (4.15) compatible with an assigned voltage or current at each line end.

Remarks

(i) As the line is dissipative, the general solutions (4.22) and (4.24) are analytic in a half-plane of the complex plane containing the imaginary axis. Therefore, there is a sinusoidal steady-state solution whose phasor form[2], $\bar{V}(x)$, $\bar{I}(x)$ can be obtained by simply replacing s with $i\omega$ in Eqs. (4.22) and (4.24).

(ii) The general solutions (4.22) and (4.24) hold also in the inductiveless limit $L \to 0$ (or in the capacitiveless limit $C \to 0$). In this case we have $s_R \to -\infty$ (or $s_C \to -\infty$). The inductiveless limit $L \to 0$ is very important from the historical point of view. The first successful submarine cable to transmit telegraph signals between England and France (1851) raised the possibility of a transatlantic cable between Europe and the United States. However, as the signal amplitude had been observed to fall off sharply with increasing length of the cable, Lord Kelvin (1855) studied the electrical transients in long cables assuming that the magnetic effects, described through the per-unit-length self-inductance L, were negligible. By using the circuit theory and Kirchhoff's laws he derived a diffusion equation for the voltage for which Fourier (1822) had given solutions. In 1857, Kirchhoff extended the long-line theory to include self-inductance effects and deduced the finite velocity of propagation of the electrical signals. "However, Kelvin's theory of the cable dominated the thinking of everyone, perhaps because the extended theory looked unapproachable to physical interpretation"

[2]The sinusoidal steady-state solution with frequency ω of the transmission line equations can be represented in its phasor form

$$v(x,t) = Re\{\bar{V}(x)\exp(i\omega t)\}, \quad i(x,t) = Re\{\bar{I}(x)\exp(i\omega t)\},$$

where $\bar{V}(x)$ and $\bar{I}(x)$ are two complex functions of the real variable x.

(Ernst Weber). In 1881, Heaviside reexamined the effect of self-induction and determined what is now known as the "traveling wave" solution. The reader is referred to the critical and historical introduction by Ernst Weber to *Electromagnetic Theory* by Oliver Heaviside (Heaviside, 1893) for a complete and comprehensive treatment of this argument.

(iii) Observe that Eqs. (4.14) and (4.15) are general, as they describe lossy transmission lines with parameters depending on the frequency too. We shall deal with these lines in the next chapter. The general solution represented by Eqs. (4.22) and (4.24) hold whatever the expressions of $Z(s)$ and $Y(s)$ are, and thus they also represent the general solution of transmission lines with frequency-dependent parameters. ◇

4.3. THE PROPAGATION ALONG A LOSSY TRANSMISSION LINE

Before tackling the problem of the characterization of lossy two-conductor lines as two-ports, we have to fully understand the influence of the losses on the propagation along the line.

Let us introduce the *local propagation operator*

$$Q(x;s) \equiv e^{-k(s)x}. \tag{4.28}$$

Immediately we observe that the terms

$$B^{+}(x;s) = Q(x - x^{+};s)V^{+}(s) \quad \text{and} \quad B^{-}(x;s) = Q(-x + x^{-};s)V^{-}(s) \tag{4.29}$$

appearing in the expression (4.22) bear a resemblance to the forward and backward voltage waves, respectively, of a lossless line in the Laplace domain. In fact for lossless lines we have $k(s) = s/c$, and by performing the inverse Laplace transform of the expressions (4.22) and (4.23), the general solutions (2.6) and (2.7), respectively, are obtained.

Contrary to what has been done in the previous two chapters, here and in the next three chapters we shall choose the constants x^{+} and x^{-} in such a way that V^{+} represents the amplitude of B^{+} at the left line end $x = 0$, while V^{-} represents the amplitude of B^{-} at the right line end $x = d$. Consequently, we must choose

$$x^{+} = 0, \quad x^{-} = d. \tag{4.30}$$

This different choice is convenient as we are considering lines initially at rest. Obviously, the results to which we shall arrive do not depend on the choice of the constants x^+ and x^-. Therefore, expressions (4.22) and (4.24) become

$$V(x, s) = Q(x; s)V^+(s) + Q(d - x; s)V^-(s), \tag{4.31}$$

$$I(x, s) = \frac{1}{Z_c(s)}[Q(x; s)V^+(s) - Q(d - x; s)V^-(s)]. \tag{4.32}$$

The voltage and current distributions along the line are uniquely determined by the functions $V^+(s)$ and $V^-(s)$ and vice versa, because the local propagation operator is always different from zero for $0 \leqslant x \leqslant d$. Thus, $V^+(s)$ and $V^-(s)$ may be assumed as the state variables of the line (in the Laplace domain) as was done for ideal transmission lines.

Let us assume that both the functions $V^+(s)$ and $V^-(s)$ are given. Starting from the expression (4.31) and using the convolution theorem, for the time domain voltage distribution along the line we obtain

$$v(x, t) = \{q(x, \cdot) * v^+(\cdot)\}(t) + \{q(d - x, \cdot) * v^-(\cdot)\}(t), \tag{4.33}$$

where $q(x, t)$, $v^+(t)$ and $v^-(t)$ are the inverse Laplace transforms of $Q(x; s)$, $V^+(s)$, and $V^-(s)$, respectively; the symbol $(u * v)(t)$ indicates the time convolution product. Obviously, the functions $v^+(t)$ and $v^-(t)$ are equal to zero for $t < 0$ because the line is assumed to be initially at rest.

For lossy lines with frequency-independent parameters the *impulse response* q may be expressed analytically. The inverse Laplace transform of

$$Q(x; s) = \exp\left[-\frac{xs}{c}\sqrt{(1 + \mu/s)^2 - (\nu/s)^2}\right], \tag{4.34}$$

is given by

$$q(x, t) = q_p(x, t) + q_r(x, t), \tag{4.35}$$

where

$$q_p(x, t) = e^{-\mu x/c}\delta(t - x/c), \tag{4.36}$$

$$q_r(x, t) = \nu^2 a(x, t)u[t - (x/c)], \tag{4.37}$$

and (see Appendix B)

$$a(x, t) = e^{-\mu t}\left(\frac{x}{c}\right)\frac{I_1[\nu\sqrt{t^2 - (x/c)^2}]}{\nu\sqrt{t^2 - (x/c)^2}}; \tag{4.38}$$

$I_1(y)$ is the modified first-order Bessel function (e.g., Abramowitz and Stegun, 1972) and $u(t)$ is the Heaviside unit step function. As for $t < x/c$ we have $q_r(x,t) = 0$, we are only interested in the case $t > x/c$, and hence the argument of the Bessel function is always real. Furthermore, $I_1 = I_1(y)$ is an odd function for real y, and hence the term q_r is independent of the sign of the parameter ν.

Two terms form the time domain local propagation function $q(x,t)$ — a *principal term* $q_p(x,t)$ and a *remainder* $q_r(x,t)$, which would vanish if there were no losses. The principal term $q_p(x,t)$ is a delayed Dirac pulse with amplitude $e^{-\mu t}$ acting at $t = x/c$, whereas the remainder $q_r(x,t)$ is a *bounded* piecewise-continuous function, which is equal to zero for $t < x/c$; $f(x)$ is piecewise continuous on an interval I iff it is continuous throughout I except for a finite number of discontinuities of the first kind.

For ideal transmission lines we have $\mu = \nu = 0$; hence Q is reduced to the *ideal delay operator* $\exp(-sx/c)$, and $q(x,t) = q_p(x,t) = \delta(t - x/c)$. Consequently, expression (4.33) reduces to $v^+(t - x/c) + v^-(t - T + x/c)$, where $T = d/c$, namely, to the d'Alembert solution equation (2.6) with $\alpha^+ = 0$ and $\alpha^- = -T$. The term $v^+(t - x/c)$ represents an undistorted forward wave with constant velocity c, and the term $v^-(t - T + x/c)$ represents an undistorted backward wave.

When there are losses and the Heaviside condition is satisfied, $\mu \neq 0$ and $\nu = 0$, the remainder q_r is again equal to zero, and

$$q(x,t) = e^{-\mu x/c}\delta(t - x/c). \tag{4.39}$$

The damping factor $e^{-\mu x/c}$ takes into account the attenuation introduced by the distributed losses along the line. In this case the expression (4.33) reduces to

$$v(x,t) = e^{-\mu x/c}v^+(t - x/c) + e^{-\mu(T - x/c)}v^-(t - T + x/c). \tag{4.40}$$

The first term always represents a voltage wave that propagates in the forward direction, with constant velocity c, damping factor $\exp(-\mu t)$, yet "relatively" undistorted. It is a *damped* "undistorted" *voltage forward wave*. Similarly, the second term is a damped "undistorted" *backward voltage wave*. Therefore, the principal term of the local propagation operator describes the undistorted damped propagation that we would have if the Heaviside condition were satisfied.

Now we shall consider the most general case for which $\nu \neq 0$. By substituting Eq. (4.35) in Eq. (4.33), for the inverse Laplace transform of $B^+(x;s)$ and $B^-(x;s)$ we obtain, respectively:

$$b^+(x,t) = [e^{-\mu x/c} v^+(t - x/c) + v^2 \int_{0^-}^{t-x/c} v^+(\tau) a(x, t-\tau) d\tau] u(t - x/c), \quad (4.41)$$

$$b^-(x,t) = [e^{-\mu(d-x)/c} v^-(t - T + x/c)$$
$$+ v^2 \int_{0^-}^{t-T+x/c} v^-(\tau) a(d-x, t-\tau) d\tau] u(t - T + x/c). \quad (4.42)$$

Let us consider the case in which $v^+ = \delta(t)$, $v^- = 0$, and hence $v(x=0,t) = \delta(t)$. The term b^+ reduces to the function q and $b^- = 0$. The Dirac pulse, which at the time instant $t = 0$ is centered at $x = 0$, propagates damped and undistorted toward the right with velocity c, as in the Heaviside case, and leaves a "wake" behind it. The wake is due to the dispersion introduced by the losses and is described by the bounded term q_r. Similar considerations may be made when v^+ is a generic function.

Obviously, $q_r(x,t)$ is equal to zero for $t < x/c$. Its value at $t = (x/\varepsilon)/c$, where ε is an arbitrary small positive number, is given by

$$q_r(x,t)|_{t=(x+\varepsilon)/c} = \frac{v^2}{2}\frac{x}{c} e^{-\mu x/c}. \quad (4.43)$$

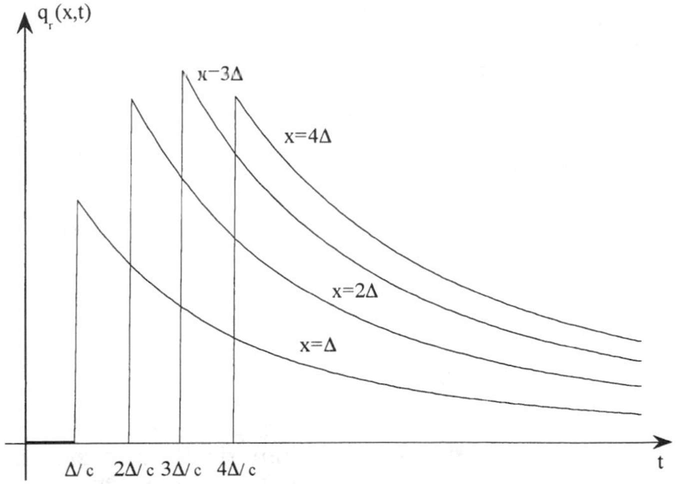

Figure 4.1. Qualitative behavior of $q_r(x,t)$ versus time at different spatial positions.

As regards the asymptotic behavior of $q_r(x,t)$ for $t \to \infty$, we obtain

$$q_r(x,t) \approx \frac{\nu^2}{\sqrt{2\pi}} \frac{x}{c} \frac{e^{-(\mu-|\nu|)t}}{(|\nu|t)^{3/2}} \quad \text{for } t \to \infty. \tag{4.44}$$

When $\mu \neq |\nu|$, the function $q_r(x,t)$ goes exponentially to zero for $t \to \infty$ with the time constant $1/(\mu - |\nu|)$, whatever the (finite) value of x. Instead, in the limit case $\mu = |\nu|$, the function q_r goes more slowly to zero for $t \to \infty$, as $1/(|\nu|t)^{3/2}$. Figure 4.1 shows the qualitative behavior of $q_r(x,t)$ as function of time for several values of the spatial variable x: $x = \Delta$, 2Δ, 3Δ, and 4Δ. The maximum value that each curve attains is given by Eq. (4.43).

In general, the term b^+ describes a *forward voltage wave* that leaves the end $x = 0$ and propagates with a wavefront moving toward the right with velocity c. By contrast, the term b^- describes a *backward voltage wave* that leaves the end $x = d$ and propagates with a wavefront moving toward the left with velocity c. Observe that the factor $\exp(-sx/c)$ in the expression of Q given by Eq. (4.34) describes the delay introduced by the finite value of the propagation velocity.

By performing the inverse Laplace transform of the expression (4.32), the general solution for the current distribution in the time domain is obtained. Placing

$$Q_i(x;s) = \frac{e^{-k(s)x}}{Z_c(s)}, \tag{4.45}$$

and applying the convolution theorem, from the expression (4.32) we obtain

$$i(x,t) = \{q_i(x,\cdot) * \nu^+(\cdot)\}(t) - \{q_i(d-x,\cdot) * \nu^-(\cdot)\}(t), \tag{4.46}$$

where $q_i(x,t)$ is the inverse Laplace transform of $Q_i(x,s)$. For lossy lines with frequency-independent parameters we have

$$Q_i(x;s) = \frac{s+\beta}{R_c} \frac{\exp\left[-\frac{x}{c}\sqrt{(s+\mu)^2 - \nu^2}\right]}{\sqrt{(s+\mu)^2 - \nu^2}}. \tag{4.47}$$

The reader may now calculate the analytical expression of q_i. The inverse Laplace transform of Q_i may be evaluated analytically by using the properties of the Laplace transform and the Laplace transform table given in Appendix B.

4.4. A SEMI-INFINITE LOSSY LINE CONNECTED TO AN IDEAL CURRENT SOURCE

It is instructive to study the dynamics of a semi-infinite lossy line, initially at rest, and fed by an independent current source $i_1(t)$ at the end $x = 0$ (Fig. 4.2). Let us indicate with $I_1(s)$ the line current at the end $x = 0$ in the Laplace domain.

For the general solution of a line of infinite length let us consider the expression

$$V(x, s) = Q(x; s)V^+(s) + Q(-x; s)V^-(s), \tag{4.48}$$

$$I(x, s) = Z_c^{-1}(s)[Q(x; s)V^+(s)] - Q(-x; s)V^-(s)]. \tag{4.49}$$

As the right end of the line is at $x = +\infty$ and the line is initially at rest, the amplitude of the backward voltage wave is equal to zero, $V^- = 0$. Instead, the function V^+, representing the amplitude of the forward wave at $x = 0$, is determined by imposing the boundary condition at $x = 0$

$$I(x = 0; s) = I_1(s). \tag{4.50}$$

Placing $V^- = 0$ in the expressions (4.48) and (4.49) and imposing the boundary condition (4.50), we obtain

$$V^+(s) = Z_c(s)I_1(s). \tag{4.51}$$

The relation between the line voltage at the end $x = 0$, that is, $V_1(s) = V(x = 0; s)$, and the current $I_1(s)$ is

$$V_1(s) = Z_c(s)I_1(s). \tag{4.52}$$

This is an interesting result that extends what we obtained in §2.2.2 for ideal two-conductor transmission lines. The behavior of a semi-infinite line initially at rest at the end $x = 0$ can be represented in the

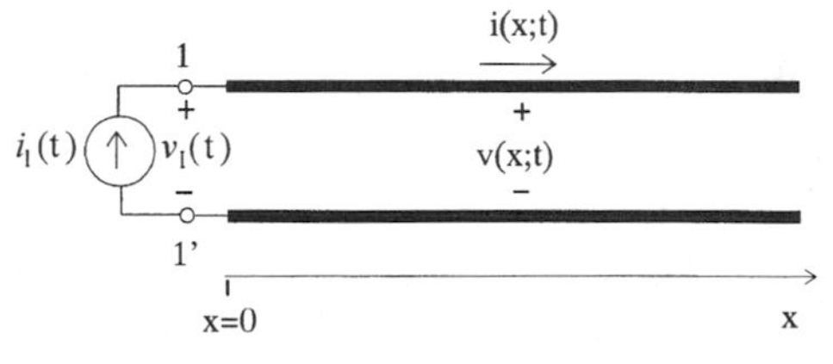

Figure 4.2. Semi-infinite line connected to an independent current source.

Laplace domain by a one-port with impedance $Z_c(s)$. By applying the convolution theorem, Eq. (4.52) in the time domain becomes

$$v_1(t) = \{z_c * i_1\}(t), \tag{4.53}$$

where $z_c(t)$ is the inverse Laplace transform of $Z_c(s)$. The function $z_c(t)$ is the current-controlled impulse response of the one-port representing the line behavior at the end $x = 0$.

4.4.1. Properties of the Characteristic Impedance $Z_c(s)$ and of the Impulse Response $z_c(t)$

Because in the limit of lossless transmission line $Z_c(s) \to R_c$, it is useful to separate this term from the one that takes into account the effects of the losses. Thus we rewrite $Z_c(s)$ as

$$Z_c(s) = Z_{cp}(s) + Z_{cr}(s), \tag{4.54}$$

where

$$Z_{cp}(s) = R_c \tag{4.55}$$

is the *principal part* and

$$Z_{cr}(s) = R_c\left(\sqrt{\frac{1 + (\mu + \nu)/s}{1 + (\mu - \nu)/s}} - 1\right) \tag{4.56}$$

is the *remainder*. The impulse response $z_c(t)$ is given by

$$z_c(t) = R_c\delta(t) + z_{cr}(t), \tag{4.57}$$

where $z_{cr}(t)$ is the inverse Laplace transform of $Z_{cr}(s)$ and is given by (see Appendix B)

$$z_{cr}(t) = \nu R_c e^{-\mu t}[I_0(\nu t) + I_1(\nu t)]u(t); \tag{4.58}$$

$I_0(\nu t)$ and $I_1(\nu t)$ are the modified Bessel functions of zero- and first-order, respectively (e.g., Abramowitz and Stegun, 1972). As $I_0(y)$ is an even function and $I_1(y)$ is an odd function for real y, the term z_{cr} depends on the sign of the parameter ν. Two terms form the impulse response z_c, that is, a Dirac pulse acting at $t = 0$ and the *bounded* piecewise-continuous function $z_{cr}(t)$ given by Eq. (4.58), which is equal to zero for $t < 0$.

Remark

Equation (4.54) is an *asymptotic expression* of $Z_c(s)$. "*An asymptotic expression for a function is an expression as the sum of a simpler function and of a remainder that tends to zero at infinity, or (more generally) which tends to zero after multiplication by some power."* (Lighthill, 1958). In fact for the remainder Z_{cr} we have the following asymptotic behavior:

$$Z_{cr}(s) \approx R_c \frac{\nu}{s} + O(1/s^2) \quad \text{for } s \to \infty, \tag{4.59}$$

whereas the principal part Z_{cp} is a constant. The inverse Laplace transform of Z_{cr} is a bounded piecewise-continuous function because Z_{cr} goes to zero as $1/s$ for $s \to \infty$ (see §5.5.1), whereas the inverse Laplace transform of the principal part is a Dirac function. This will be a very important point in the next three chapters, whenever the inverse Laplace transform of a given function $F(s)$ can not be performed analytically. Once the asymptotic expression of the function $F(s)$ is known, we can separate the part that is numerically transformable in the time domain from that which can be transformed analytically. ◇

By substituting Eq. (4.57) in the convolution equation (4.53) we obtain

$$v_1(t) = R_c i_1(t) + \{z_{cr} * i_1\}(t). \tag{4.60}$$

The first term is the voltage we should have at the end $x = 0$ if the line were lossless or the Heaviside condition were satisfied. The other term describes the wake produced by the dispersion due to the losses.

The value of z_{cr} at $t = 0^+$ is $z_{cr}(t = 0^+) = \nu R_c$. As regards the asymptotic behavior of z_{cr} for $t \to \infty$, we obtain

$$z_{cr}(t) \approx \frac{2\nu R_c}{\sqrt{2\pi}} \frac{e^{-(G/C)t}}{\sqrt{|\nu| t}} \quad \text{for } \nu > 0, \tag{4.61}$$

and

$$z_{cr}(t) \approx \frac{\nu R_c}{2\sqrt{2\pi}} \frac{e^{-(R/L)t}}{(|\nu| t)^{3/2}} \quad \text{for } \nu < 0. \tag{4.62}$$

If $\mu \neq |\nu|$ the function z_{cr} goes exponentially to zero for $t \to \infty$ with the time constant $1/(\mu - |\nu|)$. By contrast, in the limit case $\mu = |\nu|$, z_{cr} goes more slowly to zero for $t \to \infty$, as $1/\sqrt{\nu t}$ for $\nu > 0$ and as $1/(|\nu| t)^{3/2}$ for $\nu < 0$. Figure 4.3 shows the qualitative behavior of the function $z_{cr}(t)$ for several values of μ/ν.

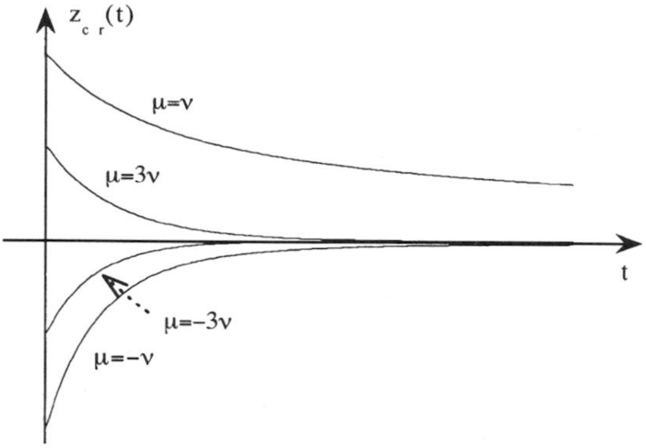

Figure 4.3. Qualitative behavior of the function $z_{cr}(t)$ for different values of μ/ν.

Remark

The asymptotic behavior of $z_{cr}(t)$ for $t \to \infty$ depends only on the branch point of $Z_c(s)$ nearer to the imaginary axis (see Appendix B).

For $\nu > 0$ we have $R/L > G/C$; thus, the branch point of polar type $-G/C$ is that nearer to the imaginary axis, and hence the factor $1/\sqrt{s + G/C}$ determines completely the asymptotic behavior of $z_{cr}(t)$.

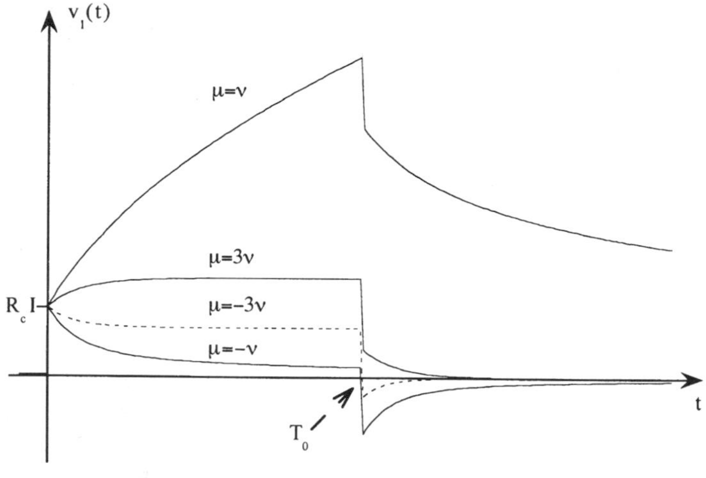

Figure 4.4. Voltage waveform at the end $x = 0$ of the semi-infinite line when i_1 is a rectangular pulse for different values of μ/ν.

By using the inverse Laplace transform of $1/\sqrt{s}$ and the property of the translation of the image function (see Appendix B), we immediately obtain that the inverse Laplace transform of $1/\sqrt{s+G/C}$ is the function $e^{-(G/C)t}/\sqrt{\pi t}$.

For $\nu = 0$ the two branch points coincide, and hence the function $Z_c(s)$ becomes a constant. However, for $\nu < 0$ we have $R/L < G/C$, thus the branch point of regular type $-R/L$ is that nearer to the imaginary axis, and hence the factor $\sqrt{s+R/L}$ determines completely the asymptotic behavior of $z_{cr}(t)$. By using the inverse Laplace transform of $\sqrt{s}$ and the linear transformation property (see Appendix B), we immediately obtain that the inverse Laplace transform of $\sqrt{s+R/L}$ asymptotically behaves as $e^{-(R/L)t}/[\Gamma(-1/2)t^{3/2}]$, where $\Gamma(y)$ is the gamma function (e.g., Abramowitz and Stegun, 1972). ◇

Figure 4.4 shows the qualitative behavior of the voltage dynamics at the end $x = 0$, when i_1 is a rectangular pulse of time length T_0 and amplitude I.

4.4.2. A Fast Convolution Algorithm

In general, the convolution integral

$$y(t) = \{z_{cr} * j\}(t) = \int_0^t z_{cr}(t-\tau)j(\tau)d\tau \tag{4.63}$$

can only be evaluated approximately through numerical integration rules. This can be done using standard techniques because the *kernel* $z_{cr}(t)$ is bounded, continuous for $t > 0$, and has a discontinuity of the first kind at $t = 0$; we are assuming that the function $j(t)$ is also piecewise continuous.

Several approximation techniques are available. Here, we shall use the very simple procedure based on trapezoidal rule (e.g., Linz, 1985; Brunner and van der Houwen, 1986). Using a homogeneous time step Δt, an approximation of the convolution equation (4.63) becomes

$$\begin{aligned} y_0 &= 0 \\ y_1 &= \frac{\Delta t}{2} z_{cr}^{(0)} j_1 + \frac{\Delta t}{2} z_{cr}^{(1)} j_0 \\ y_n &= \frac{\Delta t}{2} z_{cr}^{(0)} j_n + \Delta t \sum_{p=1}^{n-1} z_{cr}^{(p)} j_{n-p} + \frac{\Delta t}{2} z_{cr}^{(n)} j_0 \quad n \geqslant 2, \end{aligned} \tag{4.64}$$

where $z_{cr}^{(p)} = z_{cr}(t_p)$, $y_p = y(t_p)$, $j_p = j(t_p)$, the discrete time variable is given by $t_p = p\Delta t$, and $p = 0, 1, 2, \ldots$. At each discrete time instant t_n, we have to perform $n + 1$ products. Therefore, to evaluate the samples $y_1, y_2, \ldots, y_m$ we have to perform $\Sigma_{p=1}^{m}(p + 1) \cong m^2/2$ multiplications if $m \gg 1$. As this numerical calculation has a high computational cost, it is convenient to introduce a technique through which we can calculate the convolution integrals in a recursive way. This can be realized through a suitable approximation of the kernel. In the literature, two main different methods have been proposed: One method uses exponential functions (e.g., Semlyen and Dabuleanu, 1975; Gordon, Blazeck, and Mittra, 1992), and the other uses polynomial functions (e.g., Chang and Kang, 1996).

Here, we shall briefly survey the method based on the exponential approximation. Let us assume that the kernel z_{cr} may be adequately approximated by

$$\bar{z}_{cr}(t) = \sum_{i=1}^{N} \alpha_i e^{-\beta_i t}, \tag{4.65}$$

where for a given N the parameters α_i and β_i are determined by minimizing the "distance" between $z_{cr}(t)$ and $\bar{z}_{cr}(t)$ with the constraint $\beta_i > 0$ for every i. These parameters can also be evaluated by approximating the remainder $Z_{cr}(s)$ through a rational function of s using the Padé approximation technique (e.g., Lin and Kuh, 1992).

Substituting the expression (4.65) in the integral (4.63), we obtain the approximate expression for the function y

$$\bar{y}(t) = \sum_{i=1}^{N} \alpha_i I_i(t) \tag{4.66}$$

where

$$I_i(t) = \int_0^t e^{-\beta_i \tau} j(t - \tau) d\tau. \tag{4.67}$$

Note that the functions $I_i(t)$ are the solutions of the first-order ordinary differential equations

$$\frac{dI_i}{dt} + \beta_i I_i = j(t) \quad \text{with } i = 1, 2, \ldots, N. \tag{4.68}$$

with $I_i(t = 0) = 0$. Thus we are approximating a system of infinite dimension by using the finite dimension system whose state equations are given by Eq. (4.68).

The integral I_i is split into two parts,

$$I_i(t) = \int_0^{\Delta t} e^{-\beta_i \tau} j(t-\tau)d\tau + \int_{\Delta t}^{t} e^{-\beta_i \tau} j(t-\tau)d\tau. \tag{4.69}$$

Then, performing a variable change in the second integral, I_i may be rewritten as

$$I_i(t) = \int_0^{\Delta t} e^{-\beta_i \tau} j(t-\tau)d\tau + e^{-\beta_i \Delta t} I_i(t-\Delta t). \tag{4.70}$$

Equation (4.70) represents the recursive formula that we were seeking. Using a discrete approach we obtain the so-called *fast convolution algorithm*

$$\begin{cases} \bar{y}_n = \sum_{i=1}^{N} \alpha_i I_i(t_n), \\ I_i(t_n) = e^{-\beta_i \Delta t} I_i(t_{n-1}) + \dfrac{\Delta t}{2}(j_n + e^{-\beta_i \Delta t} j_{n-1}), \end{cases} \tag{4.71}$$

where $n \geqslant 1$; $I_i(t_0) = 0$ for every i.

In order to compare the computational cost of the two different approaches we observe that by using the algorithm (4.71) to evaluate the samples $y_1, y_2, \ldots, y_m$ we should perform $3N(m-1)$ multiplications. Therefore, the fast convolution algorithm is more efficient than the "crude" one when $m \gg 6N$. The accuracy of the computation in both methods is the same, depending only on the time-step size, and the error goes to zero as Δt^2 for $\Delta t \rightarrow 0$.

Remark

The kernel $z_{cr}(t)$ goes to zero exponentially with a time constant given by $1/(\mu - |\nu|)$; for $\nu > 0$ we have $\mu - |\nu| = G/C$ and for $\nu < 0$ we have $\mu - |\nu| = R/L$. Therefore, a large number of exponential functions would be required to approximate the kernel when $1/(\mu - |\nu|)$ is much less than the greatest characteristic time of the problem. It is evident that if this is the case, the fast convolution algorithm that we have described is no longer useful because the computational cost may become too high. In these situations, convolution algorithms based on polynomial functions may be more convenient. ◇

4.5. REPRESENTATION OF LOSSY TWO-CONDUCTOR LINES AS TWO-PORTS

In general, the values of the voltages and the currents at the line ends are not known, but they are actually unknowns of the problem and they depend on the actual network to which the lines are connected at their ends.

We have already seen when dealing with ideal two-conductor lines that the solution of a network composed of transmission lines and lumped circuits can be simplified considerably if we first determine the voltage and current at the line ends by representing the line behavior through equivalent circuits of Thévenin or Norton type (see §2.4). Then, once the voltages or the currents at the line ends are known, we can also calculate the voltage and current distribution all along the line by using the convolution relations (4.33) and (4.46).

The same approach can be adopted to study lossy transmission lines of finite length connecting generic lumped circuits. To determine the characteristic relations of the two-port we shall proceed as follows (Fig. 4.5):

(i) first we characterize it in the Laplace domain where relations are purely algebraic; and
(ii) then the representation in the time domain is obtained by applying the convolution theorem.

In the Laplace domain the voltages and the currents at the line ends are related to the voltage and current distributions along the line through the following:

$$V_1(s) = V(x = 0; s), \quad I_1(s) = I(x = 0; s), \tag{4.72}$$

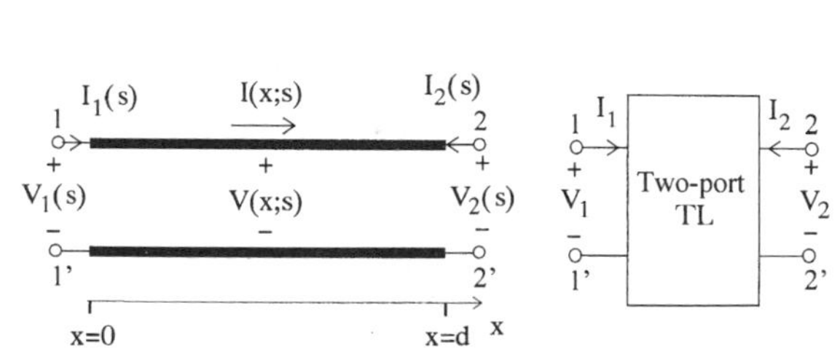

Figure 4.5. Two-conductor transmission line as a two-port element in the Laplace domain.

and

$$V_2(s) = V(x = d; s), \quad I_2(s) = -I(x = d; s). \tag{4.73}$$

The reference directions are always chosen according to the normal convention for the two-ports corresponding to the two ends of the line. To determine the characteristic relations of the two-port representing the line we have to particularize the general solution of the line equations at the line ends.

4.5.1. Terminal Behavior of the Line in the Laplace Domain: the Describing Functions $Z_c(s)$ and $P(s)$

Specifying the expressions (4.31) and (4.32) at the end $x = 0$ and recalling that $Q(0; s) = 1$ we obtain the relations

$$V_1(s) = V^+(s) + P(s)V^-(s), \tag{4.74}$$

$$Z_c(s)I_1(s) = V^+(s) - P(s)V^{-1}(s), \tag{4.75}$$

where

$$P(s) = Q(d; s) = \exp[-sT\sqrt{(1 + \mu/s)^2 - (\nu/s)^2}], \tag{4.76}$$

and

$$T = \frac{d}{c}. \tag{4.77}$$

By contrast, specifying the expressions (4.31) and (4.32) at the end $x = d$ we obtain relations

$$V_2(s) = P(s)V^+(s) + V^-(s), \tag{4.78}$$

$$-Z_c(s)I_2(s) = P(s)V^+(s) - V^-(s). \tag{4.79}$$

The *global propagation operator* $P(s)$ defined by Eq. (4.76) will play an important part in the theory developed in this book. It links the amplitude of the forward voltage wave at the line end $x = d$, that is, PV^+, with the one at the line end $x = 0$, that is, V^+. As the line is uniform, P is also the operator linking the amplitude of the backward voltage wave at the line end $x = 0$, that is, PV^-, with the one at the line end $x = d$, that is, V^-. For nonuniform transmission lines, in general, they are different, as we shall see in Chapter 7.

Subtracting Eqs. (4.74) and (4.75) termwise, and summing Eqs. (4.78) and (4.79) termwise, we have, respectively,

$$V_1(s) - Z_c(s)I_1(s) = 2P(s)V^-(s), \tag{4.80}$$

$$V_2(s) - Z_c(s)I_2(s) = 2P(s)V^+(s). \tag{4.81}$$

If the state of the line in the Laplace domain, represented by V^+ and V^-, were completely known, these equations would completely determine the terminal behavior of the line. Actually, as for the ideal two-conductor transmission lines, the voltage waves V^+ and V^- are themselves unknowns (see §2.4).

As for ideal transmission lines, different formulations of the equations governing the state are possible.

By using Eq. (4.74) it is possible to express the amplitude of the outcoming forward wave at $x = 0$, namely, V^+, as a function of the voltage and of the amplitude of the incoming forward wave at the same end. In the same way, by using Eq. (4.78) it is possible to express the amplitude of the outcoming backward wave at $x = d$, namely, V^-, as a function of the voltage and the amplitude of the incoming forward wave at the same end. Therefore, from Eqs. (4.74) and (4.78) we immediately obtain

$$V^+(s) = V_1(s) - P(s)V^-(s), \tag{4.82}$$

$$V^-(s) = V_2(s) - P(s)V^+(s). \tag{4.83}$$

By contrast, summing Eqs. (4.74) and (4.75) termwise, and subtracting Eqs. (4.78) and (4.79) termwise, we have

$$2V^+(s) = V_1(s) + Z_c(s)I_1(s), \tag{4.84}$$

$$2V^-(s) = V_2(s) + Z_c(s)I_2(s). \tag{4.85}$$

The state equations (4.82) and (4.83) describe in *implicit form* the relation between the state of the line and the electrical variables at the line ends, whereas the state equations (4.84) and (4.85) provide the same relation, but in an *explicit form*.

The state equations in implicit form similar to Eqs. (4.82) and (4.83) involving the terminal variables I_1 and I_2 instead of V_1 and V_2, may be obtained from Eqs. (4.75) and (4.79). As these equations lead to a time domain model that requires two convolution products per iteration, we shall only consider the implicit formulation based on Eqs. (4.82) and (4.83).

Equations (4.80) and (4.81), joined to the state equations, give in implicit form the relations between the voltages V_1, V_2 and the currents I_1, I_2. In particular, substituting the expression of V^+ given by Eq. (4.84) in Eq. (4.81), and the expression of V^- given by Eq. (4.85) in Eq. (4.80), we obtain two linearly independent equations in terms of the terminal variables V_1, V_2, I_1, and I_2,

$$V_1(s) - Z_c(s)I_1(s) - P(s)[V_2(s) + Z_c(s)I_2(s)] = 0, \tag{4.86}$$

$$V_2(s) - Z_c(s)I_2(s) - P(s)[V_1(s) + Z_c(s)I_1(s)] = 0. \tag{4.87}$$

Remark

We immediately observe from Eq. (4.84) that V^+ would be equal to zero if the line were connected at the left end to a one-port with impedance $Z_c(s)$,

$$V_1(s) = -Z_c(s)I_1(s), \tag{4.88}$$

—*perfectly matched line at the left end*—and hence its inverse Laplace transform $v^+(t)$ should be equal to zero. The same considerations would hold for the backward wave if the line were *perfectly matched at the right end*,

$$V_2(s) = -Z_c(s)I_2(s). \tag{4.89}$$

In general, the matching conditions (4.88) and (4.89) are not satisfied. Then, a forward wave with amplitude V^+ is generated at the left line end $x = 0$, and propagates toward the other end $x = d$, where its amplitude is PV^+. Likewise, the backward wave excited at $x = d$ with amplitude V^- propagates toward the left line end, where its amplitude is PV^-. ◇

By operating in the Laplace domain, we have found for the lossy two-conductor lines the same results as those we found for the ideal two-conductor lines; see §2.4. The system of equations (4.80) to (4.83) and the system of equations (4.86) and (4.87) are two fundamental results of considerable importance as they are two different mathematical models describing the two-port representing the line. The set of equations (4.80) to (4.83) describes the internal state of the line, represented by V^+ and V^-, as well as the terminal properties. It is the *internal* or *input-state-output description* of the line in the Laplace domain. Equations (4.82) and (4.83) govern the behavior of the state, whereas Eqs. (4.80) and (4.81) describe the terminal properties. Instead the system of Eqs. (4.86) and (4.87) describes only the terminal property of the line. It is the *external* or *input-output description* of the line in the Laplace domain. Kuznetsov and Schutt-Ainé make the same distinction (Kuznetsov and Schutt-Ainé, 1996): They call *closed-loop characterization* the input-output description and *open-loop characterization* the input-state-output description.

Both descriptions are completely characterized by the two functions $Z_c(s)$ and $P(s)$, which we call the *describing functions* of the two-port representing the line behavior in the Laplace domain. The properties of the describing function $Z_c(s)$ and the corresponding impulse response $z_c(t)$ have been outlined in §4.4.1. In what follows, we shall study the properties of the describing function $P(s)$ and the corresponding impulse response $p(t)$.

4.5.2. Properties of the Global Propagation Operator $P(s)$ and the Impulse Response $p(t)$

It is useful to rewrite the propagation operator P as follows:

$$P(s) = e^{-(s+\mu)T}\hat{P}(s), \tag{4.90}$$

where the function $\hat{P}(s)$ is given by

$$\hat{P}(s) = \exp\left\{T(s+\mu)\left[1-\sqrt{1-\left(\frac{\nu}{\mu+s}\right)^2}\right]\right\}. \tag{4.91}$$

The factor $\exp[-(s+\mu)T]$, which represents a damped ideal delay operator with delay T, oscillates for $s \to \infty$, whereas the factor $\hat{P}(s)$ tends to 1. Therefore, the function $\hat{P}(s)$ has the following asymptotic expression:

$$\hat{P}(s) = 1 + \hat{P}_r(s), \tag{4.92}$$

where the *remainder* $\hat{P}_r(s)$ has the following asymptotic behavior:

$$\hat{P}_r(s) = \frac{\nu^2 T}{2}\frac{1}{s+\mu} + O(1/s^2) \quad \text{for } s \to \infty. \tag{4.93}$$

The first term of $\hat{P}(s)$, that is, 1, is that which we should have if the line were lossless or if the Heaviside condition were satisfied. The other term $\hat{P}_r(s)$ describes the wake produced by the dispersion due to the losses.

The operator P would coincide with the *ideal delay operator* $\exp(-sT)$ if the line were lossless. For lossy lines, the operator P reduces to the product of the ideal delay operator $\exp(-sT)$ with the decaying factor $\exp(-\mu T)$ when the Heaviside condition is satisfied. When $\nu \neq 0$, besides the ideal delay and the decaying factor, there is also the factor $(1 + \hat{P}_r)$.

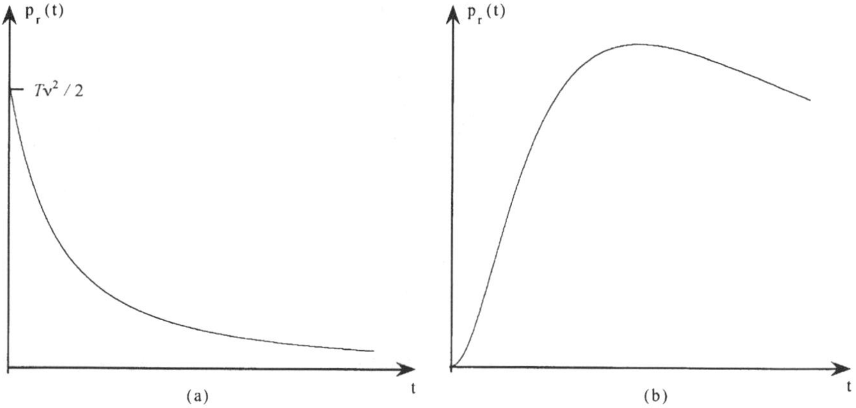

Figure 4.6. Qualitative behavior of the function $p_r(t)$ for (a) $T = T^*$, and (b) $T = 10T^*$, with $\mu = \nu$.

The inverse Laplace transform of P may be expressed as

$$p(t) = e^{-\mu T}\delta(t - T) + p_r(t - T), \tag{4.94}$$

where the function

$$p_r(t) = T\nu e^{-\mu(t+T)} \frac{I_1[\nu\sqrt{(t+T)^2 - T^2}\,]}{\sqrt{(t+T)^2 - T^2}} u(t) \tag{4.95}$$

is the inverse Laplace transform of the function $P_r(s) = e^{-\mu T}\hat{P}_r(s)$. Therefore, the function $p(t)$ is equal to zero for $t < T$. This is the manifestation of the delay introduced by the finite value of the propagation velocity. According to the asymptotic behavior of $\hat{P}_r$ given by Eq. (4.93), the function $p_r(t)$ is a bounded, piecewise-continuous function. As $I_1(y)$ is an odd function for real y, the function $p_r(t)$ does not depend on the sign of ν.

The first term on the right-hand side of Eq. (4.94) is the impulse response $p(t)$ that we should have if the Heaviside condition were satisfied. The remainder p_r describes the wake produced by the dispersion due to the losses. Figure 4.6 shows the qualitative behavior of the function $p_r(t)$ versus the time for two values of T.

The value of $p_r(t)$ at time $t = 0^+$ is $p_r(0^+) = T\nu^2/2$. For $t \to \infty$ the function $p_r(t)$ behaves as follows:

$$p_r(t) \approx \frac{\nu^2 T}{\sqrt{2\pi}} \frac{e^{-(\mu - |\nu|)t}}{(|\nu| t)^{3/2}}. \tag{4.96}$$

Thus, $p_r(t)$ vanishes exponentially for $t \to \infty$ with the time constant $1/(\mu - |\nu|)$ if $\mu \neq |\nu|$. However, it vanishes more slowly if $\mu = |\nu|$, as $1/(|\nu|t)^{3/2}$. Remember that for $\nu > 0$ we have $\mu - |\nu| = G/C$ and for $\nu < 0$ we have $\mu - |\nu| = R/L$.

Remark

As for $z_{cr}(t)$ (see §4.4.1), the asymptotic behavior of $p_r(t)$ for $t \to \infty$ depends only on the branch point of $P(s)$ that is nearer to the imaginary axis (see Appendix B). Unlike the branch points of $Z_c(s)$, both the branch points of $P(s)$ are regular.

For $\nu > 0$ we have $R/L > G/C$; the branch point $-G/C$ is that nearer to the imaginary axis, and hence the factor $\sqrt{s + G/C}$ determines completely the asymptotic behavior of $p_r(t)$. By using the inverse Laplace transform of $\sqrt{s}$ and the linear transformation property (see Appendix B), we immediately obtain that the inverse Laplace transform behaves asymptotically as $e^{-(G/C)t}/[\Gamma(-1/2)t^{3/2}]$.

For $\nu = 0$ the two branch points coincide, and hence they disappear. However, for $\nu < 0$ we have $R/L < G/C$; the branch point of regular type $-R/L$ is that nearer to the imaginary axis, and hence the factor $\sqrt{s + R/L}$ determines completely the asymptotic behavior of $p_r(t)$. By using the inverse Laplace transform of $\sqrt{s}$ and the linear transformation property we immediately obtain that the inverse Laplace transform behaves asymptotically as $e^{-(R/L)t}/[\Gamma(-1/2)t^{3/2}]$. ◇

4.6. THE INPUT-STATE-OUTPUT DESCRIPTION

In this section we shall deal with the input-state-output description based on the system of equations (4.80) to (4.83).

4.6.1. Laplace Domain Equivalent Circuits of Thévenin and Norton Type

Because in Eqs. (4.80) and (4.81) the state of the line appears through $2P(s)V^-(s)$ and $2P(s)V^+(s)$, we rewrite these equations as

$$V_1(s) - Z_c(s)I_1(s) = W_1(s), \tag{4.97}$$

$$V_2(s) - Z_c(s)I_2(s) = W_2(s), \tag{4.98}$$

where the "source" terms are given by

$$W_1(s) = 2P(s)V^-(s), \tag{4.99}$$

$$W_2(s) = 2P(s)V^+(s). \tag{4.100}$$

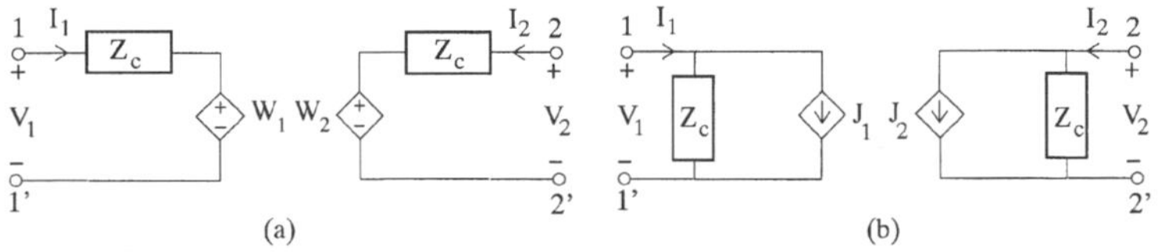

Figure 4.7. Laplace domain equivalent circuits of (a) Thevenin type and (b) Norton type for a uniform lossy two-conductor line.

As the functions V^+, V^- are uniquely related to W_1, W_2, the latter functions can also be regarded as state variables. Note that, except for the factor 2, $W_1(s)$ and $W_2(s)$ are, respectively, the backward voltage wave amplitude at the end $x = 0$ and the forward voltage wave amplitude at the end $x = d$.

From the state equations (4.82) and (4.83) we obtain the equations for W_1 and W_2:

$$W_1(s) = P(s)[2V_2(s) - W_2(s)], \tag{4.101}$$

$$W_2(s) = P(s)[2V_1(s) - W_1(s)]. \tag{4.102}$$

Let us now consider Eqs. (4.97) and (4.98). Equation (4.97) states that, in the Laplace domain, the voltage at the line end $x = 0$, that is, V_1, is equal to the sum of two terms, as in the case of ideal two-conductor lines. If the line were perfectly matched at end $x = d$, it would behave as a one-port with impedance Z_c at $x = 0$ and V_1 would be equal to $Z_c I_1$. By contrast, W_1 is the voltage that would be at the end $x = 0$ if the line were connected to an open circuit at that end. Similar considerations can be made for Eq. (4.98). Consequently, the behavior of each port of a lossy line may be represented through an equivalent Thévenin circuit (see Fig. 4.7a). Each port of the transmission line behaves as an impedance $Z_c(s)$ connected in series with a controlled voltage source as in the lossless case (see §2.6). The governing laws of these controlled sources are simply the state equations (4.101) and (4.102).

The Thévenin equivalent circuits of Fig. 4.7a and the governing laws (4.101) and (4.102) for the voltage-controlled sources still hold for lines with frequency dependent parameters (see Chapter 5).

Figure 4.7b shows the line-equivalent circuit of Norton type in the Laplace domain. The controlled current sources $J_1(s)$ and $J_2(s)$ are related to the controlled voltage sources of the circuit in Fig. 4.7a through the relations

$$J_1 = -\frac{W_1}{Z_c} \quad \text{and} \quad J_2 = -\frac{W_2}{Z_c}. \tag{4.103}$$

The control laws of these sources may be obtained from the state equations (4.101) and (4.102). We obtain

$$J_1(s) = P(s)[-2I_2(s) + J_2(s)], \tag{4.104}$$

$$J_2(s) = P(s)[-2I_1(s) + J_1(s)]. \tag{4.105}$$

4.6.2. Time Domain Thévenin Description

The equivalent two-ports representing a lossy two-conductor line in the time domain may be obtained from those shown in Fig. 4.7 by using the convolution theorem.

The time domain output equations of the two-port are obtained from Eqs. (4.97) and (4.98). They are

$$v_1(t) - \{z_c * i_1\}(t) = w_1(t), \tag{4.106}$$

$$v_2(t) - \{z_c * i_2\}(t) = w_2(t). \tag{4.107}$$

The voltages w_1 and w_2 depend only on the initial conditions for $0 \leqslant t \leqslant T$: $w_1(t) = w_2(t) = 0$ for $0 \leqslant t \leqslant T$ because we are considering transmission lines initially at rest. For $t > T$, they are both unknowns of the problem and are related to the voltages at the line ends through the control laws obtained from Eqs. (4.101) and (4.102),

$$w_1(t) = \{p * (2v_2 - w_2)\}(t), \tag{4.108}$$

$$w_2(t) = \{p * (2v_1 - w_1)\}(t). \tag{4.109}$$

As $p(t) = 0$ for $0 \leqslant t < T$, Eqs. (4.108) and (4.109) reduce to

$$w_1(t) = u(t-T)\int_{0^-}^{t^+-T} p(t-\tau)[2v_2(\tau) - w_2(\tau)]d\tau, \tag{4.110}$$

$$w_2(t) = u(t-T)\int_{0^-}^{t^+-T} p(t-\tau)[2v_1(\tau) - w_1(\tau)]d\tau. \tag{4.111}$$

Therefore, the voltages $w_1(t)$ and $w_2(t)$ depend, respectively, only on the values assumed by w_1 and w_2 and by v_1 and v_2 in the time interval $(0, t-T)$. Consequently, if the solution for $0 \leqslant t \leqslant iT$, with

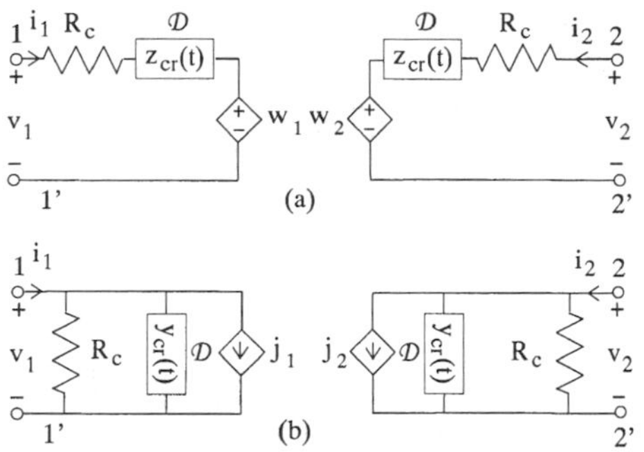

Figure 4.8. Time domain line-equivalent circuit of (a) Thevenin type and (b) Norton type for a uniform lossy two-conductor line.

$i = 1, 2, \ldots,$ is known, both w_1 and w_2 are known for $iT \leqslant t \leqslant (i+1)T$. This fact allows the controlled sources w_1 and w_2 to be treated as "independent" sources if the problem is resolved by means of iterative procedures.

Unlike the lossless lines, due to the wake caused by the dispersion, $w_1(t)$ depends on the whole history of functions w_2 and v_2 in the interval $(0, t - T)$, and $w_2(t)$ depends on the whole history of functions w_1 and v_1 in the interval $(0, t - T)$.

Equations (4.106) and (4.107), describing the terminal behavior of the line, and Eqs. (4.110) and (4.111), governing the controlled voltage sources, still hold for lines with frequency-dependent parameters (see Chapter 5).

By substituting Eq. (4.57) into Eqs. (4.106) and (4.107), we obtain the linear convolution equations

$$v_1(t) - R_c i_1(t) - \{z_{cr} * i_1\}(t) = w_1(t), \tag{4.112}$$

$$v_2(t) - R_c i_2(t) - \{z_{cr} * i_2\}(t) = w_2(t). \tag{4.113}$$

These relations suggest the time domain equivalent circuit of Thévenin type shown in Fig. 4.8a. The dynamic one-port $\mathcal{D}$ is characterized by the current-based impulse response $z_{cr}(t)$. It takes into account the wake due to the dispersion. When the Heaviside condition is satisfied we have $z_{cr}(t) = 0$, and the Thévenin equivalent circuit of Fig. 4.8a reduces to that of Fig. 2.11a relevant to lossless lines.

By substituting the expression of p given by Eq. (4.94) in Eqs. (4.110) and (4.111), for the governing laws of the state we obtain the linear difference-convolution equations

$$w_1(t) = [2v_2(t-T) - w_2(t-T)]e^{-\mu T}u(t-T)$$
$$+ u(t-T)\int_0^{t-T} p_r(t-T-\tau)[2v_2(\tau) - w_2(\tau)]d\tau, \quad (4.114)$$
$$w_2(t) = [2v_1(t-T) - w_1(t-T)]e^{-\mu T}u(t-T)$$
$$+ u(t-T)\int_0^{t-T} p_r(t-T-\tau)[2v_1(\tau) - w_1(\tau)]d\tau. \quad (4.115)$$

When the Heaviside condition is satisfied we have $p_r(t) = 0$, and the state equations reduce to the linear difference equations

$$w_1(t) = [2v_2(t-T) - w_2(t-T)]e^{-\mu T}u(t-T), \quad (4.116)$$
$$w_2(t) = [2v_1(t-T) - w_1(t-T)]e^{-\mu T}u(t-T). \quad (4.117)$$

These equations, except for the damping factor $e^{-\mu T}$, coincide with the state equations obtained for the lossless lines (see §2.6).

4.6.3. Time Domain Norton Description

Figure 4.8b shows the time domain equivalent two-port of Norton type. In this circuit the dynamic one-port $\mathscr{D}$ is characterized by the voltage-based impulse response $y_{cr}(t)$. It is the bounded part of the inverse Laplace transform of the characteristic line-admittance operator

$$Y_c(s) = \sqrt{\frac{G+sC}{R+sL}} = \frac{1}{R_c}\sqrt{\frac{1+(\mu-\nu)/s}{1+(\mu+\nu)/s}}. \quad (4.118)$$

The admittance Y_c can be rewritten as follows:

$$Y_c(s) = \frac{1}{R_c} + Y_{cr}(s), \quad (4.119)$$

where the function Y_{cr} is given by

$$Y_{cr}(s) = \frac{1}{R_c}\left(\sqrt{\frac{1+(\mu-\nu)/s}{1+(\mu+\nu)/s}} - 1\right) \quad (4.120)$$

and has the property

$$Y_{cr}(s) \approx -\frac{1}{R_c}\frac{\nu}{s} + O(1/s^2) \quad \text{for } s \to \infty. \quad (4.121)$$

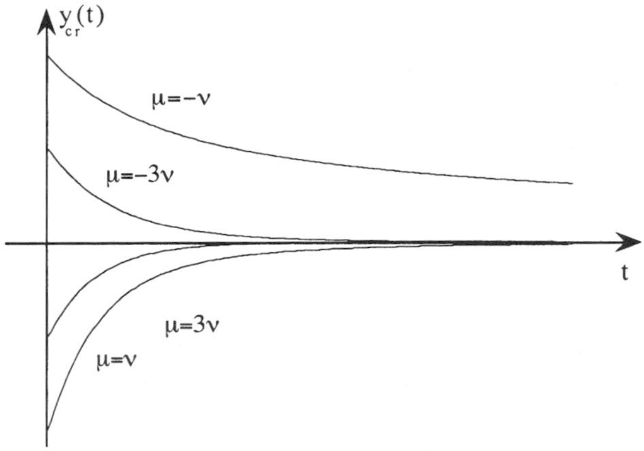

Figure 4.9. Qualitative behavior of the function $y_{cr}(t)$ for different values of μ/ν.

The inverse Laplace transform of $Y_c(s)$ is given by

$$y_c(t) = \frac{1}{R_c}\delta(t) + y_{cr}(t), \tag{4.122}$$

where $y_{cr}(t)$ is the inverse Laplace transform of Y_{cr} and it is given by (see Appendix B)

$$y_{cr}(t) = \frac{\nu}{R_c} e^{-\mu t}[I_1(\nu t) - I_0(\nu t)]u(t). \tag{4.123}$$

As I_0 is an even function and I_1 is an odd function (on the real axis), the term y_{cr} depends on the sign of the parameter ν. Because of the asymptotic behavior (4.121), the inverse Laplace transform of $Y_{cr}(s)$ is a bounded piecewise-continuous function. The reader may study the qualitative behavior of the function $y_{cr}(t)$ as we have done in §4.4.1 for the function $z_{cr}(t)$. Figure 4.9 shows the qualitative behavior of the function $y_{cr}(t)$ for different values of μ/ν. Note that the sign of $y_{cr}(t)$ is the opposite of that for ν, unlike what happens for $z_{cr}(t)$ (see Fig. 4.3).

The governing laws of the controlled current sources are

$$j_1(t) = [-2i_2(t-T) + j_2(t-T)]e^{-\mu T}u(t-T) + \{p_r * (-2i_2 + j_2)\}(t-T), \tag{4.124}$$

$$j_2(t) = [-2i_1(t-T) + j_1(t-T)]e^{-\mu T}u(t-T) + \{p_r * (-2i_1 + j_1)\}(t-T). \tag{4.125}$$

Remark

An input-state-output description that uses as state variables the current and the voltage distribution along the line also is possible (e.g., Roychowdhury, Newton, and Pederson, 1994). ◇

4.7. INPUT-OUTPUT DESCRIPTION IN EXPLICIT FORM

The general uniform two-conductor transmission line has an input-output Laplace domain description in the implicit form described by Eqs. (4.86) and (4.87). We shall study these relations here.

As we know from the circuit theory, there are six possible explicit representations of two of the four variables V_1, V_2, I_1, and I_2 in terms of the remaining two (e.g., Chua, Desoer, and Kuh, 1987).

4.7.1. The Impedance Matrix

The *current-controlled* representation expresses V_1 and V_2 as functions of I_1 and I_2,

$$V_1(s) = Z_{11}(s)I_1(s) + Z_{12}(s)I_2(s), \tag{4.126}$$

$$V_2(s) = Z_{21}(s)I_1(s) + Z_{22}(s)I_2(s). \tag{4.127}$$

Here, Z_{11} is the driving-point impedance at the port $x = 0$ when $I_2 = 0$, that is, when the port at $x = d$ is kept open-circuited; Z_{22} is the driving-point impedance at the port $x = d$ when, $I_1 = 0$, that is, when the port at $x = 0$ is kept open-circuited. They are equal because of the left-right symmetry of uniform lines, and are given by

$$Z_{11} = Z_{22} = Z_d = Z_c \frac{1 + P^2}{1 - P^2}; \tag{4.128}$$

Z_{12} is the *transfer impedance* when $I_1 = 0$, that is, when the port at $x = 0$ is kept open-circuited; Z_{21} is the *transfer impedance* when $I_2 = 0$, that is, when the port at $x = d$ is kept open-circuited. They are equal because of the reciprocity property shown in Chapter 1 (see §1.9), and are given by

$$Z_{12} = Z_{21} = Z_t = Z_c \frac{2P}{1 - P^2}. \tag{4.129}$$

The matrix

$$\mathrm{Z} = \begin{bmatrix} Z_d & Z_t \\ Z_t & Z_d \end{bmatrix} \tag{4.130}$$

is the *impedance matrix* of the two-conductor transmission line.

Remarks

(i) It is clear that to analyze sinusoidal or periodic steady states we can use the impedance matrix evaluated along the imaginary axis $Z(i\omega)$ to characterize the terminal behavior of the line.

(ii) Note that, unlike the describing functions $Z_c(s)$ and $P(s)$, characterizing the input-state-output description, the impedances $Z_d(s)$ and $Z_t(s)$ have an infinite number of poles that are the solutions of the equation $1 - P^2(s) = 0$, $s_h^{\pm} = -\mu \pm \sqrt{\nu^2 - (h\pi/T)^2}$ for $h = 0, 1, 2, \ldots$. This is due to the multiple reflections arising at the line ends when the line is characterized through the impedance matrix. It is evident that for $h = 0$ we have the pole with the largest real part, which is real, that is, either $s_0^- = s_L = -R/L$ or $s_0^+ = s_G = -G/C$. Let us indicate with σ_c the pole with the largest real part. For $Re\{s\} > \sigma_c$ the impedance matrix $Z(s)$ is an analytic function. ◇

By using the convolution theorem, in the time domain Eqs. (4.126) and (4.127) become

$$v_1(t) = \{z_d * i_1\}(t) + \{z_t * i_2\}(t), \tag{4.131}$$

$$v_2(t) = \{z_t * i_1\}(t) + \{z_d * i_2\}(t), \tag{4.132}$$

where $z_d(t)$ and $z_t(t)$ are the inverse Laplace transform of $Z_d(s)$ and $Z_t(s)$, respectively.

Equations (4.128) and (4.129) are easily inversely transformed if we represent $1/(1 - P^2)$ by the geometric series in the region $Re\{s\} > \sigma_c$ (where $|P^2| < 1$),

$$\frac{1}{1 - P^2} = \sum_{i=0}^{\infty} P^{2i}. \tag{4.133}$$

Thus the functions (4.128) and (4.129) can be expressed as

$$Z_d(s) = Z_c(s)[1 + P^2(s)] \sum_{i=0}^{\infty} P^{2i}(s), \tag{4.134}$$

$$Z_t(s) = 2Z_c(s) \sum_{i=0}^{\infty} P^{2i+1}(s). \tag{4.135}$$

The generic term of the infinite sums (4.134) and (4.135) is of the type

$$F_n(s) = Z_c(s)P^n(s) = R_c(s + \mu + \nu)\frac{\exp[-nT\sqrt{(s + \mu)^2 - \nu^2}]}{\sqrt{(s + \mu)^2 - \nu^2}}, \tag{4.136}$$

and its Laplace inverse transform is given by (see Appendix B)

$$f_n(t) = L^{-1}\{F_n(s)\}$$
$$= R_c e^{-nT\mu}\delta(t - nT)$$
$$+ \nu R_c e^{-\mu t}\left\{I_0[\nu\sqrt{t^2-(nT)^2}] + t\frac{I_1[\nu\sqrt{t^2-(nT)^2}]}{\sqrt{t^2-(nT)^2}}\right\}u(t-nT). \tag{4.137}$$

Therefore, for $z_d(t)$ and $z_t(t)$ we obtain

$$z_d(t) = f_0(t) + 2\sum_{i=1}^{\infty} f_{2i}(t), \tag{4.138}$$

$$z_t(t) = 2\sum_{i=0}^{\infty} f_{2i+1}(t). \tag{4.139}$$

As $f_n(t) = 0$ for $t < nT$, for any fixed t only a finite number of terms contribute to the sums that appear in Eqs. (4.138) and (4.139). By placing

$$N(t) = \text{int}(t/2T), \tag{4.140}$$

Eqs. (4.138) and (4.139) become

$$z_d(t) = f_0(t) + 2\sum_{i=1}^{N(t)} f_{2i}(t), \tag{4.141}$$

$$z_t(t) = 2\sum_{i=0}^{N(t+T)} f_{2i+1}(t). \tag{4.142}$$

Remarks

(i) The generic term f_n contains a Dirac pulse of amplitude $\exp(-n\mu T)$ acting at the time instant $t = nT$, and a bounded term acting only for $t > nT$. The bounded term vanishes for $\nu = 0$.

(ii) The number of nonzero terms and, in particular, of the Dirac pulses in the impulse responses $z_d(t)$, and $z_t(t)$, increases as the time increases in accordance with $N(t)$. This is due to the successive reflections that one should to take into account when describing the line through the impedance matrix. For instance, $z_d(t)$ is the voltage of the left end when a current unit Dirac pulse is applied to the left end and the right end is open-circuited ($i_2 = 0$). As for $t > 0$ both the ends behave

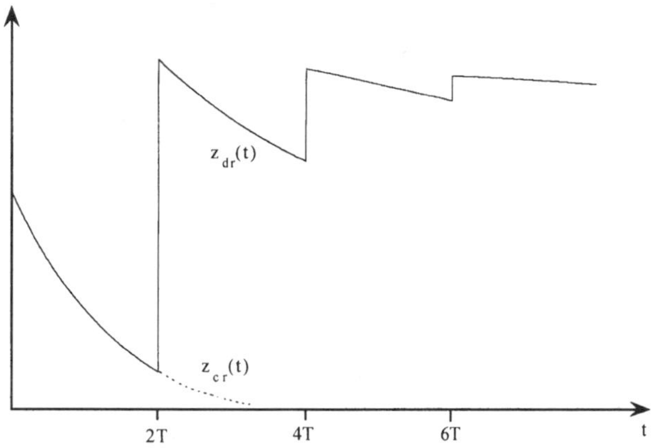

Figure 4.10. Qualitative behavior of the nonimpulsive part $z_{dr}(t)$ of the impulse response $z_d(t)$, compared with $z_{cr}(t)$.

as if they were open-circuited, the line is mismatched at both ends, and an infinite series of reflections takes place. This is the major drawback of the input-output description. Figure 4.10 shows the qualitative behavior of the nonimpulsive part $z_{dr}(t)$ of the impulse response $z_d(t)$ together with that of $z_{cr}(t)$ defined as in (4.58). Note that we have $z_{dr}(t) = z_{cr}(t)$ for $0 \leqslant t < 2T$.

(iii) Relations (4.131) and (4.132) express the voltages $v_1(t)$ and $v_2(t)$ as functions of the history of the "input variables" i_1 and i_2 on the whole time interval $(0, t)$. Numerical simulators of transmission lines based on a description of this type are time-consuming if evolution times much longer than T are needed because of the convolution products. This problem is absent in the input-state-output description based on the forward and backward voltage waves, because the impulse responses $z_c(t)$ and $p(t)$ have a duration much shorter than that of the impulse responses $z_d(t)$ and $z_t(t)$. Kuznetsov and Schutt-Ainé state that: "*Closed-loop characterizations* (namely, the input-output descriptions) (....) *include reflections from the terminations and lead to complicated oscillating transfer functions and transient characteristics.... The open-loop characterization* (namely, the input-state-output description) (....) *separates forward and backward waves and results in the simplest transfer functions and transient characteristics....*" (Kuznetsov and Schutt-Ainé, 1996). It is clear that the input-state-output description is more suitable for transient analysis. To deal with sinusoidal and periodic steady-state operating conditions in the frequency domain, the input-output description is preferable. In this case, the terminal behavior of the line may be directly characterized through the impedance matrix evaluated along the imaginary axis $Z(i\omega)$. ◇

4.7.2. The Admittance Matrix

The *voltage-controlled* representation expresses I_1 and I_2 as functions of V_1 and V_2,

$$I_1(s) = Y_{11}(s)V_1(s) + Y_{12}(s)V_2(s), \tag{4.143}$$

$$I_2(s) = Y_{21}(s)V_1(s) + Y_{22}(s)V_2(s). \tag{4.144}$$

The admittances Y_{11} (*driving-point admittance* at the end $x = 0$ when the other end is short-circuited) and Y_{22} (*driving-point admittance* at the end $x = d$ when the other end is short-circuited) are equal and are given by

$$Y_{11} = Y_{22} = Y_d = Y_c \frac{1 + P^2}{1 - P^2} = Y_c(1 + P^2) \sum_{i=0}^{\infty} P^{2i}, \tag{4.145}$$

where Y_c is the characteristic admittance matrix. The admittances Y_{12} (*transfer admittance* at the end $x = d$ when the other end is short-circuited) and Y_{21} (*transfer admittance* at the end $x = 0$ when the other end is short-circuited) are equal and are given by

$$Y_{12} = Y_{21} = Y_t = -Y_c \frac{2P}{1 - P^2} = -2Y_c P \sum_{i=0}^{\infty} P^{2i}; \tag{4.146}$$

Y_{11} is equal to Y_{22} because the line is uniform; Y_{12} is equal to Y_{21} because of the reciprocity property shown in Chapter 1 (§1.9). The matrix

$$\mathrm{Y} = \begin{bmatrix} Y_d & Y_t \\ Y_t & Y_d \end{bmatrix} \tag{4.147}$$

is the *admittance matrix* of the two-conductor transmission line.

At this point it is easy to obtain the corresponding time domain equations for this representation, thus we shall leave this as an exercise. However, all the considerations we have made for the impedance matrix still hold.

4.7.3. The Hybrid Matrices

The two hybrid representations express V_1 and I_2 as functions of I_1 and V_2 and the converse. The hybrid 1 representation is given by

$$V_1(s) = H_{11}(s)I_1(s) + H_{12}(s)V_2(s), \tag{4.148}$$

$$I_2(s) = H_{21}(s)I_1(s) + H_{22}(s)V_2(s). \tag{4.149}$$

Here, H_{11} is the *driving-point impedance* at the end $x = 0$ with the end $x = d$ short-circuited,

$$H_{11} = \frac{1}{Y_d} = Z_c\frac{1 - P^2}{1 + P^2} = Z_c(1 - P^2)\sum_{i=0}^{\infty}(-P^2)^i; \tag{4.150}$$

H_{22} is the *driving-point admittance* at the end $x = d$ with the end $x = 0$ open-circuited,

$$H_{22} = \frac{1}{Z_d} = Y_c\frac{1 - P^2}{1 + P^2} = Y_c(1 - P^2)\sum_{i=0}^{\infty}(-P^2)^i; \tag{4.151}$$

H_{21} is the forward current transfer function and H_{12} is the reverse voltage transfer function given by

$$H_{12} = -H_{21} = \frac{Z_t}{Z_d} = \frac{2P}{1 + P^2} = 2P\sum_{i=0}^{\infty}(-P^2)^i; \tag{4.152}$$

H_{12} is equal to $-H_{21}$ because of the reciprocity property shown in Chapter 1. The matrix

$$\mathrm{H} = \begin{bmatrix} \dfrac{1}{Y_d} & \dfrac{Z_t}{Z_d} \\ -\dfrac{Z_t}{Z_d} & \dfrac{1}{Z_d} \end{bmatrix} \tag{4.153}$$

is called the *hybrid 1 matrix* of the two-conductor transmission line.

A dual treatment can be given to the *hybrid 2 matrix* expressing I_1 and V_2 as functions of V_1 and I_2. Due to the spatial uniformity, the two ends are symmetric and thus we have

$$V_2(s) = H_{11}(s)I_2(s) + H_{12}(s)V_1(s), \tag{4.154}$$

$$I_1(s) = H_{21}(s)I_2(s) + H_{22}(s)V_1(s). \tag{4.155}$$

At this point it is easy to obtain the corresponding time domain equations for these representations. We shall leave this as an exercise, with all the considerations we have made for the impedance matrix still holding.

4.7.4. The Transmission Matrices

The *forward transmission* matrix representation expresses I_2 and V_2 as functions of I_1 and V_1,

$$V_2(s) = T_{11}(s)V_1(s) + T_{12}(s)I_1(s), \tag{4.156}$$

$$-I_2(s) = T_{21}(s)V_1(s) + T_{22}(s)I_1(s), \tag{4.157}$$

where

$$T_{11} = T_{22} = \frac{1 + P^2}{2P}, \tag{4.158}$$

$$T_{12} = Y_t^{-1} = -Z_c \frac{1 - P^2}{2P}, \tag{4.159}$$

$$T_{21} = -Z_t^{-1} = -Y_c \frac{1 - P^2}{2P}. \tag{4.160}$$

The matrix

$$\mathrm{T} = \begin{bmatrix} T_{11} & T_{12} \\ T_{21} & T_{22} \end{bmatrix} \tag{4.161}$$

is called the *forward transmission matrix* of the two-conductor transmission line.

A dual treatment can be given to the *backward transmission matrix* expressing I_1 and V_1 as functions of I_2 and V_2,

$$V_1(s) = T'_{11}(s)V_2(s) + T'_{12}(s)[-I_2(s)], \tag{4.162}$$

$$I_1(s) = T'_{21}(s)V_2(s) + T'_{22}(s)[-I_2(s)], \tag{4.163}$$

where $T'_{11} = T'_{22} = T_{11}$, $T'_{12} = -T_{12}$ and $T'_{21} = -T_{21}$.

Remark

The time domain representations based on the transmission matrices require particular attention as we shall see now. Let us refer, for the sake of simplicity, to the ideal line case, for which:

$$T_{11} = T_{22} = \tfrac{1}{2}(e^{sT} + e^{-sT}), \tag{4.164}$$

$$T_{12} = -\frac{R_c}{2}(e^{sT} - e^{-sT}), \tag{4.165}$$

$$T_{21} = -\frac{1}{2R_c}(e^{sT} - e^{-sT}). \tag{4.166}$$

By performing the inverse Laplace transform of Eqs. (4.164) to (4.166) and by applying the convolution theorem to Eqs. (4.156) and (4.157) we obtain

$$v_2(t) = \tfrac{1}{2}[v_1(t+T) - R_c i_1(t+T)]u(t+T) + \tfrac{1}{2}[v_1(t-T) + R_c i_1(t-T)]u(t-T). \quad (4.167)$$

$$-R_c i_2(t) = \tfrac{1}{2}[R_c i_1(t+T) - v_1(t+T)]u(t+T) + \tfrac{1}{2}[R_c i_1(t-T) + v_1(t-T)]u(t-T). \quad (4.168)$$

It is evident that these relations are useless. Actually, the time domain forward transmission matrix representation is not an explicit form of the input-output descriptions, as it links the values of v_2 and i_2 at the present time instant t to the values of v_1 and i_1 at the future time instant $t + T$. Similar considerations hold for the other transmission matrix. This is the main drawback of the characterization based on the transmission matrices.

However, there is another way to read these relations. Both Eqs. (4.167) and (4.168) express the value of $(v_1 - R_c i_1)$ at the future time instant $t + T$ as function of the value of $(v_1 + R_c i_1)$ at the past time instant $t - T$ and of the value of v_2 and i_2 at the present time instant t. In other words, we are assuming the value of $(v_1 - R_c i_1)$ at the time instant $t + T$ as the unknown. By using Eq. (4.162) (or Eq. (4.163)) we obtain the analog relations expressing $(v_2 + R_c i_2)$ at the time instant $t + T$ in terms of $(v_2 - R_c i_2)$ at the time instant $t - T$ and of v_1 and i_1 at the present time instant t. In this way we again obtain the input-state-output representation. Note that, by summing Eqs. (4.167) and (4.168) termwise, we eliminate the terms depending on the "future" and obtain the time domain equation corresponding to Eq. (4.87). To obtain the time domain equation corresponding to Eq. (4.86) we have to use the other transmission matrix. ◇

The transmission matrix characterization is particularly useful to analyze in the Laplace domain cascade connections of uniform transmission lines with different per-unit-length parameters. Let us consider a nonuniform transmission line consisting of a cascade on n uniform lines, and let us designate the transmission matrix of the ith uniform tract by $\mathrm{T}^{(i)}$. The resulting two-port is characterized by a *forward* transmission matrix T that is equal to the product $\mathrm{T}^{(n)}\mathrm{T}^{(n-1)}\ldots\mathrm{T}^{(2)}\mathrm{T}^{(1)}$.

Remark

All the input-out representations described in this paragraph may be easily obtained by using the standing wave representation for the general solution of the line equations. We leave this as an exercise. ◇

4.8. A LOSSY TRANSMISSION LINE CONNECTING TWO LINEAR RESISTIVE ONE-PORTS

Here, we apply the state variable description introduced in §4.6 to study the network shown in Fig. 4.11. The transmission line has losses, is initially at rest, and connects two linear resistive circuits. In Chapter 2 we have studied a similar circuit for a lossless line.

The unknowns of this circuit are the electrical variables at the transmission line ends and the voltages w_1 and w_2. It is straightforward to determine the time domain equations for the electrical variables at the line ends if the line is represented through its equivalent circuit of Thévenin type (see Fig. 4.8a). The equivalent network obtained in this way is shown in Fig. 4.12.

The characteristic equations of the two-terminal one-ports are

$$v_1(t) = -R_1 i_1(t) + e_1(t), \tag{4.169}$$

$$v_2(t) = -R_2 i_2(t). \tag{4.170}$$

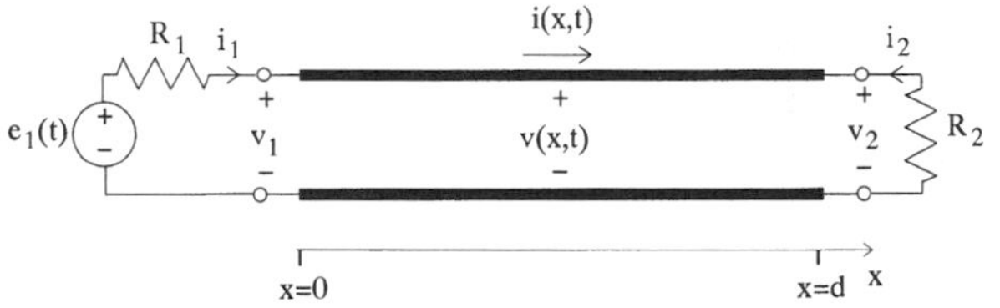

Figure 4.11. Lossy line connecting two linear resistive one-ports.

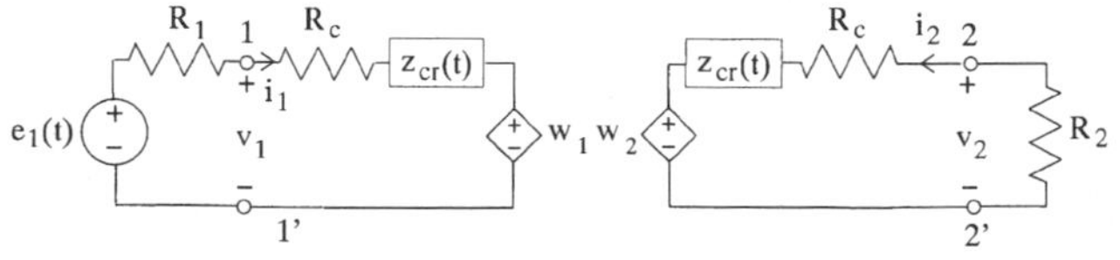

Figure 4.12. Equivalent circuit of the network shown in Fig. 4.11.

By using Eqs. (4.112) and (4.113) we obtain the linear integral equations

$$v_1(t) + \frac{1}{R_1 + R_c}\int_0^\tau z_{cr}(t-\tau)v_1(\tau)d\tau = \frac{R_1}{R_1 + R_c}w_1(t) + g(t), \quad (4.171)$$

$$v_2(t) + \frac{1}{R_2 + R_c}\int_0^\tau z_{cr}(t-\tau)v_2(\tau)d\tau = \frac{R_2}{R_2 + R_c}w_2(t), \quad (4.172)$$

where $g(t)$ is a known function given by

$$g(t) = \frac{1}{R_1 + R_c}\int_0^t z_{cr}(t-\tau)e_1(\tau)dt + \frac{R_c}{R_1 + R_c}e_1(t) \quad (4.173)$$

Equations (4.171) and (4.172) are *linear Volterra integral equations of the second kind* (e.g., Linz, 1985; Brunner and van der Houwen, 1986); they are integral equations of convolution type because they involve the difference kernel $z_{cr}(t-\tau)$.

Equations (4.171) and (4.172) together with the control laws equations (4.114) and (4.115) of the voltage sources w_1 and w_2 determine the solution of the problem we are examining. It is evident that the dispersion introduced by the losses, and described by the kernels $z_{cr}(t-\tau)$ and $p_r(t-T-\tau)$, makes the solution of the problem more difficult: A system of difference-Volterra integral equations has to be solved.

The solution of the examined circuit could be evaluated analytically as was done in Chapter 2 for the ideal line case (see §2.7): First, the circuit is solved in the Laplace domain and then the inverse Laplace transform is performed to obtain the time domain solution. In this case the inverse Laplace transform may be performed analytically by using the properties and formulas described in Appendix B. We shall leave this as an exercise. If the lumped circuits also contain dynamic elements, an analytical inverse transform is not possible. In these cases numerical inversion algorithms have to be used. If the terminal circuits contain time-varying and/or nonlinear lumped elements, the problem has to be necessarily solved in the time domain. The solution of such a problem can only be obtained numerically. Henceforth, a simple, accurate and reliable numerical procedure is presented through which we can tackle the most general case. We introduce this method by referring to the particular circuit considered.

Equations (4.114) and (4.115) govern the dynamics of w_1 and w_2. As we have already stressed, in the generic time interval $[nT, (n+1)T]$, with $n = 1, 2, \ldots$, the voltages $w_1(t)$ and $w_2(t)$ depend, respectively, only on the time history of v_2 and w_2 and of v_1 and w_1 on the interval $(0, nT)$. Consequently, if the solution for $t \leqslant nT$ is known, both $w_1(t)$ and $w_2(t)$ can be determined for $nT \leqslant t \leqslant (n+1)T$ by using Eqs. (4.114) and (4.115); for $0 \leqslant t \leqslant T$, $w_1(t)$ and $w_2(t)$ depend solely on the initial conditions, and in the case under examination they are zero. This property allows us to solve the problem iteratively: At the nth step (for $nT \leqslant t \leqslant (n+1)T$) we have to solve the following system of integral equations:

$$v_1(t) + \frac{1}{R_1 + R_c} \int_{nT}^{t} z_{cr}(t-\tau) v_1(\tau) d\tau = g_1^{(n)}(t) \tag{4.174}$$

$$v_2(t) + \frac{1}{R_2 + R_c} \int_{nT}^{t} z_{cr}(t-\tau) v_2(\tau) d\tau = g_2^{(n)}(t) \tag{4.175}$$

where

$$g_1^{(n)}(t) = -\frac{1}{R_1 + R_c} \int_0^{nT} z_{cr}(t-\tau) v_1(\tau) d\tau + \frac{R_1}{R_1 + R_c} w_1(t) + g(t), \tag{4.176}$$

$$g_2^{(n)}(t) = -\frac{1}{R_2 + R_c} \int_0^{nT} z_{cr}(t-\tau) v_2(\tau) d\tau + w_2(t). \tag{4.177}$$

The values of the functions $g_1^{(n)}$ and $g_2^{(n)}$ for $t \leqslant (n+1)T$ depend solely on the values of v_2 and w_2 and of v_1 and w_1 on the interval $(0, nT)$. Therefore, they are known if the solution is known for $t \leqslant nT$.

We operate as follows. First, v_1 and v_2 are determined for $0 \leqslant t \leqslant T$ by solving the equation system (4.174) and (4.175) with $n = 0$ and $g_1^{(0)} = g(t)$, $g_2^{(0)} = 0$ (for $0 \leqslant t \leqslant T$, $w_1 = 0$ and $w_2 = 0$). Once v_1 and v_2 are known on $(0, T)$, then w_1 and w_2 are evaluated for $T \leqslant t \leqslant 2T$ by using relations (4.114) and (4.115), and $g_1^{(1)}$ and $g_2^{(1)}$ are evaluated from Eqs. (4.176) and (4.177). At this point, by solving again the equation systems (4.174) and (4.175) with $n = 1$ we determine v_1 and v_2 for $T \leqslant t \leqslant 2T$, and so on. The continuous differentiability of v_1 and v_2 ensures that of $g_1^{(n)}$ and $g_2^{(n)}$ if the initial conditions are compatible with the boundary conditions. In this way, the solution of the problem is reduced to the solution of a sequence of two linear Volterra integral equations of the second kind with the kernel $z_{cr}(t-\tau)$.

This is the main idea on which the numerical method for the analysis of a generic network composed of transmission lines and lumped circuits discussed in Chapter 9 is based.

For lossless lines or for lossy Heaviside lines we have $z_{cr} = 0$, and the integral equations (4.174) and (4.175) reduce to the linear algebraic equations (see §2.7)

$$v_1(t) = \frac{R_1}{R_1 + R_c} w_1(t) + g(t), \tag{4.178}$$

$$v_2(t) = \frac{R_2}{R_2 + R_c} w_2(t). \tag{4.179}$$

In this case the state dynamics are described by the linear difference equations (4.116) and (4.117).

4.8.1. Numerical Solution of Volterra Integral Equations of the Second Kind

The kernel $z_{cr}(t - \tau)$ of the Volterra integral equations (4.174) and (4.175) is absolutely integrable. Therefore, these equations can be solved by using well-known approximation methods (e.g., Linz, 1985; Brunner and van der Houwen, 1986). An approximate solution of Eqs. (4.174) and (4.175) can be computed by the following iterative process. For the sake of simplicity, let us refer to the case $n = 0$.

Let us divide the interval $(0, T)$ into M parts of length $\Delta t = T/M$, and indicate with

$$V_1^{(h)} = v_1(t_h), \quad V_2^{(h)} = v_2(t_h) \tag{4.180}$$

the samples of v_1 and v_2 at the time instants $t_h = h\Delta t$, with $h = 0, 1, \ldots, M$. Suppose that we know v_1 and v_2 at time instants t_k for $k = 0, 1, \ldots, m-1$ with $m \leqslant M$. An approximation of $V_1^{(m)}$ and $V_2^{(m)}$ may then be computed by replacing the integrals on the left-hand side of Eqs. (4.174) and (4.175) by means of a numerical integration rule, using the samples of the integrands at $t_0, t_1, \ldots, t_{m-1}$, and solving the resulting linear algebraic equations for $V_1^{(m)}$ and $V_2^{(m)}$. By evaluating the expressions (4.174) and (4.175) at the time instant $t = 0$, we obtain

$$v_1(t = 0) = \frac{R_c}{R_1 + R_c} e_1(t = 0), \tag{4.181}$$

$$v_2(t = 0) = 0. \tag{4.182}$$

Zero initial conditions are compatible with the boundary conditions at $t = 0$ if $e_1(t = 0) = 0$. Let us assume, therefore, that $V_1^{(0)} = 0$ and $V_2^{(0)} = 0$.

Here, we shall use the *trapezoidal rule* to approximate the integral operators. The results that we shall show later also hold for multistep integration rules (e.g., Linz, 1985). By using the trapezoidal rule, we obtain[3], for $k = 1, 2$:

$$\int_0^{t_m} z_{cr}(t_m - \tau)v_k(\tau)d\tau \cong \frac{\Delta t}{2} z_{cr}^{(0)} V_k^{(m)} + \Delta t \sum_{p=1}^{m-1} z_{cr}^{(m-p)} V_k^{(p)} + \frac{\Delta t}{2} z_{cr}^{(m)} V_k^{(0)} \tag{4.183}$$

where

$$z_{cr}^{(0)} = z_{cr}(0^+), \quad z_{cr}^{(h)} = z_{cr}(t_h) \quad \text{for } h \geqslant 1. \tag{4.184}$$

By using Eq. (4.183) to replace the integrals in the set of equations (4.174) and (4.175), we obtain the "approximate" equations for $V_1^{(m)}$ and $V_2^{(m)}$

$$\left(1 + \frac{\Delta t}{2} \frac{z_{cr}^{(0)}}{R_1 + R_c}\right) V_1^{(m)} = \Phi_1^{(m)} \quad \text{for } m = 1, 2, \ldots, \tag{4.185}$$

$$\left(1 + \frac{\Delta t}{2} \frac{z_{cr}^{(0)}}{R_2 + R_c}\right) V_2^{(m)} = \Phi_2^{(m)} \quad \text{for } m = 1, 2, \ldots, \tag{4.186}$$

where

$$\Phi_1^{(m)} = g_1^{(0)}(t_m) - \frac{\Delta t}{R_1 + R_c} \left[\sum_{p=1}^{m-1} z_{cr}^{(m-p)} V_1^{(p)} + \frac{z_{cr}^{(m)}}{2} V_1^{(0)} \right], \tag{4.187}$$

$$\Phi_2^{(m)} = g_2^{(0)}(t_m) - \frac{\Delta t}{R_2 + R_c} \left[\sum_{p=1}^{m-1} z_{cr}^{(m-p)} V_2^{(p)} + \frac{z_{cr}^{(m)}}{2} V_2^{(0)} \right], \tag{4.188}$$

Note that $\Phi_1^{(m)}$ and $\Phi_2^{(m)}$ depend only on the time sequence $V_1^{(0)}$, $V_2^{(0)}$, $V_1^{(1)}$, $V_2^{(1)}, \ldots, V_1^{(m-1)}$, $V_2^{(m-1)}$ and on the voltage source e_1. The convergence properties of this numerical algorithm will be discussed in Chapter 9.

The approximation (4.183) also is used to evaluate $g_1^{(0)}(t_k)$ and $g_2^{(0)}(t_k)$; for $0 \leqslant t \leqslant T$ the functions w_1 and w_2 are equal to zero. Furthermore, by means of the approximation (4.183) we can also calculate the convolution integrals in Eqs. (4.114) and (4.115) because the kernel $p_r(t - T - \tau)$ is absolutely integrable.

4.9. THE MATCHING PROBLEM FOR LOSSY LINES

The matching concept is a major point of interest in the lumped and distributed circuit design. In §4.5 it has been shown that a lossy line is perfectly matched at one end if a one-port with impedance Z_c is

[3]Henceforth, we shall use the symbol $\Sigma_{i=m}^{n} f_i$, assuming that $\Sigma_{i=m}^{n} f_i = 0$ when $n < m$.

connected to that end. The characteristic impedance Z_c can be written as a sum of two terms (see Eq. (4.54)), that is, a constant term R_c, and a remainder Z_{cr}, depending on s and decaying as $1/s$ for $s \to \infty$. The latter term shows two branch points, one of regular type and the other of polar type. It is clear that the synthesis (even approximated) of Z_{cr} presents major difficulties, especially in the case $\mu = |\nu|$.

What happens when the line is connected to a resistor with resistance R_c? To analyze this case, let us refer to the circuit of Fig. 4.11. By using the Laplace method we obtain the following equations for W_1 and W_2:

$$W_1 - \Gamma_2 P W_2 = 0, \tag{4.189}$$

$$-\Gamma_1 P W_1 + W_2 = P \frac{2Z_c}{R_1 + Z_c} E_1, \tag{4.190}$$

where

$$\Gamma_1(s) = \frac{R_1 - Z_c(s)}{R_1 + Z_c(s)}, \quad \Gamma_2(s) = \frac{R_2 - Z_c(s)}{R_2 + Z_c(s)}. \tag{4.191}$$

The reflection coefficients Γ_1 and Γ_2 may be rewritten as follows:

$$\Gamma_1(s) = \Gamma_{10} + \Gamma_{1r}(s), \quad \Gamma_2(s) = \Gamma_{20} + \Gamma_{2r}(s), \tag{4.192}$$

where

$$\Gamma_{10} = \frac{R_1 - R_c}{R_1 + R_c}, \quad \Gamma_{20} = \frac{R_2 - R_c}{R_2 + R_c}, \tag{4.193}$$

and the remainders Γ_{1r} and Γ_{2r} have the following asymptotic behavior:

$$\Gamma_{1r}(s) \approx O(1/s) \quad \text{and} \quad \Gamma_{2}(s) \approx O(1/s) \quad \text{for } s \to \infty. \tag{4.194}$$

Let us consider the case in which $e_1(t)$ is a Dirac pulse with a unitary amplitude, and hence $E_1 = 1$. We obtain

$$W_1(s) = \Gamma_2 \frac{P^2}{1 - \Gamma_1 \Gamma_2 P^2} \frac{2Z_c}{Z_c + R_1}, \tag{4.195}$$

$$W_2(s) = \frac{P}{1 - \Gamma_1 \Gamma_2 P^2} \frac{2Z_c}{Z_c + R_1}. \tag{4.196}$$

It is evident that

$$P^{-2} W_1(s) \approx \Gamma_{20} \frac{2R_c}{R_c + R_1} + O(1/s) \quad \text{for } s \to \infty. \tag{4.197}$$

Therefore, when $\Gamma_{20} \neq 0$ the backward voltage wave contains a Dirac pulse centered at $t = 2T$ due to the reflections introduced by the resistor connected at the end $x = d$, whereas when $\Gamma_{20} = 0$ the Dirac pulse no longer exists. In conclusion we can say that by choosing $R_2 = R_c$, the transmission line is "matched" at the end $x = d$ with respect to the Dirac pulses, thus in this way it is possible to reduce the "intensity" of the reflected signals (see §8.3.3). This is the *quasi-matched* condition used by Djordjevic, Sarkar, and Harrington (1986) to achieve an input-output description with shorter impulse responses.

4.10. LOSSY TRANSMISSION LINES WITH DISTRIBUTED SOURCES

In this section the two-port model obtained in §4.5 is extended to lines with distributed sources and nonzero initial conditions. The time domain equations for a lossy two-conductor line with distributed sources are

$$\frac{\partial v}{\partial x} + L\frac{\partial i}{\partial t} + Ri = e_s, \tag{4.198}$$

$$\frac{\partial i}{\partial x} + C\frac{\partial v}{\partial t} + Gv = j_s, \tag{4.199}$$

where $e_s(x, t)$ and $j_s(x, t)$ are assigned functions defined for $0 \leqslant x \leqslant d$ and $t \geqslant 0$. In the Laplace domain, Eqs. (4.198) and (4.199) become

$$\frac{dV}{dx} = -Z(s)I + E, \tag{4.200}$$

$$\frac{dI}{dx} = -Y(s)V + J, \tag{4.201}$$

where

$$E(x; s) = Li_0(x) + L\{e_s(x, t)\}, \tag{4.202}$$

$$J(x; s) = Cv_0(x) + L\{j_s(x, t)\}. \tag{4.203}$$

To characterize the line at the ends in the Laplace domain we have to determine the general solution of Eqs. (4.200) and (4.201) as in the ideal case. The general solution of these equations may be expressed as

$$V(x, s) = Q(x; s)V^+(s) + Q(d - x; s)V^-(s) + V_p(x; s), \tag{4.204}$$

$$I(x, s) = \frac{1}{Z_c(s)}[Q(x; s)V^+(s) - Q(d - x; s)V^-(s)] + I_p(x; s), \tag{4.205}$$

where $V_p(x; s)$ and $I_p(x; s)$ are a particular solution of the system of equations (4.200) and (4.201).

4.10.1. A Particular Solution of Line Equations with Distributed Sources

The way to determine a particular solution of Eqs. (4.200) and (4.201) is the same as that used in Chapter 2 for ideal lines. A particular solution is one that satisfies the boundary conditions of the line as if it is perfectly matched. This ensures a particular solution that in the time domain has a short duration. To determine this solution we shall use Green's function method.

By applying the property of sampling of the Dirac function, we rewrite Eqs. (4.200) and (4.201) as

$$\frac{dV}{dx} = -Z(s)I + \int_{x=0^-}^{x=d+} \delta(x-\xi)E(\xi;s)d\xi, \tag{4.206}$$

$$\frac{dI}{dx} = -Y(s)V + \int_{x=0^-}^{x=d+} \delta(x-\xi)J(\xi;s)d\xi. \tag{4.207}$$

Let us indicate with $G_v^{(e)}(x;\xi,s)$ and $G_i^{(e)}(x;\xi,s)$ the solution of the "auxiliary problem" defined by the equations

$$\frac{dG_v^{(e)}}{dx} = -Z(s)G_i^{(e)} + \delta(x-\xi), \tag{4.208}$$

$$\frac{dG_i^{(e)}}{dx} = -Y(s)G_v^{(e)}, \tag{4.209}$$

with the boundary conditions

$$G_v^{(e)}(x=0;\xi,s) = -Z_cG_i^{(e)}(x=0;\xi,s), \tag{4.210}$$

$$G_v^{(e)}(x=d;\xi,s) = Z_cG_i^{(e)}(x=d;\xi,s). \tag{4.211}$$

Moreover, let $G_v^{(j)}(x;\xi,s)$ and $G_i^{(j)}(x;\xi,s)$ be the solution of the equations

$$\frac{dG_v^{(j)}}{dx} = -Z(s)G_i^{(j)}, \tag{4.212}$$

$$\frac{dG_i^{(j)}}{dx} = Y(s)G_v^{(j)} + \delta(x-\xi), \tag{4.213}$$

satisfying the same boundary conditions of $G_v^{(e)}(x;\xi,s)$ and $G_i^{(e)}(x;\xi,s)$.

Due to the spatial uniformity of the line parameters it must be that

$$G_v^{(e)}(x;\xi,s) = G_v^{(e)}(x-\xi;s), \quad G_i^{(e)}(x;\xi,s) = G_i^{(e)}(x-\xi;s), \tag{4.214}$$

$$G_v^{(j)}(x;\xi,s) = G_v^{(j)}(x-\xi;s), \quad G_i^{(j)}(x;\xi,s) = G_i^{(j)}(x-\xi;s), \tag{4.215}$$

Using the superposition property, a particular solution of Eqs. (4.206) and (4.207) can be expressed in the form

$$V_p(x;s) = \int_{0^-}^{d^+} [G_v^{(e)}(x-\xi;s)E(\xi,s) + G_v^{(j)}(x-\xi;s)J(\xi;s)]d\xi, \tag{4.216}$$

$$I_p(x;s) = \int_{0^-}^{d^+} [G_i^{(e)}(x-\xi;s)E(\xi,s) + G_i^{(j)}(x-\xi;s)J(\xi;s)]d\xi. \tag{4.217}$$

This particular solution satisfies by construction the perfect matching condition at the line ends.

Let us resolve the first auxiliary problem. By integrating both sides of Eqs. (4.208) and (4.209) on the spatial interval $(\xi - \varepsilon, \xi + \varepsilon)$ we obtain

$$G_v^{(e)}(0^+;s) - G_v^{(e)}(0^-;s) = 1, \tag{4.218}$$

$$G_i^{(e)}(0^+;s) = G_i^{(e)}(0^-;s). \tag{4.219}$$

For $0 \leqslant x < \xi$ and $\xi < x \leqslant d$, Eqs. (4.208) and (4.209) are homogeneous, therefore in these intervals the solution may be represented through expressions such as (4.22) and (4.23),

$$G_v^{(e)} = \begin{cases} A^+(s)e^{-k(s)x} + A^-(s)e^{k(s)(x-\xi)} & \text{for } 0 \leqslant x < \xi, \\ B^+(s)e^{-k(s)(x-\xi)} + B^-(s)e^{k(s)x} & \text{for } \xi < x \leqslant d, \end{cases} \tag{4.220}$$

$$G_i^{(e)} = \frac{1}{Z_c(s)} \begin{cases} A^+(s)e^{-k(s)x} - A^-(s)e^{k(s)(x-\xi)} & \text{for } 0 \leqslant x < \xi, \\ B^+(s)e^{-k(s)(x-\xi)} - B^-(s)e^{k(s)x} & \text{for } \xi < x \leqslant d. \end{cases} \tag{4.221}$$

By imposing the boundary conditions (4.210) and (4.211) we obtain,

$$A^+ = B^- = 0, \tag{4.222}$$

whereas by imposing the initial conditions (4.218) and (4.219), we obtain

$$A^- = -1/2, \quad B^+ = 1/2. \tag{4.223}$$

Thus the functions $G_v^{(e)}$ and $G_i^{(e)}$ are given by

$$G_v^{(e)}(x-\xi;s)=\tfrac{1}{2}\,\mathrm{sgn}(x-\xi)e^{-k(s)|\xi-x|}, \qquad (4.224)$$

$$G_i^{(e)}(x-\xi;s)=\frac{1}{2Z_c(s)}e^{-k(s)|\xi-x|}. \qquad (4.225)$$

Proceeding in the same way, we determine the solution of the other auxiliary problem. We obtain

$$G_v^{(j)}(x-\xi;s)=\tfrac{1}{2}e^{-k(s)|\xi-x|}, \qquad (4.226)$$

$$G_i^{(j)}(x-\xi;s)=\frac{1}{2Z_c(s)}\,\mathrm{sgn}(x-\xi)e^{-k(s)|\xi-x|}. \qquad (4.227)$$

Therefore, Eqs. (4.216) and (4.217) become

$$V_p(x;s)=\frac{1}{2}\int_{0^-}^{d+}\exp[-k(s)|\xi-x|\,][\,\mathrm{sgn}(x-\xi)E(\xi,s)+J(\xi;s)]d\xi, \qquad (4.228)$$

$$I_p(x;s)=\frac{1}{2Z_c}\int_{0^-}^{d+}\exp[-k(s)|\xi-x|][E(\xi,s)+\mathrm{sgn}(x-\xi)J(\xi;s)]d\xi. \qquad (4.229)$$

4.10.2. Two-Port Characterization

Specifying Eqs. (4.204) and (4.205) at the end $x=0$ and recalling that $Q(0;s)=1$ and $Q(d;s)=P(s)$ we obtain

$$V_1(s)=V^+(s)+P(s)V^-(s)+V_p(x=0;s), \qquad (4.230)$$

$$Z_c(s)I_1(s)=V^+(s)-P(s)V^-(s)+Z_cI_p(x=0;s); \qquad (4.231)$$

whereas specifying them at the end $x=d$, we obtain the relations

$$V_2(s)=P(s)V^+(s)+V^-(s)+V_p(x=d;s), \qquad (4.232)$$

$$-Z_c(s)I_2(s)=P(s)V^+(s)-V^-(s)+Z_c(s)I_p(x=d;s). \qquad (4.233)$$

By subtracting Eqs. (4.230) and (4.231) termwise, and summing Eqs. (4.232) and (4.233) termwise, we have, respectively:

$$V_1(s)-Z_c(s)I_1(s)=2P(s)V^-(s)+E_1(s), \qquad (4.234)$$

$$V_2(s)-Z_c(s)I_2(s)=2P(s)V^+(s)+E_2(s), \qquad (4.235)$$

where

$$E_1(s) = V_p(x = 0; s) - Z_c(s)I_p(x = 0; s), \qquad (4.236)$$

$$E_2(s) = V_p(x = d; s) + Z_c(s)I_p(x = d; s). \qquad (4.237)$$

Thus, it is clear that a line with distributed sources can be represented with the equivalent two-port of Thévenin type shown in Fig. 4.8a, provided that we insert two independent voltage sources E_1 and E_2 in series with the controlled voltage sources W_1 and W_2, respectively.

The control laws for the voltage sources W_1 and W_2 are in the Laplace domain

$$W_1(s) = P(s)[2V_2(s) - W_2(s) - 2V_p(x = d; s)], \qquad (4.238)$$

$$W_2(s) = P(s)[2V_1(s) - W_1(s) - 2V_p(x = 0; s)]. \qquad (4.239)$$

The line can also be represented by the Norton two-port shown in Fig. 4.8b, provided that suitable independent current sources are inserted in parallel with the controlled current sources.

The inverse transform of $V_p(x; s)$ and $I_p(x; s)$, and hence those of E_1 and E_2, may be determined by means of the convolution theorem and Laplace transform formulas given in Appendix B.

4.11. CHARACTERIZATION OF THE TERMINAL BEHAVIOR OF THE LINE THROUGH THE SCATTERING PARAMETERS

We adopt the following form for the relationship between the voltages and currents at the line ends (e.g., Schutt-Ainé and Mittra, 1988; Collin, 1992; Maio, Pignari, and Canavero, 1994):

$$V_p(s) = A_p + B_p \quad \text{for } p = 1, 2, \qquad (4.240)$$

$$I_p(s) = \frac{1}{Z_r(s)}[A_p(s) - B_p(s)] \quad \text{for } p = 1, 2, \qquad (4.241)$$

where

$$A_p(s) = \tfrac{1}{2}[V_p(s) + Z_r(s)I_p(s)] \quad \text{for } p = 1, 2, \qquad (4.242)$$

$$B_p(s) = \tfrac{1}{2}[V_p(s) - Z_r(s)I_p(s)] \quad \text{for } p = 1, 2, \qquad (4.243)$$

and $Z_r(s)$ is the reference impedance. It is immediate that

$$V_1(s) - Z_r(s)I_1(s) = 2B_1(s), \tag{4.244}$$

$$V_2(s) - Z_r(s)I_2(s) = 2B_2(s). \tag{4.245}$$

By using the expressions (4.74), (4.75), (4.78), and (4.79) for the terminal voltages and currents, we obtain for the auxiliary voltages waves $A_1(s)$, $A_2(s)$, $B_1(s)$, and $B_2(s)$:

$$A_1 = \tfrac{1}{2}\alpha_+ V^+ + \tfrac{1}{2}\alpha_- PV^-, \tag{4.246}$$

$$B_1 = \tfrac{1}{2}\alpha_- V^+ + \tfrac{1}{2}\alpha_+ PV^-, \tag{4.247}$$

$$A_2 = \tfrac{1}{2}\alpha_- PV^+ + \tfrac{1}{2}\alpha_+ V^-, \tag{4.248}$$

$$B_2 = \tfrac{1}{2}\alpha_+ PV^+ + \tfrac{1}{2}\alpha_- V^-, \tag{4.249}$$

where

$$\alpha_\pm = 1 \pm \frac{Z_r}{Z_c}. \tag{4.250}$$

The auxiliary voltages waves $A_1(s)$, $A_2(s)$, $B_1(s)$, and $B_2(s)$ depend on the choice of the reference impedance $Z_r(s)$.

The governing equations of the auxiliary "voltage waves" B_1 and B_2 may be written as

$$B_1(s) = S_{11}(s)A_1(s) + S_{12}(s)A_2(s), \tag{4.251}$$

$$B_2(s) = S_{21}(s)A_1(s) + S_{22}(s)A_2(s), \tag{4.252}$$

where S_{ij} are the so-called *scattering parameters*. For the symmetry of the line we have $S_{11} = S_{22}$ and $S_{12} = S_{21}$.

Remark

If we choose the reference impedance equal to the line characteristic impedance

$$Z_r = Z_c, \tag{4.253}$$

for the auxiliary voltages waves $A_1(s)$, $A_2(s)$, $B_1(s)$, and $B_2(s)$ we obtain

$$A_1 = V^+, \tag{4.254}$$

$$B_1 = PV^-, \tag{4.255}$$

$$A_2 = V^-, \tag{4.256}$$

$$B_2 = PV^+. \tag{4.257}$$

Consequently, we have

$$S_{11} = S_{22} = 0, \tag{4.258}$$

$$S_{12} = S_{21} = P. \tag{4.259}$$

Therefore, if we choose the reference impedance equal to the line characteristic impedance, the characterization based on the scattering parameters coincides with the input-state-output characterization described in §4.6. ◇

Let us now consider the more general case in which $Z_r \neq Z_c$. It is important to verify that the scattering parameters are related to the describing functions $Z_c(s)$ and $P(s)$ through the relations

$$S_{11} = S_{22} = \alpha_+\alpha_- \frac{1 - P^2}{\alpha_+^2 - \alpha_-^2 P^2}, \tag{4.260}$$

$$S_{12} = S_{21} = P \frac{\alpha_+^2 - \alpha_-^2}{\alpha_+^2 - \alpha_-^2 P^2}. \tag{4.261}$$

The scattering parameters, unlike the describing functions $Z_c(s)$ and $P(s)$, have an infinite number of poles for $Z_r \neq Z_c$. Therefore, the characterization based on the scattering parameters has the same drawbacks of the input-output descriptions discussed in §4.7.

CHAPTER 5

Lossy Two-Conductor Transmission Lines with Frequency-Dependent Parameters

5.1. INTRODUCTION

In Chapter 4 we have shown how to model lossy two-conductor transmission lines as dynamic two-ports in the time domain, and several representations of these two-ports have been considered. They may be divided into two groups, namely, those based on an *input-state-output* description and those based on an *input-output* description. Additionally, in Chapter 4 we have already stressed that these representations are more general and may also be used to describe lossy transmission lines with frequency-dependent parameters. In fact, the system of equations (4.14) and (4.15) describes transmission lines (initially at rest and in the absence of distributed sources) characterized in the Laplace domain by generic per-unit-length impedance and admittance functions $Z(s)$ and $Y(s)$.

The network functions describing a generic (uniform) two-conductor line as a two-port in the Laplace domain are: the line *characteristic impedance*

$$Z_c(s) = \sqrt{\frac{Z(s)}{Y(s)}}; \tag{5.1}$$

the line *global propagation operator*

$$P(s) = \exp[-d\sqrt{Z(s)Y(s)}]; \tag{5.2}$$

and the line *characteristic admittance* $Y_c(s)$ given by $Y_c = 1/Z_c$. The corresponding functions in the time domain, namely, the impulse responses $p(t) = L^{-1}\{P(s)\}$, $z_c(t) = L^{-1}\{Z_c(s)\}$, and $y_c(t) = L^{-1}\{Y_c(s)\}$ are needed to describe the line as a two-port in the time domain (see the previous chapter).

The per-unit-length impedance and admittance functions depend on the electromagnetic behavior of the guiding structure modeled by the transmission line. Evaluated along the imaginary axis of the s-plane $Z(i\omega)$ and $Y(i\omega)$, they describe the electrical behavior of the line in the frequency domain. Due to physical reasons (lossy lines are passive systems), the convergence region of the Laplace transforms for lossy lines contains the imaginary axis.

Ideal transmission lines are characterized by the per-unit-length impedance and admittance $Z(i\omega) = i\omega L$ and $Y(i\omega) = i\omega C$, whereas lossy two-conductor transmission lines are characterized by the per-unit-length impedance and admittance $Z(i\omega) = R + i\omega L$ and $Y(i\omega) = G + i\omega C$. The parameters L, C, R, and G are constant. The expression "transmission lines with frequency-independent parameters" simply indicates that the parameters L, C, R, and G do not depend on the frequency.

In the quasi-TEM approximation there are two reasons why the line parameters depend on the frequency, and they are profoundly different in nature. The line parameters L, C, R, and G depend on both the time dispersion of the materials and the transverse distribution of the electromagnetic field.

Due to the losses in the dielectrics and the conductors, the transverse distribution of the electromagnetic field varies as the frequency varies (even if the medium is not dispersive in time), and hence the parameters L, C, R, and G also vary.

When the transverse configuration of the electromagnetic field does not vary with the frequency, L, C, R, and G depend on the frequency only if the medium is dispersive in time. Actually, in the frequency ranges in which transmission line models are applicable, the embedding dielectric is dispersive in time, and the influence of the dielectric and the conductor losses on the transverse distribution of the electromagnetic field depends on the frequency.

In the general formulation of the transmission line model, which includes these effects, $Z(i\omega)$ and $Y(i\omega)$ are expressed as a combination of the per-unit-length frequency-dependent parameters

$$Z(i\omega) = \mathcal{R}(\omega) + i\omega\mathcal{L}(\omega), \tag{5.3}$$

$$Y(i\omega) = \mathcal{G}(\omega) + i\omega\mathcal{C}(\omega). \tag{5.4}$$

The functions $\mathcal{R}$, $\mathcal{G}$, $\mathcal{L}$, and $\mathcal{C}$ are real, positive definite, and even. Furthermore, they have to satisfy the Kronig-Kramers relations in order to ensure causality. These are their general properties.

The actual expressions of $Z(i\omega)$ and $Y(i\omega)$ depend on how the guiding structures modeled by transmission lines are made, hence we can not know their general expressions. Nevertheless, we know their qualitative behavior, which allows us to foresee the main properties of the describing functions P, Z_c, and Y_c, and hence the main properties of the impulse responses p, z_c, and y_c.

In the following two sections we shall recall the general behavior of the admittance and impedance functions $Y(i\omega)$ and $Z(i\omega)$ for transmission lines modeling guiding structures of interest in digital and communication networks. These lines have dielectric dispersive in time, frequency-dependent dielectric losses, and skin effect. We shall also touch on transmission lines modeling power lines above a finite conductivity ground and superconducting guiding structures.

However, the transmission line model can still be applied when the quasi-TEM hypothesis is no longer satisfied. In ideal guiding structures the behavior of each higher propagation mode (TE or TM) can be described in terms of an equivalent transmission line, whose parameters L and C are frequency dependent and can be obtained by solving Maxwell equations (e.g., Franceschetti, 1997). In the most general case of open lossy interconnections with a nonhomogeneous dielectric, operating at high frequencies, an equivalent frequency-dependent transmission-line model can still be found by using a full-wave approach (e.g., Brews, 1986; Collin, 1992; Olyslager, De Zutter, and de Hoop, 1994).

The inverse transforms of P, Z_c, and Y_c for lines with parameters depending on the frequency can not be performed analytically because $Z(i\omega)$ and $Y(i\omega)$ are complicated functions, which in most cases are not known in analytical form. The impulse responses p, z_c, and y_c could be evaluated by performing the inverse Fourier transform of $P(i\omega)$, $Z_c(i\omega)$, and $Y_c(i\omega)$ numerically. Unfortunately, this can not be done because of their asymptotic behavior for $\omega \to \infty$. In fact, in Chapter 4 we have found that the impulse responses p, z_c, and y_c of a lossy line contain Dirac impulse functions.

The impulse responses p, z_c, and y_c for lines with frequency-dependent parameters contain highly irregular *terms* that are "unbounded" and can not be handled numerically, such as Dirac pulses, terms that numerically behave as Dirac pulses, and functions of the type $1/t^{\alpha}$ with $0 < \alpha < 1$. These are the basic difficulties arising when the time domain two-port representations described in Chapter 4 are extended to lossy transmission lines with frequency-dependent parameters. However, these difficulties may be overcome if the asymptotic expressions of P, Z_c, and Y_c are known. This is a very important point. Once the asymptotic expressions of P, Z_c, and Y_c are known, we can separate the parts that are numerically transformable in the time domain from those that can only be transformed analytically (Maffucci and Miano, 2000a). This is the main reason why it is important to know the asymptotic behavior of the per-unit-length admittance and impedance of the line. This problem will be dealt with in the rest of this chapter.

5.2. FREQUENCY BEHAVIOR OF THE PER-UNIT-LENGTH ADMITTANCE $Y(s)$

The reader is assumed to be acquainted with the basic elements of the dispersion and the loss in the guiding structures for electromagnetic fields. Thus, it is obvious that the short notes presented in this and in the following section are merely intended to familiarize the reader with those general properties of $Y(i\omega)$ and $Z(i\omega)$ that are most frequently utilized throughout the book and that determine the asymptotic behavior of the describing functions P, Z_c, and Y_c. The reader is referred to the literature for a more thorough and comprehensive treatment of the subject (e.g., Collin, 1992; Paul, 1994; Matick, 1995).

The frequency behavior of the per-unit-length admittance depends on the distribution in the transverse plane of the transverse components of the electric and the current density fields.

We say that a dielectric is *ideal* if it is not dispersive in time and has zero electrical conductivity, otherwise we say that it is *imperfect*.

Let us consider a transmission line (without distributed sources) composed of two parallel conductors embedded in an imperfect dielectric, and let us assume that the conductor length is much greater than their distance. Figure 5.1 shows a longitudinal section. We shall operate in the frequency domain.

Equation (1.14), involving the per-unit-length admittance $Y(i\omega)$, is obtained by applying the law of electrical charge conservation (e.g.,

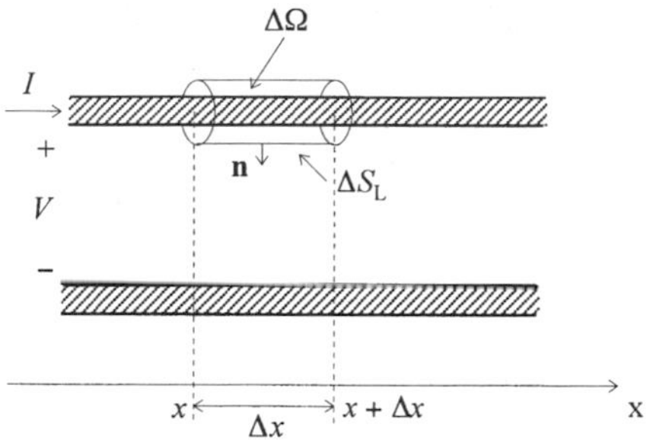

Figure 5.1. Longitudinal section of a transmission line.

Paul, 1994; Franceschetti, 1997) to the volume $\Delta\Omega$ (see Fig. 5.1) and taking the limit $\Delta x \to 0$,

$$\lim_{\Delta x \to 0} \frac{I(x + \Delta x; i\omega) - I(x; i\omega)}{\Delta x} = -K_{\perp}(x; i\omega) - i\omega Q(x; i\omega). \quad (5.5)$$

The term $\Delta x Q(x; i\omega)$ is the electrical charge inside the volume $\Delta\Omega$, lying on the conductor surface (we are implicitly supposing that the conductors are homogeneous), and the term $\Delta x K_{\perp}(x; i\omega)$ is the transverse electrical current flowing through the lateral surface ΔS_L of the volume $\Delta\Omega$; $I(x; i\omega)$ is the current flowing along the transmission line at the position x.

The per-unit-length transverse electrical current $K_{\perp}(x; i\omega)$ is related to the transverse part $\mathbf{J}_{\perp}$ of the current density field through the relation

$$K_{\perp}(x; i\omega) = \lim_{\Delta x \to 0} \frac{1}{\Delta x} \iint_{\Delta S_L} \mathbf{J}_{\perp}(x, \mathbf{r}_{\perp}; i\omega) \cdot \mathbf{n}\, dS, \quad (5.6)$$

where $\mathbf{r}_{\perp}$ indicates the transverse part of the position vector $\mathbf{r}$ (x is its longitudinal component). The per-unit-length electrical charge $Q(x; i\omega)$ is related to the transverse part $\mathbf{D}_{\perp}$ of the electric displacement field through the relation

$$Q(x; i\omega) = \lim_{\Delta x \to 0} \frac{1}{\Delta x} \iint_{\Delta S_L} \mathbf{D}_{\perp}(x, \mathbf{r}_{\perp}; i\omega) \cdot \mathbf{n}\, dS. \quad (5.7)$$

If we indicate with ε and σ the dielectric constant and the electrical conductivity of the material embedding the two conductors, we obtain

$$K_{\perp}(x; i\omega) = \lim_{\Delta x \to 0} \frac{1}{\Delta x} \iint_{\Delta S_L} \sigma \mathbf{E}_{\perp}(x, \mathbf{r}_{\perp}; i\omega) \cdot \mathbf{n}\, dS, \tag{5.8}$$

$$Q(x; i\omega) = \lim_{\Delta x \to 0} \frac{1}{\Delta x} \iint_{\Delta S_L} \varepsilon \mathbf{E}_{\perp}(x, \mathbf{r}_{\perp}; i\omega) \cdot \mathbf{n}\, dS, \tag{5.9}$$

where $\mathbf{E}_{\perp}$ is the transverse part of the electric field. If the material embedding the two conductors is not homogeneous transversally, ε and σ depend on the transverse spatial coordinate. In the guiding structure of interest in digital and communication networks the dielectric may be dispersive in time, thus ε may also depend on the frequency. By contrast, the conductivity σ is substantially independent of frequency in the ranges of interest.

In the quasi-TEM approximation, the dynamics of the fields $\mathbf{D}_{\perp}$, $\mathbf{E}_{\perp}$, and $\mathbf{J}_{\perp}$ are not greatly influenced by the longitudinal components D_x, E_x, and J_x and may be described through the electroquasistatic approximation of Maxwell equations (e.g., Haus and Melcher, 1989). In particular, the electric field $\mathbf{E}_{\perp}$ can be expressed as

$$\mathbf{E}_{\perp}(x, \mathbf{r}_{\perp}; i\omega) = \mathbf{e}_{\perp}(\mathbf{r}_{\perp}; i\omega) V(x; i\omega), \tag{5.10}$$

where the field $\mathbf{e}_{\perp}(\mathbf{r}_{\perp}; i\omega)$ depends only on the geometry and the electrical properties of the material embedding the two conductors; $V(x; i\omega)$ is the voltage between the two wires at the position x. By substituting Eq. (5.10) in Eqs. (5.8) and (5.9), we obtain

$$K_{\perp}(x; i\omega) + i\omega Q(x; i\omega) = Y(i\omega) V(x; i\omega), \tag{5.11}$$

where

$$Y(i\omega) = \lim_{\Delta x \to \infty} \frac{1}{\Delta x} \iint_{\Delta S_{\perp}} [\sigma(\mathbf{r}_{\perp}) + i\omega\varepsilon(\mathbf{r}_{\perp}, i\omega)] \mathbf{e}_{\perp}(\mathbf{r}_{\perp}; i\omega) \cdot \mathbf{n}\, dS, \tag{5.12}$$

is the per-unit-length transverse admittance of the two wires. Substituting the expression (5.11) in Eq. (5.5), we obtain Eq. (1.14) (with $J_s = 0$).

If the material embedding the two conductors is homogeneous, the free electrical charges are present only on the conductor surfaces, and the field $\mathbf{e}_{\perp}$ does not depend on the frequency, even if the dielectric is dispersive in time (e.g., Haus and Melcher, 1989). In these cases the functions $\mathscr{G}$ and $\mathscr{C}$, defined by the expression (5.4), depend on the frequency only if the medium is dispersive in time, according to the function $\varepsilon(i\omega)$.

When the material embedding the two wires is not homogeneous, the configuration of the field $\mathbf{e}_\perp$ varies as the frequency varies, due to the presence of the losses (e.g., Haus and Melcher, 1989), even if ε does not depend on the frequency. In such cases the free electrical charges are also present in the regions where the material is not homogeneous (for instance, on the separation surfaces between materials with different electrical conductivities). The physical phenomenon governing their behavior is the *charge relaxation*. This mechanism depends strongly on the frequency. Therefore, in the presence of losses, the functions $\mathscr{C}$ and $\mathscr{G}$ may depend on the frequency, even if the dielectric is not dispersive in time.

5.2.1. Homogeneous Embedding Medium

Let us now consider transmission lines with homogeneous embedding materials. For these lines, according to the foregoing remarks, the frequency behavior of the per-unit-length admittance depends only on the frequency behavior of the embedding dielectric. The per-unit-length admittance may be expressed as

$$Y(i\omega) = G_{dc} + i\omega[\bar{\varepsilon}(i\omega)C_{dc}], \tag{5.13}$$

where G_{dc} represents the per-unit-length *dc* conductance due to the bulk conductivity of the dielectric material and the surface leakages, C_{dc} is the per-unit-length static capacitance, and $\bar{\varepsilon}$ is the complex dielectric constant of the material $\varepsilon(i\omega)$ normalized to its value at $\omega = 0$, $\bar{\varepsilon} \equiv \varepsilon/\varepsilon(0)$. We have $\bar{\varepsilon} = 1$ for dielectrics that are not dispersive.

In transmission lines of practical interest, in practice, the function $\varepsilon(i\omega)$ is constant and dielectric losses are negligible for frequencies up to the low gigahertz range. In fact, the dielectrics normally used in ribbon cables and printed circuit boards have a loss tangent of the order of 10^{-4}. These losses may be represented satisfactorily by a constant conductance whose value is not considerably larger than 10^{-3} S/m, and then they are often neglected (e.g., Paul, 1994). Instead, for frequencies around and above 100 MHz, dispersion in time and losses in the dielectric material can no longer be negligible.

It is well known that, in general, the real part ε' of the complex dielectric constant $\varepsilon = \varepsilon' - i\varepsilon''$ of a dielectric material decreases with increasing the frequency, eventually approaching that of free space. However, ε' does not behave monotonically as ω increases, but it shows some extrema in correspondence with the absorption bands. The qualitative behavior of ε' as a function of the frequency for a hypothetical dielectric material is shown in Fig. 5.2 (e.g., Kittel, 1966).

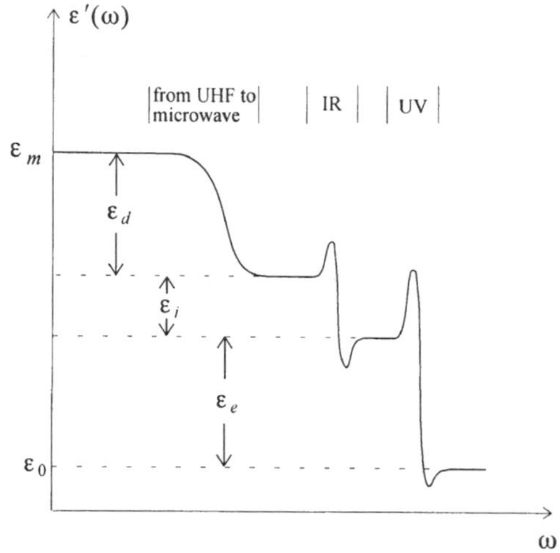

Figure 5.2. Different contributions to the real part of the complex dielectric constant versus frequency.

There are three major contributions to ε', namely dipolar, ionic, and electronic polarization. The contribution due to the electrons comes from the displacement of the electronic orbit with respect to the atomic nucleus. The contribution due to the ions comes from the displacement and deformation of a single ion with respect to all others. The dipolar contribution is due to the molecules that have a permanent dipole moment.

Because transmission lines are usually designed for operating far below 100 GHz, we need only describe how the dipolar contribution varies as a function of the frequency. The material may be considered as an ideal dielectric composed of randomly oriented dielectric dipoles. The electric field tends to line up the dipoles, at variance with the thermal agitation. A simple expression of ε taking into account the relaxation due to the thermal agitation of the electric dipoles is given by (e.g., Gordon, Blazeck, and Mittra, 1992; Franceschetti, 1997)

$$\varepsilon(i\omega) = (\varepsilon_m - \varepsilon_d) + \frac{\varepsilon_d}{1 + i\omega\tau}, \tag{5.14}$$

where the constants ε_m, ε_d, and the relaxation time constant τ characterize the material, and the latter is clearly temperature dependent. The function $\bar{\varepsilon}$ is related to ε through the relation $\bar{\varepsilon} = \varepsilon/\varepsilon_m$.

The real part $\varepsilon'(\omega)$ decreases monotonically as ω increases, that is, for $\omega \to 0$, ε' tends to ε_m, whereas for $\omega \to \infty$, ε' tends to $\varepsilon_m - \varepsilon_d$.

The imaginary part $\varepsilon''(\omega)$ is always negative, according to the requirement that the real part of $Y(i\omega)$ must always be positive, due to the passivity of the physical structure modeled by the line. It has a minimum at $\omega = 1/\tau$ and tends to zero monotonically for $\omega \to 0$ and $\omega \to \infty$.

In the high-frequency range $\omega \gg 1/\tau$, only the electronic and ionic contributions remain.

By substituting the expression (5.14) in the expression (5.13) we obtain

$$Y(i\omega) = \left[G_{dc} + (1-\eta)C_{dc} \frac{\omega^2\tau}{1+\omega^2\tau^2} \right] + i\omega \left(\frac{1+\eta\omega^2\tau^2}{1+\omega^2\tau^2} \right) C_{dc}, \qquad (5.15)$$

where η is given by

$$\eta = \frac{\varepsilon_m - \varepsilon_d}{\varepsilon_m}; \qquad (5.16)$$

the parameter η is always less than 1. The function $\mathscr{G}(\omega)$, which represents the real part of $Y(i\omega)$, increases as the frequency increases, whereas the function $\mathscr{C}(\omega)$, which represents the imaginary part of $Y(i\omega)/\omega$, decreases.

The real part of $Y(i\omega)$ tends to G_{dc} for $\omega \to 0$, and to

$$G_\infty = \left(G_{dc} + \frac{1-\eta}{\tau} C_{dc} \right) \qquad (5.17)$$

for $\omega \to \infty$. The imaginary part behaves as ωC_{dc} for $\omega \to 0$ and as ωC_∞ for $\omega \to \infty$, where

$$C_\infty = \eta C_{dc}. \qquad (5.18)$$

Therefore, the asymptotic expression (see §4.4.1) of $Y(i\omega)$ is of the type

$$Y(i\omega) = (i\omega C_\infty + G_\infty) + Y_r(i\omega), \qquad (5.19)$$

where the *remainder* $Y_r\,(i\omega)$ has the following asymptotic behavior:

$$Y_r(i\omega) \approx O(1/\omega) \quad \text{for } \omega \to \infty. \qquad (5.20)$$

By analytically extending Eqs. (5.13) and (5.14) into the complex plane, we obtain

$$Y(s) = G_{dc} + s\bar{\varepsilon}(s)C_{dc}, \qquad (5.21)$$

where

$$\bar{\varepsilon}(s) = \eta + \frac{1-\eta}{1+\tau s}. \tag{5.22}$$

The function $Y(s)$ is everywhere analytic except at $s = -1/\tau$ and infinity, where it has two poles.

Actually, the electrical behavior of many dielectric materials used in printed circuit boards is governed by a continuum relaxation mechanism with several relaxation times rather than by a single one. To allow for this, we have to introduce other terms in the expression of ε with different relaxation time constants τ and weighting constants ε_d (e.g., Gordon, Blazeck, and Mittra, 1992).

5.2.2. Nonhomogeneous Embedding Medium

In actual guiding structures the embedding medium is not homogeneous transversally. For example, integrated circuit interconnections are embedded into an insulating layer above a finite conductivity substrate (Fig. 5.3).

The distribution of the transverse component of the electric field depends strongly on the frequency because of the losses and the nonhomogeneity in the transverse section. In particular, for such cases, the frequency behavior of the per-unit-length admittance is dominated by the effect of the finite conductivity of the conducting substrate rather than by the frequency behavior of the dielectric constant. A strong dependence on frequency of the admittance $Y(i\omega)$ has been shown when low substrate conductivity $\sigma < \omega\varepsilon_0\varepsilon_{r2}$ is considered (Braunisch and Grabinski, 1998). In this case the frequency dependency of the admittance can not be neglected even for the low gigahertz range, where the variation in frequency of the dielectric constant is negligible.

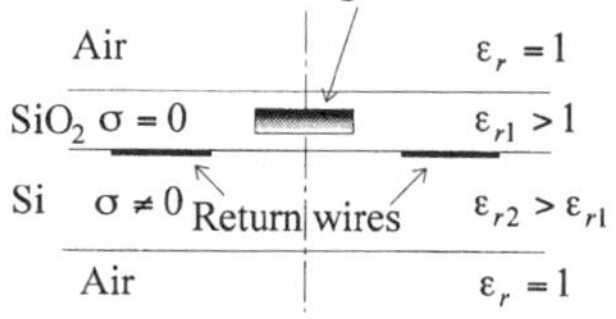

Figure 5.3. Sketch of an interconnection nonhomogeneous in the transverse section.

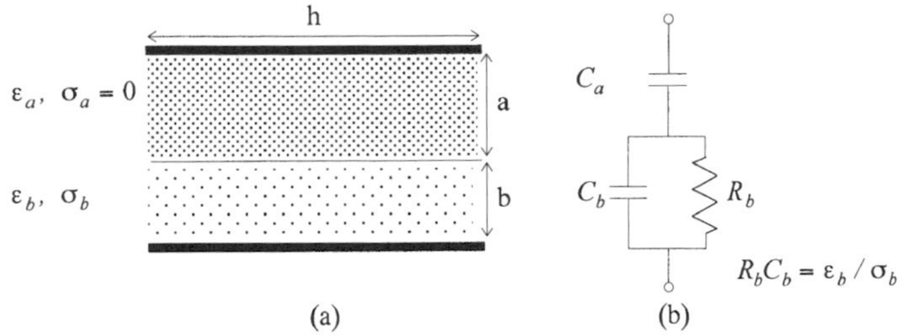

Figure 5.4. (a) Strip line with material having nonuniform ε and σ; (b) transverse equivalent circuit.

To gain an insight into this question we consider a simple guiding structure for which an approximated expression of the per-unit-length transverse admittance can be given analytically; see Fig. 5.4a.

If $a \approx b \ll h$, an accurate approximation of the per-unit-length transverse admittance is given by the admittance of the equivalent circuit shown in Fig. 5.4b (e.g., Haus and Melcher, 1989),

$$Y(i\omega) = i\omega C_\infty \frac{i\omega + \alpha_b}{i\omega + \alpha}, \tag{5.23}$$

where

$$C_\infty = \frac{C_a C_b}{C_a + C_b}, \quad \alpha_b = 1/(C_b R_b) \quad \text{and} \quad \alpha = 1/[(C_a + C_b)R_b]. \tag{5.24}$$

We immediately obtain that

$$\mathcal{G}(\omega) = (\alpha_b - \alpha) C_\infty \frac{\omega^2}{\omega^2 + \alpha^2}, \tag{5.25}$$

$$\mathcal{C}(\omega) = C_\infty \frac{\omega^2 + \alpha\alpha_b}{\omega^2 + \alpha^2}. \tag{5.26}$$

The real part of $Y(i\omega)$ vanishes for $\omega \to 0$ and tends to

$$G_\infty = \frac{1}{R_b} \frac{C_a^2}{(C_a + C_b)^2}, \tag{5.27}$$

increasing monotonically for $\omega \to \infty$. The imaginary part behaves as ωC_a for $\omega \to 0$, and as ωC_∞ for $\omega \to \infty$. Note that $\mathcal{C}(\omega)$ decreases monotonically as ω increases because $C_\infty < C_a$. The qualitative behavior of $\mathcal{G}(\omega)$ and $\mathcal{C}(\omega)$ obtained from this simple guiding structure explains the quantitative numerical results obtained in the literature (e.g., Grotelüschen, Dutta, and Zaage, 1994; Braunisch and Grabinski, 1998).

The asymptotic expression of $Y(i\omega)$ is of the same type as that given by Eq. (5.19). By prolonging analytically expression (5.23) into the complex plane, we obtain

$$Y(s) = sC_\infty \frac{s + \alpha_b}{s + \alpha}. \tag{5.28}$$

5.2.3. Asymptotic Expression of $Y(s)$

In general, the expression of $Y(i\omega)$, and hence of $Y(s)$, depends on the actual geometry and material properties of the guiding structure modeled by the line. Therefore, it is not possible to express it analytically. However, from the foregoing remarks we argue that for all cases of interest, the asymptotic expression of $Y(s)$ is of the type

$$Y(s) = (sC_\infty + G_\infty) + Y_r(s), \tag{5.29}$$

where C_∞ and G_∞ are two constants, and the remainder $Y_r(s)$ has the following asymptotic behavior:

$$Y_r(s) \approx O(s^{-1}) \quad \text{for } s \to \infty. \tag{5.30}$$

The term $(sC_\infty + G_\infty)$ represents the leading term of the admittance for $s \to \infty$. Furthermore, we expect that the remainder is analytic at infinity, where it has a zero, and all its singular points are of the polar type. Thus, the behavior of $Y_r(s)$ at infinity may be represented through a Taylor series in terms of s^{-1}.

The expression of $Y_r(s)$, or $Y_r(i\omega)$, may be evaluated analytically or numerically once the geometry and the material properties of the guiding structure are given. Note that the remainder $Y_r(s)$ vanishes in the limit of ideal dielectric, that is, a dielectric without losses and time dispersion.

Remark

In microstrip transmission lines of practical interest the propagation is quasi-TEM below a few gigahertz. At higher frequencies the electric field becomes more confined to the region between the microstrip and the ground plane. In the high-frequency limit the electric field is guided substantially by the dielectric substrate on the ground plane even if the conducting strip is moved to infinity (surface-wave mode). This results in an increase in the effective dielectric constant defined as $\varepsilon_e = C/C_v$, where C_v is the per-unit-length capacitance of the air-filled line (Collin, 1992). Typically, the effective dielectric constant ε_e increases as the frequency increases and tends to the relative dielectric constant of the substrate for $\omega \to \infty$. For a comprehensive treatment of this question the reader is referred to Collin (1992). ◇

5.3. FREQUENCY BEHAVIOR OF THE PER-UNIT-LENGTH IMPEDANCE $Z(s)$

The frequency behavior of the per-unit-length impedance depends on the distribution in the transverse plane of the longitudinal component J_x of the electric current density field inside the conducting materials.

We say that a conductor is *ideal* if it has zero electrical resistivity and if it is not dispersive in time, otherwise we say that it is *imperfect*.

Let us consider a transmission line (without distributed sources) composed of two parallel imperfect conductors and let us assume that the conductor length is much greater than their distance. Figure 5.5 shows a longitudinal section. We shall operate in the frequency domain.

Equation (1.13) involving the per-unit-length impedance $Z(i\omega)$ is obtained by applying the Faraday-Neumann law (e.g., Paul, 1994; Franceschetti, 1997) to the closed path *ABCD* (see Fig. 5.5) and taking the limit $\Delta x \to 0$,

$$\lim_{\Delta x \to 0} \frac{V(x + \Delta x; i\omega) - V(x; i\omega)}{\Delta x} = -\mathscr{V}_{||}(x; i\omega) - i\omega\Phi(x; i\omega). \quad (5.31)$$

The term $\Delta x\Phi(x; i\omega)$ is the flux of the magnetic field linked with the closed path *ABCD* according to the normal unit vector $\mathbf{n}$, and the term $\Delta x\mathscr{V}_{||}(x; i\omega)$ is the voltage along the path *AB* plus the voltage along the path *CD*; $V(x; i\omega)$ is the voltage across the transmission line at the position x.

The per-unit-length magnetic flux $\Phi(x; i\omega)$ is related to the transverse part $\mathbf{B}_\perp$ of the magnetic field through the relation

$$\Phi(x; i\omega) = \lim_{\Delta x \to 0} \frac{1}{\Delta x} \iint_{\Delta S} \mathbf{B}_\perp(x, \mathbf{r}_\perp; i\omega) \cdot \mathbf{n}\, dS, \quad (5.32)$$

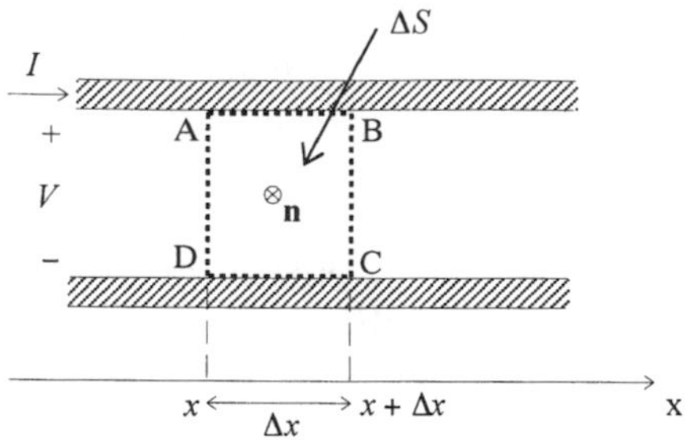

Figure 5.5. Longitudinal section of a transmission line.

where ΔS is the plane surface whose contour is the closed path $ABCD$; see Fig. 5.5. The per-unit-length voltage $\mathscr{V}_{||}(x; i\omega)$ is related to the longitudinal component of the electric field through the relation

$$\mathscr{V}_{||}(x; i\omega) = E_x(x, \mathbf{r}_{\perp A}; i\omega) - E_x(x, \mathbf{r}_{\perp D}; i\omega). \tag{5.33}$$

In the quasi-TEM approximation the behavior of the fields $\mathbf{B}_\perp$, E_x, and J_x is not greatly influenced by the transverse components of the electric and current density fields. Furthermore, below the gigahertz range the displacement current density inside the conductors is negligible compared with the conduction current density J_x. Therefore, the behavior of $\mathbf{B}_\perp$, E_x, and J_x may be described through the magnetoquasistatic approximation of Maxwell equations (e.g., Haus and Melcher, 1989).

In general, the transverse distribution of the longitudinal component of the current density is not uniform and is an unknown of the problem. Once J_x is known, it is easy to evaluate $\mathbf{B}_\perp$ and E_x.

The physical phenomenon governing the behavior of J_x is *diffusion*. This mechanism depends strongly on the frequency. In contrast to the effects due to the dispersion and the losses of an imperfect dielectric, the effects due to an imperfect conductor may be significant even for frequencies below the low gigahertz range. These effects occur substantially by means of a penetration of the electromagnetic field inside the conductors, depending on the frequency (e.g., Paul, 1994; Matick, 1995).

In an ideal conductor, the electromagnetic field can not penetrate and the electric currents flow only on the surface. In actual conductors the currents do not flow on the conductor surface but are distributed over the conductor cross sections. The current density distribution depends appreciably on the frequency. In the low-frequency limit, the distribution is almost uniform, whereas for high frequencies the current concentrates in proximity to the conductor surface.

The magnetic field $\mathbf{B}_\perp$ and the electric field E_x can be expressed as

$$\mathbf{B}_\perp(x, \mathbf{r}_\perp; i\omega) = \mathbf{b}_\perp(\mathbf{r}_\perp; i\omega)I(x; i\omega), \tag{5.34}$$

$$E_x(x, \mathbf{r}_\perp; i\omega) = e_x(\mathbf{r}_\perp; i\omega)I(x; i\omega), \tag{5.35}$$

where $\mathbf{b}_\perp(\mathbf{r}_\perp; i\omega)$ and $e_x(\mathbf{r}_\perp; i\omega)$ depend on the geometry and the electrical properties of the materials, and $I(x; i\omega)$ is the current flowing along the wire at the position x. By substituting Eqs. (5.34) and (5.35) in Eqs. (5.33) and (5.32), we obtain

$$\mathscr{V}_{||}(x; i\omega) + i\omega\Phi(x; i\omega) = Z(i\omega)I(x; i\omega), \tag{5.36}$$

where

$$Z(i\omega) = [e_x(\mathbf{r}_{\perp A}; i\omega) - e_x(\mathbf{r}_{\perp D}; i\omega)] + i\omega \lim_{\Delta x \to \infty} \frac{1}{\Delta x} \iint_{\Delta S} \mathbf{b}_\perp(\mathbf{r}_\perp; i\omega) \cdot \mathbf{n}\, dS \tag{5.37}$$

is the per-unit-length transverse impedance of the two wires. Substituting Eq. (5.36) in Eq. (5.31) we obtain Eq. (1.13) (with $E_s = 0$).

If the conductors were ideal, the first term in the expression of $Z(i\omega)$ would vanish, the second term would be given by $i\omega L_e$, where L_e is the per-unit-length external inductance, and the current would flow only on the surfaces. Note that L_e does not coincide with the value that the inductance assumes in the static limit L_{dc} because in this operating condition the current is distributed uniformly over the cross section $L_{dc} = L_e + L_i$, where L_i is the internal inductance (e.g., Paul, 1994).

The first term in the expression of $Z(i\omega)$ is the so-called per-unit-length *surface impedance* of the conductor pair. It is defined as the ratio between the per-unit-length voltage drop along the surface of the conductor pair and the total current flowing in the conductor labeled with the sign + (see Fig. 5.5).

Remark

In the model based on the magnetoquasistatic approximation describing the transverse behavior of J_x, the voltage between the conductor pair, far from the conductor ends, does not depend on the end points of the transverse path along which it is defined, provided that they belong to the conductor pair. ◇

5.3.1. Skin Effect

We shall describe the general behavior of the per-unit-length impedance by referring, for simplicity, to transmission lines made by two identical conductors, and assuming that the distance between the conductors is much greater than their transverse dimensions. In this case the per-unit-length impedance may be modeled as

$$Z(i\omega) = 2Z_s(i\omega) + i\omega L_e, \tag{5.38}$$

where L_e is the per-unit-length "external" self-inductance and $Z_s(i\omega)$ is the per-unit-length surface impedance of the single conductor (see, Paul, 1994; Matick, 1995).

Typically, the current density field in the conductors is related to the electric field through Ohm's law. Let us introduce the skin depth

$$\delta(\omega) = \sqrt{2/\omega\mu\sigma}, \tag{5.39}$$

where σ is the *dc* electrical conductivity of the conductor and μ is its magnetic permeability. If δ is less than the radius r_w of the conductors, the electromagnetic fields and current density decay rapidly in the conductor and are essentially confined at the surface to a layer of thickness equal to a few skin depths. This is the so-called *skin-effect* phenomenon. In this case, an approximated expression for the surface impedance is given by

$$Z_s(i\omega) = \tfrac{1}{2}(R_{dc} + K\sqrt{i\omega}), \tag{5.40}$$

where $R_{dc}/2$ is the per-unit-length *dc* longitudinal resistance of the single conductor and K $(\Omega \cdot s^{1/2}/m)$ is a constant depending on the geometry and the physical parameters of the conductors. Typically, the constant K may be expressed approximately as (e.g., Matick, 1995)

$$K = W\sqrt{\mu/\sigma}, \tag{5.41}$$

where $W(m^{-1})$ is a positive constant depending on the geometry of the conductors (e.g., for wires with circular cross section, $W \cong 1/\pi r_w$ where r_w is the conductor radius). The function $Z_s(i\omega)$ given by Eq. (5.40) may be rewritten as

$$Z_s(i\omega) = R_s(\omega) + i\omega L_s(\omega), \tag{5.42}$$

where the term R_s, referred to as the per-unit-length surface resistance, is given by

$$R_s(\omega) = \frac{1}{2}(R_{dc} + K\sqrt{\omega/2}) = \frac{1}{2}\left[R_{dc} + \frac{W}{\sigma\delta(\omega)}\right], \tag{5.43}$$

and the term L_s, referred to as the per-unit-length surface inductance, is given by

$$L_s(\omega) = W\frac{\mu_0\delta(\omega)}{4}. \tag{5.44}$$

Observe that the skin depth decreases as $1/\sqrt{\omega}$ as the frequency increases. Therefore, as frequency increases, the surface resistance increases as $\sqrt{\omega}$, whereas the surface inductance decreases as $1/\sqrt{\omega}$. When the frequencies are not very low, the per-unit-length internal inductance is almost equal to the per-unit-length surface inductance L_s.

By substituting Eq. (5.40) in Eq. (5.38) and by prolonging analytically the obtained expression into the complex plane, we obtain

$$Z(s) = R_{dc} + sL_e + K\sqrt{s}. \tag{5.45}$$

This function is multivalued and has two branches, one of which has a positive real part. The branch points are at $s = 0$ and $s = \infty$. At $s = 0$ the branch is of regular type, whereas at $s = \infty$ it is of polar type; $Z(s)$ also has an isolated singular point of polar type at $s = \infty$, see Appendix B.

For physical reasons we have to consider only the branch of $Z(s)$ that has a positive real part. This branch is defined in the region obtained by cutting the s-plane along the negative real axis.

For $s \to 0$, the impedance function $Z(s)$ tends to the "static" resistance R_{dc}; sL_e is the leading term for $s \to \infty$, whereas the remainder behaves as $K\sqrt{s}$.

In general, Eq. (5.40), and hence Eq. (5.45), describes correctly the frequency behavior of $Z_s(i\omega)$ only for high frequencies or for $\omega \to 0$ (expression (5.40)) is exact only for plane conductors). In the intermediate frequency range between dc and high frequency, $Z_s(i\omega)$ is a complicated function of the parameters, depending on the actual geometry of the conductors.

For interconnections of practical interest, the actual frequency behavior of the per-unit-length impedance is, of course, much more complicated. The geometries of the conductors are not elementary and may differ from each other, as, for instance, in printed circuit boards and integrated circuit interconnections. In addition, the side and the proximity effects may be not negligible. Furthermore, for particular technologies, other effects can influence the frequency behavior. For instance, in integrated circuit interconnections the frequency dependence of the transmission line parameters is completely dominated by the substrate effects (e.g., Braunisch and Grabinski, 1998). Skin and proximity effects in the signal lines in general are negligible due to the smallness of the transverse conductor dimension compared to the skin depth. However, in printed circuit boards, the skin and proximity effects are significant because the transverse dimensions are much larger.

Therefore, in all cases of interest, $Z_s(s)$ may be represented as

$$Z(s) = R_{dc} + sL_e + K\sqrt{s}f(\sqrt{s}), \tag{5.46}$$

where the function $f(u)$ is a monodrome analytic function, depending on the particular transverse shape of the line conductors, having the following asymptotic expression:

$$f(u) \approx 1 + f_1 u^{-1} + O(u^{-2}) \quad \text{for } u \to \infty \tag{5.47}$$

(f_1 is a constant). The expression (5.46) may be represented as

$$Z(s) = sL_\infty + K\sqrt{s} + R_\infty + Z_r(s), \tag{5.48}$$

where $L_\infty = L_e$, $R_\infty = R_{dc} + Kf_1$, and the remainder Z_r has the following asymptotic behavior:

$$Z_r(s) \approx O(s^{-1/2}). \tag{5.49}$$

The term $(sL_\infty + K\sqrt{s} + R_\infty)$ represents the leading term of the impedance function as $s \to \infty$; at infinity, Z_r may be represented through a Laurent series in terms of $1/\sqrt{s}$.

Remark

It is not surprising that the general expression of $Z(s)$ depends on s only through $\sqrt{s}$. This is a consequence of the fact that the basic mechanism governing the transverse behavior of J_x is diffusion. In fact, by rescaling the diffusion equation it is simple to show that J_x depends on the frequency only through the skin depth $\delta(\omega)$ given by Eq. (5.39). The situation may be more complicated if conductors with different skin depths are present.

5.3.2. Anomalous Skin Effect

At very low temperatures and high frequency, Ohm's law no longer holds as the electron mean free path Λ, that is, the mean distance between two collisions, is equal or greater than the skin depth ($\Lambda/\sigma \cong 6.5 \cdot 10^{-16}\,\Omega\text{m}^2$ for copper). This is called the *anomalous skin effect*, and the expression (5.40) becomes

$$Z_s(i\omega) = \tfrac{1}{2}[R_{dc} + K(i\omega)^{2/3}], \tag{5.50}$$

where the constant K now is given approximately by (e.g., Matick, 1995)

$$K \cong 0.6526\mu^{2/3}(\Lambda/\sigma)^{1/3}. \tag{5.51}$$

In this case, as the frequency increases, the surface resistance increases as $\omega^{2/3}$, and the surface inductance decreases as $1/\sqrt[3]{\omega}$. The function $Z(s)$ has three branches, one of which has a positive real part.

5.3.3. Superconducting Transmission Lines

An interesting case of a transmission line with frequency-dependent parameters is provided by superconducting striplines. By using the London equation for the current density in a superconductor, we can model the per-unit-length impedance function through (e.g., Matick, 1995)

$$Z(s) = \frac{2}{W}\frac{\mu s}{\sqrt{\mu\sigma_n s + 1/\lambda_L^2}}\coth\left[\left(\frac{1}{\lambda^2} + \mu\sigma_n s\right)^{1/2}\Delta\right] + sL_e, \tag{5.52}$$

where L_e is the per-unit-length external inductance, W is the strip-line width, Δ is the superconductor thickness, σ_n is the dc conductivity characterizing the current due to the "normal electrons," and λ_L is the London constant that characterizes the current contribution due to the "superconducting electrons." The expression (5.52) holds only when the displacement current density is negligible compared with the conduction current. The parameters λ_L and σ_n depend on the superconducting material. Typically, we have $\lambda_L \approx 10 \div 100$ nm; σ_n is smaller than that of ohmic conductors, such as, for example, copper.

The function $Z(s)$ has two branches because of the square root $\sqrt{\mu\sigma_n s + 1/\lambda_L^2}$; the branch points at $s = -1/(\lambda_L^2\mu\sigma_n)$ and $s = \infty$ are of polar type. The branch of $Z(s)$ with positive real part is defined in the region obtained by cutting the s-plane along the negative real axis on the left of the branch point $s = -1/(\lambda_L^2\mu\sigma_n)$. Note that in this case, $Z(s)$ at infinity may be represented through a Laurent series in terms of power of s.

5.3.4. A Single Wire Above a Finite Conductivity Ground Plane

The per-unit-length impedance of a single wire above a ground plane with conductivity σ_g and dielectric constant ε_g is given approximately by (e.g., Sunde, 1968)

$$Z(s) = R_{dc} + s\frac{\mu}{2\pi} ln\left[1 + \frac{1}{h\sqrt{s\mu(\sigma_g + s\varepsilon_g)}}\right] + sL_e, \qquad (5.53)$$

where h is the height of the wire, L_e is the per-unit-length external inductance, and R_{dc} is the per-unit-length dc longitudinal resistance of the wire. The ground conductivity is very low compared with that of normal conductors, thus the contribution of the displacement current may be significant even at low frequencies.

The term $1/[h\sqrt{s\mu(\sigma_g + s\varepsilon_g)}]$ in the argument of the logarithmic function has two branches. The branch points are $s = 0$ and $s = -\varepsilon_g/\sigma_g$ and they are of polar type. The branch of Z with a positive real part is defined in the region obtained by cutting the s-plane along the segment of the real axis with ends at $s = -\varepsilon_g/\sigma_g$ and $s = 0$. With this choice the overall function results clearly single valued. Note also in this case that $Z(s)$ at infinity may be represented through a Laurent series in terms of power of s. The function $Z(s)$ tends to R_{dc} for $s \to 0$ (the "dc resistance"), whereas sL_e is the leading term for $s \to \infty$,

$$Z(s) \approx sL_e + O(1) \quad \text{for } s \to \infty. \qquad (5.54)$$

For a detailed analysis of the influence of finite-conductivity grounds on the line parameters see the thorough study in D'Amore and Sarto (1996a, 1996b).

5.4. PROPERTIES OF THE DESCRIBING FUNCTIONS $P(s)$, $Z_c(s)$, AND $Y_c(s)$

The terminal behavior of a two-conductor transmission line in the time domain may be described through the input-output descriptions or the input-state-output description examined in the previous chapter. All the descriptions are completely determined once the *impulse responses* $p(t)$, $z_c(t)$, and $y_c(t)$ are known. Once the functions $Z(s)$ and $Y(s)$ are given, the impulse responses $p(t)$, $z_c(t)$, and $y_c(t)$ are obtained by performing the inverse Laplace transforms of $P(s)$, $Z_c(s)$, and $Y_c(s)$, respectively.

For an ideal two-conductor line (see Chapter 2) we have:

$$z_c(t) = R_c\delta(t); \tag{5.55}$$

$$p(t) = \delta(t - T); \tag{5.56}$$

and

$$y_c(t) = \frac{1}{R_c}\delta(t). \tag{5.57}$$

A two-conductor line with frequency-independent losses is characterized by (see Chapter 4)

$$z_c(t) = R_c\delta(t) + \nu R_c e^{-\mu t}[I_0(\nu t) + I_1(\nu t)]u(t), \tag{5.58}$$

$$p(t) = e^{-\mu T}\delta(t - T) + \nu T e^{-\mu t}\frac{I_1[\nu\sqrt{t^2 - T^2}]}{\sqrt{t^2 - T^2}}u(t - T), \tag{5.59}$$

$$y_c(t) = \frac{1}{R_c}\delta(t) + \nu\frac{1}{R_c}e^{-\mu t}[I_1(\nu t) - I_0(\nu t)]u(t). \tag{5.60}$$

In the general case of lines with parameters depending on the frequency these inverse transfoms can not be evaluated analytically, and hence we need to resort to numerical inversion. On the other hand, inversion procedures entirely based on numerical methods can be effectively applied only when the resulting time domain functions are sufficiently regular. Hence a preliminary and accurate study of the regularity properties of $p(t)$, $z_c(t)$, and $y_c(t)$ is required.

The impulse responses $p(t)$, $z_c(t)$, and $y_c(t)$ of ideal lines are composed of Dirac pulses as direct consequence of the basic properties of the electromagnetic propagation phenomenon. In fact, in the time domain, ideal matched or semi-infinite lines must behave at each end as resistive one-port with resistance R_c (see Chapter 2). Thus the impulse response $z_c(t)$ relating the voltages and the currents at the same end is a Dirac function centered at $t = 0$, as in expression (5.55). Moreover, a signal applied to a given end will act on the other one only after a certain finite time due to the delay due to the finite propagation velocity. Therefore, the impulse response $p(t)$, which relates the amplitudes of the voltage forward wave (voltage backward wave) at the two line ends, is a delayed Dirac function as in the expression (5.56). These properties are a direct consequence of the hyperbolic nature of the transmission line equations (see Chapter 8).

What happens when the losses are considered? In the expressions (5.58) and (5.59) the Dirac pulses that are related to the intrinsic properties of the electromagnetic propagation phenomenon are still present. The losses merely give rise to the phenomena of damping and dispersion, both described through regular functions: The positive parameter μ is the *damping rate* of the propagation, and the parameter ν characterizes the dispersion due to the losses. The structure of the lossy transmission line equations still remains hyperbolic (see Chapter 8).

As we shall show in the next section, the impulse responses $p(t)$, $z_c(t)$, and $y_c(t)$ of a line with parameters depending on the frequency contain similar "irregular" terms (Maffucci and Miano, 2000a). Even if it is not possible to express $p(t)$, $z_c(t)$, and $y_c(t)$ analytically, it is possible to predict all their irregular terms once the asymptotic behaviors of $P(s)$, $Z_c(s)$, and $Y_c(s)$ around $s = \infty$ are known. Once the asymptotic expressions of P, Z_c, and Y_c are known, we can separate the parts that are numerically transformable in the time domain, for example, through a FFT inversion, from those that can only be transformed analytically. The exact knowledge of the irregular terms of the impulse responses is fundamental for the correct evaluation of the line dynamics. For instance, if we miss the Dirac pulses in the impulse line responses, or we do not accurately evaluate them, the behavior of the line is described with great inaccuracy.

Now we shall study the main properties of the describing functions P, Z_c, and Y_c by referring to the per-unit-length admittance and impedance functions whose asymptotic expressions are given by Eqs. (5.29) and (5.48), respectively. By using these properties we shall foresee the qualitative behavior of the impulse responses p, z_c, and y_c, which will also allow us to evaluate all the irregular terms that are present in them.

By substituting Eqs. (5.29) and (5.48) in the expressions of Z_c and P, given by Eqs. (5.1) and (5.2), we obtain

$$Z_c(s) = \sqrt{\frac{(sL_\infty + K\sqrt{s} + R_\infty) + Z_r(s)}{(sC_\infty + G_\infty) + Y_r(s)}}, \tag{5.61}$$

$$P(s) = \exp\{-d\sqrt{[(sL_\infty + K\sqrt{s} + R_\infty) + Z_r(s)][(G_\infty + sC_\infty) + Y_r(s)]}\}. \tag{5.62}$$

In the complex plane cut along the negative part of the real axis, the real part of $Z_c(s)$ is everywhere positive, whereas the real part of $P(s)$

is positive for $Re\{s\} > 0$. The imaginary axis, without the origin, belongs to the region where $Z_c(s)$ and $P(s)$ are analytic.

Now we shall determine the asymptotic expressions of Z_c, Y_c, and P as we have done in Chapter 4 (see §4.4.1, 4.5.2, and 4.6.3).

5.4.1. Asymptotic Expression of Z_c

The impedance function Z_c, given by Eq. (5.61), can be rewritten as

$$Z_c(s) = R_c + Z_{cr}(s), \tag{5.63}$$

where the constant term R_c is given by

$$R_c = \sqrt{L_\infty / C_\infty}, \tag{5.64}$$

the function $Z_{cr}(s)$ is given by

$$Z_{cr}(s) = R_c \left[\sqrt{\frac{1 + N(s)}{1 + M(s)}} - 1 \right], \tag{5.65}$$

and

$$N(s) = [K\sqrt{s} + R_\infty + Z_r(s)]/sL_\infty, \tag{5.66}$$

$$M(s) = [G_\infty + Y_r(s)]/sC_\infty; \tag{5.67}$$

$Z_{cr}(s)$ would vanish if the line were ideal.

For $s \to \infty$, the functions $N(s)$ and $M(s)$ tend to zero as $1/\sqrt{s}$ and as $1/s$, respectively; thus

$$Z_{cr}(s) \approx R_c \frac{K}{2L_\infty} \frac{1}{\sqrt{s}} + O(s^{-1}) \quad \text{for } s \to \infty. \tag{5.68}$$

Therefore, the term R_c is the principal part of Z_c and $Z_{cr}(s)$ is the remainder. These results suggest rewriting the expression of Z_c as

$$Z_c(s) = R_c \left(1 + \frac{\theta}{\sqrt{s}} \right) + Z_{cb}(s), \tag{5.69}$$

where the constant θ is given by

$$\theta = \frac{K}{2L_\infty}, \tag{5.70}$$

and the function Z_{cb}, given by

$$Z_{cb}(s) = Z_{cr}(s) - R_c \frac{\theta}{\sqrt{s}}, \tag{5.71}$$

has the asymptotic behavior

$$Z_{cb}(s) \approx O(s^{-1}) \quad \text{for } s \to \infty. \tag{5.72}$$

As we shall show in the next section, the inverse transform of Z_{cb}, which cannot be evaluated analytically, can be performed numerically because of its asymptotic behavior. The inverse Laplace transform of the term

$$Z_{ci}(s) = R_c\left(1 + \frac{\theta}{\sqrt{s}}\right), \tag{5.73}$$

which can be evaluated analytically, as we shall show in the next section, can not be performed numerically because the corresponding time domain function is highly "irregular."

Remark

Note that the value of R_c depends only on the high-frequency limit of the per-unit-length inductance L_∞ and capacitance C_∞. ◇

5.4.2. Asymptotic Expression of *P*

The global propagation operator $P(s)$ can be rewritten as

$$P(s) = e^{-sT}\exp[-\lambda(s)], \tag{5.74}$$

where the function $\lambda(s)$, which tends to 1 in the limit of an ideal line, is given by

$$\lambda(s) = sT_0[\sqrt{[1 + N(s)][1 + M(s)]} - 1], \tag{5.75}$$

and

$$c = \frac{1}{\sqrt{L_\infty C_\infty}}, T = \frac{d}{c}. \tag{5.76}$$

The function $\lambda(s)$ has the asymptotic behavior

$$\lambda(s) \approx \alpha\sqrt{s} + \mu + O(s^{-1/2}) \quad \text{for } s \to \infty, \tag{5.77}$$

where the constants α and μ are given by

$$\alpha = \frac{Kd}{2R_c}, \quad \mu = \frac{1}{2}\left[\frac{R_\infty}{L_\infty} + \frac{G_\infty}{C_\infty} - \left(\frac{K}{2L_\infty}\right)^2\right]. \tag{5.78}$$

This result suggests rewriting P as

$$P(s) = e^{-sT}e^{-\mu T}[e^{-\alpha\sqrt{s}} + \hat{P}_r(s)], \tag{5.79}$$

where the function $\hat{P}_r$, given by

$$\hat{P}_r(s) = \exp[(s+\mu)T][P(s) - e^{-\alpha\sqrt{s}}], \tag{5.80}$$

has the asymptotic behavior

$$\hat{P}_r(s) \approx O(e^{-\alpha\sqrt{s}}/\sqrt{s}) \quad \text{for } s \to \infty. \tag{5.81}$$

Remark

Note that the value of T depends only on the high-frequency limit of the per-unit-length inductance L_∞ and capacitance C_∞. ◇

5.4.3. Asymptotic Expression of Y_c

The admittance $Y_c = 1/Z_c$ may be rewritten as

$$Y_c(s) = \frac{1}{R_c}\left(1 - \frac{\theta}{\sqrt{s}}\right) + Y_{cb}(s), \tag{5.82}$$

where the function Y_{cb}, given by

$$Y_{cb}(s) = \frac{1}{R_c}\left[\sqrt{\frac{1+M(s)}{1+N(s)}} - 1 + \frac{\theta}{\sqrt{s}}\right], \tag{5.83}$$

has the asymptotic behavior

$$Y_{cb}(s) \approx O(s^{-1}) \quad \text{for } s \to \infty. \tag{5.84}$$

Remark

Similar asymptotic expressions may be obtained for the other per-unit-length impedances described in the sections 5.3.2, 5.3.3, and 5.3.4.

For lines with anomalous skin effect (see §5.3.2) the functions $Z_c(s)$, $\lambda(s)$, and $Y_c(s)$ may be expanded into Laurent series in the neighborhood of $s = \infty$ by powers of $1/\sqrt[3]{s}$ (Maffucci and Miano,

2000a). The asymptotic expressions of $Z_c(s)$, $Y_c(s)$, and $P(s)$ are

$$Z_c(s) = R_c\left(1 + \frac{\theta_1}{s^{1/3}} + \frac{\theta_2}{s^{2/3}}\right) + Z_{cb}(s), \tag{5.85}$$

$$Y_c(s) = \frac{1}{R_c}\left(1 - \frac{\theta_1}{s^{1/3}} - \frac{\theta_2}{s^{2/3}}\right) + Y_{cb}(s), \tag{5.86}$$

$$P(s) = e^{-(s+\mu)T}[e^{-\alpha_1 s^{1/3} - \alpha_2 s^{2/3}} + \hat{P}_r(s)], \tag{5.87}$$

where $Z_{cb}(s) \approx O(1/s)$, $Y_{cb}(s) \approx O(1/s)$, and $\hat{P}_r(s) \approx O(1/s^{1/3})$ for $s \to \infty$.

For superconducting lines and lines above a conductive ground the functions $Z_c(s)$, $\lambda(s)$, and $Y_c(s)$ may be expanded into Laurent series in the neighborhood of $s = \infty$ by powers of $1/s$ (Maffucci and Miano, 2000a). The asymptotic expressions of $Z_c(s)$, $Y_c(s)$, and $P(s)$ are

$$Z_c(s) = R_c + Z_{cr}(s), \tag{5.88}$$

$$Y_c(s) = \frac{1}{R_c} + Y_{cr}(s), \tag{5.89}$$

$$P(s) = e^{-(s+\mu)T}[1 + \hat{P}_r(s)], \tag{5.90}$$

where $Z_{cr}(s) \approx O(1/s)$, $Y_{cr}(s) \approx O(1/s)$, and $\hat{P}_r(s) \approx O(1/s)$ for $s \to \infty$. As we have shown (Maffucci and Miano, 2000a), in both cases, the inverse Laplace transform of the principal terms may be obtained analytically, whereas the inverse transform of remainders must be performed numerically. ◇

5.5. QUALITATIVE BEHAVIOR OF THE IMPULSE RESPONSES $p(t)$, $z_c(t)$, AND $y_c(t)$

The asymptotic expressions of $Z_c(s)$, $P(s)$, and $Y_c(s)$ given, respectively, by Eqs. (5.69), (5.79), and (5.82), are of notable interest because through them we may determine all the irregular terms of the corresponding impulse responses in the time domain.

5.5.1. Inverse Laplace Transform of Functions Behaving as s^{-1} for $s \to \infty$

Let us consider a complex function $D_b(s)$ of the complex variable s that is analytic for $Re\{s\} > \sigma_c$ with $\sigma_c < 0$, and let us assume that

$$D_b(s) \approx O(s^{-1}) \quad \text{for } s \to \infty. \tag{5.91}$$

The function D_b can be rewritten as

$$D_b(s) = \frac{d_1}{s} + \hat{D}(s). \tag{5.92}$$

The remainder function $\hat{D}(s)$ is such that

$$\lim_{s \to \infty} \frac{\hat{D}(s)}{1/s} = 0, \tag{5.93}$$

hence its inverse Laplace transform $\hat{d}(t)$ exists and it is bounded everywhere (e.g., Korn and Korn, 1961; Appendix B). Note that $\hat{d}(0^+) = 0$. Therefore, the inverse Laplace transform of $D_b(s)$, given by

$$d_b(t) = d_1 u(t) + \hat{d}(t), \tag{5.94}$$

is always continuous and bounded, except for a discontinuity of the first kind at $t = 0$.

Remark

In general, the function $d_b(t)$ may be determined by using the inverse Fourier transform provided that the region of convergence of $D_b(s)$ contains the imaginary axis

$$d_b(t) = \frac{1}{2\pi} \int_{-\infty}^{+\infty} D_b(i\omega) e^{i\omega t} \, d\omega. \tag{5.95}$$

This integral can be evaluated numerically through FFT algorithms. Because the velocity of convergence increases with the increase of the decaying velocity of $|D_b(i\omega)|$ for $\omega \to \infty$, the computational cost of the numerical inversion depends on the decaying velocity of $|D_b(i\omega)|$ as $\omega \to \infty$. Once the expansion coefficient d_1 is known, the numerical inversion can be accelerated by evaluating analytically the inverse Laplace transform of the first term in the expression of D_b. The cost of the numerical inversion of $\hat{D}$ is less than that of the overall function D_b because $\hat{D}$ goes to zero for $\omega \to \infty$ more quickly than D_b: The quicker the decay as $\omega \to \infty$, the less the computational cost. By considering asymptotic expressions of D_b with higher-order remainders we can reduce the cost of the numerical inversion even more.

5.5.2. Qualitative Behavior of the Impulse Responses z_c and y_c

By using the inversion formula (see, Doetsch, 1974)

$$L^{-1}\left\{\frac{\Gamma(1+\eta)}{s^{1+\eta}}\right\} = t^{\eta} u(t), \tag{5.96}$$

where $\Gamma(y)$ is the Euler gamma function (e.g., Abramowitz and Stegun, 1972), from the asymptotic expression (5.69) we obtain

$$z_c(t) = z_{ci}(t) + z_{cb}(t), \tag{5.97}$$

where

$$z_{ci}(t) = R_c \delta(t) + R_c \frac{\theta}{\sqrt{\pi t}} u(t), \tag{5.98}$$

and $z_{cb}(t)$ indicates the inverse Laplace transform of the remainder Z_{cb} given by Eq. (5.71), which behaves as $1/s$ for $s \to \infty$. Thus the function z_{cb} is always continuous and bounded, except for a discontinuity of the first kind at $t = 0$, and hence it can be evaluated numerically as we have seen in the foregoing.

Similarly, for the impulse response y_c we have

$$y_c(t) = y_{ci}(t) + y_{cb}(t), \tag{5.99}$$

where

$$y_{ci}(t) = \frac{1}{R_c} \delta(t) - \frac{1}{R_c} \frac{\theta}{\sqrt{\pi t}} u(t), \tag{5.100}$$

and $y_{cb}(t)$ indicates the inverse Laplace transform of the remainder $Y_{cb}(s)$ given by Eq. (5.83), which also behaves as $1/s$ for $s \to \infty$. Thus the function $y_{cb}(t)$ is always continuous and bounded, except for a discontinuity of the first kind at $t = 0$, and hence it can be evaluated numerically as we also have seen in the foregoing.

5.5.3. Qualitative Behavior of the Impulse Response *p*

By using the inverse Laplace transforms (e.g., Doetsch, 1974)

$$\delta(t;\alpha) = L^{-1}\{e^{-\alpha\sqrt{s}}\} = \frac{\alpha}{2\sqrt{\pi}\, t^{3/2}} e^{-\alpha^2/4t} u(t), \tag{5.101}$$

we obtain for the impulse response $p(t)$ the following:

$$p(t) = p_i(t - T) + p_b(t - T), \tag{5.102}$$

where

$$p_i(t) = e^{-\mu T} \delta(t;\alpha), \tag{5.103}$$

and $p_b(t)$ is the inverse Laplace transform of

$$P_b(s) = \exp(sT)P(s) - e^{-\mu T} e^{-\alpha\sqrt{s}}. \tag{5.104}$$

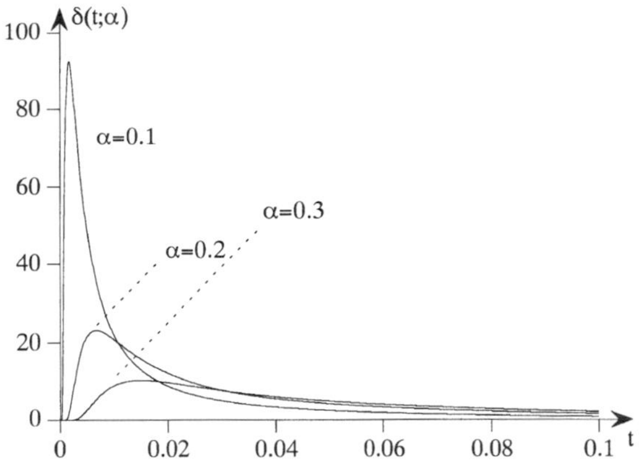

Figure 5.6. Plot of $\delta(t;\alpha)$ for different values of the parameter α in arbitrary units.

The continuously differentiable function $\delta(t;\alpha)$ has the following properties:

(i) $\delta(t \leqslant 0;\alpha) = 0,\ \delta(t = \infty;\alpha) = 0 \quad \forall\alpha \neq 0$;
(ii) $\delta(t;\alpha)$ attains its maximum value

$$\delta_m = \sqrt{\frac{54}{\pi}}\, e^{-3/2}\frac{1}{\alpha^2} \quad \text{at } t_m = \alpha^2/6; \quad \text{and}$$

(iii) $\int_{-\infty}^{+\infty} \delta(t;\alpha)dt - 1 \quad \forall\alpha.$

Figure 5.6 shows $\delta(t;\alpha)$ versus time for different values of the parameter α. These plots show some properties outlined in the foregoing. Thus the function $\delta(t;\alpha)$ approximates the unit Dirac impulse as $\alpha \to 0$, namely,

$$\lim_{\alpha\to 0} \delta(t;\alpha) = \delta(t). \tag{5.105}$$

Remarks

Some considerations are in order. By comparing, respectively, Eqs. (5.97), (5.99), and (5.102) with Eqs. (5.58), (5.60), and (5.59) we can deduce the changes introduced in the "irregular part" of the impulsive responses z_c, y_c, and p by the dispersion and the losses in the dielectric material and by the skin effect in the conductors.

(i) The delayed Dirac pulse $e^{-\mu T}\delta(t-T)$ appearing in the impulse response p of a line with frequency-independent losses (see Eq. (5.59)) is replaced by the continuously differentiable function $e^{-\mu T}\delta(t-T;\alpha)$ because of the "strong dispersion" introduced by the skin effect: The more negligible the skin effect, the more $\delta(t;\alpha)$ behaves as a Dirac pulse. We recall that the skin effect is the result of the diffusion of the electrical current in the transverse cross sections of the conductors. More precisely, we can consider the parameter α^2 as the *characteristic time* of the strong distortion phenomenon due to the skin effect. It is proportional to the square of the ratio between the line length and the transverse characteristic dimension of the conductors (see Eqs. (5.41) and (5.78)): For all cases of technical interest, α^2 is of the order of $10^{-3}(d/r_w)^2$ to $10^{-7}(d/r_w)^2$ femtoseconds, where r_w is the radius of the conducting wire. When the smallest characteristic time t_c of the electrical variables of the system satisfies the condition $t_c \gg \alpha^2$, $\delta(t;\alpha)$ behaves in practice as a Dirac pulse. As the line length increases, so does the damping and the dispersion of the propagating impulse. This result gives a theoretical explanation to the behavior observed in numerical computations of the impulse responses of the line with frequency-dependent losses (Gordon, Blazeck, and Mittra, 1992).

(ii) The dispersion and the losses in the dielectric material and the skin effect in the conductors do not affect the Dirac pulse appearing in z_c and y_c because it is related to the *instantaneous response* of the line. However, another irregular term of the type $1/\sqrt{t}$ arises due to the strong dispersion mechanism related to the skin effect.

(iii) The losses and the dispersion of the dielectric do not have a direct influence on the shape of the impulse responses $p_i(t)$, $z_{ci}(t)$, and $y_{ci}(t)$: They influence only the parameters R_c, μ, and T. In fact, the parameter α is different from zero even if the dielectric is ideal. ◇

5.5.4. A Numerical Experiment

Now we apply the method described in this chapter to determine the impulse responses of a two-conductor transmission line with pronounced skin effect, namely, a *RG-21 two-conductor line cable*

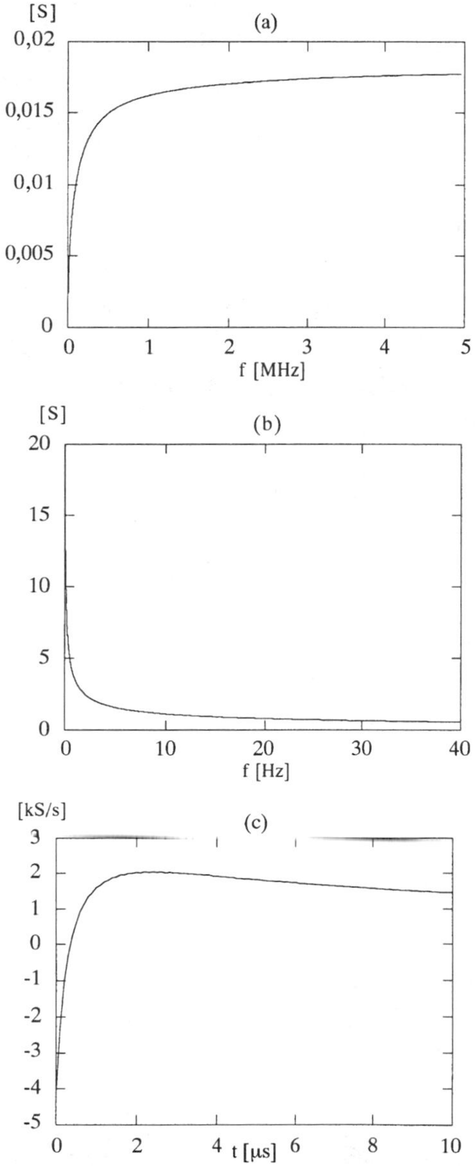

Figure 5.7. (a) Plots of $|Y_c(i\omega)|$, (b) its remainder $|Y_{cb}(i\omega)|$ versus frequency, and (c) $y_{cb}(t)$ versus time for the RG-21 cable.

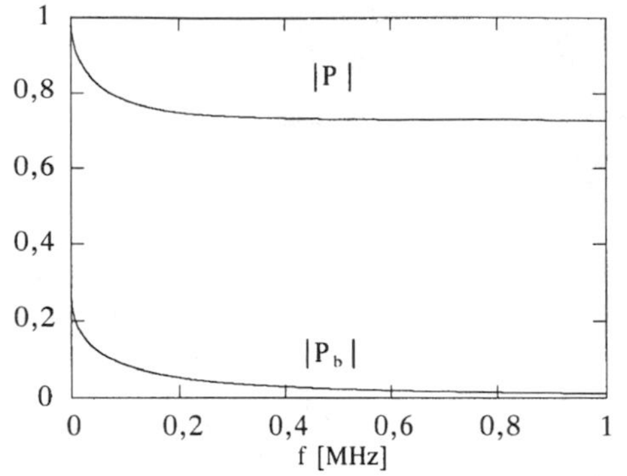

Figure 5.8. Plots of $|P(i\omega)|$ and its remainder $|P_b(i\omega)|$ versus frequency for the RG-21 cable.

(Nahman and Holt, 1972),

$$L = 265 \quad [nH]/[m]$$

$$C = 94.3 \quad [pF]/[m]$$

$$G = 0$$

$$R = 0.35 \quad [\Omega]/[m]$$

$$K = 0.25 \quad [m\Omega] \cdot [s^{-1/2}]/[m].$$

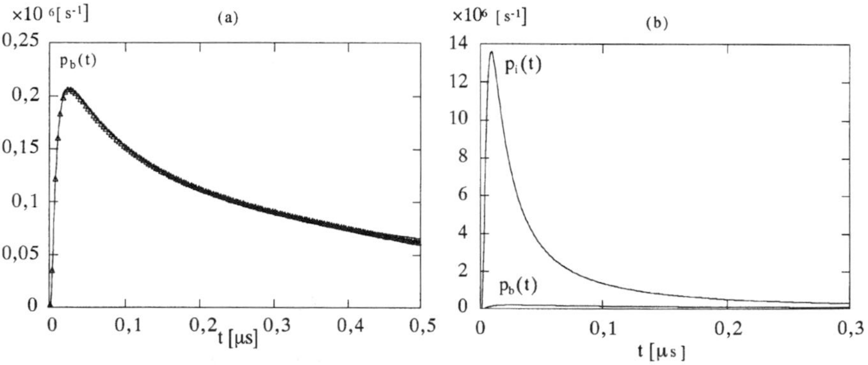

Figure 5.9. Two evaluations of (a) $p_b(t)$ with —— 2000 and △ 10 sample points in frequency for the RG-21 cable; (b) $p_i(t)$ is compared with $p_b(t)$.

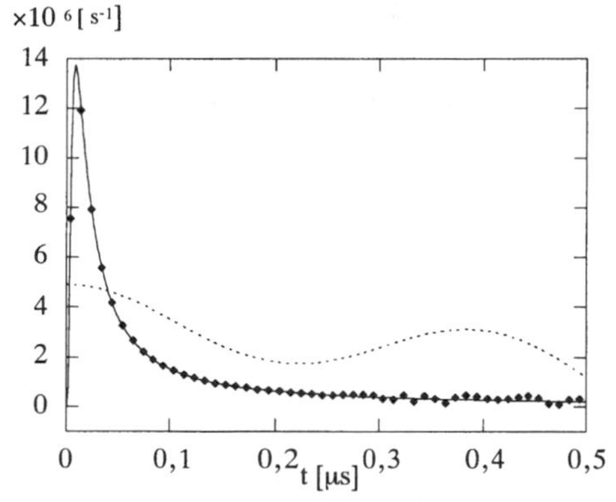

Figure 5.10. Comparison of $p(t)$ evaluated by: —— the formula (5.102) with 10 points, brute force numerical inversion with ◆ 2000 and ···· 10 sample points in frequency.

We have considered a line 96.8 m long; in this case the one-way transit time is $T = 484$ ns and the damping rate is $\mu = 5.49 \cdot 10^5\,\mathrm{s}^{-1}$.

In Figs. 5.7a and 5.7b we show, respectively, the amplitude spectra of the characteristic admittance Y_c and of its remainder Y_{cb}, given by Eq. (5.83); in Fig. 5.7c the corresponding time-domain remainder y_{cb} is reported. The computation of the time-domain remainder has been easy due to the asymptotic behavior of the remainder in the frequency domain as shown in Fig. 5.7b. A direct numerical inversion of the overall characteristic admittance would be meaningless.

In Fig. 5.8 both the amplitude spectra of the propagation function P and of its remainder P_b (see Eq. (5.104)) are plotted. In Fig. 5.9a two evaluations of the time-domain remainder $p_b(t)$ are reported. They are obtained, respectively, by means of (full line) 2000 and (triangles) 10 sample points for the evaluation of the Fourier integral. As in the case of the characteristic admittance, the computation of the remainder $p_b(t)$ appears to be very easy. In particular, it can be performed by using a few sample points in frequency due to the asymptotic behavior of P_b. In Fig. 5.9b the leading term $p_i(t)$ of the impulse response $p(t)$, given by Eq. (5.103), is compared with the remainder $p_b(t)$.

The time domain impulse response $p(t)$ could also be evaluated by a numerical direct inversion of $P(s)$ (remember that when there is the skin effect the Dirac pulse in $p(t)$ is replaced by the pseudoimpulsive function $\delta(t;\alpha)$). Figure 5.10 shows the comparison of the impulse response $p(t)$ obtained by using Eq. (5.102) and that obtained by applying a *brute force* numerical inversion to $P(s)$. The full line plot refers to the result obtained through Eq. (5.102) and by using 10 sample points in frequency to evaluate the remainder. The *brute force* evaluation has been performed by using (rhombi) 2000 and (dot line) 10 sample points. The improvement due to the analytical extraction of the "irregular" term is impressive.

CHAPTER 6

Lossy Multiconductor Transmission Lines

6.1. INTRODUCTION

This chapter discusses the characterization of uniform lossy multiconductor transmission lines as multiports in the time domain. First, we shall deal with lossy lines with frequency-independent parameters, and then we shall extend the analysis to the more general case of parameters depending on the frequency.

In Chapter 3 we have determined the general solution of the ideal multiconductor line equations in the time domain by decoupling them through the similarity transformations (3.5) and (3.6). In the more general situation of lossy multiconductor lines this is not possible except for certain special cases that exhibit structural symmetry.

The equations for lossy multiconductor lines with frequency-independent parameters are (see Chapter 1),

$$-\frac{\partial \mathbf{v}}{\partial x} = \mathrm{L}\frac{\partial \mathbf{i}}{\partial t} + \mathrm{R}\mathbf{i}, \tag{6.1}$$

$$-\frac{\partial \mathbf{i}}{\partial x} = \mathrm{C}\frac{\partial \mathbf{v}}{\partial t} + \mathrm{G}\mathbf{v}. \tag{6.2}$$

By applying the linear transformations (3.5) and (3.6) to Eqs. (6.1) and

(6.2) we obtain

$$-\frac{\partial \mathbf{u}}{\partial x} = \mathrm{T}_{\mathrm{v}}^{-1}\mathrm{L}\mathrm{T}_{\mathrm{i}}\frac{\partial \mathbf{f}}{\partial t} + \mathrm{T}_{\mathrm{v}}^{-1}\mathrm{R}\mathrm{T}_{\mathrm{i}}\mathbf{f}, \tag{6.3}$$

$$-\frac{\partial \mathbf{f}}{\partial x} = \mathrm{T}_{\mathrm{i}}^{-1}\mathrm{C}\mathrm{T}_{\mathrm{v}}\frac{\partial \mathbf{u}}{\partial t} + \mathrm{T}_{\mathrm{i}}^{-1}\mathrm{G}\mathrm{T}_{\mathrm{v}}\mathbf{u}. \tag{6.4}$$

By choosing T_{v} equal to the modal matrix of LC, and T_{i} equal to the modal matrix of CL, both the matrices $\mathrm{T}_{\mathrm{v}}^{-1}\mathrm{L}\mathrm{T}_{\mathrm{i}}$ and $\mathrm{T}_{\mathrm{i}}^{-1}\mathrm{C}\mathrm{T}_{\mathrm{v}}$ become diagonal (see relations (3.56)), but the matrices $\mathrm{T}_{\mathrm{v}}^{-1}\mathrm{R}\mathrm{T}_{\mathrm{i}}$ and $\mathrm{T}_{\mathrm{i}}^{-1}\mathrm{G}\mathrm{T}_{\mathrm{v}}$ are not diagonal except for certain special cases that we shall deal with separately in the next section.

The problem of the solution of the Eqs. (6.1) and (6.2) may be approached in another way. Assuming that the line is initially at rest, in the Laplace domain Eqs. (6.1) and (6.2) become

$$-\frac{d\mathbf{V}}{dx} = Z(s)\mathbf{I}, \tag{6.5}$$

$$-\frac{d\mathbf{I}}{dx} = \mathrm{Y}(s)\mathbf{V}, \tag{6.6}$$

where $\mathbf{V}(x; s)$ and $\mathbf{I}(x; s)$ are respectively the Laplace transforms of $\mathbf{v}(x, t)$ and $\mathbf{i}(x, t)$, and the per-unit-length impedance and admittance matrix functions Z and Y are given by

$$\mathrm{Z}(s) = \mathrm{R} + s\mathrm{L}, \tag{6.7}$$

$$\mathrm{Y}(s) = \mathrm{G} + s\mathrm{C}. \tag{6.8}$$

Nonzero initial conditions can be dealt with by using equivalent distributed sources along the line. The more general situation, in which the Laplace domain line equations have distributed sources, may be dealt with as in the two-conductor case (see §4.10).

Except for its vector nature, the structure of the system of equations (6.5) and (6.6) is formally the same as that of the system of equations (4.14) and (4.15) relevant to lossy two-conductor lines. This suggests reasoning in the same way. In fact, we shall show in this chapter how it is possible to extend to multiconductor lines all the line representations as multiports introduced in Chapter 4. For multiconductor lines the describing functions, which characterize the representations, are matrix operators whose spectral decomposition depends on s in a complicated way that cannot be represented analytically. This is the difficulty introduced by the vector nature of the problem.

As with lossy two-conductor lines, the impulse responses, characterizing the equivalent multiport in the time domain, contain highly irregular *terms* that are "unbounded" and cannot be handled numerically. As was done for lossy two-conductor lines with frequency-dependent parameters (see previous chapter), these difficulties may be overcome if the asymptotic expressions of the describing operator in the Laplace domain are known. A general method that is simple and elegant has been proposed by the authors to evaluate the asymptotic expressions of the describing operators (Maffucci and Miano, 1999a; Maffucci and Miano, 1999b). It is based on the perturbation theory of the spectrum of symmetric matrices and can be easily and effectively applied to the most general case of frequency-dependent lossy multiconductor lines. In this chapter, we shall describe this method in detail and apply it to obtain the analytical expressions of the irregular parts of the impulse responses.

6.2. LOSSY MULTICONDUCTOR LINES EXHIBITING A STRUCTURAL SYMMETRY

The system of the equations (6.1) and (6.2) may be diagonalized when the lines exhibit a particular structure symmetry leading to symmetric tridiagonal Toeplitz parameter matrices (e.g., Romeo and Santomauro, 1987; Canavero *et al.*, 1988; Corti *et al.*, 1997). Let M be any of the matrices R, L, G, or C. Let M have the following structure:

$$M_{ij} = \begin{cases} a & \text{for } i = j, \\ b & \text{for } i = j \pm 1, \\ 0 & \text{otherwise.} \end{cases} \tag{6.9}$$

Such matrices describe a particular condition, which is however a good approximation for many cases of practical interest: It is assumed that the conductors are identical and equally spaced, that each conductor is only coupled to the closest one on the right and on the left, and that the side effects are negligible.

The eigenvalues of the matrix M are given by

$$\lambda_{M,k} = a - 2b\cos\left(\frac{k\pi}{n+1}\right) \quad \text{for } k = 1, 2, \ldots, n, \tag{6.10}$$

while the eigenvectors do not depend on the matrix entries, but only on the matrix dimension n. In fact, the entries of the modal matrix T are given by

$$T_{ij} = \gamma_j \Phi_{i-1}\left[-2\cos\left(j\frac{\pi}{n+1}\right)\right] \quad \text{for } i, j = 1, 2, \ldots, n, \tag{6.11}$$

where $\Phi_k(y)$ is the polynomial defined by the recursive equation

$$\Phi_k(y) = y\Phi_{k-1}(y) - \Phi_{k-2}(y) \quad \text{for } k = 2, 3, \ldots, n, \tag{6.12}$$

with

$$\Phi_0(y) = 1 \quad \text{and} \quad \Phi_1(y) = y. \tag{6.13}$$

Furthermore, γ_j is an arbitrary factor that can be chosen to normalize the eigenvectors.

The matrix T diagonalizes the matrices L, C, R, and G simultaneously,

$$\mathrm{T}^{-1}\mathrm{LT} = \mathrm{diag}(\lambda_{\mathrm{L},1}, \ldots, \lambda_{\mathrm{L},n}), \tag{6.14}$$

$$\mathrm{T}^{-1}\mathrm{CT} = \mathrm{diag}(\lambda_{\mathrm{C},1}, \ldots, \lambda_{\mathrm{C},n}), \tag{6.15}$$

$$\mathrm{T}^{-1}\mathrm{RT} = \mathrm{diag}(\lambda_{\mathrm{R},1}, \ldots, \lambda_{\mathrm{R},n}), \tag{6.16}$$

$$\mathrm{T}^{-1}\mathrm{GT} = \mathrm{diag}(\lambda_{\mathrm{G},1}, \ldots, \lambda_{\mathrm{G},n}), \tag{6.17}$$

where the eigenvalues are obtained according to Eq. (6.10). Now, by applying the linear transformations

$$\mathbf{v}(x, t) = \mathbf{T}\mathbf{u}(x, t) \quad \text{and} \quad \mathbf{i}(x, t) = \mathbf{T}\mathbf{f}(x, t), \tag{6.18}$$

the line Eqs. (6.1) and (6.2) become, for $k = 1, \ldots, n$,

$$-\frac{\partial u_k}{\partial x} = \lambda_{\mathrm{L},k}\frac{\partial f_k}{\partial t} + \lambda_{\mathrm{R},k} f_k, \tag{6.19}$$

$$-\frac{\partial f_k}{\partial x} = \lambda_{\mathrm{C},k}\frac{\partial u_k}{\partial t} + \lambda_{\mathrm{G},k} u_k. \tag{6.20}$$

In this way we recast the vector problem (6.1) and (6.2) in terms of n scalar problems (6.19) and (6.20). Each of these scalar problems is the same problem as that dealt with in Chapter 4, where lossy two-conductor lines are studied.

Remarks

(i) The matrix T also diagonalizes the products LC and CL, as

$$\begin{aligned}\mathrm{T}^{-1}\mathrm{LCT} &= (\mathrm{T}^{-1}\mathrm{LT})(\mathrm{T}^{-1}\mathrm{CT}) \\ &= \mathrm{diag}(\lambda_{\mathrm{L},1}\lambda_{\mathrm{C},1}, \ldots, \lambda_{\mathrm{L},n}\lambda_{\mathrm{C},n}).\end{aligned} \tag{6.21}$$

(ii) We may choose two different transformation matrices for the current and the voltages, as was done in Chapter 3. Additionally, we may choose a set of eigenvectors such that they are normalized with respect the scalar product $\langle \mathbf{a}, \mathbf{b} \rangle_{\mathrm{L}} = \mathbf{a}^{\mathrm{T}} \mathrm{L}^{-1} \mathbf{b}$. In this way we obtain the matrix T_{v}. This can be done by carefully choosing the arbitrary factors γ_j in Eq. (6.11). In the same way, we may choose the matrix T by imposing a unitary norm to the eigenvectors with respect to the scalar product $\langle \mathbf{a}, \mathbf{b} \rangle_{\mathrm{C}} = \mathbf{a}^{\mathrm{T}} \mathrm{C}^{-1} \mathbf{b}$. ◇

6.3. LOSSY MULTICONDUCTOR LINE EQUATIONS IN THE LAPLACE DOMAIN

In the general case, we have to solve the system of equations (6.5) and (6.6). Starting from these equations, by derivation and substitution, we obtain the vector second-order differential equation for $\mathbf{V}(x; s)$

$$\frac{d^2\mathbf{V}}{dx^2} - s^2 \Lambda(s) \mathbf{V} = \mathbf{0}, \tag{6.22}$$

where the matrix Λ is

$$\Lambda(s) = \frac{\mathrm{Z}(s)\mathrm{Y}(s)}{s^2}. \tag{6.23}$$

The current vector $\mathbf{I}$ satisfies a similar equation where, instead of the matrix Λ, there is the matrix Π, defined as

$$\Pi(s) = \frac{\mathrm{Y}(s)\mathrm{Z}(s)}{s^2}. \tag{6.24}$$

However, once the expression of the general solution for the voltage is known, from Eq. (6.5) we shall evaluate the expression of the general solution for the current. Note that in the lossless limit, $\Lambda = \mathrm{LC}$ and $\Pi = \mathrm{CL}$.

To determine the general solution of the vector equation (6.22), a preliminary study is needed of the properties of the matrix Λ.

6.3.1. The Eigenvalues and Eigenvectors of the Matrices $\Lambda = \mathbf{ZY}/s^2$ and $\Pi = \mathbf{YZ}/s^2$

Let us consider the eigenvalue problem

$$\Lambda(s)\mathbf{e}(s) = \beta(s)\mathbf{e}(s). \tag{6.25}$$

The matrices Λ and ZY have the same eigenvectors. The generic eigenvalue β_h and the corresponding eigenvectors $\mathbf{e}_h$ are monodrome functions of the variable s, analytic through the s-plane, except for a finite number of isolated singularities.

The matrix Λ has n linearly independent eigenvectors $\mathbf{e}_1(s)$, $\mathbf{e}_2(s), \ldots, \mathbf{e}_n(s)$, whether or not the corresponding eigenvalues are distinct. The demonstration is very simple. The matrices Z and Y are symmetric and, moreover, they become positive definite on the positive real axis because of the general properties of the matrices R, L, G, and C (see §1.3). Thus, the matrix Λ for s belonging to the positive real axis $s = \sigma > 0$, has all the properties of the matrix LC shown in §3.1.1. In particular, $\Lambda(\sigma)$ always has n linearly independent eigenvectors $\mathbf{e}_1(\sigma)$, $\mathbf{e}_2(\sigma), \ldots, \mathbf{e}_n(\sigma)$. The same results extend to the complex matrix $\Lambda(s)$ by analytic continuation of $\mathbf{e}_1(\sigma)$, $\mathbf{e}_2(\sigma), \ldots, \mathbf{e}_n(\sigma)$ in the s-plane. Furthermore, the eigenvalues of Λ evaluated along the positive real axis $\beta_1(\sigma)$, $\beta_2(\sigma), \ldots, \beta_n(\sigma)$ are all real and positive. Indeed, for the eigenvalues of Λ a more general property holds because of the properties of Z and Y: The real parts of all the eigenvalues of Λ are positive in the region $Re\{s\} > \sigma_0$ with $\sigma_0 < 0$; the actual value of σ_0 depends only on the structure of Z(s) and Y(s).

The matrix $\mathrm{T_v}(s)$

$$\mathrm{T_v}(s) = |\mathbf{e}_1(s) \vdots \mathbf{e}_2(s) \vdots \cdots \vdots \mathbf{e}_n(s)|, \tag{6.26}$$

is the modal matrix of Λ, that is,

$$\mathrm{T_v^{-1}} \Lambda \mathrm{T_v} = \mathrm{B}(s), \tag{6.27}$$

where B is the diagonal matrix of the eigenvalues of $\Lambda(s)$,

$$\mathrm{B}(s) = \mathrm{diag}(\beta_1(s), \beta_2(s), \ldots, \beta_n(s)). \tag{6.28}$$

Note that

$$\mathrm{T_v^{-1}}(\mathrm{ZY})\mathrm{T_v} = s^2\mathrm{B}. \tag{6.29}$$

Two different eigenvectors $\mathbf{e}_i$ and $\mathbf{e}_j$ satisfy the notable property (see §3.1.1)

$$\mathbf{e}_i^{\mathrm{T}}(s)(\mathrm{R}/s + \mathrm{L})^{-1}\mathbf{e}_j(s) = 0 \tag{6.30}$$

Furthermore, we can always normalize the generic eigenvector $\mathbf{e}_i$ in such a way that

$$\mathbf{e}_i^{\mathrm{T}}(s)(\mathrm{R}/s + \mathrm{L})^{-1}\mathbf{e}_i(s) = 1. \tag{6.31}$$

As the matrices Z and Y are symmetric, the matrix YZ is the adjoint matrix of ZY (see §3.1.2). Therefore, the eigenvalues of $\Pi(s)$ coincide with those of the matrix $\Lambda(s)$. As for the eigenvectors $\mathbf{g}_1(s)$,

$\mathbf{g}_2(s), \ldots, \mathbf{g}_n(s)$ of $\Pi(s)$, they are different from those of $\Lambda(s)$, but have the same properties. With

$$\mathrm{T_i}(s) = |\mathbf{g}_1(s) \vdots \mathbf{g}_2(s) \vdots \cdots \vdots \mathbf{g}_n(s)|. \tag{6.32}$$

we indicate the modal matrix of $\Pi(s)$.

Two different eigenvectors $\mathbf{g}_i$ and $\mathbf{g}_j$ satisfy the notable property (see §3.1.1)

$$\mathbf{g}_i^{\mathrm{T}}(s)(\mathrm{G}/s + \mathrm{C})^{-1}\mathbf{g}_j(s) = 0. \tag{6.33}$$

Furthermore, we can always normalize the generic eigenvector $\mathbf{g}_i$ in such a way that

$$\mathbf{g}_i^{\mathrm{T}}(s)(\mathrm{G}/s + \mathrm{C})^{-1}\mathbf{g}_i(s) = 1. \tag{6.34}$$

Remark

If the eigenvectors of $\Lambda(s)$ and $\Pi(s)$ satisfy the normalization conditions (6.31) and (6.34), they may always be chosen in such a way that

$$\mathbf{g}_h(s) = s\sqrt{\beta_h(s)}\,\mathrm{Z}^{-1}(s)\mathbf{e}_h(s), \tag{6.35}$$

$$\mathbf{e}_h(s) = s\sqrt{\beta_h(s)}\,\mathrm{Y}^{-1}(s)\mathbf{g}_h(s). \tag{6.36}$$

Because the generic eigenvalue β_h evaluated along the positive real axis is always real and positive, the function $\sqrt{\beta_h(s)}$ can not have branch points there. With $\sqrt{\beta_h(s)}$ we always indicate the monodrome function corresponding to the branch of $\sqrt{\beta_h(s)}$ that has a positive real part for $Re\{s\} > 0$, that is, we always cut the complex plane along the negative real axis. From the relations (6.35) and (6.36) we find that the modal matrices $\mathrm{T_i}$ and $\mathrm{T_v}$ are related through

$$\mathrm{T_i} = s\mathrm{Z}^{-1}\mathrm{T_v}\mathrm{A}, \tag{6.37}$$

$$\mathrm{T_v} = s\mathrm{Y}^{-1}\mathrm{T_i}\mathrm{A}, \tag{6.38}$$

where

$$\mathrm{A} = \sqrt{\mathrm{B}} = \mathrm{diag}(\sqrt{\beta_1}, \sqrt{\beta_2}, \ldots, \sqrt{\beta_n}). \quad \diamond \tag{6.39}$$

6.3.2. The General Solution

Now, in order to solve Eq. (6.22), we shall use the concept of the function of a matrix given in Appendix A. Let us consider the matrix

$$\mathrm{D}(x) = \exp(\mathrm{H}x), \tag{6.40}$$

where H is a constant matrix. From the definition of the exponential of a matrix, we obtain the notable identity

$$\frac{d\mathrm{D}}{dx} = \mathrm{H}\exp(\mathrm{H}x) = \exp(\mathrm{H}x)\mathrm{H}. \qquad (6.41)$$

The matrices H and exp(Hx) commutate because they have the same eigenvectors. The identity (6.41) suggests looking for a solution of Eq. (6.22) of the type

$$\mathbf{V}(x; s) = \exp[\mathrm{H}(s)(x - x_0)]\mathbf{U}(s), \qquad (6.42)$$

where $\mathbf{U}$ is a vector and x_0 a scalar constant. By substituting the expression (6.42) in Eq. (6.22) and using the identity (6.41), we obtain

$$(\mathrm{H}^2 - s^2\Lambda)\{\exp[\mathrm{H}(x - x_0)]\mathbf{U}\} = \mathbf{0}. \qquad (6.43)$$

Since the vector $\mathbf{U}$ and the scalar x are arbitrary, the matrix H must satisfy the equation

$$\mathrm{H}^2 = s^2\Lambda. \qquad (6.44)$$

Equation (6.44) has the two solutions:

$$\mathrm{H}_{\pm}(s) = \pm\mathrm{K}(s), \qquad (6.45)$$

where

$$\mathrm{K}(s) = s\sqrt{\Lambda(s)} = s\mathrm{T}_{\mathrm{v}}(s)\mathrm{A}(s)\mathrm{T}_{\mathrm{v}}^{-1}(s). \qquad (6.46)$$

Thus the general solution of Eq. (6.22) may be expressed as

$$\mathbf{V}(x; s) = \mathrm{Q}(x; s)\mathbf{V}^{+}(s) + \mathrm{Q}(d - x; s)\mathbf{V}^{-}(s), \qquad (6.47)$$

where the local propagation matrix operator $\mathrm{Q}(x; s)$ is given by

$$\mathrm{Q}(x; s) = \exp[-\mathrm{K}(s)x], \qquad (6.48)$$

and $\mathbf{V}^{+}(s)$ and $\mathbf{V}^{-}(s)$ are two arbitrary vector functions of the complex variable s, depending only on the boundary conditions at the line ends. Expression (6.47) is formally the same as that obtained in the two-conductor case; see Eq. (4.31). Note that $\mathrm{Q}(0; s) = \mathrm{I}$, where I is the identity matrix.

The functions $\mathrm{Q}(x; s)\mathbf{V}^{+}(s)$ and $\mathrm{Q}(d - x; s)\mathbf{V}^{-}(s)$, respectively, bear a resemblance to the forward and backward waves of an ideal multiconductor line in the Laplace domain. Contrary to the choice that has been made in Chapter 3, here $\mathbf{V}^{+}(s)$ represents the amplitudes of the forward voltage waves at the end $x = 0$, and $\mathbf{V}^{-}(s)$ represents the amplitudes of the backward voltage waves at the end $x = d$.

As in the two-conductor case, the voltage and the current distributions along the line are uniquely determined by the functions $\mathbf{V}^{+}(s)$ and $\mathbf{V}^{-}(s)$ and vice versa because the local propagation operator is always invertible. Thus $\mathbf{V}^{+}$ and $\mathbf{V}^{-}$ may be assumed as the state variables of the line (in the Laplace domain), as we have done for ideal transmission lines.

By substituting the expression (6.47) in Eq. (6.5) and using the property (6.41), for the current distribution we obtain

$$\mathbf{I}(x;s) = \mathrm{Z}_c^{-1}(s)\{\mathrm{Q}(x;s)\mathbf{V}^{+}(s) - \mathrm{Q}(d-x;s)\mathbf{V}^{-}(s)\}, \tag{6.49}$$

where Z_c is the *characteristic impedance matrix* of the line given by

$$\mathrm{Z}_c(s) = \sqrt{(\mathrm{Z}(s)\mathrm{Y}(s))^{-1}}\,\mathrm{Z}(s) = \mathrm{K}^{-1}(s)\mathrm{Z}(s). \tag{6.50}$$

Remember that, in general, $\sqrt{\mathrm{ZY}} \neq \sqrt{\mathrm{YZ}}$.

For physical reasons, the matrix operators $\mathrm{K}(s)$, $\mathrm{Z}_c(s)$, and $\mathrm{Y}_c(s)$ are certainly analytic for $Re\{s\} > \sigma_c$ with $\sigma_c < 0$; the actual value of σ_c depends on the structure of $\mathrm{Z}(s)$ and $\mathrm{Y}(s)$. Applying the reciprocity property of §1.10.4, we immediately find that the characteristic impedance and admittance matrices are symmetric. Furthermore, applying the Poynting theorem equation (1.104) from the passivity of the line we obtain that the real part of $\mathrm{Z}_c(i\omega)$ and the real part of $\mathrm{Y}_c(i\omega)$ are positive definite.

Remark

From Eqs. (6.37) and (6.38) we easily obtain

$$\mathrm{T}_{\mathrm{v}}^{-1}\mathrm{Z}\mathrm{T}_{\mathrm{i}} = \mathrm{T}_{\mathrm{i}}^{-1}\mathrm{Y}\mathrm{T}_{\mathrm{v}} = s\mathrm{A}. \tag{6.51}$$

From Eqs. (6.38) and (6.50) we also obtain

$$\mathrm{T}_{\mathrm{v}}(s) = \mathrm{Z}_c(s)\mathrm{T}_{\mathrm{i}}(s). \tag{6.52}$$

Furthermore, the eigenvectors of Λ and the eigenvectors of Π are orthogonal because Π is the adjoint matrix of Λ (see §3.1.2). Consequently, we have also the notable relation

$$\mathrm{T}_{\mathrm{i}}^{\mathrm{T}}\mathrm{T}_{\mathrm{v}} = \mathrm{A}. \tag{6.53}$$

By applying the properties (6.51), we can directly diagonalize the system of equations (6.5) and (6.6) with the similarity transformations

$$\mathbf{V}(x;s) = \mathrm{T}_{\mathrm{v}}(s)\mathbf{U}(x;s), \tag{6.54}$$

$$\mathbf{I}(\mathrm{x};s) = \mathrm{T}_{\mathrm{i}}(s)\mathbf{U}(x;s), \tag{6.55}$$

obtaining

$$-\frac{d\mathbf{U}}{dx} = s\mathrm{A}(s)\mathbf{F}, \tag{6.56}$$

$$-\frac{d\mathbf{F}}{dx} = s\mathrm{A}(s)\mathbf{U}. \tag{6.57}$$

In the lossless case $\Lambda = \mathrm{LC}$, and the corresponding time domain representation of Eqs. (6.56) and (6.57) reduces to Eqs. (3.57) and (3.58) written in the Laplace domain. In the general case, however, although this diagonalization is a simple and elegant way to recast the problem, the difficulty in evaluating the solution is the same as we shall find in evaluating the general solution (6.47) and (6.49). In both cases the main difficulty consists in the fact that the eigenvalues and the eigenvectors are complicated functions of s that can not be evaluated analytically.

6.4. LOSSY MULTICONDUCTOR TRANSMISSION LINES AS MULTIPORTS IN THE LAPLACE DOMAIN

In general, the values of the voltages and the currents at the line ends are not known, but they actually are also unknowns of the problem, that is, they depend on the actual network to which the line is connected.

We have already seen how lossy two-conductor lines can be characterized as two-ports in the time domain (see Chapters 4 and 5). The same approach can be adopted to represent lossy multiconductor transmission lines of finite length as multiports.

Let us consider a multiconductor line consisting of $n + 1$ conductors (Fig. 6.1). The voltages and the currents at the line ends are related to the voltage and the current distributions along the line through the relations

$$\mathbf{V}_1(s) = \mathbf{V}(x = 0; s), \tag{6.58}$$

$$\mathbf{I}_1(s) = \mathbf{I}(x = 0; s), \tag{6.59}$$

and

$$\mathbf{V}_2(s) = \mathbf{V}(x = d; s), \tag{6.60}$$

$$\mathbf{I}_2(s) = -\mathbf{I}(x = d; s). \tag{6.61}$$

The reference directions are always chosen according to the normal convention for the ports corresponding to the two ends of the line.

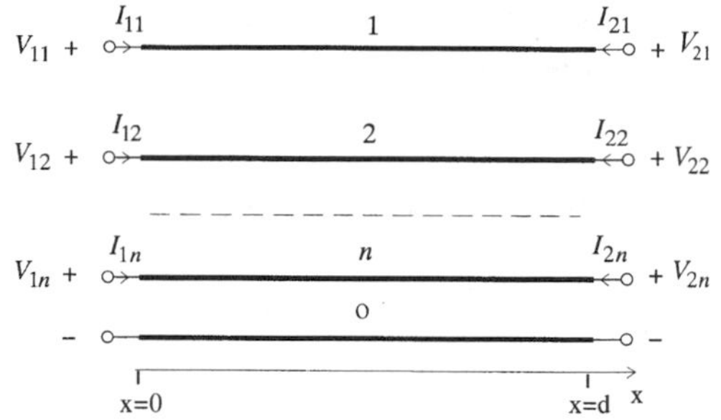

Figure 6.1. Sketch of a multiconductor transmission line of finite length.

To determine the characteristic relations of the multiport representing the line we have to particularize the general solution of the line equations at the line ends.

Let us introduce the *global propagation operator*

$$\mathrm{P}(s) = \exp[-\mathrm{K}(s)d]. \tag{6.62}$$

This plays an important part in the theory developed in this chapter. As with the *global propagation operator* (4.76) introduced in Chapter 4 to describe two-conductor lossy lines, the matrix operator $\mathrm{P}(s)$ links the amplitudes of the forward voltage waves at the line end $x = d$ with those at the line end $x = 0$; $\mathbf{V}^+$ represents the amplitudes of the forward voltage waves at $x = 0$ and $\mathrm{P}\mathbf{V}^+$ represents their amplitudes at $x = d$. As the line is uniform, P is also the operator linking the amplitudes of the backward voltage waves at the line end $x = 0$ with those at the line end $x = d$; $\mathbf{V}^-$ represents the amplitudes of the backward voltage wave at $x = d$ and $\mathrm{P}\mathbf{V}^-$ represents their amplitudes at $x = 0$.

Specifying the expressions (6.47) and (6.49) at the end $x = 0$, and recalling that $\mathrm{Q}(0; s) = \mathrm{I}$, we obtain

$$\mathbf{V}_1(s) = \mathbf{V}^+(s) + \mathrm{P}(s)\mathbf{V}^-(s), \tag{6.63}$$

$$\mathrm{Z}_c(s)\mathbf{I}_1(s) = \mathbf{V}^+(s) - \mathrm{P}(s)\mathbf{V}^-(s), \tag{6.64}$$

whereas specifying them at the end $x = d$ we obtain

$$\mathbf{V}_2(s) = \mathrm{P}(s)\mathbf{V}^+(s) + \mathbf{V}^-(s), \tag{6.65}$$

$$-\mathrm{Z}_c(s)\mathbf{I}_2(s) = \mathrm{P}(s)\mathbf{V}^+(s) - \mathbf{V}^-(s). \tag{6.66}$$

Subtracting Eqs. (6.63) and (6.64) member by member and adding Eqs. (6.65) and (6.66) member by member, we have, respectively,

$$\mathbf{V}_1(s) - \mathrm{Z}_c(s)\mathbf{I}_1(s) = 2\mathrm{P}(s)\mathbf{V}^{-1}(s), \tag{6.67}$$

$$\mathbf{V}_2(s) - \mathrm{Z}_c(s)\mathbf{I}_2(s) = 2\mathrm{P}(s)\mathbf{V}^{+}(s). \tag{6.68}$$

As in the two-conductor case dealt with in Chapter 4, if the state of the line represented by $\mathbf{V}^+(s)$ and $\mathbf{V}^-(s)$ were completely known, these equations would completely determine the terminal behavior of the line. Actually, as with two-conductor transmission lines, the voltage waves $\mathbf{V}^+(s)$ and $\mathbf{V}^+(s)$ are themselves unknowns.

As with two-conductor lines (see Chapter 4), different formulations of the equations governing the state are possible. By using Eq. (6.63) it is possible to express the amplitudes of the outcoming forward waves at $x = 0$, $\mathbf{V}^+$, as a function of the voltages and the amplitudes of the incoming backward waves at that end,

$$\mathbf{V}^+(s) = \mathbf{V}_1(s) - \mathrm{P}(s)\mathbf{V}^-(s). \tag{6.69}$$

In the same way, by using Eq. (6.65) it is possible to express the amplitudes of the outcoming backward waves at $x = d$, $\mathbf{V}^-$, as a function of the voltages and the amplitudes of the incoming forward waves at that end,

$$\mathbf{V}^-(s) = \mathbf{V}_2(s) - \mathrm{P}(s)\mathbf{V}^+(s). \tag{6.70}$$

By contrast, summing Eqs. (6.63) and (6.64) termwise and subtracting Eqs. (6.65) and (6.66), termwise, we have

$$2\mathbf{V}^+(s) = \mathbf{V}_1(s) + \mathrm{Z}_c(s)\mathbf{I}_1(s), \tag{6.71}$$

$$2\mathbf{V}^-(s) = \mathbf{V}_2(s) + Z_c(s)\mathbf{I}_2(s). \tag{6.72}$$

The state equations (6.69) and (6.70) describe in *implicit form* the relation between the state of the line and the electrical variables at the line ends, whereas the state equations (6.71) and (6.72) provide the same relation, but in *explicit form*.

Remark

We observe from Eq. (6.71) that $\mathbf{V}^+$ should be equal to zero if the line were connected at the left end to a multiport with impedance $\mathrm{Z}_c(s)$,

$$\mathbf{V}_1(s) = -\mathrm{Z}_c(s)\mathbf{I}_1(s), \tag{6.73}$$

— *perfectly matched line at the left end* — and hence

$$\mathbf{v}^{+}(t) = L^{-1}\{\mathbf{V}^{+}(s)\}$$

should be equal to zero. The same considerations hold for the backward wave if the line is *perfectly matched at the right end*,

$$\mathbf{V}_2(s) = -\mathrm{Z}_c(s)\mathbf{I}_2(s). \tag{6.74}$$

In general, these matching conditions are not satisfied. Thus, a forward wave with amplitude $\mathbf{V}^{+}$ is generated at the left line end $x = 0$ and propagates toward the other end $x = d$ where its amplitude is $\mathrm{P}\mathbf{V}^{+}$. Similarly, the backward wave excited at $x = d$ with amplitude $\mathbf{V}^{-}$ propagates toward the left line end, where its amplitude is $\mathrm{P}\mathbf{V}^{-}$. ◇

State equations in implicit form similar to Eqs. (6.69) and (6.70) involving the terminal variables $\mathbf{I}_1$ and $\mathbf{I}_2$ instead of $\mathbf{V}_1$ and $\mathbf{V}_2$, may be obtained from Eqs. (6.64) and (6.66). Because these equations lead to a time domain model that requires more convolution products per iteration than those required for Eqs. (6.69) and (6.70), we shall only consider the implicit formulation based on the latter equations.

Substituting the expression of $\mathbf{V}^{+}$ given by Eq. (6.71) into Eq. (6.68), and the expression of $\mathbf{V}^{-}$ given by Eq. (6.72) into Eq. (6.67), we obtain two linearly independent equations in terms of $\mathbf{V}_1$, $\mathbf{V}_2$, $\mathbf{I}_1$, and $\mathbf{I}_2$:

$$\mathbf{V}_1(s) - \mathrm{Z}_c(s)\mathbf{I}_1(s) - \mathrm{P}(s)[\mathbf{V}_2(s) + \mathrm{Z}_c(s)\mathbf{I}_2(s)] = \mathbf{0}, \tag{6.75}$$

$$\mathbf{V}_2(s) - \mathrm{Z}_c(s)\mathbf{I}_2(s) - \mathrm{P}(s)[\mathbf{V}_1(s) + \mathrm{Z}_c(s)\mathbf{I}_1(s)] = \mathbf{0}. \tag{6.76}$$

As in the case of two-conductor lines, Eqs. (6.67) to (6.70) and Eqs. (6.75) and (6.76) are a fundamental result of considerable importance. Specifically, they are two different mathematical models to describe the multiport representing the line. The set of equations (6.67) to (6.70) govern the behavior of the internal state of the line represented by $\mathbf{V}^{+}$ and $\mathbf{V}^{-}$, as well as the terminal properties. They are the *internal* or *input-state-output description* of a multiconductor transmission line in the Laplace domain: Equations (6.69) and (6.70) govern the behavior of the state (*state equations*), whereas Eqs. (6.67) and (6.68) describe the terminal properties (*output equations*).

Equation (6.67) states that in the Laplace domain, the voltage at the line end $x = 0$, $\mathbf{V}_1$, is equal to the sum of two terms as in the case of ideal multiconductor transmission lines. If the line were perfectly matched at the end $x = d$, it would behave as an n-port with impedance Z_c at $x = 0$ and $\mathbf{V}_1$ would be equal to $\mathrm{Z}_c\mathbf{I}_1$. By contrast, $2\mathrm{P}(s)\mathbf{V}^{-}(s)$

is the voltage that would be at end $x = 0$ if the line were connected to an open circuit at that end.

If the backward voltage wave $\mathbf{V}^{-}(s)$ were known, the behavior of the line at the left end would be determined entirely by Eq. (6.67), as in the semiinfinite line case (see §4.4). However, for transmission lines of finite length, the backward voltage wave depends on what is connected both to the right and left ends of the line due to the reflections, and thus it is an unknown of the problem. Similar considerations hold for Eq. (6.68). Therefore, the behavior of each end of a lossy multiconductor line can be represented through a Thévenin- or Norton-type equivalent representation by using suitable controlled sources. The governing laws for $\mathbf{V}^{-}(s)$ and $\mathbf{V}^{+}(s)$ are the state equations (6.69) and (6.70).

The set of equations (6.75) and (6.76) describes only the terminal properties of the line. They are the *external* or *input-output description* of a multiconductor line in the Laplace domain. As in the two-conductor case, the *input-state-output description* and the *input-output description* are not equivalent from the computational point of view (see §4.7). In the following two sections we shall analyze in detail these two descriptions. Both descriptions are completely characterized by the two matrix operators $\mathrm{Z}_c(s)$ and $\mathrm{P}(s)$, which we call *describing functions* of the multiport representing the line behavior in the Laplace domain.

The terminal behavior of a multiconductor transmission line may also be characterized through the scattering parameters as done in the two-conductor case (see §4.11). We shall leave the extension of this characterization to multiconductor lines as an exercise.

6.5. THE INPUT-STATE-OUTPUT DESCRIPTION AND THE EQUIVALENT REPRESENTATIONS OF THÉVENIN AND NORTON TYPE

In this section we shall deal with the input-state-output description based on the system of equations (6.67) to (6.70), and we shall extend all the results obtained in §4.6.1.

6.5.1. Laplace Domain

Because in Eqs. (6.67) and (6.68) the state of the line appears through $2\mathrm{P}(s)\mathbf{V}^{-}(s)$ and $2\mathrm{P}(s)\mathbf{V}^{+}(s)$, we rewrite these equations as

$$\mathbf{V}_1(s) - \mathrm{Z}_c(s)\mathbf{I}_1(s) = \mathbf{W}_1(s), \tag{6.77}$$

$$\mathbf{V}_2(s) - \mathrm{Z}_c(s)\mathbf{I}_2(s) = \mathbf{W}_2(s), \tag{6.78}$$

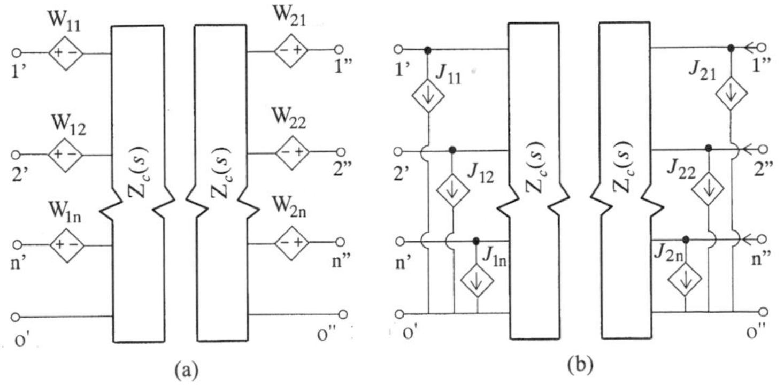

Figure 6.2. Laplace domain equivalent circuits of (a) Thévenin and (b) Norton type for a lossy multiconductor transmission line.

where

$$\mathbf{W}_1 = 2\mathrm{P}\mathbf{V}^-, \tag{6.79}$$

$$\mathbf{W}_2 = 2\mathrm{P}\mathbf{V}^+. \tag{6.80}$$

Because the functions $\mathbf{V}^+$ and $\mathbf{V}^-$ are uniquely related to $\mathbf{W}_1$ and $\mathbf{W}_2$, the latter functions can also be regarded as state variables. Note that $\mathbf{W}_1$ and $\mathbf{W}_2$, except for the factor 2, are, respectively, the backward voltage wave amplitude at the end $x = 0$ and the forward voltage wave amplitude at the end $x = d$.

From the state equations (6.69) and (6.70) we obtain the equations for $\mathbf{W}_1$ and $\mathbf{W}_2$,

$$\mathbf{W}_1(s) = \mathrm{P}(s)[2\mathbf{V}_2(s) - \mathbf{W}_2(s)], \tag{6.81}$$

$$\mathbf{W}_2(s) = \mathrm{P}(s)[2\mathbf{V}_1(s) - \mathbf{W}_1(s)]. \tag{6.82}$$

The behavior of each port of a lossy multiconductor line may be represented in the Laplace domain through an equivalent Thévenin circuit (see Fig. 6.2a). The equivalent circuit consists of two linear multiports, each of which is characterized by the matrix impedance $Z_c(s)$ and controlled voltage sources. Let us indicate with $W_{1h}(s)$ and $W_{2h}(s)$ the hth component of the vectors $\mathbf{W}_1$ and $\mathbf{W}_2$, respectively. A controlled voltage source of the value $W_{1h}(s)$ ($W_{2h}(s)$) is inserted in series at the h-terminal of the resistive multiport on the left (right). As we shall see in §6.10, the amplitudes of the controlled sources at the generic time $t > T_1$ are dependent only on the history of the voltages at the ends of the line in the interval $(0, t - T_1)$, where T_1 would be the one-way transit time associated with the fastest propagation mode of the line if it were lossless. This fact will allow the controlled sources representing the line to be treated as independent sources if the problem is resolved by means of iterative procedures as in the two-conductor case. The governing laws of these controlled sources are simply the state equations (6.81) and (6.82).

Figure 6.2b shows an equivalent circuit of Norton type in the Laplace domain. The controlled current sources $\mathbf{J}_1(s)$ and $\mathbf{J}_2(s)$ are related to the controlled voltage sources of the circuit shown in Fig. 6.2a through the relations

$$\mathbf{J}_1 = -\mathrm{Y}_c\mathbf{W}_1 \quad \text{and} \quad \mathbf{J}_2 = -\mathrm{Y}_c\mathbf{W}_2, \tag{6.83}$$

where

$$\mathrm{Y}_c = \mathrm{Z}_c^{-1} \tag{6.84}$$

is the *characteristic admittance matrix* of the line. The control laws of these sources are

$$\mathbf{J}_1(s) = \mathrm{Z}^{-1}(s)\mathrm{P}(s)\mathrm{Z}(s)[-2\mathbf{I}_2(s) + \mathbf{J}_2(s)], \tag{6.85}$$

$$\mathbf{J}_2(s) = \mathrm{Z}^{-1}(s)\mathrm{P}(s)\mathrm{Z}(s)[-2\mathbf{I}_1(s) + \mathbf{J}_1(s)]. \tag{6.86}$$

6.5.2. Time Domain

The equivalent multiport representing a lossy multiconductor line in the time domain may be obtained from the equivalent multiports shown in Fig. 6.2 by using the convolution theorem. The time domain output equations of the multiport are obtained from Eqs. (6.77) and (6.78). They are

$$\mathbf{v}_1(t) - \{z_c * \mathbf{i}_1\}(t) = \mathbf{w}_1(t), \tag{6.87}$$

$$\mathbf{v}_2(t) - \{z_c * \mathbf{i}_2\}(t) = \mathbf{w}_2(t), \tag{6.88}$$

where the impulse response matrix $z_c(t)$ is the inverse Laplace transform of the describing function matrix $\mathrm{Z}_c(s)$.

The voltages $\mathbf{w}_1(t)$ and $\mathbf{w}_2(t)$ are both unknowns of the problem and are related to the voltage at the line ends through the control laws

$$\mathbf{w}_1(t) = \{p * (2\mathbf{v}_2 - \mathbf{w}_2)\}(t), \tag{6.89}$$

$$\mathbf{w}_2(t) = \{p * (2\mathbf{v}_1 - \mathbf{w}_1)\}(t), \tag{6.90}$$

where the impulse response matrix $p(t)$ is the inverse Laplace transform of the describing function matrix $\mathrm{P}(s)$. These equations describe in implicit form the relations between the source voltages $\mathbf{w}_1(t)$ and $\mathbf{w}_2(t)$ and the voltages $\mathbf{v}_1(t)$ and $\mathbf{v}_2(t)$ at the line ends. Similar relations involving the currents $\mathbf{j}_1(t)$ and $\mathbf{j}_2(t)$ and the impulse response $y_c(t)$, which is the inverse Laplace transform of the describing function matrix $\mathrm{Y}_c(s)$, are obtained for the time domain Norton representation.

Section 6.10 will consider the problem of the evaluation of the impulse responses $z_c(t)$, $y_c(t)$, and $p(t)$.

6.6. INPUT-OUTPUT DESCRIPTIONS IN EXPLICIT FORM

Multiconductor transmission lines have input-output Laplace domain descriptions in the implicit form described by Eqs. (6.75) and (6.76). In this section we shall study these relations. As in the two-conductor case (see Chapter 4) there are six possible explicit representations of two of the four vector variables $\mathbf{V}_1$, $\mathbf{V}_2$, $\mathbf{I}_1$, and $\mathbf{I}_2$ in terms of the remaining two.

6.6.1. The Impedance Matrix

The *current-controlled* representation expresses $\mathbf{V}_1$ and $\mathbf{V}_2$ as functions of $\mathbf{I}_1$ and $\mathbf{I}_2$,

$$\mathbf{V}_1(s) = \mathrm{Z}_{11}(s)\mathbf{I}_1(s) + \mathrm{Z}_{12}(s)\mathbf{I}_2(s), \tag{6.91}$$

$$\mathbf{V}_2(s) = \mathrm{Z}_{21}(s)\mathbf{I}_1(s) + \mathrm{Z}_{22}(s)\mathbf{I}_2(s). \tag{6.92}$$

Here, Z_{11} is the driving-point impedance matrix at the end $x = 0$ when $\mathbf{I}_2 = \mathbf{0}$, that is, all the ports at $x = d$ are kept open-circuited; Z_{22} is the driving-point impedance matrix at the end $x = d$ when $\mathbf{I}_1 = \mathbf{0}$, that is, all the ports at $x = 0$ are kept open-circuited. They are equal because of the left-right symmetry of uniform lines, and are symmetric because of the reciprocity property shown in §1.10.4; Z_{12} is the *transfer impedance* matrix when $\mathbf{I}_1 = \mathbf{0}$, and Z_{21} is the *transfer impedance* when $\mathbf{I}_2 = \mathbf{0}$. They are equal because of the reciprocity property shown in §1.10.1. Furthermore, Z_{12}, and hence Z_{21}, are symmetric because of the left-right symmetry of the line and reciprocity property.

From Eqs. (6.75) and (6.76) we obtain

$$\mathrm{Z}_{11} = \mathrm{Z}_{22} = \mathrm{Z}_d = (\mathrm{I} - \mathrm{P}^2)^{-1}(\mathrm{I} + \mathrm{P}^2)\mathrm{Z}_c, \tag{6.93}$$

and

$$Z_{12} = Z_{21} = Z_t = 2(\mathrm{I} - \mathrm{P}^2)^{-1}\mathrm{P}\mathrm{Z}_c. \tag{6.94}$$

Note that it is quite cumbersome to show directly the symmetry property of the matrices Z_d and Z_t starting from their analytical expressions (6.93) and (6.94).

By using the convolution theorem, in the time domain we obtain

$$\mathbf{v}_1(t) = \{z_d * \mathbf{i}_1\}(t) + \{z_t * \mathbf{i}_2\}(t), \tag{6.95}$$

$$\mathbf{v}_2(t) = \{z_t * \mathbf{i}_1\}(t) + \{z_d * \mathbf{i}_2\}(t), \tag{6.96}$$

where $z_d(t)$ and $z_t(t)$ are the inverse Laplace transforms of Z_d and Z_t, respectively.

The expressions (6.95) and (6.96) are easily inversely transformed if we represent $(\mathrm{I} - \mathrm{P}^2)^{-1}$ by the geometric series

$$(\mathrm{I} - \mathrm{P}^2)^{-1} = \sum_{i=0}^{\infty} \mathrm{P}^{2i}. \tag{6.97}$$

This series converges in a region of the complex plane $Re\{s\} > \sigma_0$ containing the imaginary axis because of the properties of the eigenvalues of A and hence of P. Thus, Eqs. (6.93) and (6.94) can be expressed as

$$\mathrm{Z}_d = \left[(\mathrm{I} + \mathrm{P}^2) \sum_{i=0}^{\infty} \mathrm{P}^{2i}\right] \mathrm{Z}_c, \tag{6.98}$$

$$\mathrm{Z}_t = 2\left(\sum_{i=0}^{\infty} \mathrm{P}^{2i+1}\right) \mathrm{Z}_c. \tag{6.99}$$

The generic term of the infinite sums (6.98) and (6.99) is of the type $\mathrm{P}^m \mathrm{Z}_c$.

6.6.2. The Admittance Matrix

The voltage-controlled representation expresses $\mathbf{I}_1$ and $\mathbf{I}_2$ as functions of $\mathbf{V}_1$ and $\mathbf{V}_2$,

$$\mathbf{I}_1(s) = \mathrm{Y}_{11}(s)\mathbf{V}_1(s) + \mathrm{Y}_{12}(s)\mathbf{V}_2(s), \tag{6.100}$$

$$\mathbf{I}_2(s) = \mathrm{Y}_{21}(s)\mathbf{V}_1(s) + \mathrm{Y}_{22}(s)\mathbf{V}_2(s). \tag{6.101}$$

The admittance matrices Y_{11}(*driving-point admittance matrix* at the end $x = 0$ when all the ports at the other end are short-circuited), and Y_{22} (*driving-point admittance matrix* at the end $x = d$ when all the ports at the other end are short-circuited) are equal and are given by

$$\mathrm{Y}_{11} = \mathrm{Y}_{22} = \mathrm{Y}_d = \mathrm{Y}_c(\mathrm{I} - \mathrm{P}^2)^{-1}(\mathrm{I} + \mathrm{P}^2) = \mathrm{Y}_c(\mathrm{I} + \mathrm{P}^2) \sum_{i=0}^{\infty} \mathrm{P}^{2i}, \tag{6.102}$$

where Y_c is the characteristic admittance matrix of the line; and Y_{11} is equal to Y_{22} because the line is uniform. Furthermore, they are symmetric because of the reciprocity property shown in §1.10.4. The admittance matrices Y_{12} (*transfer admittance matrix* at the end $x = d$ when the other end is short-circuited) and Y_{21} (*transfer admittance matrix* at the end $x = 0$ when the other end is short-circuited) are equal and are given by

$$Y_{12} = Y_{21} = Y_t = -2Y_c(I - P^2)^{-1}P = -2Y_c \sum_{i=0}^{\infty} P^{2i+1}. \tag{6.103}$$

Here, Y_{12} is equal to Y_{21} because of the reciprocity property shown in §1.10.2.

At this point as it is easy to obtain the corresponding time domain equations for this representation, we shall leave this as an exercise. Note that to evaluate the inverse transforms of Y_d and Y_t we need the inverse transforms of the functions $Y_c P^m$.

6.6.3. The Hybrid Matrices

The two hybrid representations express $\mathbf{V}_1$ and $\mathbf{I}_2$ as functions of $\mathbf{I}_1$ and $\mathbf{V}_2$ and the converse. The *hybrid*-1 representation is

$$\mathbf{V}_1(s) = H_{11}(s)\mathbf{I}_1(s) + H_{12}(s)\mathbf{V}_2(s), \tag{6.104}$$

$$\mathbf{I}_2(s) = H_{21}(s)\mathbf{I}_1(s) + H_{22}(s)\mathbf{V}_2(s). \tag{6.105}$$

Here, H_{11} is the *driving-point impedance matrix* at the end $x = 0$ with the end $x = d$ short-circuited,

$$H_{11} = Y_d^{-1} = (I + P^2)^{-1}(I - P^2)Z_c = \left[(I - P^2) \sum_{i=0}^{\infty} (-P^2)^i\right] Z_c; \tag{6.106}$$

H_{22} is the *driving-point admittance matrix* at the end $x = d$ with the end $x = 0$ open-circuited,

$$H_{22} = Z_d^{-1} = Y_c(1 + P^2)^{-1}(I - P^2) = Y_c(I - P^2) \sum_{i=0}^{\infty} (-P^2)^i; \tag{6.107}$$

H_{21} is the *forward current transfer matrix* and H_{12} is the *reverse voltage transfer matrix* given by

$$H_{12} = Z_t Z_d^{-1} = 2(I + P^2)^{-1}P = 2P \sum_{i=0}^{\infty} (-P^2)^i \tag{6.108}$$

H_{21} is equal to $-H_{12}^T$ because of the reciprocity property shown in §1.10.3. This property may be shown in another way. Since $H_{21} = -Z_d^{-1}Z_t$, from the symmetry of the matrices Z_d and Z_t it follows that $H_{21} = -H_{12}^T$.

A dual treatment can be given to the *hybrid*-2 *matrix* expressing $\mathbf{I}_1$ and $\mathbf{V}_2$ as functions of $\mathbf{V}_1$ and $\mathbf{I}_2$. Due to the spatial uniformity, the two ends are symmetric; thus we have

$$\mathbf{V}_2(s) = H_{11}(s)\mathbf{I}_2(s) + H_{12}(s)\mathbf{V}_1(s), \tag{6.109}$$

$$\mathbf{I}_1(s) = H_{21}(s)\mathbf{I}_2(s) + H_{22}(s)\mathbf{V}_1(s). \tag{6.110}$$

At this point as it is easy to obtain the corresponding time domain equations for these representations, we shall leave this as an exercise.

Remarks

(i) It is clear that to analyze sinusoidal or periodic steady states we can use the impedance, the admittance or the hybrid matrices evaluated along the imaginary axis $Z(i\omega)$, $Y(i\omega)$, and $H(i\omega)$, to characterize the terminal behavior of the line.

(ii) The foregoing input-output descriptions have the same drawbacks as in the two-conductor case. Due to the successive reflections the impedance, admittance and hybrid matrices have an infinite number of poles, and the corresponding impulse responses have much longer duration than that of the impulse responses characterizing the input-state-output description. Therefore, it is clear that the input-state-output description is more suitable for transient analysis. To deal with sinusoidal and periodic steady-state operating conditions in the frequency domain, the input-output description is preferable. In this case, the terminal behavior of the line may be directly characterized through the impedance, the admittance or the hybrid matrices. ◇

6.6.4. The Transmission Matrices

The *forward transmission* matrix representation expresses $\mathbf{I}_2$ and $\mathbf{V}_2$ as functions of $\mathbf{I}_1$ and $\mathbf{V}_1$,

$$\mathbf{V}_2(s) = T_{11}(s)\mathbf{V}_1(s) + T_{12}(s)\mathbf{I}_1(s), \tag{6.111}$$

$$-\mathbf{I}_2(s) = T_{21}(s)\mathbf{V}_1(s) + T_{22}(s)\mathbf{I}_1(s), \tag{6.112}$$

where

$$T_{11} = T_{22} = \tfrac{1}{2}(P + P^{-1}), \tag{6.113}$$

$$T_{12} = Y_t^{-1} = \tfrac{1}{2}(P - P^{-1})Z_c, \tag{6.114}$$

$$T_{21} = -Z_t^{-1} = \tfrac{1}{2}Y_c(P - P^{-1}). \tag{6.115}$$

Note that $T_{11} = T_{22}$ because of the left-right symmetry of the line.

The matrix

$$T = \begin{bmatrix} T_{11} & T_{12} \\ T_{21} & T_{22} \end{bmatrix} \tag{6.116}$$

is called the *forward transmission matrix* of the multiconductor transmission line.

The time domain transmission matrix descriptions have the same drawbacks as in the two-conductor case (see §4.7.4). The transmission matrix characterization is particularly useful in analyzing cascade connections of uniform multiconductor transmission lines with different per-unit-length parameters. Let us consider a nonuniform multiconductor transmission line consisting of a cascade on n uniform lines, and let us designate the transmission matrix of the ith uniform tract by $T^{(i)}$. The resulting multiport is characterized by a *forward* transmission matrix T that is equal to the product $T^{(n)}T^{(n-1)} \cdots T^{(2)}T^{(1)}$.

A dual treatment can be given to the *backward transmission matrix* T′ expressing $\mathbf{I}_1$ and $\mathbf{V}_1$ as function of $\mathbf{I}_2$ and $\mathbf{V}_2$,

$$\mathbf{V}_1(s) = T'_{11}(s)\mathbf{V}_2(s) + T'_{12}(s)[-\mathbf{I}_2(s)], \tag{6.117}$$

$$\mathbf{I}_1(s) = T'_{21}(s)\mathbf{V}_2(s) + T'_{22}(s)[-\mathbf{I}_2(s)]. \tag{6.118}$$

The matrix T′ is simply related to the matrix T. We shall leave the determination of this relation as an exercise.

6.7. THE PROBLEM OF THE INVERSE LAPLACE TRANSFORM OF THE MATRIX OPERATORS P(s), $Z_c(s)$, AND $Y_c(s)$

The behavior of the line as a $2n$-port in the time domain depends on the *impulse responses* $p(t)$, $z_c(t)$, and $y_c(t)$ whether an input-output or an input-state-output representation is chosen. Once the matrices Z(s) and Y(s) are given, $p(t)$, $z_c(t)$, and $y_c(t)$ are obtained by performing the inverse Laplace transforms of P(s), $Z_c(s)$, and $Y_c(s)$, respectively.

By substituting Eq. (6.46) in the expression of P(s) given by Eq. (6.62) we obtain

$$P(s) = T_v(s)\exp[-sA(s)d]T_v^{-1}(s). \tag{6.119}$$

Furthermore, from the expression of Z_c given in Eq. (6.50), we obtain

$$Z_c(s) = \frac{1}{s} T_v(s)A^{-1}(s)T_v^{-1}(s)Z(s), \tag{6.120}$$

$$Y_c(s) = sZ^{-1}(s)T_v(s)A(s)T_v^{-1}(s). \tag{6.121}$$

Remark

The global propagation matrix function P(s) may be rewritten as

$$P(s) = \sum_{i=1}^{n} \mathbf{e}_i(s)\mathbf{s}_i^T(s)\exp[-sd\sqrt{\beta_i(s)}], \tag{6.122}$$

where $\mathbf{s}_i(s)$ are the columns of the transpose matrix of T_v^{-1}. If we normalize the eigenvectors $\mathbf{e}_1, \mathbf{e}_2, \ldots, \mathbf{e}_n$ according to Eq. (6.31), then $\mathbf{s}_i(s)$ (for any i) is given by

$$\mathbf{s}_i(s) = sZ^{-1}\mathbf{e}_i(s). \quad \diamond \tag{6.123}$$

Generally it is not possible to give a closed-form expression to A(s) and $T_v(s)$, and hence to $p(t)$, $z_c(t)$, and $y_c(t)$, unless for very particular cases, for example, ideal lines and lossy lines exhibiting structural symmetries (see §6.2). In general, to perform the inverse Laplace transforms of P(s), Z_c(s), and Y_c(s) we need to resort to numerical inversion. On the other hand, inversion procedures based on numerical methods can be effectively applied only when the resulting time-domain functions are sufficiently regular.

To gain an insight into this question, it is profitable first to survey briefly two simple and very significant cases for which $p(t)$, $z_c(t)$, and $y_c(t)$ are known in analytical form.

It is educational to determine the impulse responses $p(t)$, $z_c(t)$, and $y_c(t)$ of lossless multiconductor lines, starting from the corresponding expression in the Laplace domain. These lines are characterized by the per-unit-length impedance and admittance matrices $Z(s) = sL$ and $Y(s) = sC$. In this case we have

$$A = \mathrm{diag}(1/c_1, 1/c_2, \ldots, 1/c_n), \tag{6.124}$$

$$Z_c = R_c = \sqrt{(LC)^{-1}}L = C^{-1}\sqrt{CL}, \tag{6.125}$$

$$Y_c = G_c = L^{-1}\sqrt{LC} = \sqrt{(CL)^{-1}}C, \tag{6.126}$$

where c_i is the propagation velocity of the ith mode (the ith eigenvalue of the matrix $(\sqrt{\mathrm{LC}})^{-1}$, see Chapter 3), R_c is the characteristic resistance matrix of the line, and $\mathrm{G}_c = \mathrm{R}_c^{-1}$ is the characteristic conductance matrix. In this case, the eigenvectors $\mathbf{e}_1, \mathbf{e}_2, \ldots, \mathbf{e}_n$ and the modal matrix T_v are, respectively, the eigenvectors and the modal matrix of LC, hence they are independent of s. Therefore, the impulse responses $p(t)$, $z_c(t)$, and $y_c(t)$ are given by

$$p(t) = \sum_{i=1}^{n} \mathbf{e}_i \mathbf{s}_i^{\mathrm{T}} \delta(t - T_i), \tag{6.127}$$

$$z_c(t) = \mathrm{R}_c \delta(t), \tag{6.128}$$

$$y_c(t) = \mathrm{G}_c \delta(t). \tag{6.129}$$

If we normalize the eigenvectors $\mathbf{e}_1, \mathbf{e}_2, \ldots, \mathbf{e}_n$ according to Eq. (3.15), then the generic vector $\mathbf{s}_i$ is given by Eq. (3.22). We have already found the expression (6.127) in Chapter 3, where we have dealt with ideal multiconductor lines (see (3.126)).

The impulse responses $p(t)$, $z_c(t)$, and $y_c(t)$ of a lossless multiconductor line are composed of Dirac pulses, as a direct consequence of the basic properties of the electromagnetic propagation phenomenon. In fact, in the time domain, ideal matched or semi-infinite multiconductor lines must behave as resistive multiports with characteristic resistance matrix R_c. Hence, the impulse response $z_c(t)$ (or $y_c(t)$) relating the currents and voltages at the same termination has to be a Dirac pulse as in Eq. (6.128) (or Eq. (6.129)). The propagation phenomenon is characterized by the superposition of n natural propagation modes of the type $\mathbf{b}_i g_i(x \pm c_i t)$. As a consequence, a signal applied to a given termination will act on another one only after a certain finite time due to the delay caused by the finite propagation velocities of the line modes. Therefore, the impulse response $p(t)$, which relates the forward and backward voltage waves at the two line ends, contains a weighted sum of delayed Dirac pulse as in expression (6.127): The delayed Dirac pulse $\delta(t - T_i)$ corresponds to the mode with propagation velocity c_i. From the mathematical point of view, these properties arise from the hyperbolic nature of the line equations (see Chapter 8).

For lines with symmetric tridiagonal Toeplitz parameter matrices (see §6.2), the eigenvectors $\mathbf{e}_1\, \mathbf{e}_2, \ldots, \mathbf{e}_n$ are again independent of the complex variable s according to Eq. (6.11) if we choose the arbitrary factor γ_j independent of s. In this case the generic vector $\mathbf{s}_i$ also is independent of s. The eigenvalues β_i are given by

$$\beta_i(s) = (\lambda_{\mathrm{L},i} + \lambda_{\mathrm{R},i}/s)(\lambda_{\mathrm{C},i} + \lambda_{\mathrm{G},i}/s), \tag{6.130}$$

where $\lambda_{L,i}$, $\lambda_{R,i}$, $\lambda_{C,i}$, and $\lambda_{G,i}$ are the eigenvalues of the matrices L, R, C, and G, which can be determined analytically according to Eq. (6.10). Therefore, for these lines the inverse Laplace transforms of P(s), $Z_c(s)$, and $Y_c(s)$ can be performed analytically, as in the two-conductor case.

From the expression (6.122), for the impulse response $p(t)$ we easily obtain

$$p(t) = \sum_{i=1}^{n} \mathbf{e}_i \mathbf{s}_i^T [e^{-\mu_i T_i} \delta(t - T_i) + p_{ri}(t - T_i)], \qquad (6.131)$$

where $p_{ri}(t)$ is the function defined as in Eq. (4.95) and

$$\mu_i = \frac{1}{2}\left(\frac{\lambda_{R,i}}{\lambda_{L,i}} + \frac{\lambda_{G,i}}{\lambda_{C,i}}\right), \quad \nu_i = \frac{1}{2}\left(\frac{\lambda_{R,i}}{\lambda_{L,i}} - \frac{\lambda_{G,i}}{\lambda_{C,i}}\right), \qquad (6.132)$$

$$T_i = d\sqrt{\lambda_{L,i}\lambda_{C,i}}. \qquad (6.133)$$

From the expressions (6.120) and (6.121), for the impulse responses $z_c(t)$ and $y_c(t)$, we obtain

$$z_c(t) = T_v \,\mathrm{diag}[z_{c1}(t), \ldots, z_{cn}(t)] T_v^{-1}, \qquad (6.134)$$

$$y_c(t) = T_v \,\mathrm{diag}[y_{c1}(t), \ldots, y_{cn}(t)] T_v^{-1}, \qquad (6.135)$$

where

$$z_{ci}(t) = R_{ci}\delta(t) + z_{cri}(t), \qquad (6.136)$$

$$y_{ci}(t) = \frac{1}{R_{ci}}\delta(t) + y_{cri}(t), \qquad (6.137)$$

the functions $z_{cri}(t)$ and $y_{cri}(t)$ are given by Eqs. (4.58) and (4.123), respectively, and

$$R_{ci} = \sqrt{\frac{\lambda_{L,i}}{\lambda_{C,i}}}. \qquad (6.138)$$

What happens when the general case is considered? For lines described by Toeplitz matrices the impulse responses still contain Dirac pulses. The losses merely give rise to the phenomena of damping and dispersion, both described through regular functions. The structure of the equations of lossy multiconductor lines still remains hyperbolic (see Chapter 8). For these reasons we feel that the impulse responses of lossy multiconductor lines behave as in the lossy two-conductor case, and hence we expect that $p(t)$, $z_c(t)$, and $y_c(t)$ still contain Dirac pulses, as we shall show in the following next two sections.

Because the impulse responses $p(t)$, $z_c(t)$, and $y_c(t)$ are not regular functions, a mere numerical inversion of P(s), Z$_c(s)$, and Y$_c(s)$ gives inaccurate results and generally fails as pointed out by Yu and Kuh (1996). *"The first step of the modelling process to extract a principal part is very essential to the second step,"* consisting of the numerical evaluation through a standard inversion procedure of the regular parts of the impulsive responses. *"If $z_c(t)$ or $p(t)$ is obtained by directly taking the inverse FFT from* Z$_c(s)$ *and the propagation matrix* P(s), *its curve will have severe ripples. No simple piece wise polynomial fitting can be done to approximate such a function well."* The impulse responses $z_c(t)$ and $y_c(t)$ contain a delta Dirac function at $t = 0$ (see, Lin and Kuh, 1992; Gordon, Blazeck, and Mittra, 1992). Concerning the impulse response $p(t)$, Gordon, Blazeck, and Mittra wrote: *"In computing the time-domain response $p(t)$ for lines with small losses, the number of points required by the inverse Fourier transform increases as $p(t)$ approaches a string of delayed Dirac delta functions, as in the lossless case. This may limit the accuracy of the derived responses in cases where the line losses are very small"* (Gordon, Blazeck, and Mittra, 1992). Hence, a preliminary and close study of the properties of $p(t)$, $z_c(t)$ and $y_c(t)$ is required.

Even if it is not possible to express $p(t)$, $z_c(t)$, and $y_c(t)$ by means of known functions, it is possible to predict all their irregular terms once the behavior of P(s), Z$_c(s)$, and Y$_c(s)$ around $s = \infty$ is known, as was done in Chapter 5 to deal with two-conductor lines with parameters depending on the frequency.

Some attempts have been reported in the literature to study the asymptotic behavior of P(s), Z$_c(s)$, and Y$_c(s)$ as $s \to \infty$. The impedance and admittance operators have been modeled by a *principal part* and a *remainder* (e.g., Lin and Kuh, 1992; Gordon, Blazeck, and Mittra, 1992; Yu and Kuh, 1996): the principal part is the limit for $s \to \infty$ and the remainder approaches to zero as $s \to \infty$. Then the inverse Laplace transform is composed of a Dirac pulse and a regular term that is numerically evaluated by FFT. Lin and Kuh proposed an asymptotic expansion of the matrix P(s) by representing $\Lambda(s)$ through a Padé approximation. The expansion coefficients are computed by numerically fitting the original matrix at a certain number of discrete points "near" $s = \infty$ (Lin and Kuh, 1992).

We have developed a simple, general and effective analytical technique to evaluate the asymptotic expressions of P(s), Z$_c(s)$, and Y$_c(s)$ (Maffucci and Miano, 1999a; Maffucci and Miano, 1999b). The technique is based on the perturbation theory for the spectrum of symmetric matrices. In the next section we shall describe in detail this technique for lossy multiconductor lines with frequency-independent parameters. In Section 6.11 we shall extend the same procedure to transmission lines with frequency-dependent parameters.

6.8. STUDY OF THE ASYMPTOTIC BEHAVIOR OF THE MATRIX OPERATOR $\Lambda(s)$ THROUGH THE RAYLEIGH-SCHRÖDINGER METHOD

Although for lossy multiconductor lines with frequency-independent parameters the matrices $Z(s)$ and $Y(s)$ have the simple expression given in Eqs. (6.7) and (6.8), it is not possible to obtain an analytical expression of the eigenvalues and eigenvectors of the matrix $\Lambda(s)$ defined in Eq. (6.23), and hence of the matrices $A(s)$ and $T_v(s)$. Nevertheless, the asymptotic expressions of $A(s)$ and $T_v(s)$ may be evaluated analytically by applying the same method used in quantum mechanics to evaluate perturbatively stationary solutions of the Schrödinger equation for anharmonic potentials — the so-called *Rayleigh-Schrödinger method* (e.g., Landau and Lifshitz, 1958; Nayfeh, 1973).

Because the behavior of $\Lambda(s)$ as $s \to \infty$ does not depend on how $s = \infty$ is approached, the analysis can be performed by considering $s \to \infty$ on the positive real axis. With the assumption that s belongs to the positive real axis $s = \sigma > 0$, let us rewrite the function $\Lambda(\sigma)$ as

$$\Lambda = \Lambda^{(0)} + \varepsilon\Lambda^{(1)} + \varepsilon^2\Lambda^{(2)} \tag{6.139}$$

where

$$\varepsilon = 1/\sigma \tag{6.140}$$

is a positive real quantity, and

$$\Lambda^{(0)} = LC, \quad \Lambda^{(1)} = LG + RC, \quad \Lambda^{(2)} = RG \tag{6.141}$$

are constant matrices. The matrices Z and Y are symmetric and, moreover, they remain positive definite on the positive real axis (see §6.3.1). Because $\varepsilon \to 0$ for $s \to +\infty$ on the real axis, we are interested in the behavior of Λ for $\varepsilon \to 0$.

From Eq. (6.139) we realize that as $\varepsilon \to 0$ the matrix Λ can be tackled as a "perturbation" of the matrix $\Lambda^{(0)} = LC$. In this way we are choosing the ideal line case as the reference. Let us indicate with $\beta_i^{(0)}$ and $\mathbf{e}_i^{(0)}$ the ith eigenvalue and eigenvector of the "unperturbed" matrix, LC, respectively. We have to remark that

$$\beta_i^{(0)} = 1/c_i^2, \tag{6.142}$$

where c_i is the propagation velocity of the ith mode when the losses are disregarded (see Chapter 3). The eigenvalues $\beta_i^{(0)}$ are always real and positive because L and C are symmetric and strictly positive-definite matrices. Furthermore, the matrix LC has n linearly independent eigenvectors $\mathbf{e}_1^{(0)}, \mathbf{e}_2^{(0)}, \ldots, \mathbf{e}_n^{(0)}$, whether or not the corresponding eigenvalues are degenerate (see Chapter 3). Similarly, these results still hold for the eigenvalues $\beta_i(\sigma)$ and the eigenvectors $\mathbf{e}_i(\sigma)$ of the "perturbed" matrix $\Lambda(\sigma)$ (see §6.3).

By means of the aforementioned properties of the matrices Λ and $\Lambda^{(0)}$, and by applying some relevant results of the perturbation theory of the spectrum of symmetric matrices (e.g., Deif, 1982; Lancaster and Tismenetsky, 1985), it is possible to relate the "perturbed" and "unperturbed" eigenvalues through the following expansion:

$$\beta_i = \beta_i^{(0)} + \beta_i^{(1)}\varepsilon + \beta_i^{(2)}\varepsilon^2 + O(\varepsilon^3) \quad \text{for } \varepsilon \to 0. \tag{6.143}$$

Let us consider the eigenvalue problem

$$(\Lambda - \mathrm{I}\beta_i)\mathbf{e}_i = \mathbf{0}. \tag{6.144}$$

The eigenvalues are the solution of the algebraic equation

$$D(\beta_i) = \det(\beta_i \mathrm{I} - \Lambda) = 0. \tag{6.145}$$

The expansion (6.143) is obtained by expanding $D(\beta_i)$ in Taylor series in a neighborhood of $\beta_i = \beta_i^{(0)}$. It always holds regardless of whether the eigenvalues of $\Lambda^{(0)}$ are simple because $\beta_i^{(0)}$ and $\beta_i(s)$, for $s = \sigma$, are real quantities (see Appendix A). As regards the behavior of the eigenvectors $\mathbf{e}_i$, it is obvious to conclude that

$$\mathbf{e}_i = \mathbf{e}_i^{(0)} + \mathbf{e}_i^{(1)}\varepsilon + \mathbf{e}_i^{(2)}\varepsilon^2 + O(\varepsilon^3) \quad \text{for } \varepsilon \to 0. \tag{6.146}$$

The expansion terms $\beta_i^{(k)}$ and $\mathbf{e}_i^{(k)}$ for $k = 1, 2, \ldots,$ can be evaluated analytically. By using Eq. (6.139) and substituting the asymptotic expansions (6.141) and (6.146) in Eq. (6.144), we obtain

$$[\Lambda^{(0)} + \varepsilon\Lambda^{(1)} + \cdots][\mathbf{e}_i^{(0)} + \varepsilon\mathbf{e}_i^{(1)} + \cdots] = [\beta_i^{(0)} + \varepsilon\beta_i^{(1)} + \cdots][\mathbf{e}_i^{(0)} + \varepsilon\mathbf{e}_i^{(1)} + \cdots]. \tag{6.147}$$

Now, by equating the coefficients of like powers of ε in Eq. (6.147), we obtain the equation for the "unperturbed" eigenvalues and eigenvectors $\beta_i^{(0)}$ and $\mathbf{e}_i^{(0)}$

$$[\Lambda^{(0)} - \mathrm{I}\beta_i^{(0)}]\mathbf{e}_i^{(0)} = \mathbf{0}, \tag{6.148}$$

and the equations for the "perturbed" eigenvalues $\beta_i^{(k)}$ and eigenvectors $\mathbf{e}_i^{(k)}$, for $k = 1, 2, \ldots,$

$$[\Lambda^{(0)} - \beta_i^{(0)}\mathrm{I}]\mathbf{e}_i^{(k)} = -\sum_{l=1}^{k} [\Lambda^{(l)} - \beta_i^{(l)}\mathrm{I}]\mathbf{e}_i^{(k-l)}. \tag{6.149}$$

We must distinguish between two cases, depending on whether or not the eigenvalues of $\Lambda^{(0)}$ are distinct. The first case is called the *nondegenerate* case, while the second case is called the *degenerate* case because repeated eigenvalues exist.

6.8.1. The Eigenvalues of $\Lambda^{(0)} = \mathrm{LC}$ Are Nondegenerate

If all the eigenvalues $\beta_i^{(0)}$ of the matrix LC are different, the set of all its eigenvectors is uniquely determined. Moreover, these eigenvectors remain *mutually orthogonal* with respect to the scalar product $\langle \mathbf{a}, \mathbf{b} \rangle_{\mathrm{L}} = \mathbf{a}^{\mathrm{T}} \mathrm{L}^{-1} \mathbf{b}$ (see Chapter 3).

As the eigenvectors $\mathbf{e}_i^{(0)}$ are linearly independent, we can always write the kth corrective terms $\mathbf{e}_i^{(k)}$ as

$$\mathbf{e}_i^{(k)} = \sum_{h=1}^{n} b_{i,h}^{(k)} \mathbf{e}_h^{(0)}. \tag{6.150}$$

Now substituting Eq. (6.150) in Eq. (6.149) and putting $k = 1$, we obtain the equation for $\beta_i^{(1)}$ and $b_{i,h}^{(1)}$,

$$\sum_{h=1}^{n} b_{i,h}^{(1)} [\Lambda^{(0)} - \beta_i^{(0)} \mathrm{I}] \mathbf{e}_h^{(0)} = -[\Lambda^{(1)} - \beta_i^{(1)} \mathrm{I}] \mathbf{e}_i^{(0)} \quad \text{for } i = 1, \ldots, n. \tag{6.151}$$

By using Eq. (6.148), Eq. (6.151) becomes

$$\sum_{h=1}^{n} b_{i,h}^{(1)} [\beta_h^{(0)} - \beta_i^{(0)}] \mathbf{e}_h^{(0)} = -[\Lambda^{(1)} - \beta_i^{(1)} \mathrm{I}] \mathbf{e}_i^{(0)} \quad \text{for } i = 1, \ldots, n. \tag{6.152}$$

Note that the term proportional to $\mathbf{e}_i^{(0)}$ on the left-hand side of Eq. (6.152) vanishes. In order to compute $b_{i,h}^{(1)}$ and $\beta_i^{(1)}$, we multiply both sides of Eq. (6.152) by $\mathbf{e}_p^{(0)}$, in the sense of the scalar product $\langle \mathbf{a}, \mathbf{b} \rangle_{\mathrm{L}}$. Because the eigenvectors $\mathbf{e}_p^{(0)}$ are mutually orthogonal with respect to this scalar product, for $p = 1, \ldots, n$ we obtain

$$b_{i,p}^{(1)} = \begin{cases} \dfrac{1}{\beta_i^{(0)} - \beta_p^{(0)}} \dfrac{\langle \mathbf{e}_p^{(0)}, \Lambda^{(1)} \mathbf{e}_i^{(0)} \rangle_{\mathrm{L}}}{\langle \mathbf{e}_p^{(0)}, \mathbf{e}_i^{(0)} \rangle_{\mathrm{L}}} & i \neq p, \\ d_i^{(1)} & i = p, \end{cases} \tag{6.153}$$

and

$$\beta_i^{(1)} = \frac{\langle \mathbf{e}_i^{(0)}, \Lambda^{(1)} \mathbf{e}_i^{(0)} \rangle_{\mathrm{L}}}{\langle \mathbf{e}_i^{(0)}, \mathbf{e}_i^{(0)} \rangle_{\mathrm{L}}}; \tag{6.154}$$

$d_i^{(1)}$ are arbitrary constants. The matrix $\Lambda^{(1)} = \mathrm{LG} + \mathrm{RC}$ is strictly positive definite with respect to the scalar product $\langle \mathbf{a}, \mathbf{b} \rangle_{\mathrm{L}}$, because of the properties of L, R, C, and G; hence the first-order perturbation terms $\beta_i^{(1)}$ also are positive.

To compute the generic terms $b_{i,p}^{(k)}$ and $\beta_i^{(k)}$, we apply the same procedure to Eq. (6.149) for $k \geqslant 2$. It is easy to show that for $i = 1, \ldots, n$ and $k \geqslant 2$,

$$b_{i,p}^{(k)} = \begin{cases} \dfrac{\sum_{l=1}^{k} \langle \mathbf{e}_p^{(0)}, \Lambda^{(l)} \mathbf{e}_i^{(k-l)} \rangle_{\mathrm{L}} - \sum_{l=1}^{k-1} \beta_i^{(l)} b_{i,p}^{(k-l)} \langle \mathbf{e}_p^{(0)}, \mathbf{e}_p^{(0)} \rangle_{\mathrm{L}}}{[\beta_i^{(0)} - \beta_p^{(0)}] \langle \mathbf{e}_p^{(0)}, \mathbf{e}_p^{(0)} \rangle_{\mathrm{L}}} & i \neq p, \\ d_i^{(k)} & i = p, \end{cases} \tag{6.155}$$

and

$$\beta_i^{(k)} = \frac{1}{\langle \mathbf{e}_i^{(0)}, \mathbf{e}_i^{(0)} \rangle_{\mathrm{L}}} \left[\sum_{l=1}^{k} \langle \mathbf{e}_p^{(0)}, \Lambda^{(l)} \mathbf{e}_i^{(k-l)} \rangle_{\mathrm{L}} - \sum_{l=1}^{k-1} \beta_i^{(l)} \langle \mathbf{e}_i^{(0)}, \mathbf{e}_i^{(k-l)} \rangle_{\mathrm{L}} \right], \tag{6.156}$$

where $d_i^{(k)}$ are again arbitrary constants. Note that $\Lambda^{(l)} = 0$ for $l \geqslant 3$.

From Eqs. (6.155) and (6.156) all the expansion terms can be calculated recursively because it is possible to compute the terms $\beta_i^{(k)}$ and $\mathbf{e}_i^{(k)}$ once $\beta_i^{(j)}$ and $\mathbf{e}_i^{(j)}$ are known for $0 \leqslant j \leqslant k-1$ $(k \geqslant 1)$.

Remark

The actual values of the perturbing terms $\beta_i^{(k)}$ of the eigenvalues do not depend on the arbitrary constants $d_i^{(k)}$. For example, let us evaluate the terms $\beta_i^{(2)}$ by substituting Eq. (6.150) in Eq. (6.156). We obtain

$$\beta_i^{(2)} = \frac{1}{\langle \mathbf{e}_i^{(0)}, \mathbf{e}_i^{(0)} \rangle_{\mathrm{L}}} \left[\langle \mathbf{e}_i^{(0)}, \Lambda^{(2)} \mathbf{e}_i^{(0)} \rangle_{\mathrm{L}} + \sum_{\substack{h=1 \\ h \neq i}}^{n} b_{i,h} \langle \mathbf{e}_i^{(0)}, \Lambda^{(1)} \mathbf{e}_h^{(0)} \rangle_{\mathrm{L}} \right]. \quad \diamond \tag{6.157}$$

6.8.2. The Eigenvalues of $\Lambda^{(0)}$ = LC Are Degenerate

Now a procedure is outlined to compute the perturbing terms $\beta_i^{(k)}$ and $\mathbf{e}_i^{(k)}$ $(k \geqslant 1)$ appearing in Eqs. (6.143) and (6.146) in the most general case in which the eigenvalues of $\Lambda^{(0)}$ can be degenerate. This procedure is applied in quantum mechanics to deal with the degeneration of energetic levels (e.g., Landau and Lifshitz, 1958).

For the sake of simplicity, let us suppose that the unperturbed eigenvalue $\beta_1^{(0)}$ has multiplicity $m < n$, whereas $\beta_{m+1}^{(0)}, \ldots, \beta_n^{(0)}$ are distinct. The matrix $\Lambda^{(0)}$ has m linearly independent eigenvectors corresponding to the repeated eigenvalue $\beta_1^{(0)}$ because of the properties of the matrices L and C (see Chapter 3). Although the eigenvectors corresponding to $\lambda_1^{(0)}$ are linearly independent, they are not uniquely determined due to the degeneracy. By comparison, the eigenvectors $\mathbf{e}_{m+1}^{(0)}, \ldots, \mathbf{e}_n^{(0)}$ corresponding to the nondegenerate eigenvalues are uniquely determined and mutually orthogonal with respect to the scalar product $\langle \mathbf{a}, \mathbf{b} \rangle_{\mathrm{L}}$. Furthermore, they are also orthogonal to the eigenvectors corresponding to the degenerate eigenvalue $\lambda_1^{(0)}$. Therefore, the calculation of the perturbation terms of the nondegenerate eigenvalues and of the corresponding eigenvectors may be developed as done previously in the nondegenerate case.

Now, let us consider the problem of the perturbation terms relevant to the degenerate eigenvalue and the corresponding eigenvectors. Let $\hat{\mathbf{e}}_1^{(0)}, \hat{\mathbf{e}}_2^{(0)}, \ldots, \hat{\mathbf{e}}_m^{(0)}$ be a possible set of unperturbed eigenvectors, corresponding to the degenerate unperturbed eigenvalue $\beta_1^{(0)}$. Then all the vectors

$$\mathbf{e}_j^{(0)} = q_{1,j}^{(0)}\hat{\mathbf{e}}_1^{(0)} + q_{2,j}^{(0)}\hat{\mathbf{e}}_2^{(0)} + \cdots + q_{m,j}^{(0)}\hat{\mathbf{e}}_m^{(0)} \tag{6.158}$$

are eigenvectors of the unperturbed problem (6.148), corresponding to the same eigenvalue $\beta_1^{(0)}$, where $q_{1,j}^{(0)}, q_{2,j}^{(0)}, \ldots, q_{m,j}^{(0)}$ are arbitrary coefficients. This indeterminacy disappears if we consider the effect of the first-order perturbation term in Eq. (6.149). The terms proportional to $\mathbf{e}_1^{(0)}, \mathbf{e}_2^{(0)}, \ldots, \mathbf{e}_m^{(0)}$ on the left-hand side of Eq. (6.152) do not give a contribution because of the degeneracy of the eigenvalue $\beta_1^{(0)}$. Therefore, Eq. (6.152) becomes

$$\sum_{h=m+1}^{n} b_{j,h}^{(1)}[\beta_h^{(0)} - \beta_1^{(0)}]\mathbf{e}_h^{(0)} = -[\Lambda^{(1)} - \beta_j^{(1)}\mathrm{I}]\mathbf{e}_j^{(0)} \quad \text{for } j = 1, \ldots, m. \tag{6.159}$$

Now, left multiplying Eqs (6.159) by $[\hat{\mathbf{e}}_j^{(0)}]^{\mathrm{T}}\mathrm{L}^{-1}$ for $j = 1, \ldots, m$, the left-hand side vanishes because $\hat{\mathbf{e}}_j^{(0)}$ are orthogonal to $\mathbf{e}_h^{(0)}$ with respect to the scalar product $\langle \mathbf{a}, \mathbf{b} \rangle_{\mathrm{L}}$ for $h > m$. Therefore, using Eq. (6.158) we obtain the following generalized eigenvalue problem:

$$\mathrm{E}^{(1)}\mathbf{q}_j^{(0)} = \mathrm{F}^{(0)}\beta_j^{(1)}\mathbf{q}_j^{(0)}, \tag{6.160}$$

where

$$\mathbf{q}_j^{(0)} = |q_{1,j}^{(0)} q_{2,j}^{(0)} \cdots q_{m,j}^{(0)}|^{\mathrm{T}}, \quad \mathrm{E}^{(1)} = [\hat{\mathrm{T}}_{\mathrm{v}}^{(0)}]^{\mathrm{T}}[\mathrm{L}^{-1}\Lambda^{(1)}]\hat{\mathrm{T}}_{\mathrm{v}}^{(0)}, \tag{6.161}$$

and

$$\mathrm{F}^{(0)} = [\hat{\mathrm{T}}_{\mathrm{v}}^{(0)}]^{\mathrm{T}}\mathrm{L}^{-1}\hat{\mathrm{T}}_{\mathrm{v}}^{(0)}, \quad \hat{\mathrm{T}}_{\mathrm{v}}^{(0)} = |\hat{\mathbf{e}}_1^{(0)} \vdots \hat{\mathbf{e}}_2^{(0)} \vdots \cdots \vdots \hat{\mathbf{e}}_m^{(0)}|. \tag{6.162}$$

Here, $\hat{\mathrm{T}}_{\mathrm{v}}^{(0)}$ is a rectangular $n \times m$ matrix, and $\mathrm{E}^{(1)}$ and $\mathrm{F}^{(0)}$ are square symmetric $m \times m$ matrices.

If the eigenvalues $\beta_j^{(1)}$ are not degenerate for $j=1,\ldots,m$, the degenerate eigenvalue $\beta_1^{(0)}$ splits into m distinct eigenvalues $\beta_1^{(0)} + \varepsilon\beta_j^{(1)}$. The corresponding "unperturbed" eigenvectors are given by $\mathbf{e}_j^{(0)} = \hat{\mathrm{T}}_{\mathrm{v}}^{(0)}\mathbf{e}_j^{(0)}$. Once $\beta_j^{(1)}$ and $\hat{\mathbf{e}}_j^{(0)}$ are known for $j = 1,\ldots,m$, by using Eq. (6.149) for $k = 2$ we can calculate $\beta_j^{(2)}$ and $\mathbf{e}_j^{(1)}$ for $j = 1,\ldots,m$. In the case considered, Eq. (6.149) for $j = 1,\ldots,m$ becomes,

$$\sum_{h=m+1}^{n} b_{j,h}^{(2)}[\beta_h^{(0)} - \beta_1^{(0)}]\mathbf{e}_h^{(0)} = -[\Lambda^{(2)} - \beta_j^{(2)}\mathrm{I}]\mathbf{e}_j^{(0)} - [\Lambda^{(1)} - \beta_j^{(1)}\mathrm{I}]\mathbf{e}_j^{(1)}. \tag{6.163}$$

The terms proportional to $\mathbf{e}_1^{(0)},\ldots,\mathbf{e}_m^{(0)}$ on the left-hand side of Eq. (6.163) again do not give a contribution because of the degeneracy of the eigenvalue $\beta_1^{(0)}$. Once more, left multiplying Eq. (6.163) by $[\hat{\mathbf{e}}_j^{(0)}]^{\mathrm{T}}\mathrm{L}^{-1}$ for $j = 1,\ldots,m$ and placing $\mathbf{e}_j^{(1)} = \hat{\mathrm{T}}_{\mathrm{v}}^{(0)}\mathbf{q}_j^{(1)}$ we have

$$[\mathrm{E}^{(1)} - \mathrm{F}^{(0)}\beta_j^{(1)}]\mathbf{q}_j^{(1)} = -[\mathrm{E}^{(2)} - \mathrm{F}^{(0)}\beta_j^{(2)}]\mathbf{q}_j^{(0)} \tag{6.164}$$

where

$$\mathrm{E}^{(2)} = [\hat{\mathrm{T}}_{\mathrm{v}}^{(0)}]^{\mathrm{T}}[\mathrm{L}^{-1}\Lambda^{(2)}]\hat{\mathrm{T}}_{\mathrm{v}}^{(0)}. \tag{6.165}$$

The problem (6.164) is equivalent to the problem (6.151) with non-degenerate eigenvalues and hence can be solved in the same manner. In this fashion we can calculate the terms of the expansions (6.143) and (6.146).

6.8.3. A Particular Case of Degeneracy: Lines with Transverse Homogeneous Dielectric

Finally, we deal with an interesting case of degenerate eigenvalues, namely, the case of a transversally homogeneous line for which we have (see Chapter 1)

$$\Lambda^{(0)} = \mathrm{LC} = \frac{1}{c_0^2}\mathrm{I}, \tag{6.166}$$

where c_0 is the velocity of the light in the actual dielectric. In this case all the eigenvalues are degenerate. Let us rewrite Λ as follows:

$$\Lambda = \beta^{(0)}\mathrm{I} + \varepsilon\Omega, \tag{6.167}$$

where

$$\beta^{(0)} = 1/c_0^2, \tag{6.168}$$

and

$$\Omega = \Lambda^{(1)} + \varepsilon\Lambda^{(2)} + \varepsilon^2\Lambda^{(3)} + O(\varepsilon^3) \quad \text{for } \varepsilon \to 0. \tag{6.169}$$

It is easy to show that the eigenvalues β_{A} of the matrix $\mathrm{A} = \beta\mathrm{I} + \mathrm{B}$ are related to those of the matrix B through the relation

$$\beta_{\mathrm{A}} = \beta + \beta_{\mathrm{B}}, \tag{6.170}$$

and A and B have the same eigenvectors. Therefore, once we have calculated the eigenvalues η_i and the eigenvectors $\mathbf{r}_i$ of the matrix Ω, we can obtain those of the matrix Λ through the relations

$$\beta_i = \beta^{(0)} + \varepsilon\eta_i, \tag{6.171}$$

and

$$\mathbf{e}_i = \mathbf{r}_i. \tag{6.172}$$

The matrix Ω may be considered as a perturbation of the matrix $\Lambda^{(1)}$ (or of the matrix $\Lambda^{(2)}$, if $\Lambda^{(1)} = 0$), and hence the asymptotic expressions of η_i and $\mathbf{r}_i$ for $\varepsilon \to 0$ can be computed as in the nondegenerate case.

6.9. ASYMPTOTIC EXPRESSIONS FOR THE MATRIX OPERATORS A(s) AND $\mathrm{T_V}$(s)

By continuing analytically the expansions (6.143) and (6.146) over the entire complex plane, for the asymptotic behavior of the eigenvalues $\beta_i(s)$ and eigenvectors $\mathbf{e}_i(s)$, we obtain

$$\beta_i = \beta_i^{(0)} + \beta_i^{(1)}s^{-1} + \beta_i^{(2)}s^{-2} + O(s^{-3}) \quad \text{for } s \to \infty, \tag{6.173}$$

$$\mathbf{e}_i = \mathbf{e}_i^{(0)} + \mathbf{e}_i^{(1)}s^{-1} + \mathbf{e}_i^{(2)}s^{-2} + O(s^{-3}) \quad \text{for } s \to \infty. \tag{6.174}$$

From the expansion (6.173) we obtain

$$a_i(s) = \sqrt{\beta_i(s)} = a_i^{(0)} + a_i^{(1)}s^{-1} + a_i^{(2)}s^{-2} + a_i^{(3)}s^{-3} + O(s^{-4}) \quad \text{for } s \to \infty, \tag{6.175}$$

where $a_i^{(k)}$ are related to $\beta_i^{(k)}$ through the relations

$$a_i^{(0)} = \sqrt{\beta_i^{(0)}}, \tag{6.176}$$

$$a_i^{(1)} = \frac{\beta_i^{(1)}}{2\sqrt{\beta_i^{(0)}}}, \tag{6.177}$$

$$a_i^{(2)} = \frac{1}{2\sqrt{\beta_i^{(0)}}}\left[\beta_i^{(2)} - \frac{[\beta_i^{(1)}]^2}{4\beta_i^{(0)}}\right], \tag{6.178}$$

$$a_i^{(3)} = \frac{1}{2\sqrt{\beta_i^{(0)}}}\left[\beta_i^{(3)} - \frac{\beta_i^{(1)}\beta_i^{(2)}}{2\beta_i^{(0)}} + \frac{[\beta_i^{(1)}]^3}{8[\beta_i^{(0)}]^2}\right]. \tag{6.179}$$

Therefore, the asymptotic expressions of $\mathrm{A}(s)$ and $\mathrm{T_v}(s)$ are

$$\mathrm{A}(s) = \mathrm{A}^{(0)} + \mathrm{A}^{(1)}s^{-1} + \mathrm{A}^{(2)}s^{-2} + O(s^{-3}) \quad \text{for } s \to \infty, \tag{6.180}$$

$$\mathrm{T_v}(s) = \mathrm{T}^{(0)} + \mathrm{T}^{(1)}s^{-1} + \mathrm{T}^{(2)}s^{-2} + O(s^{-3}) \quad \text{for } s \to \infty, \tag{6.181}$$

where

$$\mathrm{A}^{(k)} = \mathrm{diag}(a_1^{(k)}, a_2^{(k)}, \ldots, a_n^{(k)}), \tag{6.182}$$

$$\mathrm{T}^{(k)} = |\mathbf{e}_1^{(k)} \vdots \mathbf{e}_2^{(k)} \vdots \cdots \vdots \mathbf{e}_n^{(k)}|. \tag{6.183}$$

Furthermore, the inverse of the modal matrix $\mathrm{T_v}(s)$ has the asymptotic expression

$$\mathrm{T_v^{-1}}(s) = \mathrm{W}^{(0)} + \mathrm{W}^{(1)}s^{-1} + O(s^{-2}) \quad \text{for } s \to \infty, \tag{6.184}$$

where

$$\mathrm{W}^{(0)} = (\mathrm{T}^{(0)})^{-1} = (\mathrm{L}^{-1}\mathrm{T}^{(0)})^{\mathrm{T}} \quad \text{and} \quad \mathrm{W}^{(1)} = -\mathrm{W}^{(0)}\mathrm{T}^{(1)}\mathrm{W}^{(0)}. \tag{6.185}$$

6.10. EVALUATION OF THE IMPULSE RESPONSES FOR LOSSY MULTICONDUCTOR LINES WITH FREQUENCY-INDEPENDENT PARAMETERS

In this section, by using the asymptotic behavior of $\mathrm{A}(s)$ and $\mathrm{T_v}(s)$, we shall determine first the asymptotic expressions of the describing functions $\mathrm{P}(s)$, $\mathrm{Z_c}(s)$, and $\mathrm{Y_c}(s)$. Then we shall study the properties of the corresponding inverse transforms.

6.10.1. Asymptotic Expressions for the Describing Functions P(s), $Y_c(s)$, and $Z_c(s)$

Now we model P, Z_c, and Y_c by *principal parts* P_p, Z_{cp}, and Y_{cp} and *remainders* P_r, Z_{cr}, and Y_{cr}: The principal parts contain the leading term as $s \to \infty$ and the remainders go to zero as $s \to \infty$.

By substituting Eqs. (6.180), (6.181), and (6.184) in the expression of P(s) given by Eq. (6.119) we obtain

$$P(s) = \left(T^{(0)} + \frac{T^{(1)}}{s} + O\left(\frac{1}{s^2}\right)\right) Q(s)\left(1 - \frac{A^{(2)}}{s}d + O\left(\frac{1}{s^2}\right)\right) \times \left(W^{(0)} + \frac{W^{(1)}}{s} + O\left(\frac{1}{s^2}\right)\right), \tag{6.186}$$

where Q(s) is the diagonal matrix

$$Q(s) = \mathrm{diag}(e^{-(s+\mu_1)T_1}, e^{-(s+\mu_2)T_2}, \ldots, e^{-(s+\mu_n)T_n}), \tag{6.187}$$

and

$$T_i = da_i^{(0)} = \frac{d}{c_i}, \quad \mu_i = \frac{a_i^{(1)}}{a_i^{(0)}} = \frac{c_i^2 \beta_i^{(1)}}{2}. \tag{6.188}$$

Therefore, the leading term of P(s) for $s \to \infty$ is given by

$$P_p(s) = \sum_{i=1}^{n} \mathbf{e}_i^{(0)}(\mathbf{s}_i^{(0)})^T \exp[-(s + \mu_i)T_i], \tag{6.189}$$

where (see Eq. (3.22))

$$\mathbf{s}_i^{(0)} = L^{-1}\mathbf{e}_i^{(0)}. \tag{6.190}$$

The remainder P_r, defined as

$$P_r(s) \equiv P(s) - P_p(s), \tag{6.191}$$

has the following expression:

$$P_r(s) = \frac{1}{s}[T^{(0)}Q(s)W^{(1)} + T^{(1)}Q(s)W^{(0)} - dT^{(0)}Q(s)A^{(2)}W^{(0)}] + \Sigma(s)Q(s)\Xi(s), \tag{6.192}$$

where $\Sigma(s) \approx \Xi(s) \approx O(s^{-1})$ for $s \to \infty$,

$$-A^{(2)}d = \tfrac{1}{2}\,\mathrm{diag}(T_1 v_1^2, T_2 v_2^2, \ldots, T_n v_n^2), \tag{6.193}$$

and

$$v_i^2 = \mu_i^2 - c_i^2 \beta_i^{(2)}. \tag{6.194}$$

By substituting Eqs. (6.180), (6.181), and (6.184) in the expression of $Y_c(s)$ given by Eq. (6.120) we obtain

$$\begin{aligned} Y_c(s) = sZ^{-1}(s)\left(T^{(0)} + \frac{T^{(1)}}{s} + O\left(\frac{1}{s^2}\right)\right)\left(A^{(0)} + \frac{A^{(1)}}{s} + O\left(\frac{1}{s^2}\right)\right) \\ \times\left(W^{(0)} + \frac{W^{(1)}}{s} + O\left(\frac{1}{s^2}\right)\right). \end{aligned} \tag{6.195}$$

Because for $s \to \infty$ the matrix function $sZ^{-1}(s)$ may be developed as

$$sZ^{-1}(s) = \left(I + \frac{L^{-1}R}{s}\right)^{-1} L^{-1} = L^{-1} - \frac{L^{-1}RL^{-1}}{s} + O\left(\frac{1}{s^2}\right), \tag{6.196}$$

the leading term of $Y_c(s)$ for $s \to \infty$ is given by

$$Y_{cp} = G_c = L^{-1}\sqrt{LC}. \tag{6.197}$$

The remainder Y_{cr}, defined as

$$Y_{cr}(s) \equiv Y_c(s) - Y_{cp}, \tag{6.198}$$

has the following asymptotic expression for $s \to \infty$:

$$\begin{aligned} Y_{cr}(s) = \frac{L^{-1}}{s}[T^{(1)}A^{(0)}W^{(0)} + T^{(0)}A^{(0)}W^{(1)} + T^{(0)}A^{(1)}W^{(0)} \\ - RL^{-1}T^{(0)}A^{(0)}W^{(0)}] + O\left(\frac{1}{s^2}\right), \end{aligned} \tag{6.199}$$

where

$$A^{(1)} = \mathrm{diag}\left(\frac{\mu_1}{c_1}, \frac{\mu_2}{c_2}, \ldots, \frac{\mu_n}{c_n}\right). \tag{6.200}$$

Finally, by substituting Eqs. (6.180), (6.181), and (6.184) in the expression of $Z_c(s)$ given by Eq. (6.120), we obtain

$$Z_c(s) = \frac{1}{s}\left(T^{(0)} + \frac{T^{(1)}}{s} + O\left(\frac{1}{s^2}\right)\right)A^{-1}(s)\left(W^{(0)} + \frac{W^{(1)}}{s} + O\left(\frac{1}{s^2}\right)\right)Z(s). \tag{6.201}$$

Because

$$A^{-1}(s) = (A^{(0)})^{-1} - \frac{A^{(1)}(A^{(0)})^{-2}}{s} + O\left(\frac{1}{s^2}\right), \tag{6.202}$$

the leading term of $Z_c(s)$ for $s \to \infty$ is given by

$$Z_{cp} = R_c\sqrt{(LC)^{-1}}L. \tag{6.203}$$

The remainder Z_{cr} defined as

$$Z_{cr}(s) \equiv Z_c(s) - Z_{cp}, \tag{6.204}$$

has the following asymptotic expression for $s \to \infty$:

$$Z_{cr}(s) = \frac{1}{s}[T^{(1)}(A^{(0)})^{-1}W^{(0)} + T^{(0)}(A^{(0)})^{-1}W^{(1)} - T^{(0)}A^{(1)}(A^{(0)})^{-2}W^{(0)} + T^{(0)}(A^{(0)})^{-1}W^{(0)}RL^{-1}]L + O\left(\frac{1}{s^2}\right). \tag{6.205}$$

6.10.2. Evaluation of the Principal Parts of the Impulse Responses $p(t)$, $z_c(t)$, and $y_c(t)$

The obtained asymptotic expressions allow an accurate evaluation of the time-domain line impulse responses. In fact, the inverse Laplace transforms of the principal parts $P_p(s)$, Z_{cp}, and Y_{cp} of $P(s)$, $Z_c(s)$, and $Y_c(s)$ are straightforward. These give the analytical expressions to the principal parts $p_p(t)$, $z_{cp}(t)$, and $y_{cp}(t)$ of $p(t)$, $z_c(t)$, and $y_c(t)$:

$$p_p(t) = \sum_{i=1}^{n} \mathbf{e}_i^{(0)}(\mathbf{s}_i^{(0)})^T e^{-\mu_i T_i}\delta(t - T_i), \tag{6.206}$$

$$z_{cp}(t) = R_c\delta(t), \tag{6.207}$$

$$y_{cp}(t) = G_c\delta(t). \tag{6.208}$$

From Eqs. (6.192), (6.199), and (6.205) we note that the remainders of $P(s)$, $Z_c(s)$, and $Y_c(s)$ go to zero as s^{-1} for $s \to \infty$, and hence the time domain remainders

$$p_r(t) \equiv p(t) - \sum_{i=1}^{n} \mathbf{e}_i^{(0)}(\mathbf{s}_i^{(0)})^T e^{-\mu_i T_i}\delta(t - T_i) \tag{6.209}$$

$$z_{cr}(t) = z_c(t) - R_c\delta(t), \tag{6.210}$$

$$y_{cr}(t) = y_c(t) - G_c\delta(t), \tag{6.211}$$

are regular functions and can be simply and effectively evaluated by numerical inversion procedures whatever the required accuracy. Because all the remainders go to zero in the Laplace domain as s^{-1} for $s \to \infty$, we call them *first-order remainders.*

Remarks

(i) From the expression of Q(s) given by Eq. (6.187), we obtain that

$$p(t) = 0 \quad \text{for } 0 \leqslant t < T_1, \tag{6.212}$$

because we have arranged the eigenvalues $\beta_i^{(0)}$ in such a way that $T_1 \leqslant T_2 \leqslant \cdots \leqslant T_n$, as was done in Chapter 3. The impulse response $p(t)$ takes into account the delay due to the finite velocity of propagation of the modes: The time required for a signal to propagate from one end to the other is the one-way transit time of the fastest mode.

(ii) The time domain remainders of $p(t)$, $z_c(t)$, and $y_c(t)$ are *regular* everywhere, that is, they are bounded and continuous except at isolated points where they may have a discontinuity of the first kind. In the ideal case, namely, R $= 0$ and G $= 0$, the remainders vanish and the impulse responses reduce to those given in Eq. (6.127) to (6.129). When the losses are considered, we still find the same Dirac pulses, which correspond to the leading terms in the asymptotic expansions of P(s), Z$_c$(s), and Y$_c$(s) near $s = \infty$. The losses have two kinds of effect: They give rise to the damping factors $e^{-\mu_i T_i}$ in the weighted sum of Dirac pulses appearing in $p_p(t)$ (the damping rates μ_i are positive since $\beta_i^{(1)}$ are positive), and to the wake phenomenon described by the remainders.

(iii) The computational cost of the numerical inversion of a function given in the Laplace domain depends on the decaying velocity of the function as $s \to \infty$: The quicker the decay as $s \to \infty$, the less the computational cost. Thus, to accelerate the numerical inversion of the remainders, we could evaluate analytically the inverse Laplace transform of the s^{-1} terms in the asymptotic expressions of the remainders P$_r$, Z_{cr}, and Y_{cr}. In this way, we have to invert numerically the *second-order remainders* that go to zero as s^{-2} for $s \to \infty$. By introducing higher-order remainders we can further reduce the cost of the numerical inversion. Their evaluation is based on the evaluation of higher-order terms in the asymptotic expansions (6.143) and (6.146).

6.10.3. An Application to a Three-Conductor Line

Now we apply the method described in this chapter to determine the impulse responses of the three-conductor line of length $d = 1\,\text{m}$ characterized by the per-unit-length parameters (Maffucci and Miano, 1999b):

$$\mathrm{R} = \mathrm{diag}(41.67, 41.67, 41.67)[\Omega]/[\mathrm{m}],$$

$$\mathrm{G} = \mathrm{diag}(0.59, 0.59, 0.59)[\mathrm{mS}]/[\mathrm{m}],$$

$$\mathrm{L} = \begin{vmatrix} 2.42 & 0.69 & 0.64 \\ 0.69 & 2.36 & 0.69 \\ 0.64 & 0.69 & 2.42 \end{vmatrix} [\mu\mathrm{H}]/[\mathrm{m}],$$

$$\mathrm{C} = \begin{vmatrix} 21.0 & -12.3 & -4.01 \\ -12.3 & 26.2 & -12.3 \\ -4.01 & -12.3 & 21.0 \end{vmatrix} [\mathrm{pF}]/[\mathrm{m}].$$

The one-way transit times T_i and the corresponding damping rates μ_i are given in Table 6.1.

In Figs. 6.3a and 6.3b we show, respectively, the amplitude spectra of the first row entries of the characteristic admittance matrix Y_c and of its remainder Y_{cr}, while in Fig. 6.5a the corresponding time domain remainders are reported. As in the previous example, the computation of the time domain remainder Y_{cr} is easy due to the asymptotic behavior of the remainder in the frequency domain, as shown in Fig. 6.3b. A direct numerical inversion of the overall characteristic admittance matrix Y_c would be meaningless (see Fig. 6.3a).

In Figs. 6.4a and 6.4b the amplitude spectra of the first row entries of the propagation function matrix P and of its remainder P_r are plotted, while in Fig. 6.5b the corresponding time-domain remain-

Table 6.1.

Mode Number	T_i [ns]	$\mu_i/10^7$ [s^{-1}]
1	3.6404	8.8272
2	6.6667	2.3437
3	8.1404	1.9879

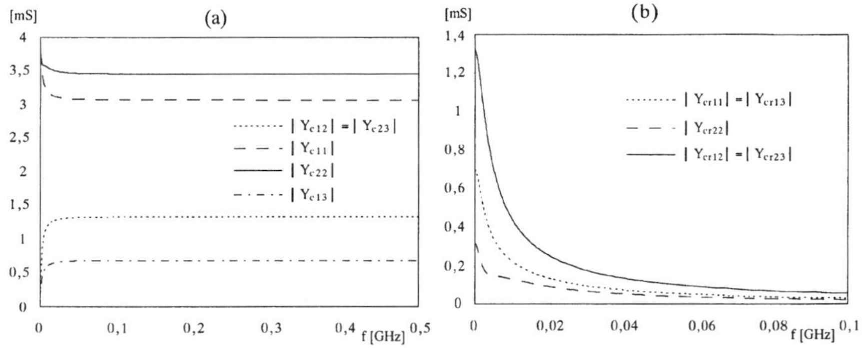

Figure 6.3. Amplitude spectra of the first row entries of (a) Y_c and (b) Y_{cr}, for the three-conductor line.

ders are reported (the origin of the time axis has been shifted to $t = T_1$). We observe in the amplitude spectra of the entries of P the following peculiar behavior:

(i) the amplitudes do not decay to zero as the frequency $f \to \infty$ because of the presence of Dirac pulses in the time domain; and

(ii) the amplitudes oscillate because of the beating between the three modes, which have different propagation velocities.

It is obvious that there is no way of numerically computing the

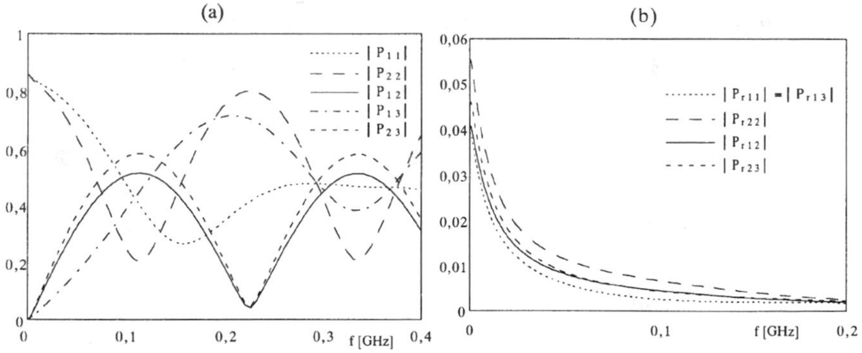

Figure 6.4. Amplitude spectra of the first row entries of (a) P and (b) P_r for the three-conductor line.

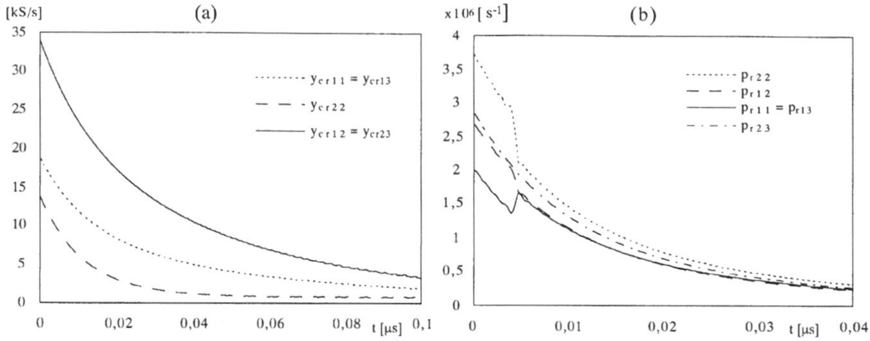

Figure 6.5. First row entries of (a) y_r and (b) p_r, versus time for the three-conductor line.

inverse transform of P. On the other hand, by extracting the principal part P_p, we obtain the remainder P_r shown in Fig. 6.4b. This can easily be numerically inverted as it is regular. The time-domain remainders show jumps corresponding to the contribution of the slower modes: For instance, Fig. 6.5b shows clearly the jumps occurring at $t = T_3 - T_1 = 4.5$ ns. These first kind of discontinuities could be computed analytically by evaluating the coefficients of the terms behaving as s^{-1} for $s \to \infty$; the resulting second-order remainder would be continuous everywhere. Note that, for the particular case we are considering, $Y_{cr11} = Y_{cr13}$, $Y_{cr12} = Y_{cr23}$ and $P_{r11} = P_{r13}$.

6.11. EVALUATION OF THE IMPULSE RESPONSES OF LOSSY MULTICONDUCTOR LINES WITH FREQUENCY-DEPENDENT PARAMETERS

The method presented in the previous sections can be applied without any restriction to multiconductor lines with arbitrary frequency dependence if the matrices $Z(s)$ and $Y(s)$ are known.

6.11.1. Multiconductor Lines with Skin Effect

These lines may be represented by the following per-unit-length impedance matrix (see §5.3.1)

$$Z(s) = R_{dc} + K\sqrt{s} + sL_e, \tag{6.213}$$

where L_e is the per-unit-length external inductance matrix, R_{dc} is the per-unit length dc longitudinal resistance matrix and $K = \chi W$. The matrix W is an $n \times n$ symmetric and strictly positive matrix depending on the geometrical parameters of the line, and $\chi = \sqrt{\mu/\sigma}$ is the *skin-effect constant* (for copper wires $\chi \approx 10^{-7}\,\Omega \cdot s^{1/2}$). The term $\sqrt{s}$ in Eq. (6.213) is the monodrome function corresponding to the branch that has a positive real part in the complex plane.

The matrix $Z(s)$ is symmetric; furthermore, it is positive definite along the positive real axis of the complex plane. Here, only for the sake of simplicity, we refer to lines with dispersionless dielectric, thus we shall develop our analysis by using the expression (6.8) for $Y(s)$. However, the asymptotic expansions for $A(s)$ and $T_v(s)$ obtained in the previous paragraph no longer hold because $s = \infty$ is a first-order branch-point of polar type of $Z(s)$. In this case the matrix $\Lambda(s)$ defined in Eqs. (6.23) has a first-order branch-point of regular type at $s = \infty$, and hence it cannot be expanded in the neighborhood of $s = \infty$ by powers of s^{-1}. This property gives rise to deep changes in the behavior of the time domain impulse responses as we have already discussed in Chapter 5 for two-conductor lines. Now we shall extend those results to multiconductor lines.

The matrix $\Lambda(s)$ can be expanded in terms of powers of $1/\sqrt{s}$ in the neighborhood of $s = \infty$:

$$\Lambda(s) = \Lambda^{(0)} + \frac{\Lambda^{(1)}}{s^{1/2}} + \frac{\Lambda^{(2)}}{s} + \frac{\Lambda^{(3)}}{s^{3/2}} + \frac{\Lambda^{(4)}}{s^2}, \tag{6.214}$$

where

$$\Lambda^{(0)} = L_e C, \quad \Lambda^{(1)} = KC, \quad \Lambda^{(2)} = L_e G + R_{dc} C, \quad \Lambda^{(3)} = KG, \quad \Lambda^{(4)} = R_{dc} G. \tag{6.215}$$

Therefore, the eigenvalues and eigenvectors of $\Lambda(s)$ can be expanded in series of powers of $1/\sqrt{s}$ near $s = \infty$:

$$\beta_i(s) = \beta_i^{(0)} + \frac{\beta_i^{(1)}}{\sqrt{s}} + \frac{\beta_i^{(2)}}{s} + O(s^{-3/2}) \quad \text{for } s \to \infty, \tag{6.216}$$

$$\mathbf{e}_i(s) = \mathbf{e}_i^{(0)} + \frac{\mathbf{e}_i^{(1)}}{\sqrt{s}} + \frac{\mathbf{e}_i^{(2)}}{s} + O(s^{-3/2}) \quad \text{for } s \to \infty, \tag{6.217}$$

where $\beta_i^{(0)}$ and $\mathbf{e}_i^{(0)}$ are, respectively, the ith eigenvalue and eigenvector of the "unperturbed" matrix $\Lambda_0 = LC$. The other coefficients of the expansions (6.216) and (6.217) can be still evaluated by studying the behavior of the eigenvalues and eigenvectors of $\Lambda(s)$ along the positive real axis $s = \sigma > 0$ because of the properties of the matrices $Z(s)$ and

Y(s). The matrix

$$\Lambda = \Lambda^{(0)} + \Lambda^{(1)}\varepsilon + \Lambda^{(2)}\varepsilon^2 + \Lambda^{(3)}\varepsilon^3 + \Lambda^{(4)}\varepsilon^4, \qquad (6.218)$$

where $\varepsilon = 1/\sqrt{\sigma}$ is a positive real quantity, can be tackled as a "perturbation" of the matrix $\Lambda^{(0)} = \mathrm{LC}$ in the neighborhood of $\varepsilon = 0$. Although the definition of ε is changed with respect to the perturbation analysis outlined in §6.8, those results are still applicable to evaluate the expansion coefficients $\mathbf{e}_i^{(k)}$ and $\beta_i^{(k)}$ for $k \geqslant 1$: They are obtained from Eqs. (6.153) and (6.155), and from Eqs. (6.154) and (6.156), respectively, with the matrices $\Lambda^{(k)}$ given in Eq. (6.215).

For the asymptotic expansions of A(s) and $\mathrm{T_v}(s)$ we obtain

$$\mathrm{A}(s) = \mathrm{A}^{(0)} + \frac{\mathrm{A}^{(1)}}{\sqrt{s}} + \frac{\mathrm{A}^{(2)}}{s} + O\left(\frac{1}{s^{3/2}}\right) \quad \text{for } s \to \infty, \qquad (6.219)$$

$$\mathrm{T_v}(s) = \mathrm{T}^{(0)} + \frac{\mathrm{T}^{(1)}}{\sqrt{s}} + \frac{\mathrm{T}^{(2)}}{s} + O\left(\frac{1}{s^{3/2}}\right) \quad \text{for } s \to \infty, \qquad (6.220)$$

where $\mathrm{A}^{(k)}$ and $\mathrm{T}^{(k)}$are always defined according to Eqs. (6.182) and (6.183); $a_i^{(k)}$ appearing in the expressions of A_k are always related to $\beta_i^{(k)}$ through Eqs. (6.176) to (6.179). Furthermore, the inverse of the modal matrix $\mathrm{T_v}(s)$ has the asymptotic expansion

$$\mathrm{T_v^{-1}}(s) = \mathrm{W}^{(0)} + \frac{\mathrm{W}^{(1)}}{\sqrt{s}} + O\left(\frac{1}{s}\right) \quad \text{for } s \to \infty, \qquad (6.221)$$

where the matrices $\mathrm{W}^{(0)}$ and $\mathrm{W}^{(1)}$ are always related to the matrices $\mathrm{T}^{(0)}$ and $\mathrm{T}^{(1)}$ through Eq. (6.185).

Now we model P(s), $\mathrm{Z_c}(s)$, and $\mathrm{Y_c}(s)$ by principal parts P_p, $\mathrm{Z_{cp}}$, and $\mathrm{Y_{cp}}$ and remainders P_r, Z_{cr}, and Y_{cr} as in the case of frequency-independent parameters: The principal parts contain the leading terms as $s \to \infty$. By substituting Eqs. (6.219) to (6.221) in the expression of P(s) given by Eq. (6.119) we obtain

$$\begin{aligned}\mathrm{P}(s) = {} & \left(\mathrm{T}^{(0)} + \frac{\mathrm{T}^{(1)}}{\sqrt{s}} + O\left(\frac{1}{s}\right)\right)\mathrm{Q}(s) \\ & \times \left(\mathrm{I} - \frac{\mathrm{A}^{(3)}}{\sqrt{s}}d + O\left(\frac{1}{s}\right)\right)\left(\mathrm{W}^{(0)} + \frac{\mathrm{W}^{(1)}}{\sqrt{s}} + O\left(\frac{1}{s}\right)\right), \end{aligned} \qquad (6.222)$$

where now Q(s) is the diagonal matrix

$$\mathrm{Q}(s) = \mathrm{diag}(e^{-(s+\mu_1)T_1 - \alpha_1\sqrt{s}}, e^{-(s+\mu_2)T_2 - \alpha_2\sqrt{s}}, \ldots, e^{-(s+\mu_n)T_n - \alpha_n\sqrt{s}}), \quad (6.223)$$

and

$$T_i = da_i^{(0)} = \frac{d}{c_i}, \quad \mu_i = \frac{a_i^{(2)}}{a_i^{(0)}}, \quad \alpha_i = da_i^{(1)}. \tag{6.224}$$

Therefore, the leading term of P(s) for $s \to \infty$ is given by

$$P_p(s) = \sum_{i=1}^{n} \mathbf{e}_i^{(0)}(\mathbf{s}_i^{(0)})^T \exp[-(s+\mu_i)T_i - \alpha_i\sqrt{s}], \tag{6.225}$$

where the vectors $\mathbf{s}_i^{(0)}$ are given by Eq. (6.190). The remainder $P_r(s)$ defined as in Eq. (6.191) is as follows:

$$P_r(s) = \frac{1}{\sqrt{s}}[T^{(0)}Q(s)W^{(1)} + T^{(1)}Q(s)W^{(0)} - d\,T^{(0)}Q(s)A^{(3)}W^{(0)}]$$
$$+ \Sigma(s)Q(s)\Xi(s), \tag{6.226}$$

where now $\Sigma(s) \approx \Xi(s) \approx O(s^{-1/2})$ for $s \to \infty$,

$$-A^{(3)}d = \mathrm{diag}(\alpha_1 \nu_1, \alpha_2 \nu_2, \ldots, \alpha_n \nu_n), \tag{6.227}$$

and

$$\nu_i = -\frac{a_i^{(3)}}{a_i^{(1)}}. \tag{6.228}$$

The expressions of the parameters $a_i^{(0)}$, $a_i^{(1)}, \ldots,$ are given by Eqs. (6.176) to (6.179).

By substituting Eqs. (6.219) to (6.221) in the expression of $Y_c(s)$ given by Eq. (6.120) we obtain

$$Y_c(s) = sZ^{-1}(s)\left(T^{(0)} + \frac{T^{(1)}}{\sqrt{s}} + O\left(\frac{1}{s}\right)\right)\left(A^{(0)} + \frac{A^{(1)}}{\sqrt{s}} + O\left(\frac{1}{s}\right)\right)$$
$$\times\left(W^{(0)} + \frac{W^{(1)}}{\sqrt{s}} + O\left(\frac{1}{s}\right)\right). \tag{6.229}$$

Because the matrix function $sZ^{-1}(s)$ may be developed as in Eq. (6.196) near $s = \infty$, the leading term of $Y_c(s)$ for $s \to \infty$ is always given by Eq. (6.197). By contrast, the remainder, Y_{cr} defined as in Eq. (6.198) has the following asymptotic expression for $s \to \infty$:

$$Y_{cr}(s) = \frac{\Sigma^{(1)}}{\sqrt{s}} + O\left(\frac{1}{s}\right), \tag{6.230}$$

where

$$\Sigma^{(1)} = L^{-1}[T^{(1)}A^{(0)}W^{(0)} + T^{(0)}A^{(1)}W^{(0)} + T^{(0)}A^{(0)}W^{(1)}] - \frac{L^{-1}KL^{-1}}{\sqrt{s}} T^{(0)}A^{(0)}W^{(0)}, \tag{6.231}$$

and

$$A^{(1)} = \frac{1}{d}\,\mathrm{diag}(\alpha_1, \alpha_2, \ldots, \alpha_n). \tag{6.232}$$

These results suggest rewriting the expression of $Y_c(s)$ as

$$Y_c(s) = \left(G_c + \frac{\Sigma^{(1)}}{\sqrt{s}}\right) + Y_{cb}(s), \tag{6.233}$$

where the function $Y_{cb}(s)$ given by

$$Y_{cb}(s) = Y_{cr}(s) - \frac{\Sigma^{(1)}}{\sqrt{s}}, \tag{6.234}$$

has the asymptotic behavior $Y_{cb}(s) \approx O(s^{-1})$ for $s \to \infty$.

Finally, by substituting Eqs. (6.219) to (6.221) in the expression of $Z_c(s)$ given by Eq. (6.121) we obtain

$$Z_c(s) = \frac{1}{s}\left(T^{(0)} + \frac{T^{(1)}}{\sqrt{s}} + O\left(\frac{1}{s}\right)\right)A^{-1}(s)\left(W^{(0)} + \frac{W^{(1)}}{\sqrt{s}} + O\left(\frac{1}{s}\right)\right)Z(s). \tag{6.235}$$

Because

$$A^{-1}(s) = (A^{(0)})^{-1} - \frac{A^{(1)}(A^{(0)})^{-2}}{\sqrt{s}} + O\left(\frac{1}{s}\right), \tag{6.236}$$

the leading term of $Z_c(s)$ for $s \to \infty$ is always given by Eq. (6.203). The remainder Z_{cr} defined as in Eq. (6.204) has the following asymptotic expression for $s \to \infty$:

$$Z_{cr}(s) = \frac{\Psi^{(1)}}{\sqrt{s}} + O\left(\frac{1}{s}\right), \tag{6.237}$$

where

$$\Psi^{(1)} = [T^{(1)}(A^{(0)})^{-1}W^{(0)} + T^{(0)}(A^{(0)})^{-1}W^{(1)} - T^{(0)}(A^{(0)})^{-2}A^{(1)}W^{(0)}]L + T^{(0)}(A^{(0)})^{-1}W^{(0)}K. \tag{6.238}$$

These results suggest rewriting the expression of $Z_c(s)$ as

$$Z_c(s) = \left(R_c + \frac{\Psi^{(1)}}{\sqrt{s}} \right) + Z_{cb}(s), \tag{6.239}$$

where the function $Z_{cb}(s)$, given by

$$Z_{cb}(s) = Z_{cr}(s) - \frac{\Psi^{(1)}}{\sqrt{s}}, \tag{6.240}$$

has the asymptotic behavior $Z_{cb}(s) \approx O(s^{-1})$ for $s \to \infty$. Note that the generic term (α_i / d) in Eq. (6.232) does not actually depend on the line length d.

The impulse responses $p(t)$, $z_c(t)$, and $y_c(t)$ may be expressed as

$$p(t) = p_p(t) + p_r(t), \tag{6.241}$$

$$z_c(t) = z_{ci}(t) + c_{cb}(t), \tag{6.242}$$

$$y_c(t) = y_{ci}(t) + y_{cb}(t), \tag{6.243}$$

where

$$p_p(t) = \sum_{i=1}^{n} \mathbf{e}_i^{(0)}(\mathbf{s}_i^{(0)})^{\mathrm{T}} e^{-\mu_i T_i} \delta(t - T_i; \alpha_i), \tag{6.244}$$

$$z_{ci}(t) = R_c \delta(t) + \frac{\Psi^{(1)}}{\sqrt{\pi t}} u(t), \tag{6.245}$$

$$y_{ci}(t) = G_c \delta(t) + \frac{\Sigma^{(1)}}{\sqrt{\pi t}} u(t), \tag{6.246}$$

and $p_r(t)$, $z_{cb}(t)$, and y_{cb} are, respectively, the inverse Laplace transforms of $P_r(s)$, $Z_{cb}(s)$, and $Y_{cb}(s)$. The function $\delta(t; \alpha)$ has already been introduced in Chapter 5, where we have dealt with two-conductor lines with skin effect.

It is possible to extend all the considerations made for two-conductor lines to multiconductor lines (see §5.5). In particular, the parameter $\alpha_{\max}^2 = \max_i(\alpha_i^2)$ can be considered as the characteristic time of the dispersive phenomenon due to the skin effects. As in the two-conductor case, it is proportional to the square of the ratio between the line length and the characteristic transverse dimension of the conductors. When the smallest characteristic time t_c of the signals satisfies the condition $t_c \gg \alpha_{\max}^2$, each $\delta(t; \alpha_i)$ behaves in practice as a Dirac pulse as in the two-conductor case. Note that the sign of the parameters μ_i may also be negative.

6.11.2. An Application to a Three-Conductor Line

Now we (Maffucci and Miano, 1999a) determine the impulse responses of a 15 cm long three-conductor line with skin effect characterized by the per-unit-length parameters

$$\mathrm{R} = \mathrm{diag}(117,\ 117,\ 117)\ [\Omega]/[\mathrm{m}], \quad \mathrm{G} = 0,$$

$$\mathrm{L} = \begin{bmatrix} 0.3277 & 0.0676 & 0.0184 \\ 0.0676 & 0.3236 & 0.0676 \\ 0.0184 & 0.0676 & 0.3277 \end{bmatrix} \mu\mathrm{H/m},$$

$$\mathrm{C} = \begin{bmatrix} 138.3 & -28.8 & -0.3 \\ -28.8 & 146.2 & -28.8 \\ -0.3 & -28.8 & 138.3 \end{bmatrix} \mathrm{pF/m},$$

$$\mathrm{W} = \begin{bmatrix} 0.802 & 0.172 & 0.036 \\ 0.172 & 0.854 & 0.172 \\ 0.036 & 0.172 & 0.802 \end{bmatrix} 10^{-2}\Omega/\mathrm{m},\ \chi = 2.$$

This is an example of a degenerate case in which all the line modes propagate with the same velocity. The one-way transit time and the damping factors are given in Table 6.2.

Figure 6.6 shows the plots of the entries of the principal diagonal of the characteristic admittance of this line, both in the frequency and time domain. Again, a direct numerical inversion of $\mathrm{Y}_c(s)$ is not possible because of the asymptotic behavior of its entries (Fig. 6.6a). The entries of the corresponding time domain function $y_c(t)$ have Dirac pulses and terms such as $1/\sqrt{t}$. The entries of the remainder $\mathrm{Y}_{cb}(s)$ defined as in Eq. (6.240) are easily numerically inverted because of their asymptotic behavior (Fig. 6.6b), and the results are shown in Fig. 6.6c.

Table 6.2.

Mode Number	T_i [ns]	$\mu_i/10^7\ [\mathrm{s}^{-1}]$
1	0.98814	−15.24
2	0.98814	−11.19
3	0.98814	−11.23

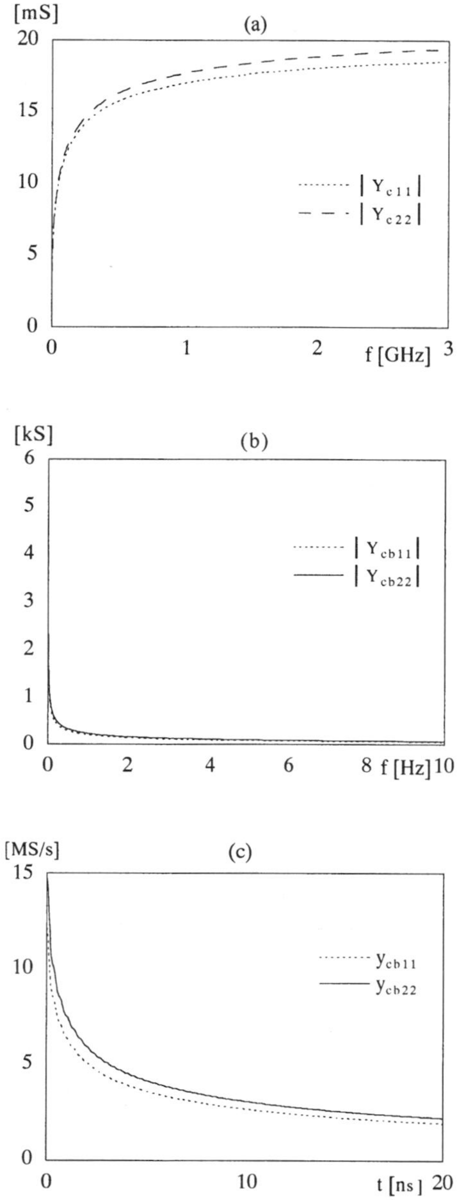

Figure 6.6. Amplitude spectra of some entries of (a) Y_c and (b) Y_{cb}; (c) the corresponding entries of $y_{cb}(t)$.

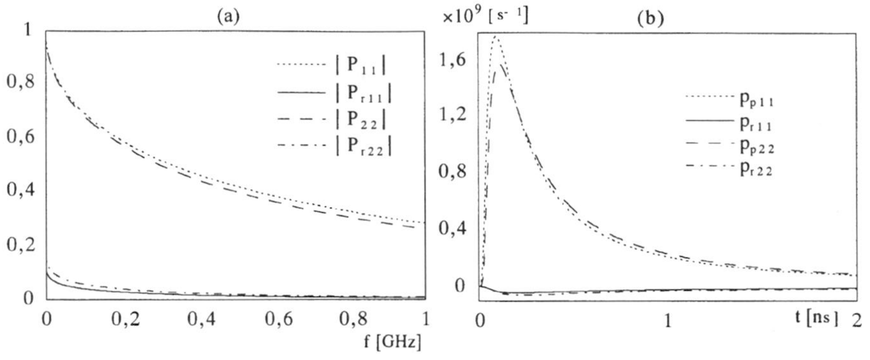

Figure 6.7. Plots of the (a) frequency domain and (b) the time domain propagation function.

Figure 6.7a shows the entries of the principal diagonal of the propagation function, both in the frequency and in the time domain. By using the method described in this chapter we have evaluated analytically the "pseudoimpulsive" terms $\delta(t; \alpha_i)$ that would make the direct numerical evaluation of the overall time domain propagation function unmanageable. Once again the regular parts are easily computed numerically (Fig. 6.7b). Note that the entries of the time domain propagation function have been shifted to $t = T$.

6.11.3. Per-Unit-Length Impedance and Admittance Matrices with Arbitrary Frequency Dependence

The same method can be applied without any restriction to multiconductor transmission lines with arbitrary per-unit length impedance and admittance matrices. To model the skin effect we have used Eq. (6.213). In other cases of practical interest the matrices $Z(s)$ and $Y(s)$ can still be expressed analytically. For instance, when the behavior of the embedding dielectric is frequency dependent, we can use a model similar to that described in Chapter 5. However, even if the matrices $Z(s)$ and $Y(s)$ are not known in analytical form, to apply the method we have described previously we only need to know their asymptotic behavior analytically.

Once the per-unit-length impedances $Z(s)$ and admittances $Y(s)$ are given, a preliminary analysis of the behavior of $\Lambda(s)$ for $s \to \infty$ is needed in order to choose the right expansion term in the Laurent series at $s = \infty$. For instance, in all those cases for which $\Lambda(s)$ is regular at $s = \infty$ and has no branches there, it can always be expanded as $\Lambda_0 + \Lambda_1/s + \Lambda_2/s^2 + \ldots$, and the analysis developed in §6.8 can be carried out. This is the case when the embedding medium is dispersive or is not homogeneous (e.g., Gordon, Blazeck, and Mittra, 1992; Grotelüsche, Dutta, and Zaage, 1994; Braunisch and Grabinski, 1998). If $\Lambda(s)$ has a first-order branch point of regular type at $s = \infty$, as in the skin-effect case (e.g., Gordon, Blazeck, and Mittra, 1992), $\Lambda(s)$ can be expanded in Laurent series in the neighborhood of $s = \infty$ by powers of $1/\sqrt{s}$. If $\Lambda(s)$ had a second-order branch point, as in the anomalous skin effect case (see §5.3.2), it should be expanded in Laurent series in the neighborhood of $s = \infty$ by powers of $1/\sqrt[3]{s}$.

CHAPTER 7

Nonuniform Transmission Lines

7.1. INTRODUCTION

Modern high-density integrated circuits are characterized by interconnections with varying cross sections and bends due to the severe geometrical constraints and to the placement of vias for signal redistribution (e.g., Schutt-Ainé, 1992). In some cases, the non-uniformity of the line is intentional, as with transmission lines designed for impedance matching (e.g., Collin, 1992), pulse shaping (Burkhart and Wilcox, 1990), and analog signal processing (Hayden and Tripathi, 1991). Whatever the case, the problem to be dealt with is very complex, as it is described by transmission line equations with *space varying* per-unit-length parameters (see §1.2).

A substantial amount of work has been devoted to the study of nonuniform transmission lines. However, only in a few particular cases is it possible to solve their equations analytically, and, moreover, only by operating in the frequency domain (e.g., Sugai, 1960; Franceschetti, 1997; Lu, 1997). The reader may consult Schutt-Ainé (Schutt-Ainé, 1992) for a comprehensive list of interesting papers on this subject.

The frequency domain analysis provides the simplest way of analyzing arbitrary profiles by approximating them through piece-wise-constant functions (e.g., Paul, 1994). In this way the problem consists in the solution of a cascade of many uniform lines. The overall transmission matrix is simply the product of the transmission matrices of each uniform tract (see §4.6.4 and §6.5.4). However, the piecewise-constant approximation, although simple, has a serious drawback. Due to the discontinuities of the parameters between two adjacent tracts, spurious multireflections arise. This problem may be overcome if a smooth approximation of the actual parameter profile is considered, provided that the line equations for each tract can be solved analytically. This has been done for a piecewise-linear approximation of a nonuniform line with a transverse homogeneous dielectric (Lu, 1997).

In many applications we only need to know the reflection coefficient of the voltage or the current along the line. It is a function of the space variable and its expression is described by a nonlinear differential equation of Riccati type, which may be solved only for particular longitudinal profiles of the per-unit-length parameters (e.g., Sugai, 1961; Collin, 1992).

The propagation along nonuniform lines may be effectively studied through the asymptotic form (for high frequencies) of the line equations (Baum, 1988).

As we shall show in Chapter 9, the most efficient method for analyzing circuits consisting of lumped parameter elements connected by tracts of transmission lines is to reduce the entire network to an equivalent circuit in which the lines are represented as multiports. Such an approach is significant provided that it is possible to express in a simple manner the characteristic equations of the multiports that represent the lines.

We are now studying the possibility of extending to nonuniform lines the *input-state-output* and *input-output* descriptions developed in Chapter 4 for uniform lossy two-conductor transmission lines (Maffucci and Miano, 1999d; 2000b). In particular here, we shall show that the terminal behavior of a lossless nonuniform two-conductor transmission line may be represented by an equivalent circuit of the kind shown in Fig. 7.1.

Ideal two-conductor lines may be characterized directly in the time domain by using d'Alembert's solution for the wave equation as described in Chapter 2. The equations of nonuniform transmission lines, can not be solved analytically in the time domain. However, they can be solved through a numerical-analytical method in the Laplace domain as we shall see in this chapter. Therefore, the approach to be followed for nonuniform lines is first to determine the characteristic relations of the two-port in the Laplace domain and then to transform them in the time domain.

How should we express the general solution of the nonuniform transmission line equations in the Laplace domain? If we wish to generalize the input-state-output and the input-output representations introduced in the previous chapters for uniform transmission

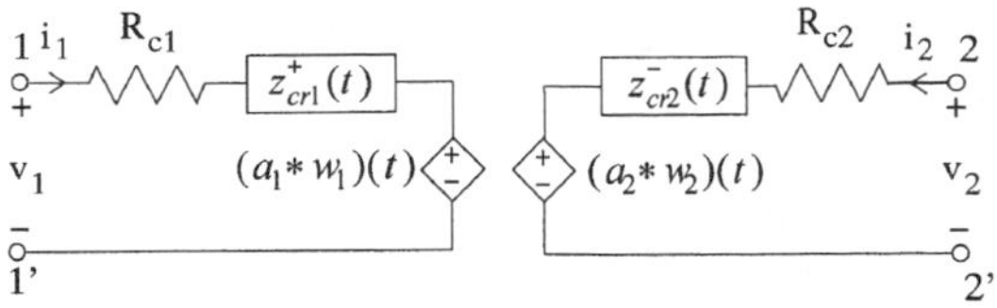

Figure 7.1. Thévenin equivalent two-port in the time domain of a two-conductor non uniform transmission line.

lines, the general solution of the line equations must be expressed through a linear combination of *traveling wave solutions* (see §4.2 and §.4.5) As we shall show in the next section, the equations of nonuniform lines admit traveling wave solutions for any kind of nonuniformity.

How could we determine the traveling wave solutions of the nonuniform line equations in the Laplace domain? In general, these equations can not be solved analytically, unless for transversally homogeneous lines and particular longitudinal profiles. For instance, for lines with an exponential longitudinal profile the general solution of the line equations is a linear combination of exponential functions of the kind $\exp(\pm k(s)x/c_0)$. For lines with linear longitudinal profiles the line equations reduce to an ordinary differential equation of Bessel type, whereas for lines with Gaussian longitudinal profiles, the line equations reduce to an ordinary differential equation, whose solutions are the Weber parabolic cylindrical functions. We shall deal with these particular and very interesting cases in Section 7.3.

In Sections 7.4 and 7.5 we shall extend the input-state-output and the input-output descriptions to these nonuniform lines.

Generally, the equations of nonuniform transmission lines even if ideal can not be solved analytically.

In Section 7.6 we shall make an attempt to catch numerically two particular solutions of the nonuniform line equations that are at the same time linearly independent and of the traveling wave type. The method is based on the WKB (Wentzel, Kramers, Brillouin) method (e.g., Bender and Orszag, 1978), and it is applicable to any type of nonuniformity.

From preliminary investigations it seems that the numerical evaluation of the traveling solutions based on this method is difficult, as we shall discuss at the end of Section 7.6. However, an equivalent circuit representation of the terminal behavior of nonuniform lines may easily be obtained numerically by means of the method of characteristics, as we shall see in Chapter 8.

7.2. EQUATIONS FOR NONUNIFORM LOSSLESS TRANSMISSION LINES

Consider a two-conductor transmission line with parameters varying in space. For the sake of simplicity, let us assume that the line is lossless, initially at rest, and that there are no distributed sources. The equations are (see §1.2)

$$-\frac{\partial v}{\partial x} = L(x)\frac{\partial i}{\partial t}, \tag{7.1}$$

$$-\frac{\partial i}{\partial x} = C(x)\frac{\partial v}{\partial t}, \tag{7.2}$$

where $L(x)$ and $C(x)$ are, respectively, the per-unit-length inductance and capacitance of the line.

Equations (7.1) and (7.2) can not be solved analytically in the time domain. They can be solved through a numerical-analytical method in the Laplace domain as we shall see in this chapter. In the Laplace domain the system of equations (7.1) and (7.2) becomes

$$\frac{dV}{dx} = -sL(x)I, \tag{7.3}$$

$$\frac{dI}{dx} = -sC(x)V. \tag{7.4}$$

The equation that describes the voltage distribution V is readily obtained from the system of equations (7.3) and (7.4):

$$\frac{d^2V}{dx^2} - \frac{d\,ln[L(x)]}{dx}\frac{dV}{dx} - s^2L(x)C(x)V = 0. \tag{7.5}$$

Equation (7.5) can be simplified by introducing the following variable transform:

$$V(x;s) = \sqrt{L(x)}U(x;s). \tag{7.6}$$

Substituting Eq. (7.6) in Eq. (7.5), we obtain the differential equation with varying coefficients in the unknown function $U(x;s)$:

$$\frac{d^2U}{dx^2} + r(x;s)U = 0, \tag{7.7}$$

where $r(x;s)$ is defined by

$$r(x;s) = -\frac{s^2}{c^2(x)} + \rho(x), \tag{7.8}$$

$c(x)$ is the "local" propagation velocity defined by

$$c^2(x) = \frac{1}{L(x)C(x)}, \tag{7.9}$$

and

$$\rho(x) = \frac{1}{2}\frac{d^2 \ln L}{dx^2} - \frac{1}{4}\left(\frac{d \ln L}{dx}\right)^2. \tag{7.10}$$

For actual transmission lines, $r(x;s)$ is an analytical function of x, and hence all the solutions of Eq. (7.7) are analytical at every x for any given s (e.g., Korn and Korn, 1961; Smirnov, 1964b).

Remark

Equation (7.7) also holds for lossy nonuniform two-conductor lines with frequency-dependent parameters if instead of $L(x)$ and $C(x)$ we consider, respectively, $Z(x;s)/s$ and $Y(x;s)/s$ in Eq. (7.6) and in the expressions (7.9) and (7.10). ◇

7.2.1. The Dyson Series

The system of equations (7.3) and (7.4) can be rewritten in the matrix form

$$\frac{d\mathbf{Y}}{dx} = -sA(x)\mathbf{Y}. \tag{7.11}$$

where $\mathbf{Y} = |V, I|^{\mathrm{T}}$ and

$$\mathrm{A}(x) = \begin{bmatrix} 0 & L(x) \\ C(x) & 0 \end{bmatrix}. \tag{7.12}$$

The solution of the equation system (7.11) is the solution of the following vector integral equation:

$$\mathbf{Y}(x) = \mathbf{Y}_0 - s\int_0^x \mathrm{A}(\zeta)\mathbf{Y}(\zeta)d\zeta, \tag{7.13}$$

where $\mathbf{Y}_0 = \mathbf{Y}(0)$. This equation can be solved by using the Picard method of successive approximation (e.g., Lefschetz, 1977). Let us start by assuming as trial solution $\mathbf{Y}^{(0)} = \mathbf{Y}_0$, and computing successive approximations to the solution $\mathbf{Y}(x)$:

$$\mathbf{Y}^{(j+1)}(x) = \mathbf{Y}_0 - s\int_0^x \mathrm{A}(\zeta)\mathbf{Y}^{(j)}(\zeta)d\zeta, \quad j = 0, 1, 2, \ldots. \tag{7.14}$$

The process converges for $j \to \infty$, subject to the Lipschitz condition

$$\|\mathrm{A}(x)(\mathbf{Y}' - \mathbf{Y}'')\| \leqslant M\|\mathbf{Y}' - \mathbf{Y}''\|, \tag{7.15}$$

where M must be bounded and independent of $\mathbf{Y}'$ and $\mathbf{Y}''$. Because in actual nonuniform lines $L(x)$ and $C(x)$ are bounded, the Lipschitz condition (7.15) is always satisfied. A formal representation of the solution obtained in this way as power of the matrix A is given by

$$\begin{aligned}\mathbf{Y}(x) = \Bigg[\mathrm{I} - \int_0^x dx_1\mathrm{A}(x_1; s) + \int_0^x dx_1 \int_0^{x_1} dx_2\mathrm{A}(x_1; s)\mathrm{A}(x_2; s) \\ - \int_0^x dx_1 \int_0^{x_1} dx_2 \int_0^{x_2} dx_3\mathrm{A}(x_1; s)\mathrm{A}(x_2; s)\mathrm{A}(x_3; s) + \cdots\Bigg]\mathbf{Y}_0,\end{aligned} \tag{7.16}$$

where $x > x_1 > x_2 > \cdots > x_n > 0$.

Let us define the ordered product of m times the matrix $\mathrm{A}(x)$, S, as the sum of all the products whose factors are arranged so that the one corresponding to x_n, namely, to the smallest space variable x, is at the right end, and the one corresponding to x_{n-1} is next on the left, and so forth. For example,

$$S\{\mathrm{A}(x)\} = \mathrm{A}(x), \qquad (m = 1) \tag{7.17}$$

$$S\{\mathrm{A}(x_1)\mathrm{A}(x_2)\} = u(x_1 - x_2)\mathrm{A}(x_1)\mathrm{A}(x_2) + u(x_2 - x_1)\mathrm{A}(x_2)\mathrm{A}(x_1), \quad (m = 2)$$

and so on, where $u(x)$ is the unit step function. The ordered space product of m times the matrix A is the sum of $m!$ terms. The definite integrals of these terms over the region $0 \leqslant x_1 \leqslant x$, $0 \leqslant x_2 \leqslant x, \ldots, 0 \leqslant x_m \leqslant x$ are equal. Then Eq. (7.16) can be rewritten formally as

$$\mathbf{Y}(x) = \left[\mathrm{I} + \sum_{m=1}^{\infty} \frac{(-s)^m}{m!} \int_0^x dx_1 dx_2 \cdots dx_m S\{\mathrm{A}(x_1) \cdots \mathrm{A}(x_m)\}\right\}\mathbf{Y}_0. \tag{7.18}$$

This infinite ordered series is known in quantum field theory as the *Dyson series* (e.g., Weinberg, 1995). It can be summed and expressed in a closed form if the matrices obtained by evaluating $\mathrm{A}(x)$ at different positions commutate. If this is the case, the sum is given by

$$\mathbf{Y}(x) = \exp\left[-s\int_0^x \mathrm{A}(\zeta)d\zeta\right]\mathbf{Y}_0. \tag{7.19}$$

In general, this condition is never verified in actual transmission lines, except in uniform ones. For transmission lines in which the nonuniformity may be dealt with as a perturbation with respect to a uniform longitudinal profile, the series (7.18) consists of an asymptotic expansion in the perturbation part (Amari, 1991).

7.3. ANALYTICAL SOLUTIONS FOR LINES WITH TRANSVERSALLY HOMOGENEOUS DIELECTRIC AND PARTICULAR PROFILES OF $L(x)$

To extend to nonuniform transmission lines the input-state-output description and the input-output description introduced for uniform two-conductor lines in Chapter 4, we have to represent the general solution of Eq. (7.7) through a superposition of two linearly independent particular solutions of the traveling wave type. A solution is of the traveling wave type if it behaves as $A(x)\exp[\pm\mu(x)s]$ for $s \to \infty$ in the region of convergence $\text{Re}\{s\} > \sigma_c$.

For uniform lines ρ is equal to zero, c is constant, and hence $r = -s^2/c^2$ also is constant. For nonuniform lines with a transversally homogeneous dielectric, c is always constant (see Chapter 1), even if L and C vary with respect to x:

$$L(x)C(x) = \frac{1}{c_0^2} = \text{constant}. \tag{7.20}$$

In this condition Eq. (7.7) can be solved analytically for some particular profiles of the per-unit-length inductance: Lines with exponential profile (e.g., Franceschetti, 1997); lines with linear profile (e.g., Lu, 1997); and lines with Gaussian profile. In this section we shall determine the traveling wave solutions of Eq. (7.7) for these particular cases.

When the condition (7.20) is satisfied, the function $c(x)$ given by Eq. (7.9) does not depend on the actual profiles of L (and C), and the expression of $r(x;s)$ reduces to

$$r(x;s) = -\frac{s^2}{c_0^2} + \frac{1}{2}\frac{d^2 \ln L}{dx^2} - \frac{1}{4}\left(\frac{d \ln L}{dx}\right)^2. \tag{7.21}$$

In this condition Eq. (7.7) can be solved analytically in terms of known functions in some cases.

7.3.1. Exponential Profile

Let us consider a transversally homogeneous transmission line with

$$L(x) = L_0 \exp(\eta x/\Delta), \tag{7.22}$$

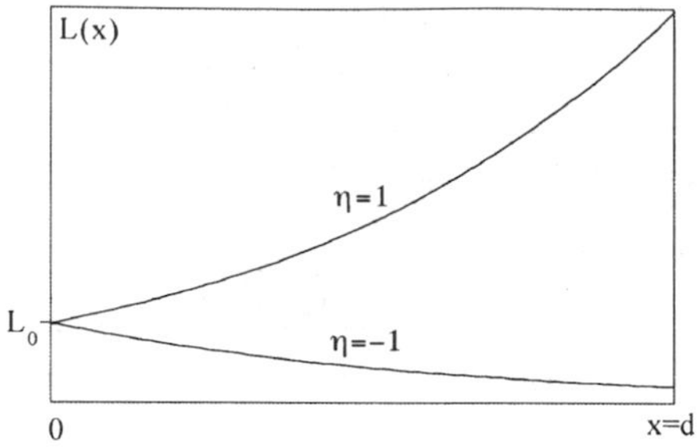

Figure 7.2. Exponential profile.

where L_0 represents the value of the per-unit-length inductance at the end $x = 0$, Δ is the characteristic length on which the inductance of the line varies ($\Delta > 0$), and η is a parameter that assumes two single values $\eta = 1$ (rising profile toward the right) and $\eta = -1$ (decaying profile toward the right) (Fig. 7.2). The corresponding capacitance profile is obtained from Eq. (7.20).

Substituting Eq. (7.22) in the expression (7.21) we find that $r(x; s)$ is constant with respect to x and its expression is given by

$$r = -\frac{1}{c_0^2}\left(s^2 + \frac{1}{\tau^2}\right), \tag{7.23}$$

where

$$\tau = \frac{2\Delta}{c_0}. \tag{7.24}$$

Let us introduce the function

$$k(s) = \frac{1}{c_0}\sqrt{s^2 + \frac{1}{\tau^2}}. \tag{7.25}$$

It is multivalued and has two branches (e.g., Smirnov, 1964b). It has two branch points of the first order along the imaginary axis, one at $s_+ = i/\tau$ and the other at $s_- = -i/\tau$; both s_- and s_+ are branch points of regular type, and in the limit of the uniform line they coincide and hence cancel each other. To render this function single valued, appropriate cuts in the complex plane are needed. It is convenient to cut the imaginary axis along the segment whose ends are at s_- and s_+. In this way we have $Re\{k(s)\} \geqslant 0$ for $Re(s) > 0$ and $Im\{k(s)\} \geqslant 0$ for $Im(s) > 0$. Therefore, in the limit of the uniform line, namely, $\Delta \to \infty$, we have $k(s) \to s/c_0$.

Two linearly independent solutions of traveling wave type of Eq. (7.7) are

$$U^{\pm}(x;s) = \exp[\mp k(s)x]. \tag{7.26}$$

The forward and backward waves represented by U^+ and U^- propagate with dispersion because of the nonuniformity. The dispersion relation is the same as that of an electromagnetic plane wave propagating in a cold nonmagnetized collisionless plasma (e.g., Franceschetti, 1997). For $\omega < 1/\tau$ both waves become of "evanescent" type.

Because $k(s) \approx s/c_0$ for $s \to \infty$, the asymptotic behavior of the solutions $U^{\pm}(x;s)$ is given by $\exp(\mp sx/c_0)$ for $s \to \infty$ and bounded x.

7.3.2. Linear Profile

Let us now consider the case where the profile of L is linear:

$$L(x) = L_0 + lx, \tag{7.27}$$

where L_0 again represents the value of the per-unit-length inductance of the line at the end $x = 0$, and l is a parameter representing the rise of L that can be either positive or negative (Fig. 7.3). The corresponding capacitance profile is obtained from Eq. (7.20).

Substituting Eq. (7.27) in the expression (7.21) we obtain

$$r(x;s) = -\left[\frac{3}{4}\frac{l^2}{(L_0 + lx)^2} + \frac{s^2}{c_0^2}\right], \tag{7.28}$$

and Eq. (7.7) becomes

$$\frac{d^2U}{dx^2} - \left[\frac{3}{4}\frac{l^2}{(L_0 + lx)^2} + \frac{s^2}{c_0^2}\right]U = 0. \tag{7.29}$$

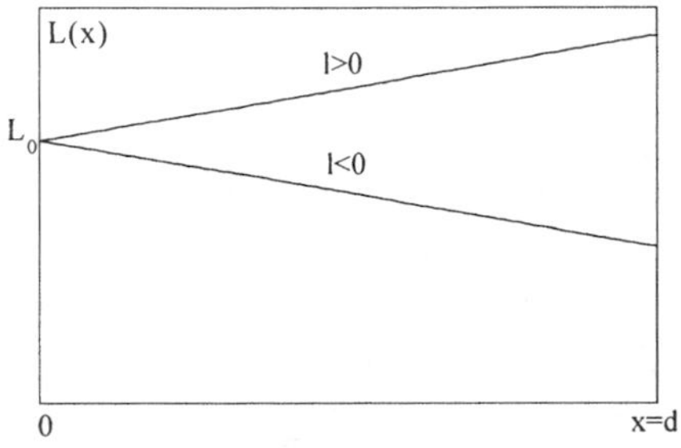

Figure 7.3. Linear profile.

Introducing the new independent variable

$$\bar{x} = \frac{L_0 + lx}{l}, \tag{7.30}$$

Eq. (7.29) becomes

$$\frac{d^2U}{d\bar{x}^2} + \left[\left(i\frac{s}{c_0}\right)^2 - \frac{3}{4}\frac{1}{\bar{x}^2}\right]U = 0. \tag{7.31}$$

This equation admits solutions of the type (e.g., Abramowitz and Stegun, 1972)

$$u = \sqrt{\bar{x}}C_1\left(i\frac{s}{c_0}\bar{x}\right), \tag{7.32}$$

where $C_1(\lambda)$ is a generic first-order Bessel function: a first-type Bessel function $J_1(\lambda)$, a second-type Bessel function $Y_1(\lambda)$, or one of the third-type Bessel functions, also known as Hankel functions, that is, $H_1^{(1)} = J_1 + iY_1$ and $H_1^{(2)} = J_1 - iY_1$. Functions $\sqrt{\lambda}J_1(\lambda)$ and $\sqrt{\lambda}Y_1(\lambda)$ behave, respectively, as the cosine and sine for $\lambda \to \infty$, while functions $\sqrt{\lambda}H_1^{(1)}(\lambda)$ and $\sqrt{\lambda}H_1^{(2)}(\lambda)$ behave, respectively, as $e^{i\lambda}$ and $e^{-i\lambda}$. Thus two linearly independent solutions of traveling wave type of Eq. (7.31) are

$$U^+(x;s) = \frac{H_1^{(1)}\left[i\frac{s}{c_0}\frac{L(x)}{l}\right]}{H_1^{(1)}\left[i\frac{s}{c_0}\frac{L(0)}{l}\right]}, \tag{7.33}$$

$$U^-(x;s) = \frac{H_1^{(2)}\left[i\frac{s}{c_0}\frac{L(x)}{l}\right]}{H_1^{(2)}\left[i\frac{s}{c_0}\frac{L(0)}{l}\right]}. \tag{7.34}$$

Therefore, U^+ and U^- behave, respectively, as $\exp(isx/c_0)$ and $\exp(-isx/c_0)$ for large s and bounded x.

7.3.3. Gaussian Profile

Finally, let us consider the case where the profile of $L(x)$ is of Gaussian type:

$$L(x) = L_0 \exp[\eta(x - d/2)^2/\Delta^2]. \tag{7.35}$$

Here, L_0 represents the inductance value at the center of the line, Δ

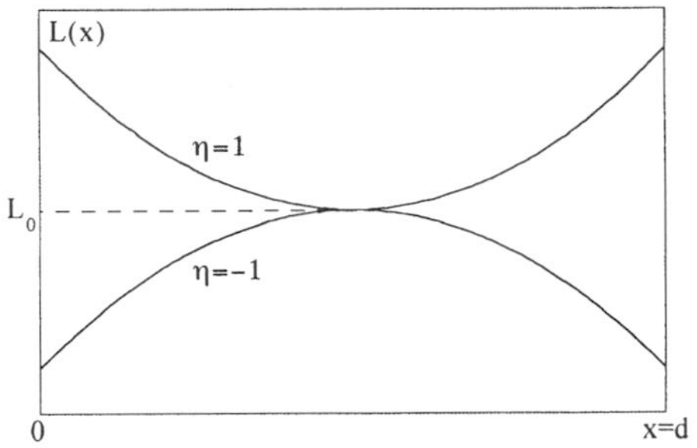

Figure 7.4. Gaussian profile.

is the characteristic length on which the inductance of the line varies ($\Delta > 0$), and η is a parameter that has only two values, namely, $\eta = 1$ (rising profile toward the right) and $\eta = -1$ (decaying profile toward the right) (Fig. 7.4). The corresponding capacitance profile is obtained from Eq. (7.20).

Substituting Eq. (7.35) in the expression of $r(x;s)$ given by Eq. (7.21), for Eq. (7.7) we obtain

$$\frac{d^2U}{dx^2} + \left\{\frac{1}{\Delta^2}\left[\eta - \left(\frac{x - d/2}{\Delta}\right)^2\right] - \frac{s^2}{c_0^2}\right\}U = 0. \tag{7.36}$$

Introducing the new spatial variable

$$\bar{x} = \sqrt{2}\left(\frac{x - d/2}{\Delta}\right), \tag{7.37}$$

Eq. (7.36) becomes

$$\frac{d^2U}{d\bar{x}^2} - \left(\frac{\bar{x}^2}{4} + b\right)U = 0, \tag{7.38}$$

where

$$b = \frac{1}{2}\left[\left(\frac{s\Delta}{c_0}\right)^2 - \eta\right]. \tag{7.39}$$

Two linearly independent solutions of Eq. (7.38) are the Weber parabolic cylindrical functions $\mathscr{U}(b, \bar{x})$ and $\mathscr{V}(b, \bar{x})$ (e.g., Abramowitz and Stegun, 1972). The functions $\mathscr{U}$ are "even," while the functions $\mathscr{V}$ are "odd," that is,

$$\mathscr{U}(b, -\bar{x}) = \mathscr{U}(b, \bar{x}), \quad \mathscr{V}(b, -\bar{x}) = -\mathscr{V}(b, \bar{x}). \tag{7.40}$$

They can be expressed through confluent hypergeometric functions (e.g., Abramowitz and Stegun, 1972), in particular, through the Kummer functions $M(\frac{1}{2}b+\frac{1}{4},\frac{1}{2},\frac{1}{2}\bar{x}^2)$ and $M(\frac{1}{2}b+\frac{3}{4},\frac{3}{2},\frac{1}{2}\bar{x}^2)$. The functions $\mathscr{U}$ and $\mathscr{V}$ behave as the cosine and the sine functions for large and negative b (e.g., Abramowitz and Stegun, 1972).

Let us introduce the functions

$$\mathscr{W}^{\pm}(b,\bar{x}) = \mathscr{U}(b,\bar{x}) \mp i\Gamma(\tfrac{1}{2}-\bar{x})\mathscr{V}(b,\bar{x}), \tag{7.41}$$

where Γ is the gamma function. The functions $\mathscr{W}^{+}$ and $\mathscr{W}^{-}$ behave as $\exp(-i\sqrt{|b|}\bar{x})$ and $\exp(i\sqrt{|b|}\bar{x})$, respectively, for large and negative b, and bounded $\bar{x}$ (e.g., Abramowitz and Stegun, 1972).

Therefore, two linearly independent solutions of the traveling wave type of Eq. (7.36) are

$$U^{+}(x;s) = \frac{\mathscr{W}^{+}\left(b,\sqrt{2}\dfrac{x-d/2}{\Delta}\right)}{\mathscr{W}^{+}\left(b,-\dfrac{d}{\Delta\sqrt{2}}\right)}, \tag{7.42}$$

$$U^{-}(x;s) = \frac{\mathscr{W}^{-}\left(b,\sqrt{2}\dfrac{x-d/2}{\Delta}\right)}{\mathscr{W}^{-}\left(b,-\dfrac{d}{\Delta\sqrt{2}}\right)}. \tag{7.43}$$

Note that the functions U^{+} and U^{-} behave as $\exp(-isx/c_0)$ and $\exp(isx/c_0)$, respectively, for large s and bounded x.

Remark

The asymptotic behavior of the traveling wave solutions we have found for these particular nonuniform transmission lines is independent of the actual profile of $L(x)$:

$$U^{\pm}(x;s) \approx \exp\left(\mp\frac{sx}{c_0}\right) \quad \text{for } s\to\infty. \tag{7.44}$$

They describe two waves, one propagating toward the right and the other toward the left, both with velocity c_0. As we shall show in Section 7.6, this result is not fortuitous, it is a general property of the traveling wave solutions of Eq. (7.7). ◇

7.3.4. General Solution of the Traveling Wave Type

According to the results we have found in the previous paragraphs, the general solution of the traveling wave type for the line equations (7.3) and (7.4) can be expressed as follows:

$$V(x;s) = Q^+(x;s)V^+(s) + Q^-(x;s)V^-(s), \tag{7.45}$$

$$I(x;s) = \frac{Q^+(x;s)}{Z_c^+(x;s)}V^+(s) - \frac{Q^-(x;s)}{Z_c^-(x;s)}V^-(s), \tag{7.46}$$

where

$$Q^+(x;s) = \sqrt{\frac{L(x)}{L(0)}}\,U^+(x;s), \tag{7.47}$$

$$Q^-(x;s) = \sqrt{\frac{L(x)}{L(d)}}\,\frac{U^-(x;s)}{U^-(d;s)}, \tag{7.48}$$

$U^+(x;s)$ and $U^-(x;s)$ are given by Eq. (7.26), or Eqs. (7.33) and (7.34), or Eqs. (7.42) and (7.43) for exponential, or linear or Gaussian per-unit-length inductance profiles, respectively, $V^+(s)$ and $V^-(s)$ are arbitrary functions of the complex variable s, depending solely on the boundary conditions at the line ends, and

$$Z_c^\pm(x;s) = \mp sL\frac{Q^\pm(x;s)}{\dfrac{dQ^\pm}{dx}}. \tag{7.49}$$

Bearing in mind Eqs. (7.47) and (7.48), for $Z_c^\pm(x;s)$ we obtain

$$Z_c^\pm(x;s) = \mp\frac{sL(x)}{\dfrac{d}{dx}\{ln[\sqrt{L(x)}\,U^\pm(x;s)]\}}. \tag{7.50}$$

The analytical expressions of $Q^\pm$ and $Z_c^\pm$ are given in Table 7.1.

Remarks

We have obtained three very interesting results.

(i) By generalizing what has been seen for uniform lossy two-conductor lines we have represented a generic voltage and current distribution by superposing a forward voltage wave and a backward one; $Q^+(x;s)V^+(s)$ represents a traveling

Table 7.1. Analytical Expressions of the Functions $Q^{\pm}$ and $Z_c^{\pm}$ for Exponential (1); Linear (2); and Gaussian (3) Profiles

	$Q^{+}(x;s)$	$Q^{-}(x;s)$	$Z_c^{\pm}(x;s)$
1	$\sqrt{\frac{L(x)}{L(0)}}\exp[-k(s)x]$	$\sqrt{\frac{L(x)}{L(d)}}\exp[k(s)(x-d)]$	$\frac{sL(x)}{k(s)\mp\eta/2\Delta}$
2	$\frac{L(x)}{L(0)}\frac{H_1^{(1)}\left[i\frac{s}{c_0}\frac{L(x)}{l}\right]}{H_1^{(1)}\left[i\frac{s}{c_0}\frac{L(0)}{l}\right]}$	$\frac{L(x)}{L(d)}\frac{H_1^{(2)}\left[i\frac{s}{c_0}\frac{L(x)}{l}\right]}{H_1^{(2)}\left[i\frac{s}{c_0}\frac{L(d)}{l}\right]}$	$\pm ic_0L(x)\frac{H_1^{(p)}\left[i\frac{s}{c_0}\frac{L(x)}{l}\right]}{H_0^{(p)}\left[i\frac{s}{c_0}\frac{L(x)}{l}\right]}$ $\begin{matrix} +\Rightarrow p=1 \\ -\Rightarrow p=2 \end{matrix}$
3	$\sqrt{\frac{L(x)}{L(0)}}\frac{W^{+}\left[b,\sqrt{2}\left(\frac{x-d/2}{\Delta}\right)\right]}{W^{+}\left(b,-\frac{d}{\sqrt{2}\Delta}\right)}$	$\sqrt{\frac{L(x)}{L(d)}}\frac{W^{-}\left[b,\sqrt{2}\left(\frac{x-d/2}{\Delta}\right)\right]}{W^{-}\left(b,\frac{d}{\sqrt{2}\Delta}\right)}$	$\frac{\mp sL(x)}{\frac{\sqrt{2}}{\Delta}\frac{d}{dy}\ln W^{\pm}(b,y)\Big\vert_{y=\sqrt{2}\left(\frac{x-d/2}{\Delta}\right)}+\eta\frac{x-d/2}{\Delta^2}}$

wave propagating toward the right with velocity c_0, whereas $Q^-(x;s)V^-(s)$ represents a traveling wave propagating toward the left with the same local propagation velocity.

(ii) The "characteristic impedance" $Z_c^+(x;s)$ that relates the amplitude of the forward voltage wave to the amplitude of the forward current wave depends on the spatial coordinate x, and is different from the "characteristic impedance" $Z_c^-(x;s)$ that links the backward voltage wave to the backward current wave. Even if both waves have the same propagation velocity, in general, nonuniform lines, unlike the uniform case, no longer show a left-right symmetry with respect to the propagation because of the asymmetric dispersion introduced by the nonuniformity. In the limit of uniform lines we obtain $Z_c^+(x;s) = Z_c^-(x;s)$.

(iii) Once the actual values of $V^+(s)$ and $V^-(s)$ have been given (or determined), the time domain expressions of the voltage and current distributions along the line may be determined from Eq. (7.45) and (7.46) by using the convolution theorem. To determine the inverse Laplace transforms of $Q^+(x;s)$, $Q^-(x;s)$, $Q^+(x;s)/Z_c^+(x;s)$, and $Q^-(x;s)/Z_c^-(x;s)$ we may apply the same procedure as that we shall describe in §7.4 to determine the inverse Laplace transforms of the line describing functions. ◇

The functions $Q^+(x;s)$ and $Q^-(x;s)$ have been chosen in a way such that

$$Q^+(0;s) = Q^-(d;s) = 1. \tag{7.51}$$

In this way, V^+ and V^- have the same formal interpretation as that given for uniform lines (see §4.3); $V^+(s)$ represents the value of the forward voltage wave at the end $x = 0$ where it is generated and $V^-(s)$ represents the value of the backward voltage wave at end $x = d$ where it is generated.

7.4. REPRESENTATION OF NONUNIFORM TRANSMISSION LINES AS TWO-PORTS IN THE LAPLACE DOMAIN

In general, although the values of the voltages and currents at line ends are not known, they actually are unknowns of the problem: They depend on the actual network to which the lines are connected at their ends.

As we have already seen when we dealt with uniform two-conductor lines, the solution of a network composed of transmission lines and lumped circuits can be greatly simplified if we first determine the voltage and the current at the line ends by representing the line behavior through equivalent circuits of Thévenin or Norton type. Then, once the voltages or the currents at the line ends are known, we can also calculate the voltage and the current distributions all along the line.

The same approach can be adopted to study nonuniform transmission lines of finite length connecting generic lumped circuits. To determine the characteristic relations of the two-port we shall proceed as in the uniform case (see §4.5):

(i) First, we characterize the two-port in the Laplace domain where the relations are purely algebraic; and
(ii) then the representation in the time domain is obtained by applying the convolution theorem.

In the Laplace domain the voltages and the currents at line ends are related to the voltage and the current distributions along the line through the relations

$$V_1(s) = V(x = 0; s), \quad I_1(s) = I(x = 0; s), \tag{7.52}$$

and

$$V_2(s) = V(x = d; s), \quad I_2(s) = -I(x = d; s). \tag{7.53}$$

The reference directions are always chosen according to the normal convention for the two-ports corresponding to the two ends of the line (see Fig. 4.5). To determine the characteristic relations of the two-port representing the line we have to particularize the general solution (7.45) and (7.46) at the line ends.

7.4.1. Terminal Behavior of the Line

Specifying the expressions (7.45) and (7.46) at the ends of the line we find

$$V_1(s) = V^+(s) + P^-(s)V^-(s), \tag{7.54}$$

$$Z_{c1}^+(s)I_1(s) = V^+(s) - \frac{Z_{c1}^+(s)}{Z_{c1}^-(s)}P^-(s)V^-(s), \tag{7.55}$$

and

$$V_2(s) = P^+(s)V^+(s) + V^-(s), \tag{7.56}$$

$$Z_{c2}^-(s)I_2(s) = -\frac{Z_{c2}^-(s)}{Z_{c2}^+(s)}P^+(s)V^+(s) + V^-(s), \tag{7.57}$$

where

$$P^+(s) = Q^+(x = d; s), \quad P^-(s) = Q^-(x = 0; s), \tag{7.58}$$

and

$$Z_{c1}^{\pm}(s) = Z_c^{\pm}(x = 0; s), \quad Z_{c2}^{\pm}(s) = Z_c^{\pm}(x = d; s). \tag{7.59}$$

Starting from Eqs. (7.54) to (7.57), we can formally extend to nonuniform lines the results obtained in Chapter 4 for uniform lines. Subtracting Eqs. (7.54) and (7.55) member by member, and subtracting Eqs. (7.56) and (7.57), we extend Eqs. (4.50) and (4.51) to nonuniform lines

$$V_1(s) - Z_{c1}^+(s)I_1(s) = A_1(s)W_1(s), \tag{7.60}$$

$$V_2(s) - Z_{c2}^-(s)I_2(s) = A_2(s)W_2(s), \tag{7.61}$$

where

$$W_1(s) = 2P^-(s)V^-(s), \quad A_1(s) = \frac{1}{2}\left[1 + \frac{Z_{c1}^+(s)}{Z_{c1}^-(s)}\right], \tag{7.62}$$

$$W_2(s) = 2P^+(s)V^+(s), \quad A_2(s) = \frac{1}{2}\left[1 + \frac{Z_{c2}^-(s)}{Z_{c2}^+(s)}\right]. \tag{7.63}$$

The auxiliary variable W_1 represents twice the value that the backward voltage wave assumes at $x = 0$ and W_2 represents twice the value that the forward voltage wave assumes at $x = d$. As in the uniform case, they may be considered as the state variables of the line in the Laplace domain.

If the state of the line in the Laplace domain, represented by W_1 and W_2, were completely known, Eqs. (7.60) and (7.61) would completely determine the terminal behavior of the line. Actually, as for ideal two-conductor transmission lines, the voltage waves W_1 and W_2 are unknowns.

As for ideal transmission lines, different formulations of the equations governing the state are possible.

By using Eq. (7.54) it is possible to express the amplitude of the outcoming forward wave at $x = 0$, V^+, as a function of the voltage and the amplitude of the incoming backward wave at the same end. In the same way, by using Eq. (7.86) it is possible to express the amplitude of the outcoming backward wave at $x = d$, V^-, as a function of the voltage and the amplitude of the incoming forward wave at the same end. From Eqs. (7.54) and (7.56) we obtain the state equations

$$W_1(s) = P^-(s)[2V_2(s) - W_2(s)], \tag{7.64}$$

$$W_2(s) = P^+(s)[2V_1(s) - W_1(s)]. \tag{7.65}$$

By multiplying both members of Eq. (7.54) by Z_{c1}^+/Z_{c1}^-, and then summing the relation obtained in this way and Eq. (7.55) member by member, we obtain

$$W_2(s) = \frac{Z_{c1}^+(s)P^+(s)}{A_1(s)}\left[\frac{V_1(s)}{Z_{c1}^-(s)} + I_1(s)\right]. \tag{7.66}$$

In the same way, from Eqs. (7.56) and (7.57) we obtain

$$W_1(s) = \frac{Z_{c2}^-(s)P^-(s)}{A_2(s)}\left[\frac{V_2(s)}{Z_{c2}^+(s)} + I_2(s)\right]. \tag{7.67}$$

The state equations (7.64) and (7.65) describe in *implicit form* the relation between the state of the line and the electrical variables at the line ends, whereas the state equations (7.66) and (7.67) provide the same relation, but in *explicit form*.

State equations in implicit form similar to Eq. (7.64) and (7.65) involving the terminal variables I_1 and I_2, instead of V_1 and V_2, may be obtained from Eqs. (7.55) and (7.57). As these equations lead to a time domain model that requires two convolution products per iteration, we shall only consider the implicit formulation based on Eqs. (7.64) and (7.65).

Equations (7.60) and (7.61), joined to the state equations, give in implicit form the relations between the voltages V_1, V_2 and the currents I_1, I_2. In particular, substituting the expression of W_1 given by Eq. (7.67) in Eq. (7.60), and the expression of W_2, given by Eq. (7.66) in Eq. (7.61), we obtain two linearly independent equations in terms of the terminal variables V_1, V_2, I_1, and I_2:

$$V_1(s) - Z_{c1}^+(s)I_1(s) = \frac{A_1(s)}{A_2(s)} Z_{c2}^-(s)P^-(s)\left[\frac{V_2(s)}{Z_{c2}^+(s)} + I_2(s)\right], \tag{7.68}$$

$$V_2(s) - Z_{c2}^-(s)I_2(s) = \frac{A_2(s)}{A_1(s)} Z_{c1}^+(s)P^+(s)\left[\frac{V_1(s)}{Z_{c1}^-(s)} + I_1(s)\right], \tag{7.69}$$

The set of equations (7.60), (7.61), (7.64), and (7.65) describe the internal state of the line represented by W_1 and W_2, as well as the terminal properties. This is the *internal* or *input-state-output description* of the line in the Laplace domain. Equations (7.64) and (7.65) govern the behavior of the state, whereas Eqs. (7.60) and (7.61) describe the terminal properties. By contrast, the system of equations (7.68) and (7.69) describes only the terminal properties of the line. This is the *external* or *input-output description* of the line in the Laplace domain. Both descriptions are completely characterized by six describing functions, that is, the four characteristic impedances Z_{c1}^{+}, Z_{c1}^{-}, Z_{c2}^{+}, and Z_{c2}^{-}, and the two propagation functions $P^{+}(s)$ and $P^{-}(s)$. In the case of uniform lines $Z_{c1}^{+} = Z_{c2}^{+} = Z_{c1}^{-} = Z_{c2}^{-} = R_c$, $A_1 = A_2 = 1$, and $P^{+}(s) = P^{-}(s) = \exp(-sT_0)$, where $T = d/c_0$, $c_0 = 1/\sqrt{LC}$ and $R_c = \sqrt{L/C}$.

Kuznetsov has proposed a model that is similar to the input-state-output description (Kuznetsov, 1998).

7.4.2. The Input-State-Output Description and an Equivalent Circuit of Thévenin Type

In this section we shall deal with the input-state-output description based on the system of equations (7.60), (7.61), (7.64), and (7.65).

As for uniform lines, we can introduce a Thévenin-type equivalent circuit to represent the behavior of the line at the terminals (Fig. 7.5). Each terminal behaves as an impedance in series with a controlled voltage source. Equations (7.64) and (7.65) govern the controlled sources.

An equivalent circuit of Norton type can easily be obtained.

We immediately observe from Eq. (7.66) that W_2, namely, the forward wave, would be equal to zero if the line were connected at the left end to a one-port with impedance $Z_{c1}^{-}(s)$. That is,

$$V_1(s) = -Z_{c1}^{-}(s)I_1(s), \tag{7.70}$$

—*perfectly matched line at the left end with respect to the incoming*

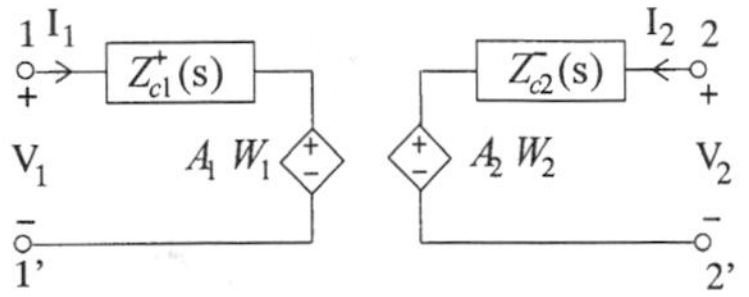

Figure 7.5. Thévenin equivalent circuit of the nonuniform line in the Laplace domain.

backward wave. Remember that $Z_{c1}^{-}(s)$ is the characteristic impedance associated with the backward wave at the left end. The same consideration holds for the backward wave, namely, W_1 would be equal to zero if the line were connected at the right end to a one-port with impedance $Z_{c2}^{+}(s)$. That is,

$$V_2(s) = -Z_{c2}^{+}(s)I_2(s), \tag{7.71}$$

—*perfectly matched line at the right end with respect to the incoming forward wave*. Remember that $Z_{c2}^{+}(s)$ is the characteristic impedance associated with the forward wave at the right end.

The impedance $Z_{c1}^{+}(s)$ is the *driving impedance* at the port $x = 0$ when the line is perfectly matched at the end $x = d$: It is the impedance seen by the outcoming forward wave at the left line end. It is different from the impedance $Z_{c1}^{-}(s)$ seen by the incoming backward wave. Dual considerations hold for the driving impedance $Z_{c2}^{-}(s)$ at the port $x = d$.

In general, the matching conditions (7.70) and (7.71) are not satisfied. Then a forward wave with amplitude V^+ generates at the left line end $x = 0$ and propagates toward the other end $x = d$, where its amplitude is P^+V^+. Similarly, the backward wave excited at $x = d$ with amplitude V^- propagates toward the left line end, where its amplitude is P^-V^-.

The value of A_1 is the ratio between the voltage V_1 and twice the amplitude of the backward voltage wave at the end $x = 0$ when $I_1 = 0$. Similar considerations are valid for A_2.

7.4.3. Input-Output Description in Explicit Form

A nonuniform two-conductor transmission line has an input-output description in the implicit form defined by Eqs. (7.68) and (7.69). As we know there are six possible explicit representations of two of the four variables V_1, V_2, I_1, and I_2 in terms of the remaining two (see §4.6).

Here, we shall consider the *current-controlled* representation expressing V_1 and V_2 as function of I_1 and I_2

$$V_1(s) = Z_{11}(s)I_1(s) + Z_{12}(s)I_2(s), \tag{7.72}$$

$$V_2(s) = Z_{21}(s)I_1(s) + Z_{22}(s)I_2(s). \tag{7.73}$$

For the driving-point impedances Z_{11} and Z_{22} we have

$$Z_{11}(s) = Z_{c1}^{+} \frac{Z_{c1}^{-}Z_{c2}^{+} + Z_{c1}^{-}Z_{c2}^{-}P^{+}P^{-}}{Z_{c1}^{-}Z_{c2}^{+} - Z_{c1}^{+}Z_{c2}^{-}P^{+}P^{-}}, \tag{7.74}$$

$$Z_{22}(s) = Z_{c2}^{-} \frac{Z_{c1}^{-}Z_{c2}^{+} + Z_{c1}^{+}Z_{c2}^{+}P^{+}P^{-}}{Z_{c1}^{-}Z_{c2}^{+} - Z_{c1}^{+}Z_{c2}^{-}P^{+}P^{-}}. \tag{7.75}$$

Note that, in general, $Z_{11} \neq Z_{22}$ because of the asymmetry introduced by the nonuniformity.

The *transfer impedances* Z_{12} and Z_{21} are given by

$$Z_{12}(s) = Z_{c2}^{+}Z_{c2}^{-}P^{-} \frac{Z_{c1}^{+} + Z_{c1}^{-}}{Z_{c1}^{-}Z_{c2}^{+} - Z_{c1}^{+}Z_{c2}^{-}P^{+}P^{-}}, \tag{7.76}$$

$$Z_{21}(s) = Z_{c1}^{+}Z_{c1}^{-}P^{+} \frac{Z_{c2}^{+} + Z_{c2}^{-}}{Z_{c1}^{-}Z_{c2}^{+} - Z_{c1}^{+}Z_{c2}^{-}P^{+}P^{-}}. \tag{7.77}$$

Due to the reciprocity property (see §1.9) it must be $Z_{12} = Z_{21}$. This is a sort of constraint imposed by the physical nature of the line on the descriptive functions.

Note that, unlike the describing functions characterizing the input-state-output description, the impedances $Z_{11}(s)$, $Z_{22}(s)$, $Z_{12}(s)$, and $Z_{21}(s)$ have an infinite number of poles, the solutions of the equation

$$Z_{c1}^{-}(s)Z_{c2}^{+}(s) - Z_{c1}^{+}(s)Z_{c2}^{-}(s)P^{+}(s)P^{-}(s) = 0. \tag{7.78}$$

This is due to the multireflections arising at the line ends when the line is characterized through the impedance matrix. This is the drawback of thc input output descriptions.

Therefore, it is clear that the input-state-output description is more suitable for transient analysis. To deal with sinusoidal and periodic steady-state operating conditions in the frequency domain the input-output description is preferable.

All the other possible input-output descriptions of explicit type may be easily obtained from Eqs. (7.68) and (7.69). We shall leave this as an exercise.

7.5. THE EQUIVALENT CIRCUIT OF THÉVENIN TYPE IN THE TIME DOMAIN

In the Laplace domain, Eqs. (7.60) and (7.61) describe the behavior of these nonuniform lines at the ends, where the dynamics of the sources W_1 and W_2 are governed by Eqs. (7.64) and (7.65). This way of representing the line as a two-port is characterized by the six describing functions $P^+(s)$, $P^-(s)$, $Z_{c1}^+(s)$, $Z_{c2}^-(s)$, $A_1(s)$, and $A_2(s)$.

In the time domain the six impulse responses

$$p^{\pm}(t) = L^{-1}\{P^{\pm}(s)\}, \quad z_{c1}^+(t) = L^{-1}\{Z_{c1}^+(s)\}, \quad z_{c2}^-(t) = L^{-1}\{Z_{c2}^-(s)\}$$

$$a_1(t) = L^{-1}\{\mathrm{A}_1(s)\}, \quad a_2(t) = L^{-1}\{\mathrm{A}_2(s)\} \tag{7.79}$$

corresponding to the six describing functions, completely characterize the terminal behavior of the line.

Using the Borel theorem, from Eqs. (7.60) and (7.61) we get, in fact,

$$v_1(t) - (z_{c1}^+ * i_1)(t) = (a_1 * w_1)(t), \tag{7.80}$$

$$v_2(t) - (z_{c2}^- * i_2)(t) = (a_2 * w_2)(t). \tag{7.81}$$

The equations that describe the dynamics of the state in the time domain are obtained by transforming Eqs. (7.64) and (7.65) as

$$w_1(t) = [p^- * (2v_2 - w_2)](t), \tag{7.82}$$

$$w_2(t) = [p^+ * (2v_1 - w_1)](t). \tag{7.83}$$

In the general case the inverse transforms of the describing functions can not be evaluated analytically, and hence we need to resort to numerical inversion. On the other hand, inversion procedures based entirely on numerical methods can be effectively applied only when the resulting time domain functions are sufficiently regular. Hence, a preliminary and accurate study of the regularity properties of $p^{\pm}(t)$, $z_{c1}^+(t)$, $z_{c2}^-(t)$, $a_1(t)$, and $a_2(t)$ is required.

As was done for uniform two-conductor lines with frequency-dependent parameters, once the asymptotic behavior of the describing functions for $s \to \infty$ is known, we can separate the parts of the impulse responses that are irregular and can be dealt with analytically from the regular ones that can be evaluated numerically, whatever the required accuracy (see Chapter 5).

7.5.1. Asymptotic Behavior of the Describing Functions

The asymptotic behavior of the describing functions $P^+(s)$, $P^-(s)$, $Z_{c1}^+(s)$, $Z_{c2}^-(s)$, $A_1(s)$, and $A_2(s)$ for $s \to \infty$ may be studied by recalling that the two independent solutions of the equations $U^+(x;s)$ and $U^-(x;s)$, have been chosen such as to satisfy the asymptotic conditions (7.44).

Let us start with the behavior of the impedances.

Let us introduce the *local characteristic resistance of the line*,

$$R_c(x) = \sqrt{\frac{L(x)}{C(x)}}. \tag{7.84}$$

It is easy to see that, for $s \to \infty$, the impedances $Z_{c1}^{\pm}$ and $Z_{c2}^{\pm}$ have an asymptotic behavior independent from the particular profile of $L(x)$: It only depends on the values of $L(x)$ at the line ends. In fact, from the expressions reported in Table 7.1, we have, for $s \to \infty$,

$$Z_{c1}^{+}(s) \approx Z_{c1}^{-}(s) \approx R_{c1} = R_c(0), \tag{7.85}$$

$$Z_{c2}^{+}(s) \approx Z_{c2}^{-}(s) \approx R_{c2} = R_c(d). \tag{7.86}$$

This behavior suggests the following asymptotic expressions for the impedances:

$$Z_{c1}^{\pm}(s) = R_{c1} + Z_{c1r}^{\pm}(s), \tag{7.87}$$

$$Z_{c2}^{\pm}(s) = R_{c2} + Z_{c2r}^{\pm}(s), \tag{7.88}$$

where the *remainders* $Z_{c1r}^{\pm}(s)$ and $Z_{c2r}^{\pm}(s)$ vanish as s^{-1} for $s \to \infty$.

The asymptotic behavior of the functions $A_1(s)$ and $A_2(s)$ is given by

$$A_1(s) \approx A_2(s) \approx 1 \quad \text{for } s \to \infty. \tag{7.89}$$

This behavior suggests the following asymptotic expressions:

$$A_1(s) = 1 + A_{1r}(s), \tag{7.90}$$

$$A_2(s) = 1 + A_{2r}(s), \tag{7.91}$$

where the *remainders* $A_{1r}(s)$ and $A_{2r}(s)$ again vanish as s^{-1} for $s \to \infty$.

Let us introduce the operator

$$P(s) = \exp(-sT), \tag{7.92}$$

where $T = d/c_0$. From the asymptotic behavior for $s \to \infty$ of the expressions of functions $Q^{\pm}$ reported in Table 7.1, we have, for $s \to \infty$,

$$P^{+}(s) \approx \sqrt{\frac{R_{c2}}{R_{c1}}} P(s) \equiv P_p^{+}(s), \tag{7.93}$$

$$P^{-}(s) \approx \sqrt{\frac{R_{c1}}{R_{c2}}} P(s) \equiv P_p^{+}(s). \tag{7.94}$$

This behavior suggests the following asymptotic expressions:

$$P^{+}(s) = P_p^{+}(s)[1 + P_r^{+}(s)], \tag{7.95}$$

$$P^{-}(s) = P_p^{-}(s)[1 + P_r^{-}(s)], \tag{7.96}$$

where the *remainders* $P_r^{+}(s)$ and $P_r^{-}(s)$ again vanish as s^{-1} for $s \to \infty$.

7.5.2. Impulse Responses

Let $p_r^+(t)$, $p_r^-(t)$, $z_{c1r}^+(t)$, $z_{c2r}^-(t)$, $a_{1r}(t)$, and $a_{2r}(t)$ indicate the Laplace inverse transforms of the remainders of the describing functions. The impulse responses can be expressed as

$$p^+(t) = \sqrt{\frac{R_{c2}}{R_{c1}}}[\delta(t-T) + p_r^+(t-T)], \tag{7.97}$$

$$p^-(t) = \sqrt{\frac{R_{c1}}{R_{c2}}}[\delta(t-T) + p_r^-(t-T)], \tag{7.98}$$

$$z_{c1}^+(t) = R_{c1}\delta(t) + z_{c1r}^+(t), \tag{7.99}$$

$$z_{c2}^-(t) = R_{c2}\delta(t) + z_{c2r}^-(t), \tag{7.100}$$

$$a_1(t) = \delta(t) + a_{1r}(t), \tag{7.101}$$

$$a_2(t) = \delta(t) + a_{2r}(t). \tag{7.102}$$

In general, the functions $p_r^+(t)$, $p_r^-(t)$, $z_{c1r}^+(t)$, $z_{c2r}^-(t)$, $a_{1r}(t)$, and $a_{2r}(t)$ can not be evaluated analytically. Because of the asymptotic behavior of the remainders, they are generally continuous and have, at most, discontinuities of the first type (see §5.5.1). Thus, they may be determined numerically with the precision required by performing the inverse Fourier transform of the remainders.

By substituting Eqs. (7.99) and (7.100) in the convolution relations (7.80) and (7.81) we obtain

$$v_1(t) - R_{c1}i_1(t) - (z_{c1r}^+ * i_1)(t) = (a_1 * w_1)(t), \tag{7.103}$$

$$v_2(t) - R_{c2}i_2(t) - (z_{c2r}^- * i_2)(t) = (a_2 * w_2)(t). \tag{7.104}$$

Remarks

(i) The result obtained allows these nonuniform transmission lines to be described by means of the Thévenin equivalent circuit in Fig. 7.1. Equations (7.82) and (7.83) govern the dynamics of w_1 and w_2. The impulse responses $z_{c1}^+(t)$ and $z_{c2}^-(t)$ that describe the behavior of the line at the ends are different because the backward and the forward waves propagate in different ways. They contain: (a) two Dirac pulses of amplitude R_{c1} and R_{c2}, respectively; and (b) regular terms, which we do not find in uniform lines, describing the wake that the impulse leaves behind it as it propagates because of the dispersion due to nonuniformity.

(ii) The impulse response $p^+(t)$ describes how a Dirac pulse propagates along the line toward the right. It contains: (a) a Dirac pulse delay of the one-way transit time T multiplied by the factor $\sqrt{R_{c2}/R_{c1}}$; (b) a regular term that we do not have in uniform lines describing the wake left by the impulse as it propagates, due to the dispersion introduced by the nonuniformity. The amplitude of the Dirac pulse is amplified if $R_{c2} > R_{c1}$, remains the same if $R_{c2} = R_{c1}$, or otherwise is damped. The impulse response $p^-(t)$, by contrast, describes how a Dirac pulse propagates toward the left. Its principal part is different from that of p^+ only by the factor $\sqrt{R_{c1}/R_{c2}}$.

(iii) As $p^+(t) = p^-(t) = 0$ for $0 \leqslant t \leqslant T$, we have $w_1(t) = w_2(t) = 0$ for $0 \leqslant t \leqslant T$ because we have assumed zero initial conditions. Moreover, for $t > T$, $w_1(t)$ depends solely on the history of w_2 and v_2 in the interval $(0, t-T)$, while $w_2(t)$ depends solely on the history of w_1 and v_1 in the interval $(0, t-T)$. Therefore, as with uniform lines, the terms $(a_1 * w_1)(t)$ and $(a_2 * w_2)(t)$ can be considered as if they were known. The fundamental difference is that in the case of ideal lines, for example, the value of w_1 at the generic instant $t > T$ depends on (and only on) the values of w_2 and v_2 at the instant $t - T$, and the impulse propagates without leaving a wake because there is no dispersion. ◇

In the special case of exponential lines (see §7.3.1), the inverse Laplace transforms of $P_r^{\pm}(s)$, $Z_{c1r}^{+}(s)$, $Z_{c2r}^{-}(s)$, $A_{1r}(s)$, and $A_{2r}(s)$ can be determined analytically by using two well-known pairs of Laplace transforms (see Appendix B). We obtain

$$z_{c1r}^{+}(t) = \frac{R_{c1}}{\tau}[F(t/\tau) + \eta]u(t), \tag{7.105}$$

$$z_{c2r}^{-}(t) = \frac{R_{c2}}{\tau}[F(t/\tau) - \eta]u(t), \tag{7.106}$$

where

$$F(\theta) = \int_0^\theta J_0(\lambda)d\lambda - J_1(\theta), \tag{7.107}$$

$J_0(\lambda)$ and $J_1(\lambda)$ are the first-type Bessel functions. The function $F(\theta)$ tends to unity for $\theta \to \infty$.

With reference to an exponential profile, with $\tau = 2T$, the amplitude spectrum of the normalized remainder Z_{cr1}^{+}/R_{c1} is plotted in

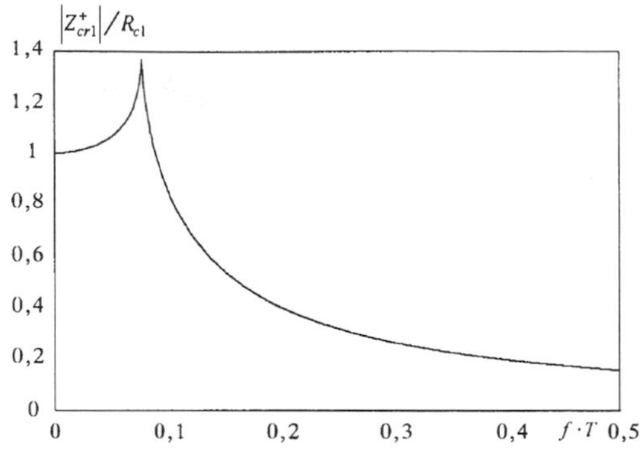

Figure 7.6. Amplitude spectrum of Z^{+}_{cr1}/R_{c1} versus $f \cdot T$ for an exponential line.

Fig. 7.6 versus the normalized frequency for the case $\eta = -1$ (rising exponential profile toward the left). Note that maximum occurs at $f = 1/(2\pi\tau)$ where the function has a branch. This branch gives rise to an oscillating behavior of the corresponding functions in the time domain (see Fig. 7.7). In Fig. 7.7 two plots are reported for the time domain remainder $z^{+}_{cr1}(t)$, referring to the rising exponential profile toward the left $\eta = 1$, and toward the right $\eta = -1$.

From the asymptotic behavior of $Z^{+}_{c1}(s)$ and $Z^{-}_{c2}(s)$ for $s \to 0$ we obtain information regarding the asymptotic behavior of $z^{+}_{c1}(t)$ and

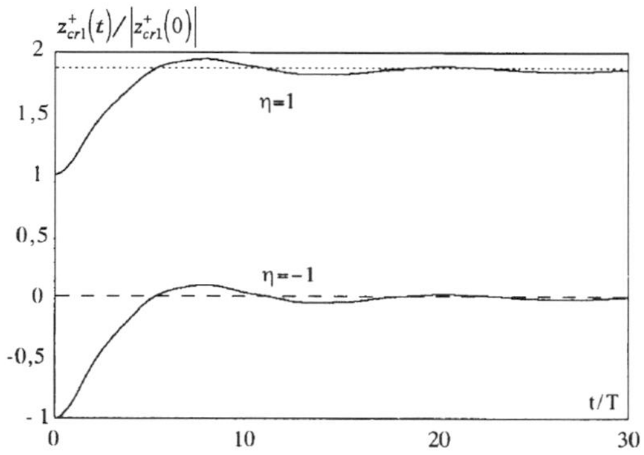

Figure 7.7. Regular part of the impulse responses $z^{+}_{c1}(t)$ normalized to $|z^{+}_{cr1}(0)|$ versus t/T for an exponential line.

$z_{c2}^{-}(t)$ for $t \to \infty$. The impedances $Z_{c1}^{\pm}(s)$ and $Z_{c2}^{\pm}(s)$ behave as

$$Z_{c1}^{\pm}(s) \approx (1 \pm \eta)\frac{R_{c1}}{\tau}\frac{1}{s} + O(s) \quad \text{for } s \to 0, \tag{7.108}$$

$$Z_{c2}^{\pm}(s) \approx (1 \pm \eta)\frac{R_{c2}}{\tau}\frac{1}{s} + O(s) \quad \text{for } s \to 0. \tag{7.109}$$

Therefore, for the impulse responses $z_{c1}^{+}(t)$ and $z_{c2}^{-}(t)$ we obtain

$$\lim_{t \to \infty} z_{c1}^{+}(t) = (1 + \eta)\frac{R_{c1}}{\tau}, \tag{7.110}$$

$$\lim_{t \to \infty} z_{c2}^{-}(t) = (1 - \eta)\frac{R_{c2}}{\tau}. \tag{7.111}$$

Consequently, the exponential line behaves at the end $x = 0$ as an integrator with respect to the forward current wave for $\eta = +1$, that is, when the inductance profile rises in the forward wave propagation direction (see Fig. 7.7). For $\eta = -1$, the line behaves as an integrator at the end $x = d$ with respect to the backward current wave. In this case the inductance profile rises in the backward wave propagation direction.

For the regular parts of the impulse responses $a_1(t)$ and $a_2(t)$ we have

$$a_{1r}(t) = \frac{1}{2\tau}\left[\frac{t}{\tau} - \eta - \eta G\left(\frac{t}{\tau}\right)\right]u(t), \tag{7.112}$$

$$a_{2r}(t) = \frac{1}{2\tau}\left[\frac{t}{\tau} + \eta + \eta G\left(\frac{t}{\tau}\right)\right]u(t), \tag{7.113}$$

where

$$G(\theta) = \int_0^{\theta} F(\lambda)d\lambda. \tag{7.114}$$

The amplitude spectrum of the remainder of the describing function $A_1(s)$ is plotted in Fig. 7.8 for $\eta = -1$. Furthermore, the describing function $A_1(s)$ ($A_2(s)$) tends to zero as s^2 for $s \to 0$ when $\eta = -1$ ($\eta = +1$), whereas it behaves as $2(\tau s)^{-2}$ when $\eta = +1$ ($\eta = -1$). In Fig. 7.9 the behavior of the remainder $a_{1r}(t)$ is plotted when $\eta = -1$. The impulse response $a_1(t)$ for $\eta = +1$ and the impulse response $a_2(t)$ for $\eta = -1$ behave as $2t/\tau^2$ for $t \to +\infty$. For this reason, in Fig. 7.10 we

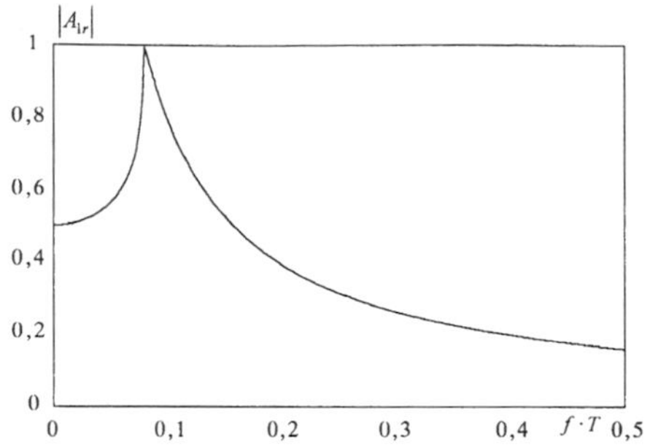

Figure 7.8. Amplitude spectra of A_{1r} versus $f \cdot T$ ($f = \omega/2\pi$) for an exponential line with $\tau = 2T$ and $\eta = -1$ (rising exponential profile toward the left); the maximum is at $f = 1/(2\pi\tau)$.

have plotted only the term

$$a_{1b}(t) = a_{1r}(t) - 2t/\tau^2, \tag{7.115}$$

for the case $\eta = +1$.

Finally, the time domain remainder $p_r^{\pm}$

$$p_r^{\pm}(t) = -\frac{T}{\tau} \frac{J_1\left[\frac{1}{\tau}\sqrt{(t+T)^2 - T^2}\right]}{\sqrt{(t+T)^2 - T^2}} u(t) \tag{7.116}$$

is plotted in Fig. 7.11.

For a linear profile we observe that the asymptotic behavior of

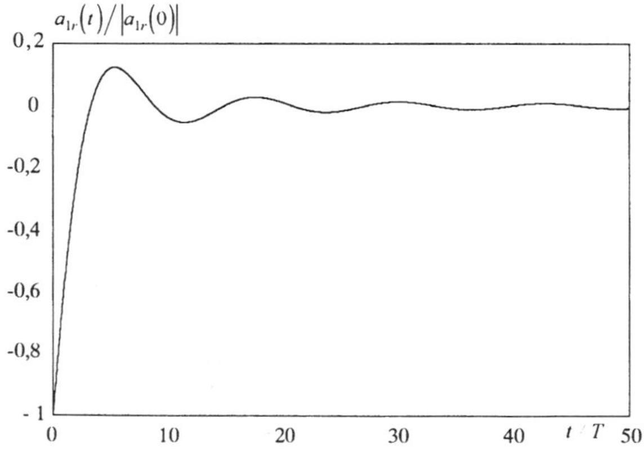

Figure 7.9. Regular part of the impulse response $a_1(t)$ normalized to $|a_{1r}(0)|$ versus t/T for an exponential line for $\eta = -1$.

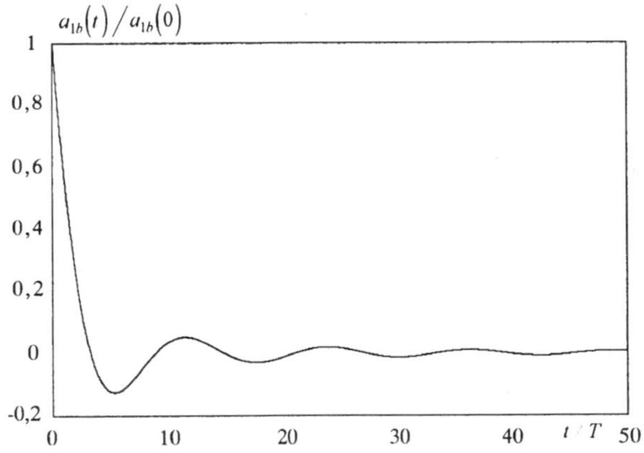

Figure 7.10. $a_{1b}(t)$ normalized to $|a_{1b}(0)|$ versus t/T for an exponential line for $\eta = 1$.

$Z_{c1}^{+}(s)$ and $Z_{c2}^{-}(s)$ for $s \to 0$ is given by

$$Z_{c1}^{\pm}(s) \approx \mp \frac{c_0^2 l}{s\, ln\left(\dfrac{is}{2c_0}\dfrac{L(0)}{l}\right)}, \tag{7.117}$$

$$Z_{c2}^{\pm}(s) \approx \mp \frac{c_0^2 l}{s\, ln\left(\dfrac{is}{2c_0}\dfrac{L(d)}{l}\right)}. \tag{7.118}$$

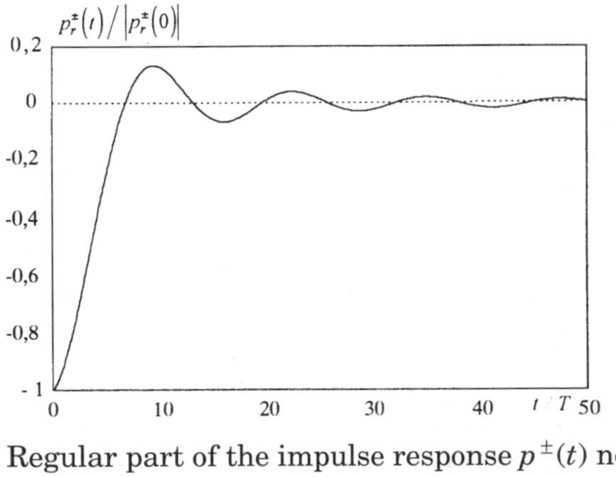

Figure 7.11. Regular part of the impulse response $p^{\pm}(t)$ normalized to $|p_r^{\pm}(0)|$ versus t/T for an exponential line with $\tau = 2T$.

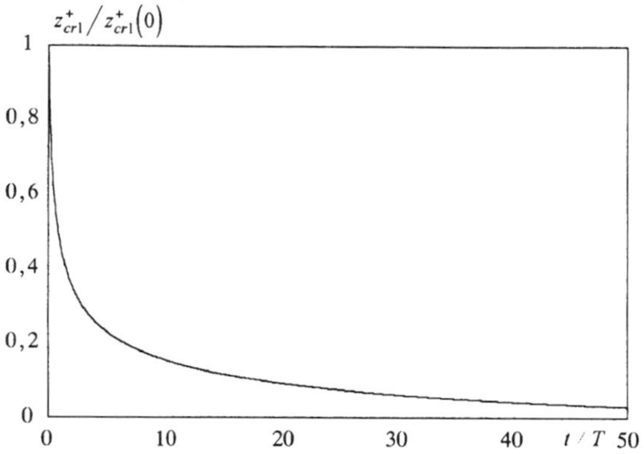

Figure 7.12. Regular part of the impulse responses $z^+_{c1}(t)$ normalized to $|z^+_{cr1}(0)|$ versus t/T for a linear profile with $l = 5L_0$.

Therefore,

$$\lim_{s\to 0}(sZ^{\pm}_{c1}(s)) = \lim_{s\to 0}(sZ^{\pm}_{c2}(s)) = 0, \tag{7.119}$$

thus the impulse responses $z^+_{c1}(t)$ and $z^-_{c2}(t)$ tend to zero for $t \to \infty$ (Fig. 7.12). The same considerations hold for the impulse responses $a_1(t)$ and $a_2(t)$.

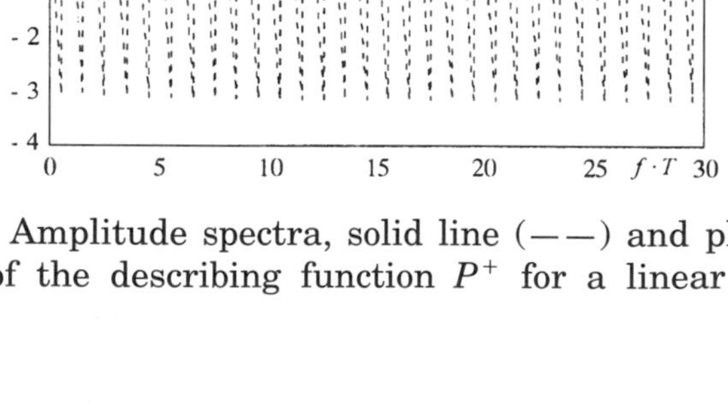

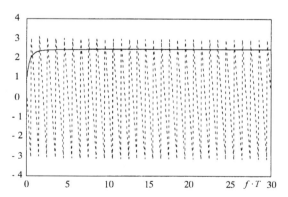

Figure 7.13. Amplitude spectra, solid line (——) and phase, dashed line (-----) of the describing function P^+ for a linear profile with $l = 5L_0$.

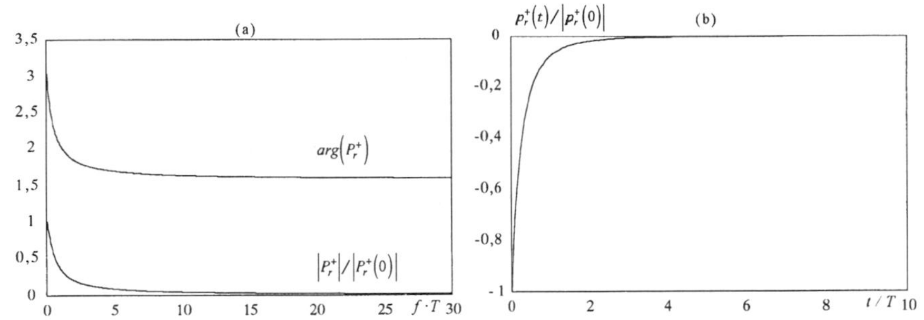

Figure 7.14. Amplitude and phase spectra of P^+ (a), and regular part of the impulse response $p_r^+(t)$ (b), normalized to $|p_r^{\pm}(0)|$ versus t/T for a linear profile with $l = 5L_0$.

Figure 7.13 shows the amplitude and phase spectra of the describing function P^+ for a linear profile. It is clear that a numerical inversion of this function fails. Figure 7.14a shows the amplitude and phase spectra of the remainder P_r^+ defined as in Eq. (7.95). This function is easily numerically inverted and the result is plotted in Fig. 7.14b. The impulse responses $p^+(t)$ and $p^-(t)$ always vanish for $t \to \infty$.

7.6. THE SOLUTION OF THE LINE EQUATIONS FOR A GENERIC PROFILE OF $L(x)$ AND $C(x)$

Although Eq. (7.7) can not be solved in a closed form, except for the particular cases we have considered in the previous section, many studies have been dedicated to it, beginning with those of Liouville and Green in 1837. In particular, it is possible to highlight the asymptotic behavior of $U(x;s)$ for fixed x and $s \to \infty$, along any straight line parallel to the imaginary axis, in the region of convergence, by means of the Liouville-Green transform (see, e.g., Nayfeh, 1973).

7.6.1. Study of the Asymptotic Behavior of the Solution Through the Liouville-Green Transformation

Let us introduce the transform of the independent variable

$$y = \phi(x), \tag{7.120}$$

and the transform of the dependent variable

$$w = \psi(x)U(x). \tag{7.121}$$

The two functions ϕ and ψ will be determined in such a way that the equation for w in the independent variable y immediately shows some interesting asymptotic behavior of the solution for fixed x and $s \to \infty$. By using the Eqs. (7.120) and (7.121), Eq. (7.7) becomes

$$\frac{d^2w}{dy^2} + \frac{1}{\phi'^2}\left(\phi'' - \frac{2\phi'\psi'}{\psi}\right)\frac{dw}{dy} + \frac{1}{\phi'^2}\left[-\frac{s^2}{c^2} + \rho(x) - \psi\left(\frac{\psi'}{\psi^2}\right)'\right]w = 0, \tag{7.122}$$

where the first- and second-order derivatives of the function $\phi(x)$ with respect to x are indicated, respectively, with ϕ' and ϕ''.

If we choose ϕ and ψ so that

$$\phi'(x) = \pm\frac{1}{c(x)}, \tag{7.123}$$

$$\frac{\psi'}{\psi} = \frac{\phi''}{2\phi'}, \tag{7.124}$$

Eq. (7.122) becomes

$$\frac{d^2w}{dy^2} - [s^2 + f(x)]w = 0, \tag{7.125}$$

where

$$f(x) = c^2(x)\left[\psi\left(\frac{\psi'}{\psi^2}\right)' - \rho(x)\right]. \tag{7.126}$$

From Eq. (7.123) we have

$$\phi(x) = \pm\int\frac{dx}{c(x)}, \tag{7.127}$$

while from Eq. (7.124) we have

$$\psi(x) = \frac{1}{\sqrt{c(x)}}. \tag{7.128}$$

Consequently, the expression of function f is

$$f(x) = c^2(x)\left[\frac{1}{4}c^4(x)\frac{d^2}{dx^2}\left(\frac{1}{c^2}\right) - \frac{5}{16}c^6(x)\left(\frac{d}{dx}\frac{1}{c^2}\right)^2 - \rho(x)\right]. \tag{7.129}$$

Because for actual transmission lines $c(x)$ is bounded, greater than zero, and twice differentiable, and $\rho(x)$ is bounded in the interval $(0, d)$, f is negligible in Eq. (7.125) with respect to s^2 for $s \to \infty$. Thus, for bounded x and $s \to \infty$, the asymptotic behavior of the solutions of Eq. (7.125) can be expressed by means of a linear combination of functions of the types $\exp(-sy)$ and $\exp(+sy)$. As a result, the asymptotic behavior of the solutions of Eq. (7.7) for bounded x and $s \to \infty$ can be represented through linear combinations of functions of the type

$$U_\infty^\pm(x;s) = \frac{\sqrt{c(x)}}{\sqrt{c(0)}} \exp\left[\mp s \int_0^x \frac{dy}{c(y)}\right]. \tag{7.130}$$

The functions U_∞^+ and U_∞^- resemble, respectively, the forward and backward waves that are excited along uniform lines and the analytical solutions we have found in the previous section for some particular cases of nonuniform transmission lines.

Remark

When condition (7.20) is not satisfied, the time delay caused by the propagation is no longer directly proportional to the distance by means of the factor $1/c_0$, but is given by $\int_0^x dy/c(y)$. For uniform lines and for nonuniform lines with a homogeneous dielectric, Eq. (7.130) becomes (see Eq. (7.44))

$$U_\infty^\pm(x;s) = \exp\left(\mp \frac{sx}{c_0}\right), \tag{7.131}$$

because in both cases $c(x)$ is constant. ◇

Let $U^+(x;s)$ and $U^-(x;s)$ indicate, respectively, two particular solutions of Eq. (7.7) whose asymptotic behavior for bounded x and $s \to \infty$ coincides, respectively, with the functions U_∞^+ and U_∞^-:

$$U^+(x;s) \approx U_\infty^+(x;s) \quad \text{and} \quad U^-(x;s) \approx U_\infty^-(x;s) \quad \text{for } s \to \infty. \tag{7.132}$$

Let us suppose for the moment that two particular solutions of this type are linearly independent. Thus, any solution of Eq. (7.7) may be expressed through a linear combination of them. Clearly, then, it is possible to express the general solution for the voltage and the current by means of a linear combination of traveling wave solutions, and thus also possible to extend the equivalent circuit models, obtained in the previous sections, to a generic nonuniform line.

7.6.2. Semianalytical Evaluation of Traveling Wave Solutions Based on the WKB Method

We shall now describe a semianalytic procedure for evaluating two particular solutions $U^+(x;s)$ and $U^-(x;s)$ of the type specified by Eq. (7.132). It is based on the WKB method (see Bender and Orszag, 1978). Furthermore, with this procedure it will be possible to obtain further details concerning the asymptotic behavior of U^+ and U^- for bounded x and $s \to \infty$.

Equations (7.130) suggest that we seek $U^+(x;s)$ and $U^-(x;s)$ in the form

$$U(x;s) = \exp\left[s \int B(x;s)dx\right]. \tag{7.133}$$

Substituting this expression in Eq. (7.7), for the new unknown function $B(x;s)$ we get the equation

$$\frac{1}{s}\frac{dB}{dx} + B^2 = \frac{1}{c^2(x)} - \frac{\rho^2(x)}{s^2}. \tag{7.134}$$

This is a Riccati equation with complex coefficients, and its solution can be given by the asymptotic expression

$$B(x;s) = \sum_{i=0}^{n-1} \frac{b_i(x)}{s^i} + B_n(x;s), \tag{7.135}$$

where n is an arbitrary positive integer number. Because $c(x)$ is bounded and greater than zero, we realize that the asymptotic behavior of the residue $B_n(x;s)$, irrespective of n, may be of the type

$$B_n(x;s) \approx O(b_n(x)/s^n) \quad \text{for } s \to \infty. \tag{7.136}$$

Let us consider the asymptotic expression of B obtained by imposing $n = 2$ in Eq. (7.135):

$$B(x;s) = b_0(x) + b_1(x)s^{-1} + B_2(x;s). \tag{7.137}$$

Substituting Eq. (7.137) in Eq. (7.133) we obtain

$$U(x;s) = \exp\left[s \int b_0(x)dx\right] \exp\left[\int b_1(x)dx\right][1 + M(x;s)], \tag{7.138}$$

where the function

$$M(x;s) = \exp\left[s \int B_2(x;s)dx\right] - 1, \tag{7.139}$$

goes to zero, at least as s^{-1} for $s \to \infty$.

The factor $\exp[s\int b_0(x)dx]$ describes the delay introduced by the propagation, the factor $\exp[\int b_1(x)dx]$ the damping or amplification introduced by the nonuniformity and, finally, $M(x;s)$ the dispersion, again due to the nonuniformity. The semianalytic procedure we propose is based essentially on the idea of analytically evaluating $b_0(x)$ and $b_1(x)$ and numerically computing $B_2(x;s)$. In this way we shall determine analytically the asymptotic behavior of the solution up to the first order for $s \to \infty$ correctly, and hence the salient features of the propagation phenomenon in the nonuniform line.

While a rigorous demonstration of the property 7.136 has been reported in Horn (1899) and Erdélyi (1956), later we shall demonstrate it for $n = 2$, in a quite informal but nonetheless rigorous, manner.

Substituting Eq. (7.137) in Eq. (7.134), we obtain the equations

$$b_0^2 = \frac{1}{c^2(x)}, \tag{7.140}$$

$$2b_0 b_1 + \frac{db_0}{dx} = 0, \tag{7.141}$$

and

$$\frac{1}{s}\frac{dB_2}{dx} + 2\left[b_0(x) + \frac{1}{s}b_1(x)\right]B_2 + B_2^2 = -\frac{1}{s^2}\left[\rho(x) + b_1^2(x) + \frac{db_1}{dx}\right]. \tag{7.142}$$

The first two equations are very easily solved and from their solution we get

$$b_0(x) = \pm\frac{1}{c(x)}, \tag{7.143}$$

$$b_1(x) = \frac{d}{dx}\left[ln\sqrt{c(x)}\right]. \tag{7.144}$$

Substituting these expressions in Eq. (7.142), we obtain the equation for the remainder B_2. Although we can not solve this equation analytically, it can be solved numerically quite easily because of the asymptotic behavior of B_2 for $s \to \infty$. In fact, because $c(x)$ is always different from zero and ρ, b_1 and db_1/dx are bounded, Eq. (7.142) has solutions whose asymptotic behavior, for fixed x and $s \to \infty$, is of the type

$$B_2(x;s) \approx b_2(x)s^{-2}, \tag{7.145}$$

where

$$b_2(x) = -\frac{\rho + b_1^2 + \dfrac{db_1}{dx}}{2b_0}. \tag{7.146}$$

It should be noted that $b_2(x)$ is a bounded function.

Let us assume, for the moment, that we are able to catch a solution of Eq. (7.142) whose asymptotic behavior for $s \to \infty$ is of the type (7.145).

As is seen from Eq. (7.140), there are two possible solutions for b_0 — one represents a forward traveling wave (that with the minus sign determination), and the other a backward traveling wave (that with the plus sign determination). Thus we have two functions, B_2, depending on which expression of b_0 we use in Eq. (7.142). Let $B_2^+(x;s)$ indicate the expression of B_2 that we have when considering the minus sign in the expression of b_0, and $B_2^-(x;s)$ the other. The expression of b_1 is independent of the sign of b_0. Therefore, two linearly independent solutions of traveling wave type are

$$U^+(x;s) = \frac{\sqrt{c(x)}}{\sqrt{c(0)}} \exp\left[-s\int_0^x dy/c(y)\right] \exp\left[s\int_0^x B_2^+(y;s)dy\right], \tag{7.147}$$

$$U^-(x;s) = \frac{\sqrt{c(x)}}{\sqrt{c(0)}} \exp\left[s\int_0^x dy/c(y)\right] \exp\left[s\int_0^x B_2^-(y;s)dy\right]. \tag{7.148}$$

Because $B_2^\pm \approx O(s^{-2})$ for $s \to \infty$, the functions U^+ and U^- given by Eqs. (7.147) and (7.148) satisfy the asymptotic condition equation (7.132). The solutions U^+ and U^- are *traveling wave* type solutions: U^+ is a forward wave type solution and U^- is a backward wave type solution. From Eqs. (7.147) and (7.148) we also get

$$\frac{d}{dx} ln[U^\pm(x;s)] = \frac{d}{dx} ln\sqrt{c(x)} \mp \frac{s}{c(x)} + sB_2^\pm(x;s). \tag{7.149}$$

To solve numerically Eq. (7.142) we have to assign the value of B_2^+ and B_2^- at a given point x, for instance, at $x = 0$. To catch the solution that tends to zero as s^{-2} for $s \to \infty$, we must impose boundary conditions $B_2^+(x = 0; s)$ and $B_2^-(x = 0; s)$ that are compatible with this requirement. The way to choose these conditions in the general case is yet an open problem. In addition, these conditions have to be chosen in such a way as to ensure that the Wronskian is different from zero everywhere in the spatial domain where the problem is defined, so that the two particular solutions U^+ and U^- are linearly independent.

The Wronskian corresponding to the functions $U^+/\sqrt{c}$ and $U^-/\sqrt{c}$ is given by

$$W(U^+/\sqrt{c}, U^-/\sqrt{c}) = s\,\frac{U^+(x;s)U^-(x;s)}{c(x)}\left[B_2^+(x;s) - B_2^-(x;s) - \frac{2}{c(x)}\right]. \tag{7.150}$$

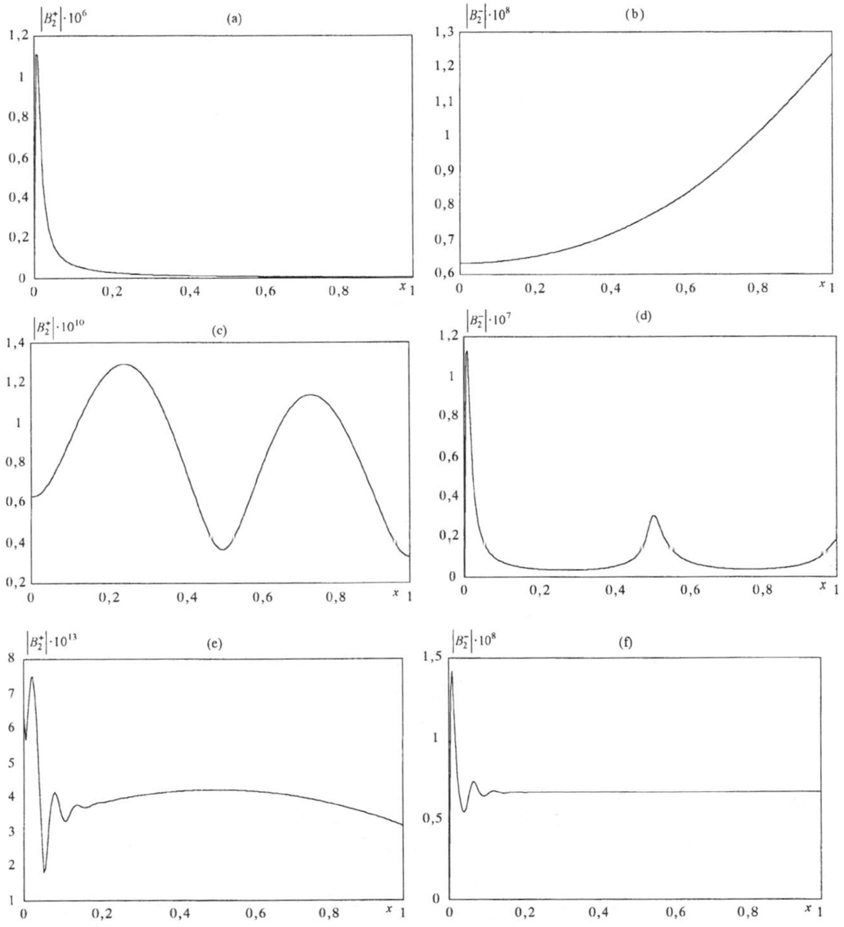

Figure 7.15. Amplitude of the functions B_2^+ and B_2^- along the line (Gaussian profile) for (a) and (b) $f = 1/(10T)$, (c) and (d) $f = 1/T$, and (e) and (f) $f = 10/T$.

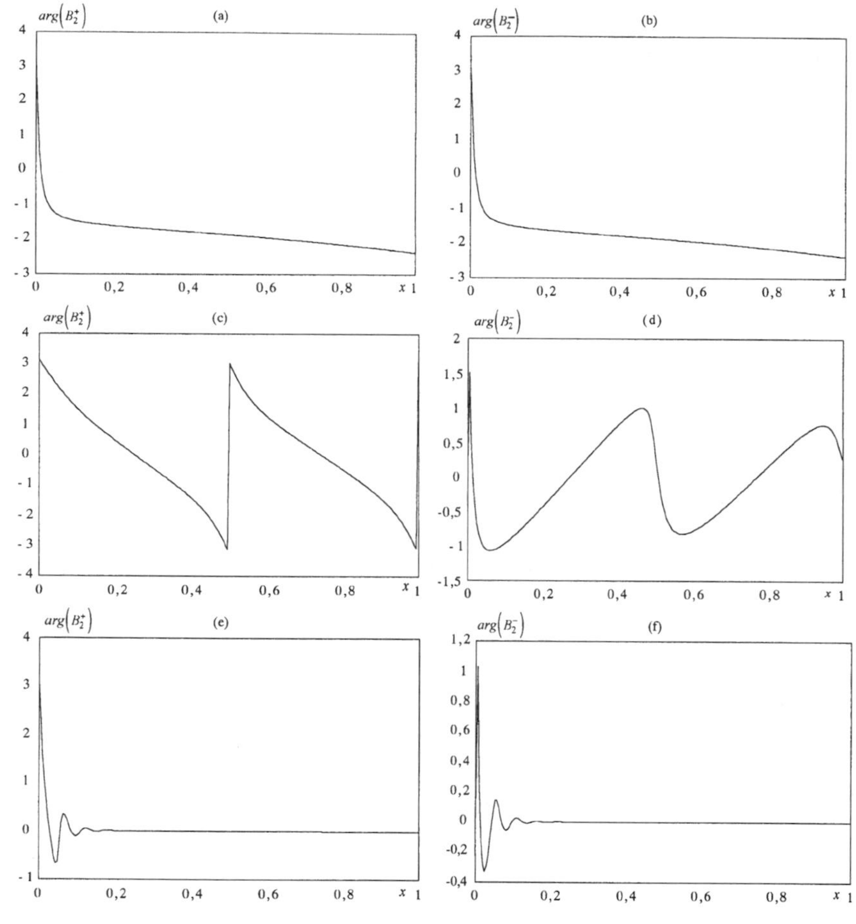

Figure 7.16. Phase of the functions B_2^+ and B_2^- along the line (Gaussian profile) for (a) and (b) $f = 1/(10T)$, (c) and (d) $f = 1/T$, and (e) and (f) $f = 10/T$.

Obviously, we do not have to solve Eq. (7.142) everywhere in the complex plane because we only need to know the solution on the imaginary axis. In Figures 7.15 and 7.16 we show some results relevant to a line transversally homogeneous with a Gaussian profile (see Eq. (7.35)). They have been obtained by solving Eq. (7.142) with the condition at $x = 0$, $B_2^{\pm}(x = 0; s) = \mp c_0 \rho(x = 0)/s^2$, for three different values of the frequency, $f = 1/(10T)$, $f = 1/T$ and $f = 10/T$ ($T = d/c_0$).

Remark

The general solution of line equations (7.3) and (7.4), for a generic profile of the line parameters, may again be expressed through traveling waves, as in Eqs. (7.45) and (7.46). Functions $Q^{\pm}$ are obtained by substituting expressions (7.147) and (7.148) in expressions (7.47) and (7.48). Thus, all the results relevant to the characterization of the terminal behavior of the line, obtained in §7.4 and 7.5, still hold.

By using identity (7.149), from the general expression of the impedances (7.50) we obtain, for $Z_{c1}^{\pm}(s)$ and $Z_{c2}^{\pm}(s)$,

$$Z_{c1}^{\pm}(s) = \mp \frac{sL(0)}{\left.\frac{1}{2}\frac{d\ln R_c(x)}{dx}\right|_{x=0} \mp \frac{s}{c(0)} + sB_2^{\pm}(0;s)}, \tag{7.151}$$

$$Z_{c2}^{\pm}(s) = \mp \frac{sL(d)}{\left.\frac{1}{2}\frac{d\ln R_c(x)}{dx}\right|_{x=d} \mp \frac{s}{c(d)} + sB_2^{\pm}(d;s)}, \tag{7.152}$$

where $R_c(x)$ is defined in (7.84).

Now, by using the above results and expressions (7.62) and (7.63), we obtain the general expressions of functions A_1 and A_2

$$A_1(s) = \frac{1}{2}\left[1 - \frac{\left.\frac{1}{2}\frac{d\ln R_c(x)}{dx}\right|_{x=0} + \frac{s}{c(0)} + sB_2^{-}(0;s)}{\left.\frac{1}{2}\frac{d\ln R_c(x)}{dx}\right|_{x=0} - \frac{s}{c(0)} + sB_2^{+}(0;s)}\right], \tag{7.153}$$

$$A_2(s) = \frac{1}{2}\left[1 - \frac{\left.\frac{1}{2}\frac{d\ln R_c(x)}{dx}\right|_{x=d} - \frac{s}{c(d)} + sB_2^{+}(d;s)}{\left.\frac{1}{2}\frac{d\ln R_c(x)}{dx}\right|_{x=d} + \frac{s}{c(d)} + sB_2^{-}(d;s)}\right]. \tag{7.154}$$

Using (7.147) and (7.148), for the propagation functions P^{+} and P^{-}, given by (7.58), we have

$$P^{+}(s) = \sqrt{\frac{R_{c2}}{R_{c1}}}\exp\left[s\int_0^d B_2^{+}(x;s)dx\right]P(s), \tag{7.155}$$

$$P^{-}(s) = \sqrt{\frac{R_{c1}}{R_{c2}}}\exp\left[s\int_0^d B_2^{-}(x;s)dx\right]P(s), \tag{7.156}$$

where $P(s)$ is defined in (7.92), with the delay time given by

$$T = \int_0^d \frac{dx}{c(x)}. \tag{7.157}$$

If $B_2^{\pm} \approx O(s^{-2})$ for $s \to \infty$, the asymptotic behavior of functions $Z_{c1}^{\pm}$, $Z_{c2}^{\pm}$, A_1, A_2, P^+, and P^- is again given by Eqs. (7.85), (7.86), (7.89), (7.93), and (7.94), respectively. The *remainders* again vanish as s^{-1} for $s \to \infty$. ◇

CHAPTER 8

Transmission Line Equations in Characteristic Form

8.1. INTRODUCTION

Let us consider a lossy two-conductor transmission line with frequency-independent parameters. The per-unit-length characteristic parameters L, C, R, and G may vary along the line. The time domain equations of these lines are (see §1.2):

$$L(x)\frac{\partial i}{\partial t} + \frac{\partial v}{\partial x} = -R(x)i + e_s(x, t), \tag{8.1}$$

$$C(x)\frac{\partial v}{\partial t} + \frac{\partial i}{\partial x} = -G(x)v + j_s(x, t). \tag{8.2}$$

These equations are partial differential equations of the *hyperbolic type*, and so they can be written in a particularly simple and expressive form that is called *characteristic form*, where only *total derivatives* appear (see, e.g., Courant and Hilbert, 1989). This chapter will be dedicated to this question.

For a system of partial differential equations the concept of *characteristic curve* can be introduced in various ways. This concept arises when one seeks a solution to satisfy conditions assigned along given curves in the space-time plane (x, t). There are partial differential equations for which there exist space-time curves along which it is not possible to assign any conditions on the solutions — these are called *characteristic curves* of the partial differential equation. In the classification of partial differential equations, those equations having this peculiarity are called *hyperbolic equations*.

To highlight what is in our view the most interesting aspect, in this chapter we shall introduce the concept of the characteristic curve in a different, more physical and, at the same time, more pragmatic way. The reader is referred to the classical texts on partial differential equations for their classification and for a more thorough and strict definition of the characteristic curves (see, e.g., Courant and Hilbert, 1989). In particular, in this chapter we shall see that there are curves in the plane (x, t), which are the characteristic curves, along which Eqs. (8.1) and (8.2) are transformed into *total differential equations*. The form assumed by the set of equations along the characteristic curves is the so-called *characteristic form*. This is the fundamental property of line equations, and is common to all wave equations.

The transmission line equations in characteristic form are interesting for at least two reasons. First, as is well known in the literature (e.g., Bobbio, Di Bello, and Gatti, 1980; Mao and Kuh, 1997), there is the fact that it is possible to solve lossy, nonuniform and time-varying line equations in characteristic form by using simple methods. Another no less important aspect is that from transmission line equations in characteristic form it is possible to understand how irregularities possibly present in the initial and/or boundary conditions propagate along a generic transmission line.

We shall also be verifying the possibility of extending the characteristic form to transmission lines with frequency-dependent parameters. In the literature, numerical algorithms have been proposed for the analysis of lines with frequency-dependent parameters, based on equations in characteristic form (e.g., Mao and Li, 1992).

8.2. A FIRST-ORDER WAVE EQUATION IN CHARACTERISTIC FORM AND THE CHARACTERISTIC CURVES

For the sake of simplicity let us first consider a single first-order partial equation, and then extend the study to the set of equations (8.1) and (8.2). Although the equation that we are now considering does not describe what happens along a transmisson line, we may use it to show, in the simplest possible way, the fundamental properties of linear hyperbolic equations.

Consider the equation

$$\frac{\partial u}{\partial t} + c(x)\frac{\partial u}{\partial x} = -a(x)u + f(x,t), \tag{8.3}$$

and assume that the solutions satisfy the initial condition

$$u(x, t = 0) = u_0(x) \quad \text{for} \ -\infty < x < \infty; \tag{8.4}$$

$a(x)$, $c(x)$, and $f(x, t)$ are known bounded functions. The parameter c is homogeneous with a velocity, the parameter a is homogeneous with the inverse of a time constant, and f represents a source term. For obvious reasons of a physical nature, $c(x)$ must always have the same sign as x varies; it is assumed that $c(x)$ is always positive.

Consider the curves in the plane (x, t) with slope

$$\frac{dx}{dt} = c(x). \tag{8.5}$$

These are the characteristic curves of the partial differential equation (8.3) (see, e.g., Courant and Hilbert, 1989). Let us indicate them with Γ. Let

$$x(t) = \mathscr{X}(t; x_0) \tag{8.6}$$

be the solution of Eq. (8.5), satisfying the initial condition

$$\mathscr{X}(t = 0; x_0) = x_0, \tag{8.7}$$

where $-\infty < x_0 < +\infty$. It is clear, then, that Eq. (8.6) represents the characteristic curve $\Gamma(P_0)$ passing through the point $P_0 = (x_0, 0)$ in the plane (x, t), see Fig. 8.1.

If $c(x)$ is a regular function, there is one, and only one, solution of Eq. (8.5) that satisfies the initial condition (8.7), and so there is one and only one, characteristic curve $\Gamma(P_0)$ passing through the point P_0. If the parameter c depended also on the unknown function u, then two

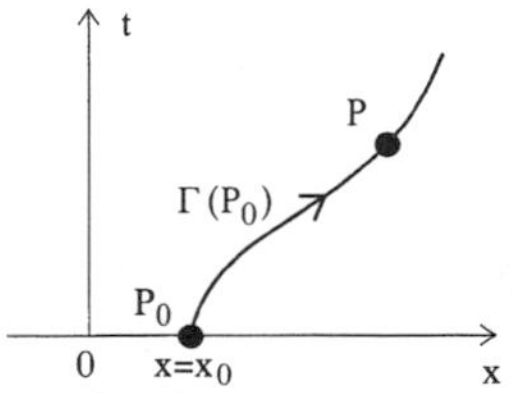

Figure 8.1. A characteristic curve of Eq. (8.3).

distinct characteristic curves could intersect each other (e.g., Whitham, 1974).

When $c(x)$ is constant, $c(x) = c_0$, and the characteristic curves are straight lines. In this case the equation of the characteristic curve passing through P_0 is

$$x = x_0 + c_0 t. \tag{8.8}$$

As x_0 varies, we have a family of parallel straight lines with slope c_0. When c is not uniform in space, we have a family of curves whose shape will obviously depend on the function $c(x)$.

Why are the characteristic curves so important? Along the characteristic curve $\Gamma(P_0)$ we can consider x as a function of the time variable t, and, consequently, $u(x, t)$ as a function of only the time variable t

$$u[\mathscr{X}(t; x_0), t] = \mathscr{V}(t; x_0). \tag{8.9}$$

The function of time $\mathscr{V}(t; x_0)$, so defined, gives the values that the solution of the problem (8.3) assumes along the characteristic curve $\Gamma(P_0)$.

Let us define the total derivative of u along the curve $\Gamma(P_0)$ as

$$\frac{Du}{Dt} = \frac{d\mathscr{V}(t; x_0)}{dt}. \tag{8.10}$$

Thus, we have immediately

$$\frac{Du}{Dt} = \frac{\partial u}{\partial t} + c(x)\frac{\partial u}{\partial x} \quad \text{along the curve } \Gamma(P_0). \tag{8.11}$$

Clearly Eq. (8.3) can be rewritten in the equivalent form:

$$\frac{Du}{Dt} + au + f = 0 \quad \text{along the curve } \Gamma(P_0). \tag{8.12}$$

Equation (8.12) is an *ordinary* differential equation and is called the *characteristic form* of Eq. (8.3). It is also equivalent to Eq. (8.3) and must be solved with the initial condition

$$\mathscr{V}(0; x_0) = u_0(x_0). \tag{8.13}$$

Because Eq. (8.12) is linear, it admits one and only one solution that satisfies the initial condition (8.13). Thus, the values of u along the characteristic curve $\Gamma(P_0)$ are determined univocally by the value of the initial condition at $x = x_0$. In consequence, no condition can be assigned to the solution along the characteristic curves because it is completely determined by the initial conditions. Moreover, having supposed that a and f are bounded, the solution is continuous. Therefore, the solutions of Eq. (8.3) are continuous along the characteristic curves in the plane (x, t).

Equation (8.12) with the initial condition (8.13) can be solved numerically using one of the many standard algorithms for ordinary differential equations. If $a = 0$ and $f = 0$ the function $u(x, t)$ is constant along $\Gamma(P_0)$, and the general solution of Eq. (8.3) is

$$u(x, t) = u_0[x - \mathscr{X}(t; x_0) + x_0], \tag{8.14}$$

where the function $u_0(x)$ describes the initial condition (see Eq. (8.4)). This solution represents a wave propagating with a nonuniform velocity $c(x)$ in the positive x direction, because $c > 0$.

In the case where the propagation velocity c is constant $c = c_0$, Eq. (8.14) reduces to the well-known d'Alembert solution for first-order wave equations

$$u(x, t) = u_0(x - c_0 t). \tag{8.15}$$

Let us turn to the general case $a \neq 0$ and $f \neq 0$. The equation in the characteristic form of Eq. (8.12) immediately shows two important properties of the solution, that will be discussed in the next paragraphs.

8.2.1. The Domain of Dependence of the Solution

The initial condition at the generic abscissa $x = x_1$ of the line affects what happens at generic abscissa $x_2 > x_1$ only beginning from the time instant

$$t^* = \mathscr{X}^{-1}(x_2; x_1), \tag{8.16}$$

where $\mathscr{X}^{-1}(\cdot; x_1)$ is the inverse of function $\mathscr{X}(t; x_1)$. Having assumed that $c > 0$, the inverse exists and is single valued. When the velocity is constant Eq. (8.16) reduces to $t^* = (x_2 - x_1)/c_0$.

8.2.2. The Transport of the Irregularities

An *irregularity* present at a certain $\hat{x}$ in the initial condition is transported in the space-time plane by moving along the characteristic curve passing through the point $(\hat{x}, t = 0)$. By irregularity we mean a discontinuity of the first or second kind, or Dirac pulses. A most interesting consequence of this property is that possible

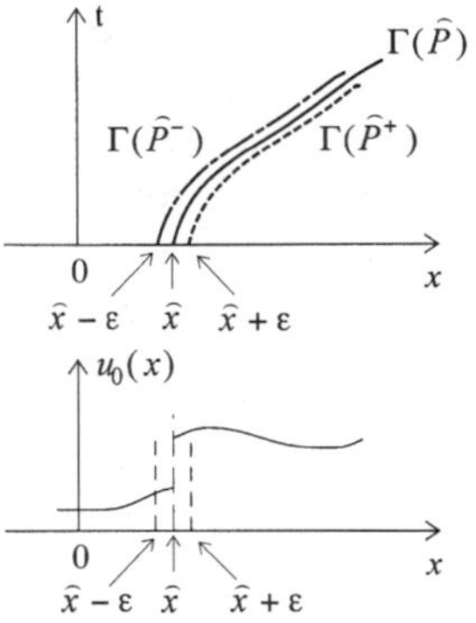

Figure 8.2. Transport of the irregularities along the characteristics.

discontinuities or Dirac pulses propagate in the physical space with finite velocity $c(x)$. Let us see why this is so.

Let us assume that the initial distribution $u_0(x)$ has, for example, a discontinuity of the first kind at the point $x = \hat{x}$ and is continuous elsewhere. Now let us show that this discontinuity moves in the plane (x, t) along the characteristic curve $\Gamma(\hat{P})$ passing through the point $\hat{P} = (\hat{x}, 0)$. From the hypotheses made on the initial distribution we have $u_0(\hat{x} + \varepsilon) \neq u_0(\hat{x} - \varepsilon)$, where ε is an arbitrarily small, positive real number. Let $\Gamma(\hat{P}^+)$ and $\Gamma(\hat{P}^-)$ indicate the curves in the plane (x, t) passing through the points $\hat{P}^+ = (\hat{x} + \varepsilon, 0)$ and $\hat{P}^- = (\hat{x} - \varepsilon, 0)$, respectively, see Fig. 8.2.

Even though the initial condition is discontinuous at $x = \hat{x}$, the generic characteristic curve $\Gamma(P_0)$ relevant to a generic x_0 varies continuously as x_0 varies, because the propagation velocity c does not depend on u. Thus, $\Gamma(\hat{P}^+)$ and $\Gamma(\hat{P}^-)$ coincide with the characteristic curve $\Gamma(\hat{P})$ passing through $\hat{P}$ for $\varepsilon \to 0$. Along the curve $\Gamma(\hat{P})$, both a and f can be expressed as functions of the time variable

$$\mathscr{A}(t; \hat{x}) = a[\mathscr{X}(t; \hat{x})], \quad \mathscr{F}(t; \hat{x}) = f[\mathscr{X}(t; \hat{x}), t]. \tag{8.17}$$

Let $\mathscr{V}^+(t) = \mathscr{V}(t; \hat{x} + \varepsilon)$ and $\mathscr{V}^-(t) = \mathscr{V}(t; \hat{x} - \varepsilon)$ indicate the solution of Eq. (8.3) with the initial condition (8.4), specified along the curves $\Gamma(\hat{P}^+)$ and $\Gamma(\hat{P}^-)$. Clearly, the functions $\mathscr{V}^+$ and $\mathscr{V}^-$ are both solutions of the ordinary differential equation

$$\frac{d\mathscr{V}}{dt} + \mathscr{A}(t; \hat{x})\mathscr{V} + \mathscr{F}(t; \hat{x}) = 0, \tag{8.18}$$

satisfying the initial conditions $\mathscr{V}^+(t=0) = u_0(\hat{x}+\varepsilon)$ and $\mathscr{V}^-(t=0) = u_0(\hat{x}-\varepsilon)$, respectively. The functions $\mathscr{V}^+(t)$ and $\mathscr{V}^-(t)$ are both continuous. Let $\Delta\hat{\mathscr{V}}(t)$ indicate the difference between $\mathscr{V}^+(t)$ and $\mathscr{V}^-(t)$. Then the equation for $\Delta\hat{\mathscr{V}}(t)$ is

$$\frac{d\Delta\hat{\mathscr{V}}}{dt} + \mathscr{A}(t;\hat{x})\Delta\hat{\mathscr{V}} = 0. \tag{8.19}$$

This is the law governing the "transport" of the discontinuity along the characteristic curve $\Gamma(\hat{P})$. It is an ordinary differential equation that must be solved with the initial condition $\Delta\hat{\mathscr{V}}(t=0) = u_0(\hat{x}+\varepsilon) - u_0(\hat{x}-\varepsilon)$. If $\Delta\hat{\mathscr{V}}(t=0) \neq 0$, then we have $\Delta\hat{\mathscr{V}}(t) \neq 0$ for any $t > 0$, so the discontinuity is transported along the characteristic curve $\Gamma(\hat{P})$, eventually vanishing or augmenting according to the sign of the coefficient a. If $a = 0$, we have $\Delta\hat{\mathscr{V}}(t) = \Delta\hat{\mathscr{V}}(t=0)$ and the discontinuity jump remains constant in time, while if $a > 0$ then $\Delta\hat{\mathscr{V}}(t) \to 0$ for $t \to \infty$; if $a < 0$, $\Delta\hat{\mathscr{V}}(t)$ diverges for $t \to \infty$.

The characteristic curve $\Gamma(\hat{P})$ separates the plane (x,t) into two regions, in each of which the solution is regular. The solution is continuous along any characteristic curve, whereas it is discontinuous when crossing the characteristic curve that passes through the point of the straight line $t = 0$ in which the initial condition is discontinuous.

It is evident that a discontinuity transmitted in the space-time plane (x,t) along a characteristic curve implies that a discontinuity propagates in the physical space in the direction of positive x. Figure 8.3 shows the profile of a signal at successive time instants for an

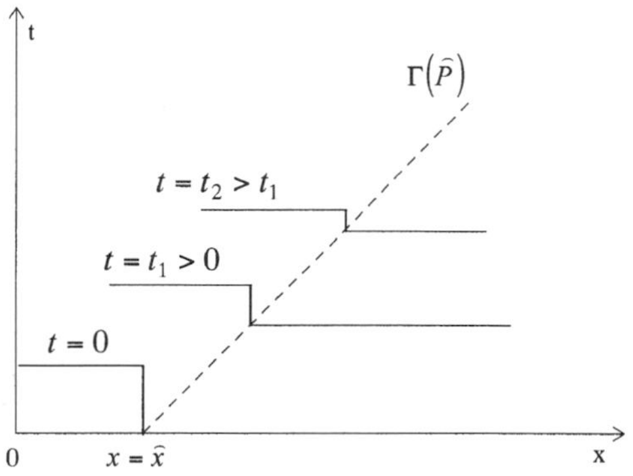

Figure 8.3. Propagation of a "front" along x.

initial step distribution, positive a and $f = 0$. The *front* of the signal, that is, the "surface" beyond which, at a given instant of time, the medium is completely at rest, propagates toward the right with velocity $c(x)$. Analogous results hold also when the initial distribution contains terms that are generalized functions, for example, Dirac pulses according to the theory of distributions (e.g., Courant and Hilbert, 1989).

Remark

We have found two very interesting results. Partial differential equation (8.3) is equivalent to the ordinary differential equation (8.12) along the characteristic curve $\Gamma(P_0)$ defined by Eq. (8.6). If there are discontinuities in the initial condition they are transported in space-time along the characteristic curves and therefore propagate in the space.

Not all partial differential equations behave in this way. Only for hyperbolic type equations do the space-time curves exist along which the equations are equivalent to ordinary differential equations and the discontinuities move. Transmission line equations are hyperbolic equations, so they admit characteristic curves, can be rewritten in characteristic form, and allow the propagation of possible irregularities present in the initial and boundary conditions. Now we shall demonstrate just this. ◇

8.3. THE CHARACTERISTIC FORM EQUATIONS FOR LINES WITH FREQUENCY-INDEPENDENT PARAMETERS

Let us now consider the problem that interests us. We have to reduce the set of equations (8.1) and (8.2) to a set of total differential equations, as was done for Eq. (8.3). In the system of equations (8.1) and (8.2) there is, however, a difficulty, as each equation contains two different unknowns and each of them is differentiated with respect to either time or space. This difficulty can nonetheless be overcome and we shall now see how.

Consider the following linear combination of Eqs. (8.1) and (8.2):

$$l_i(x)\left[L(x)\frac{\partial i}{\partial t} + \frac{\partial v}{\partial x}\right] + l_v(x)\left[C(x)\frac{\partial v}{\partial t} + \frac{\partial i}{\partial x}\right]$$

$$= l_i(x)[-R(x)i + e_s(x,t)] + l_v(x)[-G(x)v + j_s(x,t)], \qquad (8.20)$$

where $l_i(x)$ and $l_v(x)$ are arbitrary functions of x. Equation (8.20) can be rewritten as

$$Ll_i\left(\frac{\partial i}{\partial t} + \frac{l_v}{l_i L}\frac{\partial i}{\partial x}\right) + Cl_v\left(\frac{\partial v}{\partial t} + \frac{l_i}{l_v C}\frac{\partial v}{\partial x}\right) = l_i(-Ri + e_s) + l_v(-Gv + j_s). \tag{8.21}$$

Given the arbitrariness of the functions l_i and l_v, we can choose them to have

$$\frac{l_v(x)}{l_i(x)L(x)} = \frac{l_i(x)}{l_v(x)C(x)} = \chi(x), \tag{8.22}$$

so that Eq. (8.21) assumes the notable form

$$Ll_i\left(\frac{\partial i}{\partial t} + \chi\frac{\partial i}{\partial x}\right) + Cl_v\left(\frac{\partial v}{\partial t} + \chi\frac{\partial v}{\partial x}\right) = l_i(-Ri + e_s) + l_v(-Gv + j_s). \tag{8.23}$$

The condition (8.22) is equivalent to the following set of homogeneous equations in the unknowns l_i and l_v:

$$\chi L l_i - l_v = 0, \tag{8.24}$$

$$l_i - \chi C l_v = 0. \tag{8.25}$$

This set admits nontrivial solutions iff the following condition is satisfied

$$\chi^2 = \frac{1}{LC}. \tag{8.26}$$

Equation (8.26) has two solutions

$$\chi_+ = \frac{1}{\sqrt{LC}}, \tag{8.27}$$

$$\chi_- = -\frac{1}{\sqrt{LC}}. \tag{8.28}$$

Placing

$$R_c(x) = \sqrt{\frac{L(x)}{C(x)}}, \tag{8.29}$$

the solutions of the homogeneous system of equations (8.24) and

(8.25), for $\chi = \chi_+$, are all the functions l_i and l_v that satisfy the condition

$$\frac{l_i(x)}{l_v(x)} = \frac{1}{R_c(x)}, \tag{8.30}$$

and, for $\chi = \chi_-$, are all the functions l_i and l_v that satisfy the condition

$$\frac{l_i(x)}{l_v(x)} = -\frac{1}{R_c(x)}. \tag{8.31}$$

Consequently, two independent equations exist in the notable form of Eq. (8.23). They are

$$\left[\frac{\partial v}{\partial t} + c\frac{\partial v}{\partial x}\right] + R_c\left[\frac{\partial i}{\partial t} + c\frac{\partial i}{\partial x}\right] = c(e_s - Ri) + \frac{1}{C}(j_s - Gv), \tag{8.32}$$

$$\left[\frac{\partial v}{\partial t} - c\frac{\partial v}{\partial x}\right] - R_c\left[\frac{\partial i}{\partial t} - c\frac{\partial i}{\partial x}\right] = c(Ri - e_s) + \frac{1}{C}(j_s - Gv), \tag{8.33}$$

where

$$c(x) = \sqrt{\frac{1}{L(x)C(x)}}. \tag{8.34}$$

For ideal uniform lines, R_c coincides with the characteristic resistance and c the propagation velocity. The two Eqs. (8.32) and (8.33) are linearly independent because of the way in which they have been obtained, and therefore are completely equivalent to the initial set of equations (8.1) and (8.2).

Let Γ^+ and Γ^- indicate the space-time curves that have, respectively, slopes

$$\frac{dx}{dt} = c(x), \tag{8.35}$$

and

$$\frac{dx}{dt} = -c(x). \tag{8.36}$$

These are the *characteristic curves* of the transmission line equations (8.1) and (8.2). In fact, along the curves Γ^+ and Γ^- the partial differential equations (8.32) and (8.33) become ordinary differential equations

$$\frac{Dv}{Dt} + R_c \frac{Di}{Dt} = c(-Ri + e_s) + \frac{1}{C}(-Gv + j_s) \quad \text{along } \Gamma^+, \tag{8.37}$$

$$\frac{Dv}{Dt} - R_c \frac{Di}{Dt} = -c(-Ri + e_s) + \frac{1}{C}(-Gv + j_s) \quad \text{along } \Gamma^-, \tag{8.38}$$

where D/Dt is the total derivative along the characteristic curves, that is,

$$\begin{aligned} Du/Dt &\equiv \partial u/\partial t + c\partial u/\partial x \quad \text{on } \Gamma^+, \\ Du/Dt &\equiv \partial u/\partial t - c\partial u/\partial x \quad \text{on } \Gamma^-. \end{aligned} \tag{8.39}$$

We have here a result analogous to that obtained for the wave equation (8.3). Equations (8.37) and (8.38) are transmission line equations in characteristic form. We should note that the two unknown functions of Eq. (8.37) are different from the two unknown functions of Eq. (8.38) because they are the voltage and the current along two different curves of the plane (x, t). Thus the set of equations (8.37) and (8.38) can not be interpreted as a usual set of equations.

Two characteristic curves pass through each point of the plane (x, t), one with a positive slope and the other with a negative one. Because of this fact, the initial conditions at the generic point $x = x_0$ affect both what happens at the points to the right of $x = x_0$ and what happens in the points to the left (see Fig. 8.4).

In a generic transmission line, therefore, part of the signal propagates, perhaps distorting itself or vanishing, toward the positive x direction, and part propagates toward the negative x direction. Propagation in the negative x direction is described by the characteristic curve Γ^- with slope $-c(x)$, while that in the positive x direction is described by the curve Γ^+ with slope $+c(x)$.

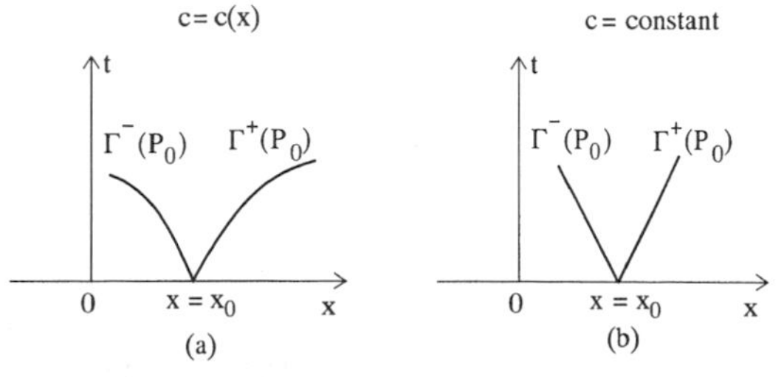

Figure 8.4. Families of characteristic curves in the plane (x, t).

If the line is uniform, c is constant. The propagation velocity c is constant even when the lines are nonuniform but the dielectric is transversally homogeneous, see §1.2 and §7.3. In all of these cases the characteristic curves are straight lines in the plane (x, t) (see Fig. 8.4b). We have two families of straight lines: the family $\Gamma^+(P_0)$ defined by

$$t = \frac{x - x_0}{c}, \tag{8.40}$$

and the family $\Gamma^-(P_0)$ defined by

$$t = -\frac{x - x_0}{c}, \tag{8.41}$$

where x_0 can be any value in the definition domain of the space coordinate.

Equations (8.37) and (8.38) reduce to

$$D(v + R_c i)/Dt = 0 \quad \text{along } \Gamma^+, \tag{8.42}$$

$$D(v - R_c i)/Dt = 0 \quad \text{along } \Gamma^-, \tag{8.43}$$

for ideal uniform transmission lines without sources. Thus, in the absence of losses and sources, we have

$$v(x, t) + R_c i(x, t) = \text{constant} \quad \text{along } \Gamma^+, \tag{8.44}$$

$$v(x, t) - R_c i(x, t) = \text{constant} \quad \text{along } \Gamma^-. \tag{8.45}$$

In this case, the expressions $(v + R_c i)/2$ and $(v - R_c i)/2$ coincide, respectively, with the forward and backward voltage waves, v^+ and v^-, introduced in Chapter 2 when dealing with ideal two-conductor lines. The amplitudes of the forward and the backward waves, therefore, are constant, respectively, along the straight lines with slopes $1/c$ and $-1/c$ in the plane (x, t), in agreement with what we found in Chapter 2.

Remark

Consider a generic point $P = (x, t)$ in the space-time and let Γ^+ and Γ^- indicate the two characteristic curves passing through it (see Fig. 8.5). Let $P^+(x - \Delta x, t - \Delta t)$ and $P^- = (x + \Delta x, t - \Delta t)$ be two points belonging, respectively, to curves Γ^+ and Γ^-, with $\Delta x = c(x)\Delta t$ and $\Delta t > 0$. By construction they are situated below P.

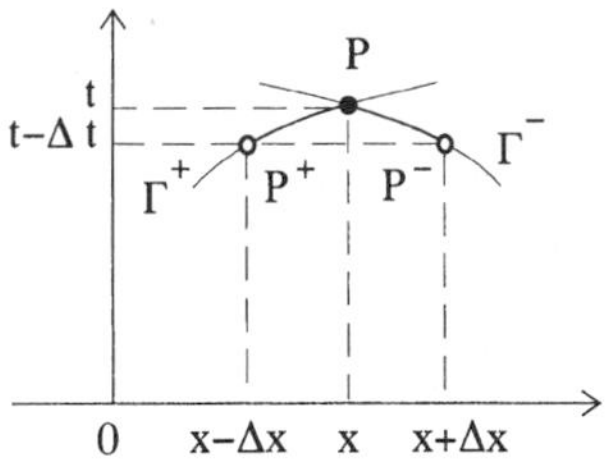

Figure 8.5. Stepwise construction of solutions using characteristics.

An important consequence of the existence of the characteristic Eqs. (8.37) and (8.38) is that the voltage and the current values at the point P depend only on the values that the voltage, the current, and the sources assume along the tracts PP^+ and PP^-. Consequently, for $\Delta t \to 0$ the solution at point P depends, at the first order, only on the values that the solution and the sources assume at points P^+ and P^-. By using this property, it is possible to realize simple iterative algorithms to solve lossy nonuniform transmission lines starting from assigned initial and boundary conditions, as we shall recall in the last section of this chapter. ◇

8.3.1. The Domain of Dependence of the Solution

In our problems it is necessary to solve the line equations (8.1) and (8.2), or their equivalents in the characteristic form (8.37) and (8.38), with assigned initial and boundary conditions.

Now we shall present some general considerations on how the information contained in the initial and boundary conditions is transported by means of the characteristic curves in the space-time plane. However, for the sake of simplicity, we shall refer to a uniform line. Nonetheless, the results that we shall show hold also for nonuniform lines. The only difference is that the propagation velocity is not uniform.

To emphasize the role played by the initial conditions we shall consider first an infinite line. In this case it is necessary to resolve the line equations with only the initial conditions

$$v(x, t = 0) = v_0(x), \quad i(x, t = 0) = i_0(x). \tag{8.46}$$

Consider a generic point $P = (x, t)$ in the space-time. Let Γ^+ and Γ^- be the two characteristic curves that pass through it, and $P_0^+ = (x^+, 0)$ and $P_0^- = (x^-, 0)$ be the points of intersection of Γ^+ and Γ^- with the

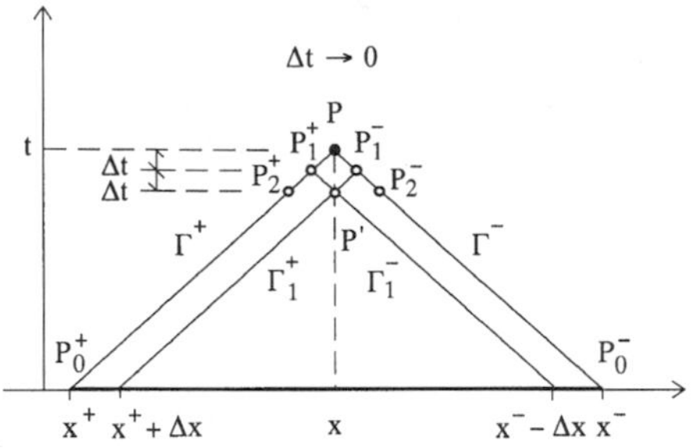

Figure 8.6. The domain of dependence of the solution.

straight line $t = 0$ (see Fig. 8.6). Unlike what happens for Eq. (8.3), the solution of the line equations at point P is affected not only by the value that the voltages and the currents assume at points P_0^+ and P_0^-, that is, by the initial values of v and i at points P_0^+ and P_0^-, but also by the values of i and v along segment $P_0^+P_0^-$ of straight line $t = 0$. This segment is the region of the space-time plane that determines univocally what happens at point P. It is called the *domain of dependence* of P.

This result is simply the consequence of the fact that, through any point on the space-time, two characteristic curves pass. The demonstration is very simple and is as follows. Consider the two points $P_1^+ = (x-\Delta x, t-\Delta t)$ and $P_1^- = (x+\Delta x, t-\Delta t)$ that belong, respectively, to the characteristic curves Γ^+ and Γ^-; let $\Delta x = c\Delta t$, where Δt (with $\Delta t > 0$) is an arbitrarily small time interval. Equations (8.37) and (8.38) simply say that the solution at point P depends, at the first order, uniquely on the values assumed by i and v, and the sources at points P_1^+ and P_1^-.

Through point P_1^+ pass the characteristic curves Γ^+ and Γ_1^-, while through point P_1^- pass the characteristic curves Γ^- and Γ_1^+. The characteristic curve Γ_1^+ intersects the straight line at $t = 0$ to the right of point P_0^+, at the point $(x^+ + \Delta x, 0)$, and the characteristic curve Γ_1^- intersects the straight line $t = 0$ to the left of point P_0^-, at the point $(x^- - \Delta x, 0)$. Consequently, the solution at P_1^+ depends also on the initial condition values at $x = x^- - \Delta x$, and the solution at P_1^- depends also on the initial condition values at point $x = x^+ + \Delta x$. In turn, the solution at P_1^+ depends on the solutions at $x = x^- - \Delta x$ and at P_2^+, and the solution at P_1^- depends on the solution at P_2^- and at $x = x^+ + \Delta x$. The solution at point P_2^- depends on the initial condition values at $x = x^+ + 2\Delta x$, and so on.

In this way, we have shown that what happens at point P is generally influenced only by the initial conditions in the space interval $x^+ \leqslant x \leqslant x^-$. This is one of the most important properties of the solution of hyperbolic equations, something that is not verified for other types of partial differential equations. For example, the solution of a diffusion equation, which is a parabolic type partial differential equation, at the generic point $P = (x, t)$ of the space-time is affected by the entire distribution of the initial values.

We can give a further explanation of this result. What happens at the generic point x_1 of the transmission line, at the generic instant t depends only on those signals that are able to reach this point in the time interval $(0, t)$. Since the signals propagate with a finite velocity c, the solution at the abscissa x_1 and at time t may be affected only by signals originating within the transmission line tract whose ends are $x_1 - ct$ and $x_1 + ct$.

In general, the signals that reach point x_1 at the instants of time antecedent to t leave a wake behind them after their passage, due to losses and nonuniformities, and continue to act at that point even afterwards, as shown in the previous chapters.

Remark

If the line is ideal, that is, uniform and lossless, the quantities $v(x,t) + R_c i(x,t)$ and $v(x,t) - R_c i(x,t)$ are constants along the straight lines Γ^+ and, Γ^-, respectively. Thus, the solution at point x_1 at the instant t depends only on the values that the voltages and the currents assume at the abscissa $x_1 - ct$ and $x_1 + ct$, that is, the domain of dependence reduces to two points. In these cases there is no wake behind the wavefront. ◇

For lines of finite length, the solution of Eqs. (8.1) and (8.2) must satisfy the boundary conditions as well as the initial conditions.

Let us consider a transmission line of finite length d, and assume that the solution must satisfy both the initial conditions (8.46) and the boundary conditions

$$v(x = 0, t) = v_1(t) \quad \text{and} \quad v(x = d, t) = v_2(t) \tag{8.47}$$

for $t \geqslant 0$, where $v_1(t)$ and $v_2(t)$ are given functions.

In this case the domain of definition $\mathscr{D}$ of the solution is the strip of the plane (x, t) bounded below by the segment S_1 whose ends are $(x = 0, t = 0)$ and $(x = d, t = 0)$ and laterally by the semistraight lines

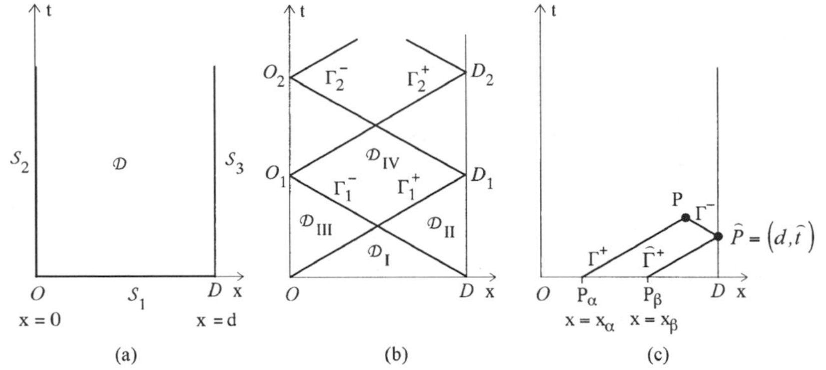

Figure 8.7. Influence of the initial and boundary conditions: the lattice diagram.

S_2 and S_3, defined, respectively, by $(x = 0, t \geqslant 0)$ and $(x = d, t \geqslant 0)$ (see Fig. 8.7a).

Domain $\mathscr{D}$ is open and its boundary $\partial\mathscr{D}$ consists of the joining of S_1, S_2, and S_3. Let us trace the characteristic curves Γ_1^+ and Γ_1^- originating from points O and D and entering region $\mathscr{D}$. They intersect the boundaries S_2 and S_3 at points O_1 and D_1, respectively. Let us now trace the characteristic curves Γ_2^+ and Γ_2^- originating from points O_1 and D_1, and always entering region $\mathscr{D}$. They intersect the boundaries S_2 and S_3 at points O_2 and D_2, respectively, and so on. In this way we shall have divided the strip $\mathscr{D}$ in the closed regions $\mathscr{D}_I$, $\mathscr{D}_{II}$, $\mathscr{D}_{III}$, and $\mathscr{D}_{IV}$ and and so on (see Fig. 8.7b).

Only the characteristic curves that originate from segment S_1 pass through region $\mathscr{D}_I$, so what happens in this region is affected only by the initial conditions and the distributed sources.

The curves passing through the points of region $\mathscr{D}_{II}$ are the characteristic curves Γ^+ originating from the segment S_1, and the characteristic curves Γ^- starting from segment DD_1 of the semistraight line S_3. Take, for example, the point $P = (x, t)$ of $\mathscr{D}_{II}$ (see Fig. 8.7c). The characteristic curve Γ^+ starting from point $P_\alpha = (x_\alpha, 0)$ of the segment S_1, with $x_\alpha = x - ct$, and characteristic curve Γ^- starting from point $\hat{P} = (d, \hat{t})$ of the tract DD_1, with $\hat{t} = t - (d - x)/c$, pass through $P = (x, t)$. Furthermore, the characteristic curve $\hat{\Gamma}^+$ that starts from point $P_\beta = (x_\beta, 0)$ of the segment S_1, with $x_\beta = d - c\hat{t}$, passes through point $\hat{P}$.

Repeating the reasoning developed a little earlier for the infinite line, it can easily be verified that the part of the boundary $\partial\mathscr{D}$ that affects what happens at point $\hat{P}$ is the piecewise straight line constituted by the horizontal tract $P_\beta D$ belonging to S_1, and by the vertical tract $D\hat{P}$ belonging to S_3. In consequence, the voltage and the current at point P depend both on the initial conditions in the interval (x_α, d), and on the values that the voltage assumes at the end $x = d$ for $0 \leqslant t \leqslant \hat{t}$. Similar considerations hold for the points in the other regions of the space-time. In particular, what happens at the points of region $\mathscr{D}_{III}$ depends also on the boundary condition at the end $x = 0$.

8.3.2. The Transportation of the Discontinuities

The characteristic curves carry information, contained in the initial and boundary conditions, from boundary $\partial\mathscr{D}$, where they originate, to the domain $\mathscr{D}$ of the space-time, wherein the problem is defined. Therefore, as in the problem described by Eq. (8.3), any possible irregularity present in the initial conditions or in the boundary conditions will move along the characteristic curves that start from the point of the space-time in which it is present, and will be transported within the domain $\mathscr{D}$.

If the initial voltage distribution along the transmission line has an irregularity at point $x = x'$, for example, a discontinuity of the first kind or a Dirac pulse, the irregularity will move along the two characteristic curves starting from point $P = (x', 0)$ of the space-time. Consequently, two waves will emerge from point $x = x'$, and travel in opposite directions, by means of which the irregularity will propagate along the line, until it reaches the ends, where it will be reflected.

An irregularity can also be present in the boundary conditions. If, for example, voltage $v_1(t)$ contains a discontinuity of the first kind or a Dirac pulse at instant $t = t'$, it will move along the characteristic curve Γ^+ starting from point $P' = (0, t')$ of the space-time. Thus, a wave leaves the end $x = 0$, so that the irregularity will propagate along the whole line toward the other end, where it will be reflected.

Note that, due to this feature, the hyperbolic equations admit generalized solutions. This is the very reason why the impulse responses of the input-state-output and input-output descriptions discussed in the previous chapters contain Dirac pulses. This is true even for transmission lines with parameters depending on the frequency, except for those with skin effect, as we shall show in §8.4.

8.3.3. The Riemann Variables

Both of the characteristic form equations (8.37) and (8.38) introduce a particular linear combination of the total differential operators of the unknown functions. In Eq. (8.37) there is the linear combination $Dv/Dt + R_c Di/Dt$, while in Eq. (8.38) there is the combination $Dv/Dt - R_c Di/Dt$. As we have already pointed out, the set of equations (8.37) and (8.38) cannot be intended as a usual set of equations, because the unknown functions in the two equations refer to two different curves of the space-time plane. Consequently, Eqs. (8.37) and (8.38) cannot be reduced to normal form, substituting, for example, the expression Di/Dt that is obtained from Eq. (8.37) in Eq. (8.38). However, if as new unknowns of the problem we introduce the values $u^+(x,t)$ and $u^-(x,t)$, so defined

$$u^{\pm} = \frac{v \pm R_c i}{2}, \tag{8.48}$$

Eqs. (8.37) and (8.38) are transformed into

$$\frac{Du^+}{Dt} + f^+(x,t;u^+,u^-) = 0 \quad \text{along } \Gamma^+, \tag{8.49}$$

$$\frac{Du^-}{Dt} + f^-(x,t;u^+,u^-) = 0 \quad \text{along } \Gamma^-, \tag{8.50}$$

where

$$f^+ = -c\left[\left(\frac{R_c' - R}{R_c}\right)(u^+ - u^-) + e_s\right] - \frac{1}{C}[-G(u^+ + u^-) + j_s], \tag{8.51}$$

$$f^- = -c\left[\left(\frac{R_c' + R}{R_c}\right)(u^+ - u^-) - e_s\right] - \frac{1}{C}[-G(u^+ + u^-) + j_s], \tag{8.52}$$

and $R_c' = dR_c/dx$.

This type of variable and this reformulation of the characteristic form of the wave equations was first introduced by Riemann in his study on wave propagation in gases (see, e.g., Whitman, 1974). The variables u^+ and u^- are known as *Riemann variables*. For ideal transmission lines without distributed sources we have $f^+ = f^- = 0$. As a consequence, u^+ and u^- are constants along the respective characteristic curves. In this case they are called *Riemann invariants*, and coincide, respectively, with the forward and backward voltage waves introduced in Chapter 2. Instead, in the more general case of nonuniform lines and/or lines with losses, variables u^+ and u^- are not invariant along the characteristic curves and do not coincide with the forward and backward voltage waves introduced in the previous chapters.

Remark

From the definitions of the Riemann invariants (8.48) we immediately deduce that we would have $u^+(x=0,t)=0$ for any t if the line were connected at the left end to a resistor with resistance $R_c(x=0)$. However, this does not imply that $u^+(x,t)=0$ for $x \neq 0$ because, in general, f^+ also depends on $u^-(x,t)$. The same considerations hold for the Riemann variable $u^-(x,t)$ if the line were to be connected at the right end to a resistor with resistance $R_c(x=d)$. We can state that the excitation of the Riemann waves is, in some sense, *distributed* along the line. Instead, for ideal lines the Riemann waves may be excited only at the line ends. When the line is connected at the end $x=0$ $(x=d)$ to a resistor with resistance $R_c(x=0)$ $(R_c(x=d))$ a Dirac pulse coming from the right (left) is completely absorbed. In these cases we can say that the transmission line is "matched" at the end $x=0$ $(x=d)$ with respect to the Dirac pulses (see, also, §4.9) $\diamond$.

8.4. THE CHARACTERISTIC FORM EQUATIONS FOR LINES WITH FREQUENCY-DEPENDENT PARAMETERS

In the literature, numerical algorithms have been proposed for the analysis of lines with frequency-dependent parameters, based on the equations in characteristic form (e.g., Orhanovic, Tripathi, and Wang, 1990; Mao and Li, 1992). It is appropriate, therefore, to present some considerations concerning the equations of these lines.

Does it make sense to speak of characteristic curves and equations in characteristic form when there is dispersion? And if so, what are the properties of the solutions?

This is a most difficult problem because it concerns, as we shall see, the properties of a set of integral-differential equations. Even if it is not possible to tackle this problem with all the rigor required, we shall try to highlight the problems that arise because of the time dispersion in the parameters, and we shall try to give an answer to the preceding questions. The considerations we shall give may provide both a stimulus to whomever wishes to face the matter in question with all the rigor due, and a minimum of theoretical support to whomever is interested only in the applied aspects of equations in characteristic form for lines with frequency-dependent parameters.

The equations for transmission lines with frequency-dependent parameters are, in the Laplace domain,

$$\frac{dV}{dx} = -Z(x;s)I, \tag{8.53}$$

$$\frac{dI}{dx} = -Y(x;s)V. \tag{8.54}$$

For the sake of simplicity, we have not taken into account possible distributed sources and have assumed the line initially at rest. The parameters $Z(x;s)$ and $Y(x;s)$ can always be expressed as (see §1.2)

$$Z(x;s) = [sL_{\infty}(x) + R_{\infty}(x)] + \hat{Z}_r(x;s), \tag{8.55}$$

$$Y(x;s) = [sC_{\infty}(x) + G_{\infty}(x)] + Y_r(x;s), \tag{8.56}$$

where R_{∞}, L_{∞}, G_{∞}, and C_{∞} do not depend on s, while $\hat{Z}_r$ and Y_r are analytic functions in the complex variable s depending on the actual physical nature of the guiding structure modeled by the line. We have stressed in Chapter 5 that, in the most used models, $\hat{Z}_r$ and Y_r have the asymptotic behavior

$$\hat{Z}_r(x;s) = O(s^{\alpha}) \quad \text{with } \alpha < 1 \quad \text{for } s \to \infty, \tag{8.57}$$

$$Y_r(x;s) = O(s^{-1}) \quad \text{for } s \to \infty. \tag{8.58}$$

For transmission lines with parameters independent of the frequency we have $\hat{Z}_r = 0$ and $Y_r = 0$.

In models of lines with skin effect the exponent α is always greater than zero; it is equal to 1/2 for *ordinary skin effect* and 2/3 for *anomalous skin effect*. Instead, in all other models it is always $\alpha \leqslant -1$. As was shown in §5.5, the difference between the models with skin effect and those without has important consequences on the properties of line equations and therefore on their behavior. We shall see now the same results, from a different point of view.

Placing

$$\hat{z}_r(x;t) = L^{-1}\{\hat{Z}_r(x;s)\}, \tag{8.59}$$

$$y_r(x;t) = L^{-1}\{Y_r(x;s)\}, \tag{8.60}$$

and using the convolution theorem, Eqs. (8.53) and (8.54) are transformed in the time domain into equations

$$L_{\infty}\frac{\partial i}{\partial t} + \frac{\partial v}{\partial x} = -R_{\infty}i - (\hat{z}_r * i)(t), \tag{8.61}$$

$$C_\infty \frac{\partial v}{\partial t} + \frac{\partial i}{\partial x} = -G_\infty v - (y_r * v)(t). \tag{8.62}$$

In §5.5 it is shown that the regularity properties of the Laplace inverse transform of an analytic function $H(s)$ depend on its asymptotic behavior for $s \to \infty$. In particular, when $H(s)$ tends to zero at least as $1/s$ for $s \to \infty$, then the inverse transform $h(t)$ is a bounded function that, at most, can have a finite number of discontinuities of the first kind, that is, it is a generally continuous function.

For lines without skin effect both $Y_r(x;s)$ and $\hat{Z}_r(x;s)$ tend to zero at least as $1/s$ for $s \to \infty$, and so $\hat{z}_r(x;t)$ and $y_r(x;s)$ are generally continuous functions (see Chapter 5).

Let us now consider lines with skin effect — in this case the function $\hat{Z}_r(x;s)$ diverges for $s \to \infty$ as s^α, with $0 < \alpha < 1$. In such cases, its inverse transform contains, as well as a generally continuous term, an irregular term that can be expressed only by means of a pseudo-function-distribution (e.g., Doetsch, 1974). Note that, for $\alpha = 1$, we have $L^{-1}\{s\} = \delta^{(1)}(t)$, where $\delta^{(1)}(t)$ is the first-order derivative of the Dirac function $\delta(t)$; for $\alpha = 0$ we have $L^{-1}\{s\} = \delta(t)$. When $0 < \alpha < 1$, the irregular term is a function that behaves as the intermediary between the Dirac pulse and its derivative. Thus, to represent this term we need to turn to the theory of distributions. This difficulty can be overcome in the following manner.

Let us introduce a new function $K(x;s)$ so that, for every s, it is

$$\hat{Z}_r(x;s) = sK(x;s). \tag{8.63}$$

Because of the property (8.57), the function $K(x;s)$ certainly tends to zero for $s \to \infty$, at least as $1/s^{(1-\alpha)}$, so its inverse Laplace transform $k(x;t)$ exists in the classical sense. The function $k(x;t)$ has a singularity at $t = 0$ of the type $1/t^\alpha$, and, for $t > 0$, it is generally continuous (see Appendix B). By using expression (8.63) we have for the convolution integral present in Eq. (8.61)

$$(\hat{z}_r * i)(t) = \left(k * \frac{\partial i}{\partial t}\right)(t). \tag{8.64}$$

As $k(x;t)$ has a singularity of the type $1/t^\alpha$, with $0 < \alpha < 1$, the convolution integral on the right-hand side of Eq. (8.64), even if improper, is convergent in the classical sense if $\partial i/\partial t$ is a generally continuous bounded function. This is a standard way to operate when the evaluation of the impulse response of a linear dynamic system is too difficult; one may evaluate the response to the step function and take the convolution with the derivative of the input function.

Remark

What would happen if exponent α were equal to 1? In this case the function K would tend to a constant for $s \to \infty$, and the function $k(x;t)$, besides a generally continuous term, say $k_r(x;t)$, would also contain a Dirac pulse at $t = 0$. Thus we would have $k(x;t) = A(x)\delta(t) + k_r(x;t)$, and, consequently, $(k * \partial i/\partial t)(t) = A(x)\partial i/\partial t + (k_r * \partial i/\partial t)(t)$. Thus, for $\alpha = 1$, a part of the convolution operator would act as a partial derivative of the first order with respect to the time variable. Therefore, we can say that for $0 < \alpha < 1$, the action of the convolution integral may be in some sense "weaker" than a first-order derivative with respect to the time. ◇

For $0 < \alpha < 1$ it is again appropriate to rewrite Eqs. (8.61) and (8.62) in characteristic form along the characteristic curves $\Gamma^+(P)$ and $\Gamma^-(P)$, defined, respectively, by Eq. (8.35) and Eq. (8.36), with $c = \sqrt{1/(L_\infty C_\infty)}$. It may be shown that the following set of ordinary differential equations is obtained,

$$\frac{Dv}{Dt} + R_c\frac{Di}{Dt} = c[-Ri - (\hat{z}_r * i)(t)] + \frac{1}{C}[-Gv - (y_r * v)(t)] \quad \text{along } \Gamma^+, \tag{8.65}$$

$$\frac{Dv}{Dt} - R_c\frac{Di}{Dt} = -c[-Ri - (\hat{z}_r * i)(t)] + \frac{1}{C}[-Gv - (y_r * v)(t)] \quad \text{along } \Gamma^-, \tag{8.66}$$

where $R_c = \sqrt{L_\infty/C_\infty}$.

These are the equations in *characteristic form* of a transmission line with parameters depending on the frequency.

We have obtained a result that is formally analogous to that obtained for equations of transmission lines with frequency-independent parameters. However, there is an important difference — in each of them there is also a convolution operator. In this case, too, we can introduce the Riemann variables and write the corresponding equations. The presence of the convolution integrals in Eqs. (8.65) and (8.66) has an immediate consequence. Points $P^+ = (x - \Delta x, t - \Delta t)$ and $P^- = (x + \Delta x, t - \Delta t)$ are considered to belong, respectively, to the two characteristic curves Γ^+ and Γ^- passing through point $P = (x, t)$ on

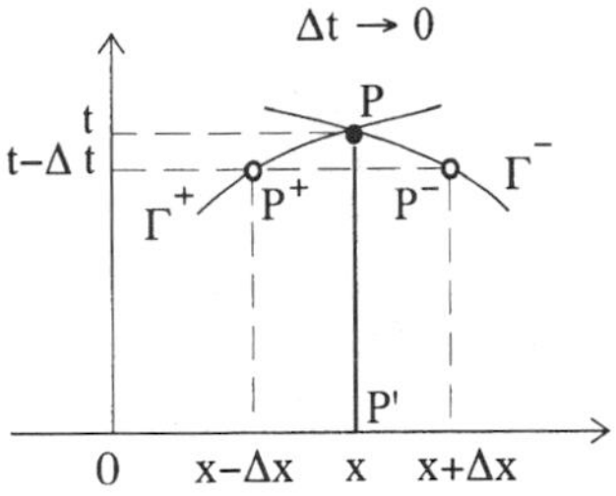

Figure 8.8. Stepwise construction of solutions using characteristics.

the plane (x, t), where $\Delta x = c(x)\Delta t$ and Δt is an arbitrarily small positive value (see Fig. 8.8).

The solution at point P depends not only on the values that i and v assume at points P^+ and P^-, but also on the values that i and v assume in segment PP', that is, the history of the current and the voltage in the time interval $(0, t)$ at the point of the line of abscissa x. This is an effect of the dependence on the frequency of the line parameters. Consequently, the presence of the convolution integrals in Eqs. (8.65) and (8.66) thwarts the possibility of using simple iterative algorithms to solve them.

One senses, however, that the dispersion introduced by the parameters varying in frequency can not make the propagation phenomenon disappear. We shall show this by referring to a uniform semi-infinite line initially at rest and connected at $x = 0$ to an ideal voltage source that imposes a nonzero voltage for $t > 0$. In this case, the space-time domain in which the problem is defined is the first quadrant of the plane (x, t) (see Fig. 8.9).

Let us indicate with Γ_0 the characteristic curve passing through the origin. In the region below this curve, both current $i(x, t)$ and

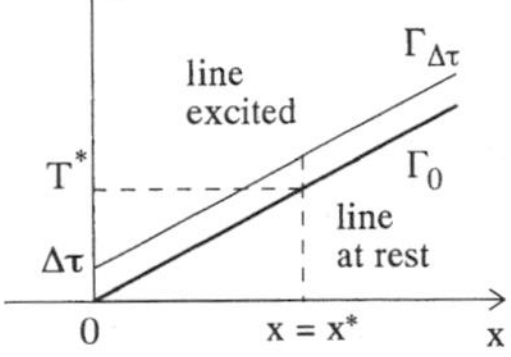

Figure 8.9. The domain of definition of the problem for a semi-infinite line.

voltage $v(x, t)$ are identically zero, because the line is initially at rest. Instead, the solution is different from zero along the straight line $\Gamma_{\Delta\tau}$ passing through point $(0, \Delta\tau)$ where there is a steady state with nonzero current and voltage because of the effect of the voltage source. Consequently, the solution at the generic point $x = x^*$ is equal to zero for $0 \leqslant t < T^*$ and different from zero for $t > T^*$, where $T^* = x^*/c$ and c is the propagation velocity. Thus, the effect of the voltage source connected to end $x = 0$ of the transmission line is manifested at the generic abscissa $x = x^*$ only after time T^*. When the initial value of the source voltage is zero, and therefore is compatible with the initial state of the line, the current and the voltage distributions are also zero along the straight line Γ_0.

What happens to an irregularity that may be present in the boundary condition at $x = 0$?

Assume, for example, that $e(t) = E_0 u(t)$, where $u(t)$ is a unit step function. In this case the voltage at the end $x = 0$ has a discontinuity of the first type at the instant $t = 0$.

In a transmission line without skin effect this discontinuity moves along the characteristic curve Γ_0, possibly attenuating, and thus propagates to the right along the line at velocity c, leaving a wake behind it. This is due to the fact that, in absence of the skin effect, the kernels of both of the convolution integrals present in Eqs. (8.65) and (8.66) are generally continuous functions. In these lines the dispersion action is so "slightly affected" by the presence of possible irregularities as not to alter their structure.

For lines with skin effect, matters are different — the discontinuity in the boundary condition is "immediately suppressed." The reason for this lies in the fact that in both equations in characteristic form there is the term $(\hat{z}_r * i)(t) = (k * \partial i/\partial t)$. The kernel k, although a generally continuous function for $t > 0$, has a $1/t^\alpha$-type singularity at $t = 0$, with $0 < \alpha < 1$. If the solution were discontinuous through the characteristic curve Γ_0, then $\partial i/\partial t$ would contain a term of the type $\delta(t - x/c)$, and hence, the integral $(\hat{z}_r * i)(t) = (k * \partial i/\partial t)(t)$ would contain a term of the type $(t - x/c)^{-\alpha} u(t - x/c)$, which can not be so. Consequently, the solution is immediately forced to be regular. The diffusion of the current in the transverse cross section of the conductors, which is at the very basis of the skin effect, immediately stops any possible irregularities present in the initial or boundary conditions from continuing. These results are in agreement with those we have found in §5.5, where we have dealt with the impulse responses of a two-conductor transmission line with skin effect.

8.5. CHARACTERISTIC EQUATIONS FOR MULTICONDUCTOR LINES

The equations for a multiconductor line (with $n + 1$ conductors) with frequency-independent parameters are (see Chapter 1)

$$\frac{\partial \mathbf{v}}{\partial x} + \mathrm{L}\frac{\partial \mathbf{i}}{\partial t} = -\mathrm{R}\mathbf{i} + \mathbf{e}_s, \tag{8.67}$$

$$\frac{\partial \mathbf{i}}{\partial x} + \mathrm{C}\frac{\partial \mathbf{v}}{\partial t} = -\mathrm{G}\mathbf{v} + \mathbf{j}_s, \tag{8.68}$$

where the characteristic $n \times n$ matrices L, C, R, and G may generally depend on the spatial coordinate x. We shall now demonstrate that these equations also can be transformed into the characteristic form as was done for two-conductor transmission lines. In Chapter 3 we have shown how, with simple linear transformations, it is in principle always possible to reduce a set of equations for an ideal multiconductor line to a set of n mutually uncoupled systems, each of which consists of two first-order coupled equations. Applying a linear transformation of the same type to Eqs. (8.67) and (8.68), we shall show how to extend the characteristic form to multiconductor transmission line equations.

By placing

$$\mathbf{u} = \begin{bmatrix} \mathbf{v} \\ \mathbf{i} \end{bmatrix}, \quad \mathrm{A} = \begin{bmatrix} 0 & \mathrm{L} \\ \mathrm{C} & 0 \end{bmatrix}, \quad \mathrm{B} = \begin{bmatrix} 0 & \mathrm{R} \\ \mathrm{G} & 0 \end{bmatrix}, \quad \mathbf{g}(x,t) = \begin{bmatrix} \mathbf{e}_s(x,t) \\ \mathbf{j}_s(x,t) \end{bmatrix}, \tag{8.69}$$

the set of equations (8.67) and (8.68) can be rewritten as

$$\mathrm{A}\frac{\partial \mathbf{u}}{\partial t} + \frac{\partial \mathbf{u}}{\partial x} = -\mathbf{B}\mathbf{u} + \mathbf{g}. \tag{8.70}$$

Let $\mathbf{U}$ be a left eigenvector of A,

$$\mathbf{U}^{\mathrm{T}}\mathrm{A} = \mathbf{U}^{\mathrm{T}}\lambda, \tag{8.71}$$

where λ is the corresponding eigenvalue. Then left multiplying

$$(\mathrm{A}\partial/\partial t + \partial/\partial x)\mathbf{u}$$

by $\mathbf{U}^{\mathrm{T}}$ yields

$$\mathbf{U}^{\mathrm{T}}\left(\mathrm{A}\frac{\partial}{\partial t} + \frac{\partial}{\partial x}\right)\mathbf{u} = \mathbf{U}^{\mathrm{T}}\left(\lambda\frac{\partial}{\partial t} + \frac{\partial}{\partial x}\right)\mathbf{u}, \tag{8.72}$$

whence Eq. (8.70) becomes

$$\mathbf{U}^{\mathrm{T}}\left(\frac{\partial}{\partial t}+\frac{1}{\lambda}\frac{\partial}{\partial x}\right)\mathbf{u}=\frac{1}{\lambda}\mathbf{U}^{\mathrm{T}}(-\mathrm{B}\mathbf{u}+\mathbf{g}). \tag{8.73}$$

We have obtained a form analogous to Eq. (8.23). Now we shall show that the $2n \times 2n$ matrix A has $2n$ linear independent eigenvectors and $2n$ real eigenvalues.

The eigenvalue problem (8.71) is equivalent to the problem

$$\mathbf{U}_a^{\mathrm{T}}\mathrm{L}=\mathbf{U}_b^{\mathrm{T}}\lambda, \tag{8.74}$$

$$\mathbf{U}_b^{\mathrm{T}}\mathrm{C}=\mathbf{U}_a^{\mathrm{T}}\lambda, \tag{8.75}$$

where we have placed

$$\mathbf{U}^{\mathrm{T}}=[\mathbf{U}_a^{\mathrm{T}} \vdots \mathbf{U}_b^{\mathrm{T}}]. \tag{8.76}$$

Combining Eqs. (8.74) and (8.75), we obtain the two eigenvalue problems

$$\mathbf{U}_a^{\mathrm{T}}\mathrm{LC}=\mathbf{U}_a^{\mathrm{T}}\lambda^2, \tag{8.77}$$

$$\mathbf{U}_b^{\mathrm{T}}\mathrm{CL}=\mathbf{U}_b^{\mathrm{T}}\lambda^2. \tag{8.78}$$

Applying the transposition operation to both sides of Eqs. (8.77) and (8.78), and using the symmetry property of L and C, we obtain the two equations

$$\mathrm{CL}\mathbf{U}_a=\lambda^2\mathbf{U}_a, \tag{8.79}$$

$$\mathrm{LC}\mathbf{U}_b=\lambda^2\mathbf{U}_b. \tag{8.80}$$

In §1 of Chapter 3 we studied the properties of the matrices CL and LC. In particular, we have shown that the eigenvalues of CL are equal to those of LC, and are real and positive. Indicating the eigenvalues of CL with

$$\beta_1 \leqslant \beta_2 \leqslant \cdots \leqslant \beta_n \tag{8.81}$$

we have

$$\begin{gathered}\lambda_1=\sqrt{\beta_1},\quad \lambda_2=\sqrt{\beta_2},\ldots,\lambda_n=\sqrt{\beta_n},\\ \lambda_{n+1}=-\sqrt{\beta_1},\quad \lambda_{n+2}=-\sqrt{\beta_2},\ldots,\lambda_{2n}=-\sqrt{\beta_n}.\end{gathered} \tag{8.82}$$

In Chapter 3 we have also shown that both CL and LC have n linearly independent eigenvectors. Having indicated with $\mathbf{g}_1, \mathbf{g}_2, \ldots, \mathbf{g}_n$ the eigenvectors of CL corresponding to the eigenvalues (8.81), and with $\mathbf{e}_1, \mathbf{e}_2, \ldots, \mathbf{e}_n$ the corresponding eigenvectors of LC, a set of $2n$ linearly independent eigenvectors of A are given by

$$\mathbf{U}_h^{\mathrm{T}} = [\mathbf{g}_h^{\mathrm{T}} \vdots \mathbf{e}_h^{\mathrm{T}}] \equiv (\mathbf{U}_h^+)^{\mathrm{T}} \quad \text{for } 1 \leqslant h \leqslant n, \tag{8.83}$$

$$\mathbf{U}_{n+i}^{\mathrm{T}} = [-\mathbf{g}_i^{\mathrm{T}} \vdots \mathbf{e}_i^{\mathrm{T}}] \equiv (\mathbf{U}_i^-)^{\mathrm{T}} \quad \text{for } 1 \leqslant i \leqslant n. \tag{8.84}$$

Then from Eq. (8.73) we have, for $1 \leqslant k \leqslant n$,

$$(\mathbf{U}_k^+)^{\mathrm{T}}\left(\frac{\partial \mathbf{u}}{\partial t} + c_k \frac{\partial \mathbf{u}}{\partial x}\right) = c_k(\mathbf{U}_k^+)^{\mathrm{T}}(-\mathbf{B}\mathbf{u} + \mathbf{g}), \tag{8.85}$$

$$(\mathbf{U}_k^-)^{\mathrm{T}}\left(\frac{\partial \mathbf{u}}{\partial t} - c_k \frac{\partial \mathbf{u}}{\partial x}\right) = c_k(\mathbf{U}_k^-)^{\mathrm{T}}(\mathbf{B}\mathbf{u} - \mathbf{g}), \tag{8.86}$$

where we have placed

$$c_k = 1/\sqrt{\beta_k} \quad \text{for } k = 1, 2, \ldots, n. \tag{8.87}$$

For nonuniform lines, c_k, $\mathbf{U}_k^+$ and $\mathbf{U}_k^-$ depend on x. Observe that c_k would coincide with the propagation velocity of the kth natural mode of the transmission line if it were uniform and without losses. Let Γ_k^+ and Γ_k^- be the curves defined, respectively, by the directions

$$\frac{dx}{dt} = c_k(x), \tag{8.88}$$

and

$$\frac{dx}{dt} = -c_k(x). \tag{8.89}$$

The curves $\Gamma_1^+, \Gamma_2^+, \ldots, \Gamma_n^+, \Gamma_1^-, \Gamma_2^-, \ldots, \Gamma_n^-$ are the *characteristic curves of the multiconductor line*. Two characteristic curves are associated to each mode, differing only in their slope sign. Thus, Eqs. (8.85) and (8.86) become

$$(\mathbf{U}_k^+)^{\mathrm{T}} \frac{D\mathbf{u}}{Dt} = c_k(\mathbf{U}_k^+)^{\mathrm{T}}(-\mathbf{B}\mathbf{u} + \mathbf{g}) \quad \text{on } \Gamma_k^+ \text{ for } 1 \leqslant k \leqslant n, \tag{8.90}$$

$$(\mathbf{U}_k^-)^{\mathrm{T}} \frac{D\mathbf{u}}{Dt} = c_k(\mathbf{U}_k^-)^{\mathrm{T}}(\mathbf{B}\mathbf{u} - \mathbf{g}) \quad \text{on } \Gamma_k^- \text{ for } 1 \leqslant k \leqslant n, \tag{8.91}$$

where the total derivative is defined as $D\mathbf{u}/Dt \equiv \partial\mathbf{u}/\partial t + c_k\partial\mathbf{u}/\partial x$ along Γ_k^+ and as $D\mathbf{u}/Dt \equiv \partial\mathbf{u}/\partial t - c_k\partial\mathbf{u}\partial x$ along Γ_k^-. We have transformed the set of $2n$ equations (8.67) and (8.68) into a set of $2n$ blocks of two ordinary differential equations of the first order. The set of equations defined by (8.90) and (8.91), for $1 \leqslant k \leqslant n$, are the equations in characteristic form for multiconductor lines.

The $2n$ scalar variables:

$$W_1^+(x,t) = (\mathbf{U}_1^+)^{\mathrm{T}}\mathbf{u}, \quad W_2^+(x,t) = (\mathbf{U}_2^+)^{\mathrm{T}}\mathbf{u}, \ldots, W_n^+(x,t) = (\mathbf{U}_n^+)^{\mathrm{T}}\mathbf{u}, \quad (8.92)$$

$$W_1^-(x,t) = (\mathbf{U}_1^-)^{\mathrm{T}}\mathbf{u}, \quad W_2^-(x,t) = (\mathbf{U}_2^-)^{\mathrm{T}}\mathbf{u}, \ldots, W_n^-(x,t) = (\mathbf{U}_n^-)^{\mathrm{T}}\mathbf{u}, \quad (8.93)$$

are the *Riemann variables* for multiconductor lines. By introducing the matrix T and the vector **W** so defined

$$\begin{aligned} \mathrm{T} &= [\mathbf{U}_1^+ \vdots \mathbf{U}_2^+ \vdots \cdots \vdots \mathbf{U}_n^+ \vdots \mathbf{U}_1^- \vdots \mathbf{U}_2^- \vdots \cdots \vdots \mathbf{U}_n^-]^{\mathrm{T}}, \\ \mathbf{W} &= [W_1^+, W_2^+, \ldots, W_n^+, W_1^-, W_2^-, \ldots, W_n^-]^{\mathrm{T}}, \end{aligned} \quad (8.94)$$

we have, for $1 \leqslant k \leqslant n$,

$$\frac{DW_k^+}{Dt} = c_k\{[(\mathbf{U}_k'^+)^{\mathrm{T}} - (\mathbf{U}_k^+)^{\mathrm{T}}\mathrm{B}]\mathrm{T}^{-1}\mathbf{W} + (\mathbf{U}_k^+)^{\mathrm{T}}\mathbf{g}\} \quad \text{on } \Gamma_k^+, \quad (8.95)$$

$$\frac{DW_k^-}{Dt} = c_k\{[-(\mathbf{U}_k'^-)^{\mathrm{T}} + (\mathbf{U}_k^+)^{\mathrm{T}}\mathrm{B}]\mathrm{T}^{-1}\mathbf{W} - (\mathbf{U}_k^-)^{\mathrm{T}}\mathbf{g}\} \quad \text{on } \Gamma_k^-, \quad (8.96)$$

where $\mathbf{U}_k' = d\mathbf{U}_k'/dx$. Summarizing, we have seen that for every propagation mode of the multiconductor transmission line we have two characteristic curves on the plane (x,t). Along the $2n$ characteristic curves the line equations are transformed into a set of ordinary differential equations. For multiconductor transmission line equations in characteristic form all the considerations made for two-conductor line equations hold.

8.6. STEPWISE INTEGRATION OF THE TRANSMISSION LINE EQUATIONS IN CHARACTERISTIC FORM

As we have already pointed out, the numerical solution of the transmission line equations in characteristic form may be obtained by imaging a construction of it at successive small time increments. Let us now see how. Only for the sake of simplicity do we refer to a uniform two-conductor transmission line.

Let us introduce a space-time discretization

$$x_i = i\Delta x \quad \text{for } i = 0,1,2,\ldots, \qquad \text{and} \qquad t_n = n\Delta t \quad \text{for } n = 0,1,2,\ldots \quad (8.97)$$

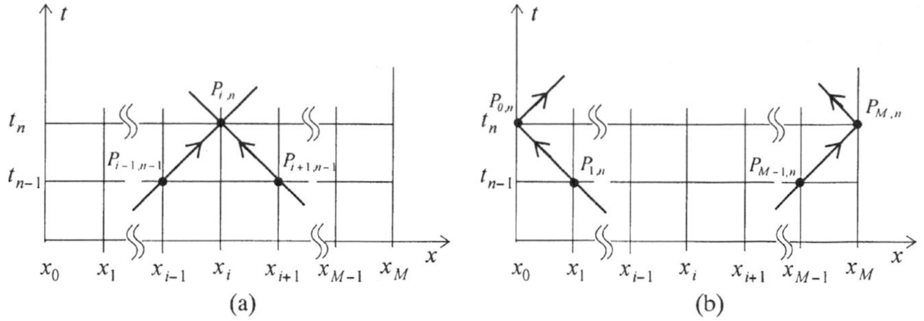

Figure 8.10. Stepwise construction of solutions using characteristics.

with the constraint

$$\Delta x = c\Delta t, \tag{8.98}$$

where c is the propagation velocity.

Let us assume that the solution is known at $t = t_{n-1}$ for every x belonging to the interval $(0, d)$. To evaluate the solution at the point $P_{i,n} = (x_i, t_n)$ of the space-time plane we should distinguish between two cases, depending on whether x_i is the coordinate of an inner point of the line $(0 < x < d)$, or of a boundary point ($x_0 = 0$ or $x_M = d$).

Let us first consider the case in which x_i is the spatial coordinate of an inner point of the line (see Fig. 8.10a). Through the point $P_{i,n}$ pass the two characteristic curves Γ^+ and Γ^- "coming from" the points $P_{i-1,n-1}$ and $P_{i+1,n-1}$, respectively. Thus, according to the results obtained in §8.3.1, the solution at point $P_{i,n}$ depends only on the values of the Riemann variables at the points $P_{i-1,n-1}$ and $P_{i+1,n-1}$. This can be exploited to build a numerical scheme by discretizing the two Eqs. (8.49) and (8.50). By using, for instance, the Euler explicit scheme, we obtain

$$u^+_{i,n} = u^+_{i-1,n-1} - \Delta t f^+(x_{i-1}, t_{n-1}; u^+_{i-1,n-1}, u^-_{i-1,n-1}), \tag{8.99}$$

$$u^-_{i,n} = u^-_{i+1,n-1} - \Delta t f^-(x_{i+1}, t_{n-1}; u^+_{i+1,n-1}, u^-_{i+1,n-1}). \tag{8.100}$$

This approximation is numerically stable (e.g., Ames, 1977).

Let us now consider the case in which x_i is the coordinate of a boundary point. We have to remark that in this case there is only one characteristic curve "incoming from the past" and passing through point $P_{i,n}$. Hence, we may use only one of the two Eqs. (8.49) and

(8.50). With reference to the left end, for instance, we have (see Fig. 8.10b)

$$u_{0,n}^{-} = u_{1,n-1}^{-} - \Delta t f^{-}(x_1, t_{n-1}; u_{1,n-1}^{+}, u_{1,n-1}^{-}). \tag{8.101}$$

The other equation is given by the boundary condition at the end $x = 0$. If the voltage $v_1(t)$ at the line end $x = 0$ were known, the equation would be

$$u_{0,n}^{+} + u_{0,n}^{-} = v_1(t_n), \tag{8.102}$$

while if the current $i_1(t)$ were known, the equation would be

$$u_{0,n}^{+} - u_{0,n}^{-} = R_c i_1(t_n). \tag{8.103}$$

In general, when the line is connected to a generic one-port, neither voltage $v_1(t)$ nor current $i_1(t)$ is known, but they are themselves unknowns of the problem. There are two relations between voltage $v_1(t)$ and current $i_1(t)$ involving the Riemann variables. These are:

$$v_1(t_n) - R_c i_1(t_n) = 2u_{0,n}^{-}, \tag{8.104}$$

$$v_1(t_n) + R_c i_1(t_n) = 2u_{0,n}^{+}. \tag{8.105}$$

From Eq. (8.101) we have that $u_{0,n}^{-}$ is independent of voltage $v_1(t_n)$ and current $i_1(t_n)$. Thus, Eq. (8.104) suggests that the behavior of the line at the left end may be described by an equivalent one-port of the type shown in Fig. 8.11. Once $v_1(t_n)$ and $i_1(t_n)$ are known, it is possible to upgrade the Riemann variable $u_{0,n}^{+}$ by using Eq. (8.105).

If we compare this result with that obtained in the previous chapters by using an input-state-output description, we may observe that the basic difference is that here there is no convolution in time but a "distributed state" appears. The driving impedance is much simpler (it reduces to a mere resistor), but the control law of the controlled voltage source $2u_{0,n}^{-}$ is much more complicated. In fact, in

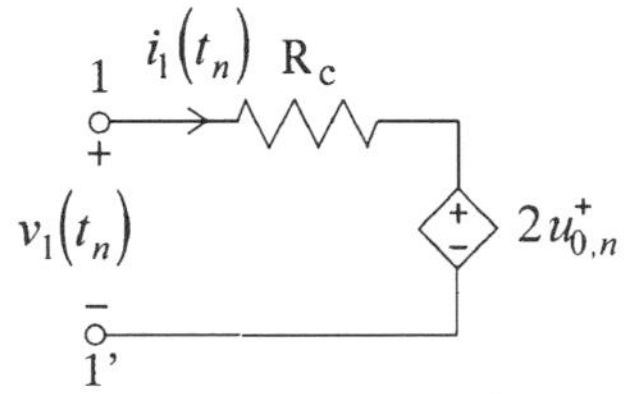

Figure 8.11. Equivalent circuit representing the line behavior at the end $x = 0$ obtained by using the characteristic method.

order to update this source, the evaluation of the dynamics and the storage of the Riemann variables at every abscissa x_j are required.

It has recently been pointed out that algorithms based on the characteristic method can be as fast as the recursive convolution method discussed in Chapter 4 (Mao and Kuh, 1997). These algorithms may be effectively applied to the most general case of multiconductor lines with parameters depending on the frequency and nonuniformity in space (Mao and Li, 1992; Orhanovic, Tripathi, and Wang, 1990).

Remark

From Eq. (8.105) we immediately deduce that $u_{0,n}^{+}$ would be equal to zero for any n if the line were connected at the left end to a resistor with resistance R_c. However, this does not imply that $u_{i,n}^{+} = 0$ for $i > 0$, because, in general, f^{+} also depends on $u_{i,n}^{-}$. ◇

The results obtained hold also for nonuniform lines, provided that a homogeneous dielectric is considered, because in this case the velocity $c(x) = c_0$ does not depend on x. Let us consider, for instance, a lossless exponential line, initially at rest, described by the per-unit-length inductance given in Eq. (7.22). A stepwise solution of the line equations may be again obtained through the numerical scheme given by Eqs. (8.99) and (8.100), relevant to the space-time discretization defined in Fig. 8.10, where $c = c_0$. In particular we have

$$f^{+} = f^{-} = -c_0 \frac{d\, ln[R_c(x)]}{dx}(u^{+} - u^{-}) - \frac{\eta c_0}{\Delta}(u^{+} - u^{-}). \quad (8.106)$$

When the dielectric is not homogeneous the velocity is not constant along the line, hence the characteristics are no longer straight lines in the (x, t) plane, but curves depending on the profile $c(x)$. Let us consider, for instance, a lossless nonuniform line, initially at rest, described by the parameters

$$L(x) = L_0(1 + \Delta Lx) \quad C(x) = C_0(1 - \Delta Cx). \quad (8.107)$$

The characteristic curves, reported in Fig. 8.12, are described by

$$\frac{dx}{dt} = \pm \frac{c_0}{\sqrt{(1 + \Delta Lx)(1 - \Delta Cx)}}, \quad (8.108)$$

where $c_0 = 1/\sqrt{L_0 C_0}$.

It is evident that a nonuniform space-time discretization is now needed, in order to follow the profile of the characteristic curves. Let

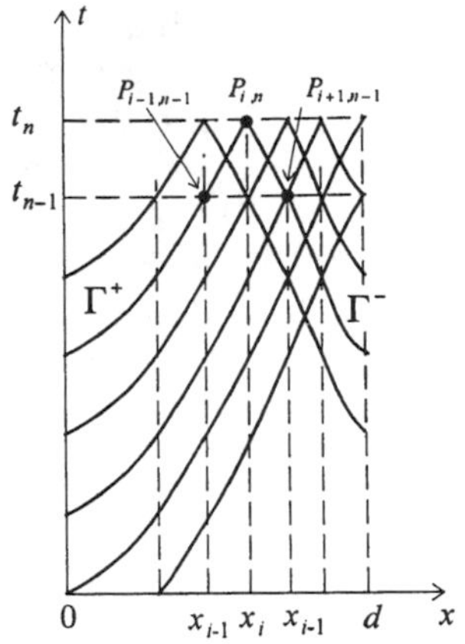

Figure 8.12. Characteristic curves for a line with nonuniform c.

us consider a uniform time discretization, with step Δt. In consequence, the space discretization has to be nonuniform. One possible choice is to use the points x_i given by the intersections in the (x, t) plane between the straight line $t = t_n$ and the characteristics (see Fig. 8.12). In this way, the points $P_{i,n}$ and $P_{i-1,n-1}$ belong to the same characteristic Γ^+, while the points $P_{i,n}$ and $P_{i+1,n-1}$ belong to Γ^-. This means that the line equations may be again solved through the numerical scheme given by Eqs. (8.99) and (8.100), where now

$$f^+ = f^- = -c_0 \frac{\Delta L + \Delta C}{2\sqrt{(1 + \Delta L x)(1 - \Delta C x)}}. \tag{8.109}$$

This particular space-discretization fits the problem well, but the drawback consists in the necessity to evaluate the positions of the segmenting points (e.g., Mao and Li, 1992). This can be avoided if the following variable transformation is introduced (El-Zein, Haque, and Chowdury, 1994)

$$\varphi(x) = \int_0^x \sqrt{L(y)C(y)}\, dy. \tag{8.110}$$

In the plane (φ, t) the characteristics are now straight lines with a slope of $\pm 45°$, and thus a uniform space-time discretization $\Delta\varphi$, Δt may be again introduced. The corresponding points on the real axis, x_i, are defined by

$$\varphi(x_i) = (i - 1)\Delta\varphi. \tag{8.111}$$

CHAPTER 9

Lumped Nonlinear Networks Interconnected by Transmission Lines

9.1. INTRODUCTION

This chapter deals with the circuit equations relevant to lumped nonlinear networks interconnected by transmission lines. The transmission lines (two or multiconductors, lossless or lossy, with parameters depending or not on the frequency, uniform or nonuniform in space) are modeled as multiports as shown in the previous chapters. The characteristic relations of these multiports are quite different in kind: for *ideal lines* they are linear algebraic difference equations; whereas for *nonideal lines* they are combinations of linear algebraic difference equations and linear convolution equations of the second kind. The characteristic relations describing lumped circuit elements are of the algebraic-differential type with ordinary derivatives, which, in general, may be nonlinear and time varying.

The literature presents many numerical methods to solve such a problem, but usually more prominence is given to the computational features rather than to the analysis of the qualitative behavior of the solutions, that is, the existence, uniqueness, asymptotic behavior, and stability. However, Lacking any prior general information about these properties, simulations *are nothing but blind groping*.

An evolution problem is said to be *well posed* if it has exactly *one* solution that satisfies the prescribed initial conditions, and the solution regularly depends on the data of the problem; otherwise it is said to be *ill posed*. Evolution problems with many solutions display an inadequacy in the model used to describe the physical system. This fact tells us that some effects, deemed insignificant during modeling, are actually crucial to the behavior of the physical system.

When a stable and consistent numerical scheme is adopted, the convergence of the approximated solution to the actual one is assured if the actual model and the numerical one are both well posed. Therefore, a preliminary study of the existence and uniqueness of the solution should be made before solving the problem numerically. These basic requirements are frequently taken for granted. The fact is, though, that they are by no means universal properties of such networks, as has been shown by Miano (1997).

This chapter first deals with the conditions that ensure the existence and uniqueness of the solutions of the equations of such networks. Once the existence and uniqueness have been established, the problem of the numerical evaluation of the solution is tackled. The next chapter will deal with the asymptotic behavior and stability of the solution.

We shall deal with these problems gradually. First, we shall consider two lumped circuits connected by a two-conductor line, and then we shall extend the analysis to the case in which the line is multiconductor. Finally, the extension to a generic network composed of lumped circuits and transmission lines will be discussed.

9.2. TIME DOMAIN FORMULATION OF THE NETWORK EQUATIONS

Let us consider a network consisting of a two-conductor line connecting two lumped circuits, as sketched in Fig. 2.1; the line may be nonuniform. The network equations consist of the characteristic equations of the single elements, lumped or distributed, and of the Kirchhoff equations describing the interaction between them. The terminal behavior of the two-conductor transmission line can be represented through either an input-state-output description, see §4.6 and §7.4.2, or an input-output description, see §4.7 and §7.4.3. We have seen that the input-state-output description is more suitable for time domain transient analysis.

As we have already pointed out in Chapters 2, 4, 5, and 7, the two-port describing the terminal behavior of the line in the time domain may be represented through an equivalent two-port of Thévenin (or Norton) type as shown in Fig. 9.1: Each port is composed

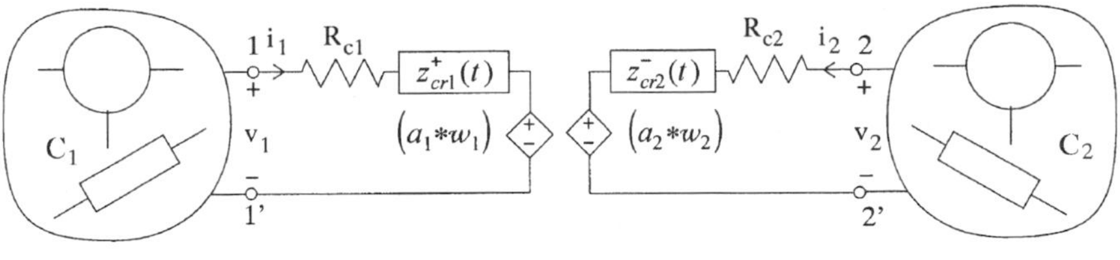

Figure 9.1. Equivalent circuit of the network shown in Fig. 2.1.

of a series connection of a linear resistor, a linear dynamic one-port, and a voltage source controlled by the terminal variables of the other port.

The equations governing the dynamics of the controlled voltage sources w_1 and w_2 are linear convolution equations with one delay (see §4.6.2 and §7.5)

$$\begin{aligned} w_1(t) &= \sqrt{R_{c1}/R_{c2}}\,\{[2v_2(t-T) - w_2(t-T)] \\ &\quad + \int_0^{t-T} p_r^-(t-T-\tau)[2v_2(\tau) - w_2(\tau)]d\tau\}u(t-T), \\ w_2(t) &= \sqrt{R_{c2}/R_{c1}}\,\{[2v_1(t-T) - w_1(t-T)] \\ &\quad + \int_0^{t-T} p_r^+(t-T-\tau)[2v_1(\tau) - w_1(\tau)]d\tau\}u(t-T), \end{aligned} \tag{9.1}$$

We recall that for uniform transmission lines it is $z^+_{cr1}(t) = z^-_{cr2}(t) = z_{cr}(t)$, $R_{c1} = R_{c2} = R_c$, $a_1(t) = a_2(t) = \delta(t)$ and $p_r^+(t) = p_r^-(t) = p_r(t)$; for ideal lines it is $z_{cr}(t) = 0$ and $p_r(t) = 0$.

From the equations governing the dynamics of the voltages w_1 and w_2 it follows that $w_1(t)$ and $w_2(t)$ depend, respectively, only on the time history of w_1 and w_2 themselves, and of v_1 and v_2, in the time interval $(0, t-T)$ where T is the one-way transit time. Therefore, both $w_1(t)$ and $w_2(t)$ are known for $iT \leqslant t \leqslant (i+1)T$ if the solution is known for $t \leqslant iT$; for $0 \leqslant t \leqslant T$ the voltages $w_1(t)$ and $w_2(t)$ depend solely on the initial conditions. Consequently, the controlled sources w_1 and w_2 may be treated as if they were "independent" sources, if the problem is resolved iteratively. Once the network of Fig. 9.1 has been solved for $0 \leqslant t \leqslant T$, $w_1(t)$ and $w_2(t)$ are "updated" through the control laws governing the dynamics of the voltage sources. Once $w_1(t)$ and $w_2(t)$

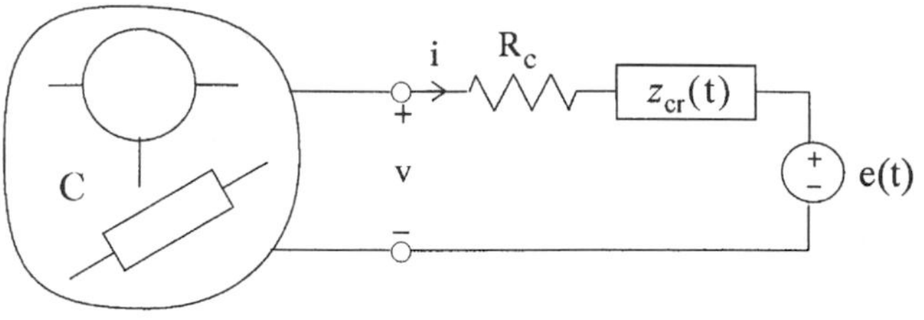

Figure 9.2. Left-end (right-end) equivalent circuit of a two-conductor line connecting a generic lumped one-port at $x = 0$ ($x = d$).

have been evaluated for $T \leqslant t \leqslant 2T$, the network is solved again, and so forth.

It is evident that at each step of the iteration the behavior of the right-hand part of the circuit shown in Fig. 9.1 does not depend on the behavior of the other part and vice versa, so they could be considered as if they were uncoupled. Consequently, we shall lose nothing in generality if we study the well posedness and the numerical solution of the equations relevant to the network shown in Fig. 9.2, where the voltage $e(t)$ is assumed to be known.

Similar considerations hold for the more general case of multiconductor transmission lines. The well posedness and the numerical solution of the circuit model may be studied by referring to the circuit shown in Fig. 9.3. The right-hand part of the circuit represents the behavior of an end of a generic multiconductor transmission line (see Chapters 3 and 6). It is composed of a linear dynamic multiport with current-based impulse response matrix $z_c(t) = R_c\delta(t) + z_{cr}(t)$, and voltage sources $e_1(t), e_2(t), \ldots, e_n(t)$. The voltage $\mathbf{e}(t) = |e_1(t), e_2(t), \ldots, e_n(t)|^T$ depends only on the time history of the electrical variables at the other line end on the interval $(0, t - T_1)$ where T_1 is the one-way transit time of the fastest line mode. For ideal multiconductor lines it is $z_{cr}(t) = 0$.

Note that the circuits of Figs. 9.2 and 9.3 may also represent semi-infinite transmission lines connected to a generic lumped circuit.

Remark

For sinusoidal and periodic steady-state analysis in the frequency domain, an input-output description is suitable to represent the terminal behavior of the line. To evaluate the sinusoidal and periodic steady-state solutions common numerical methods may be used (Chua and Lin, 1975). ◇

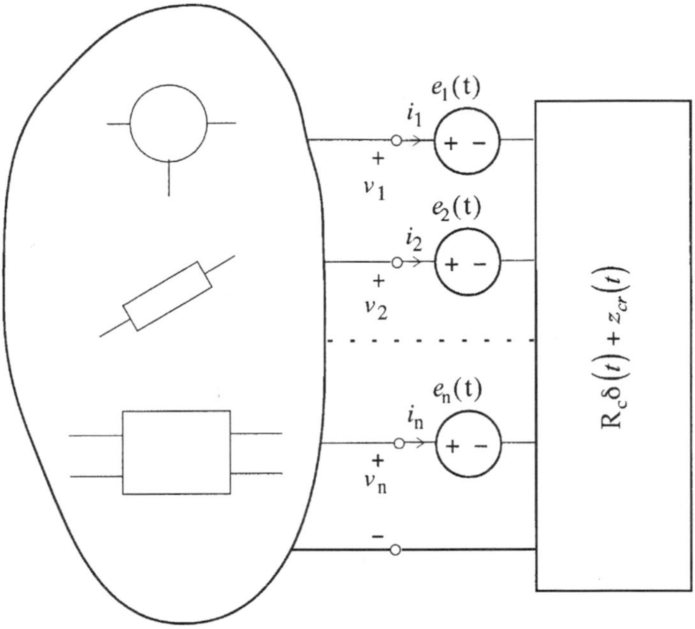

Figure 9.3. Left-end (right-end) equivalent circuit of a multiconductor line connecting a generic lumped multiport at $x = 0$ ($x = d$).

9.3. A GLIMPSE AT THE UNIQUENESS PROBLEM FOR IDEAL TWO-CONDUCTOR TRANSMISSION LINES

We shall gradually study the uniqueness problem, starting from the simple case of an ideal line connected to a nonlinear resistor and then extending the results to the more general case of imperfect lines and/or nonlinear dynamic terminations. It is worth stressing that the analysis of the simple ideal line case will lead to a uniqueness criterion that will be extended to the most general case.

Let us consider the case in which the one-port C in the circuit of Fig. 9.2 is a simple nonlinear resistor (see Fig. 9.4), which we assume, for example, to be voltage controlled, $i = -g(v)$; $g(v)$ represents the characteristic of the resistor defined according to the normal convention. Furthermore, let us assume that the function g is continuous and the resistor is weakly active.

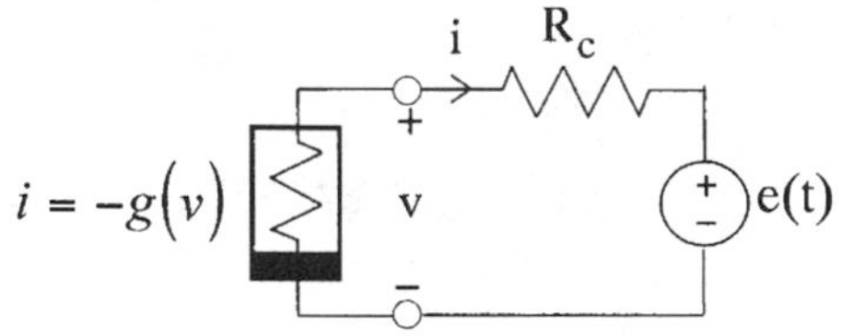

Figure 9.4. Left-end (right-end) equivalent circuit of an ideal two-conductor line connected to nonlinear resistors.

Remark: Weakly Active Resistors

A voltage-controlled resistor is *weakly active* if there exists a positive constant I such that ($i' = -i$)

$$i' = g(v) \geqslant -I \quad \text{for } v \geqslant 0 \qquad \text{and} \qquad i' = g(v) \leqslant I \quad \text{for } v \leqslant 0. \tag{9.2}$$

A weakly active voltage-controlled resistor can produce at most the electrical power $I|v|$, that is, for any voltage and current it must be $i'v \geqslant -I|v|$.

A current-controlled resistor $v = r(i')$, is weakly active if there exists a positive constant V such that

$$v = r(i') \geqslant -V \quad \text{for } i' \geqslant 0 \qquad \text{and} \qquad v = r(i') \leqslant V \quad \text{for } i' \leqslant 0. \tag{9.3}$$

It can produce at most the electrical power $V|i'|$, that is, for any voltage and current it must be $i'v \geqslant -V|i'|$. These definitions bear a resemblance to that given by Hasler and Neirynck (1986). ◇

The equations of the circuit of Fig. 9.4 are the system of algebraic equations

$$\begin{cases} i + g(v) = 0, \\ v - R_c i = e. \end{cases} \tag{9.4}$$

This system must necessarily be solved numerically, as it is nonlinear. However, before building a numerical algorithm, we must investigate both the existence and the uniqueness of the solution for given $e(t)$. Even if the circuit shown in Fig. 9.4 is a particular case, its study is of fundamental importance in the theory that we shall develop in this chapter. In fact, as we shall see, in the study of circuits with imperfect lines, one often resorts to auxiliary circuits related to the ideal case.

By eliminating i in the system (9.4) we get

$$f(v; e) \equiv v + R_c g(v) - e = 0. \tag{9.5}$$

Thus the existence and uniqueness of the solution of the circuit represented in Fig. 9.4 depend on the existence and uniqueness of the solution of Eq. (9.5) in the unknown v.

Let us assume that $f(v; e)$ satisfies the following conditions:

(i) $f(v; e)$ is continuous;
(ii) $f(v; e) \to \pm\infty$ as $v \to \pm\infty$ for any value of e; and
(iii) $f(v; e)$ is strictly increasing with respect to v for any value of e.

Then, the nonlinear equation (9.5) has one, and only one, solution. The existence follows from the *intermediate value theorem* of Bolzano (e.g., Bronshtein and Semendyayev, 1985), whereas the uniqueness statement is obvious.

Condition (i) holds because $g(v)$ is continuous, condition (ii) holds because the nonlinear resistor is weakly active, whereas condition (iii) holds if $g(v)$ satisfies the following inequality

$$\frac{dg}{dv} > -\frac{1}{R_c}. \tag{9.6}$$

Therefore, under condition (9.6), Eq. (9.5), and hence, the system (9.4) has one, and only one, solution for any assigned $e(t)$. Equation (9.5) defines implicitly the function $v = F(e)$, expressing the voltage v as a function of the supplied voltage e. Under the assumption (9.6) function F is continuous and single valued, according to the implicit function theorem (e.g., Korn and Korn, 1961). In particular, $F(e)$ is bounded for any bounded value of e. This is a direct consequence of the weakly activeness property.

Remark

In the same manner we can show that, when the nonlinear resistor is current controlled, $v = r(i')$, the uniqueness of the solution is assured by the condition

$$\frac{dr}{di'} > -R_c. \quad \diamond \tag{9.7}$$

Equation (9.6) (or Eq. (9.7)) is always satisfied if the characteristic curve of the nonlinear resistor is not decreasing. In particular, when the terminal resistor is both voltage and current controlled, and hence *locally passive* (e.g., Hasler and Neirynck, 1986), the solution is always unique. Instead, when the characteristic curve is decreasing somewhere, condition (9.6) (or condition (9.7)) may be unsatisfied in some intervals of the values of v (or i).

Remark

The inequalities (9.6) and (9.7) bear a resemblance to the conditions that we have to impose to put lumped circuit state equations in the *normal form*, that is, $\dot{\mathbf{x}} = \mathbf{f}(\mathbf{x}, t)$ where f is a single valued function. (e.g., Braiton and Moser, 1964; Chua and Rohrer, 1965). For instance, the circuit equations of a linear inductor (capacitor) in parallel (in series) with a voltage-(current) controlled resistor and a linear resistor R_c, may be put in normal form only if Eq. (9.6) (Eq. (9.7)) holds. ◇

9.3.1. An Ill-Posed Problem

The solution of Eq. (9.5), in general, is not unique when the condition (9.6) is not satisfied. Let us consider a "simple" case in which the resistor has a characteristic curve like that of a tunnel diode (see Fig. 9.5a). In the plane (i, v) the solutions of the circuit correspond, at each instant t, to the intersection points of the characteristic curve of the diode with the straight line defined by the second equation of the system (9.4). If the slope of the diode characteristic obeys Eq. (9.6), then there is one, and only one, intersection, and the function $v = F(e)$ is single valued.

If condition (9.6) is not satisfied, there may be three intersection points P_1, P_2, and P_3, and the nonlinear function $v = F(e)$ is multivalued, as shown in Fig. 9.5b. The function F bifurcates into three branches when $E^{\downarrow} \leqslant e(t) \leqslant E^{\uparrow}$, where $E^{\uparrow}$ and $E^{\downarrow}$ are the bifurcation points, that is, the values of e for which the straight line is tangent to the diode characteristic curve.

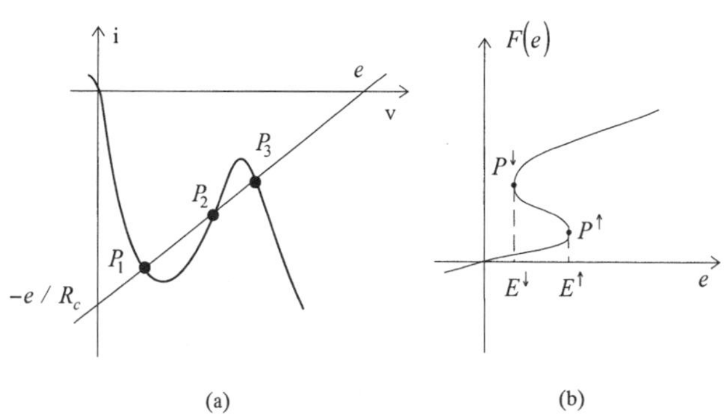

Figure 9.5. Graphic solutions of the system of equations (9.4) for $R_c dg/dv < -1$.

What happens when there is more than one solution? Voltage $v(t)$ may "jump" uninterrupted between P_1, P_2, and P_3 until $E^{\downarrow} \leqslant e(t) \leqslant E^{\uparrow}$. Remember that the transmission line equations are hyperbolic, hence they admit generalized solutions (e.g., Courant and Hilbert, 1989).

9.3.2. A Circuit with an Additional Parasitic Reactance

Since the model described by the system of equations (9.4) leads to indeterminacy when the condition (9.6) is not satisfied, from a physical point of view the considered circuit is an inadequate model of the corresponding physical system. Therefore, we have to add some effects, neglected during modeling, that are actually critical to the system's behavior. For example, when modeling the voltage-controlled nonlinear resistor, we have not taken into consideration the capacitive parasitic effects. We may model these effects by connecting a linear capacitor in parallel to the nonlinear resistor (see Fig. 9.6).

These parasitic reactances, which are often freely omitted from the circuit model, bear a resemblance to those that: (a) we have to include in the model of a lumped electrical circuit to put the network equations in normal form (Chua and Alexander, 1971); and (b) we need to include in order to correctly assess the stability of operating points of any given physical lumped circuit (Green and Wilson, 1992).

The equation describing the dynamics of the voltage v of the equivalent circuit of Fig. 9.6 is

$$R_c C_p \frac{dv}{dt} = -f(v; e), \tag{9.8}$$

where $f(v; e)$ is defined in Eq. (9.5). This equation admits one, and only one, solution compatible with the initial value of v, for any value of $C_p \neq 0$, whatever it may be (Miano, 1997), because $f(v; e)$ is *locally Lipschitz* with respect to v and the solution is bounded for any t.

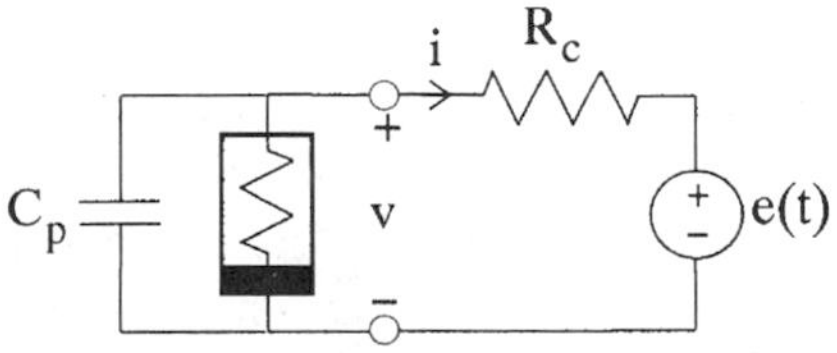

Figure 9.6. The circuit of Fig. 9.4, with a parasitic capacitor C_p.

DEFINITIONS (e.g., Korn and Korn, 1961; Lefschetz, 1977).

(*a*) *The scalar function* $h(y;t)$ *is Lipschitz with respect to* y *on an interval* $I_{a,b} = (a, b)$, *if there is a constant* k, *such that* $|h(y_1;t) - h(y_2;t)| \leqslant k|y_1 - y_2|$ *for any pair of* y_1 *and* y_2 *belonging to* $I_{a,b}$ *and for any* t.

The following properties are not difficult to prove. A function, which is Lipschitz on an interval $I_{a,b}$, is there everywhere continuous; a function, that has a bounded derivative in $I_{a,b}$, is Lipschitz there.

(*b*) *The scalar function* $h(y;t)$ *is locally Lipschitz with respect to* y *at a generic point* y^* *if there is a neighborhood of the point* y^* *in which it is Lipschitz.*

A function that has a continuous first-order derivative at a given point is locally Lipschitz at that point. In general the property providing for a function to be locally Lipschitz is more restrictive than the continuity and less restrictive than the differentiability.

(*c*) *The scalar function* $h(y;t)$ *is globally Lipschitz with respect to* y *if it is Lipschitz with respect to* y *in the whole interval* $(-\infty, +\infty)$.

A piece wise linear function is Lipschitz on the whole interval $(-\infty, +\infty)$. Note that, *in general*, a function, although being locally Lipschitz for any bounded y, may be non-Lipschitz in the real axis as a whole. An example is given by the function $h = ay^3$, or by the exponential function $h = ae^y$. In these two cases, although h is Lipschitz on each bounded interval, it is not Lipschitz in the interval $(-\infty, +\infty)$ as a whole $\diamond$.

The local Lipschitz property guarantees only the existence and the uniqueness of the solution of the ordinary differential equation $dy/dt = h(y;t)$ within a time interval of nonzero length (e.g., Korn and Korn, 1961; Zeidler, 1986). An example is given by the simple differential equation $dy/dt = ay^3 + e$ with $a > 0$. In this case the solution diverges after a finite time, and hence, cannot be indefinitely extended into the future whatever the value of the initial condition (e.g., Zeidler, 1986; Hasler and Neirynck, 1986). The reason is that h is non-Lipschitz in the whole interval $(-\infty, +\infty)$. Instead, if $h(y;t)$ were globally Lipschitz, the equation $dy/dt = h(y;e)$ would admit, for any t, one, and only one, solution satisfying the imposed initial conditions (e.g., Korn and Korn, 1961; Zeidler, 1986). If the characteristic of the nonlinear resistor is piecewise linear, then the function $f(v;e)$ is globally *Lipschitz* for any bounded $e(t)$. However, if the characteristic is a polynomial or exponential, which is the case with many ordinary models, the function $f(v;e)$ is not globally *Lipschitz*.

Actually, the solution of Eq. (9.8) exists on the whole interval $0 \leqslant t < +\infty$ and is unique even if $f(v;e)$ is not globally *Lipschitz*, but is only locally *Lipschitz*, because the resistor is supposed to be weakly active. In fact, any solution of Eq. (9.8) must remain bounded for any t, that is, it can not go to infinity after a finite time (see Appendix C). Therefore, the solution v always stays within a region where the function $f(v; e)$ is *Lipschitz*.

In conclusion, the local Lipschitz property guarantees only the existence and the uniqueness within a time interval of nonzero length, whereas the boundedness of the solution assures its existence and uniqueness on the whole interval $(0, +\infty)$ (e.g., Zeidler, 1986; Hasler and Neirynck, 1986).

To understand the effects of the parasitic capacitor, we now briefly investigate the behavior of the solution of Eq. (9.8) in the limit $C_p \downarrow 0$ when the condition (9.6) is not satisfied.

Let us consider the relation $v = F(e)$ that we would obtain if the capacitor were absent (see Fig. 9.5a). The stability of a point on the curve $v = F(e)$ is determined by the sign of dg/dv at the corresponding point on the characteristic of the diode. The branches of $F(e)$ with positive slope correspond to points of the diode characteristic that have positive slope, while the branch with negative slope corresponds to operating points on the negative slope of the tunnel diode characteristic. Thus the branches of $F(e)$ with positive slopes are stable, whereas the branch with negative slope is unstable. Therefore, if the circuit is operating initially on a stable branch, it will continue to operate on that branch because of the capacitor "inertia" (*inertia postulate*), until the solution arrives at a branch point $P^{\uparrow}$ or $P^{\downarrow}$, at which time instant it takes a vertical jump to the other stable branch (*jump postulate*) (Chua and Alexander, 1971), (see Fig. 9.7).

A repetitive application of the inertia and jump postulates will determine unambiguously the solution. Thus, in the limit case $C_p \downarrow 0$ the dynamics of the voltage at the end of the line could be directly described through a *relay hysteresis operator* (Fig. 9.7). In Fig. 9.8a the qualitative behavior of the solution of the circuit of Fig. 9.6 is shown for the voltage waveform $e(t)$ shown in Fig. 9.8b; we have assumed a homogeneous initial condition for the capacitor.

In conclusion, when the condition (9.6) is not satisfied, the solution experiences a jump only at the two bifurcation values $E^{\uparrow}$ and $E^{\downarrow}$ from a stable branch to the other stable branch in the limit $C_p \downarrow 0$ (the jump time goes to zero for $C_p \downarrow 0$). Moreover, the bifurcation value at which the jump takes place is determined by the direction in which the supplied voltage $e(t)$ varies. Note that, when the condition (9.6) is satisfied, the effects of the parasitic capacitor may be disregarded, in the limit $C_p \downarrow 0$.

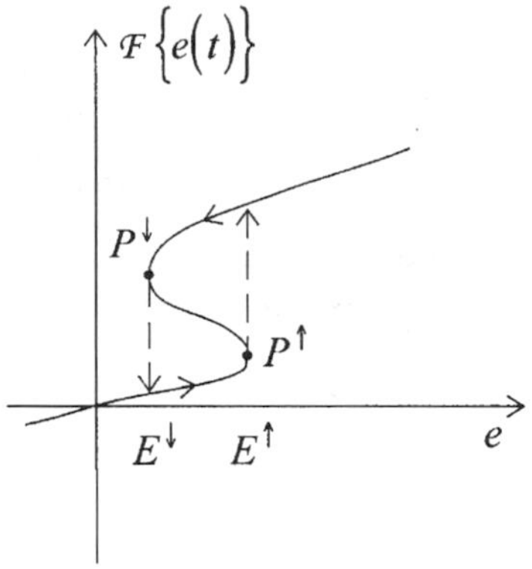

Figure 9.7. *Relay hysteresis* operator (for $R_c dg/dv < -1$, in the limit $C_p \downarrow 0$).

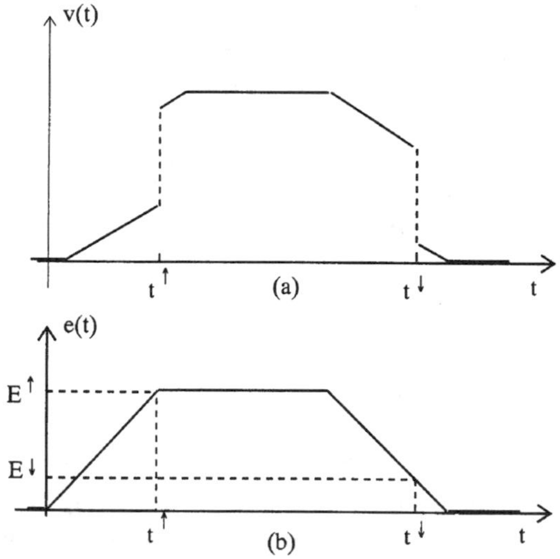

Figure 9.8. Voltage behavior (a), for the voltage waveform $e(t)$ represented in (b) (for $R_c dg/dv < -1$, in the limit $C_p \downarrow 0$).

When the condition (9.7) is not satisfied, the uniqueness can be assured by adding a parasitic inductance in series to the current controlled resistor.

9.4. A GLIMPSE AT THE UNIQUENESS PROBLEM FOR IMPERFECT TWO-CONDUCTOR TRANSMISSION LINES: ASSOCIATED RESISTIVE CIRCUIT

Let us now consider an imperfect two-conductor line connecting two nonlinear weakly active resistors. The left-end (or right-end) equivalent circuit is shown in Fig. 9.9.

The equations describing this circuit are

$$\begin{cases} i(t) + g[v(t)] = 0, \\ v(t) - R_c i(t) - v_z(t) - e(t) = 0, \\ v_z(\mathrm{t}) = \displaystyle\int_0^t z_{cr}(t-\tau) i(\tau) d\tau. \end{cases} \tag{9.9}$$

This system is constituted of a linear algebraic equation, a nonlinear algebraic equation, and a linear convolution relation. The first two equations describe the behavior of the "resistive" part of the circuit, whereas the third equation describes the dynamic one.

9.4.1. The Volterra Integral Equation of the Second Kind in Normal Form

As we shall see, because the problem (9.9) can be recast in terms of nonlinear Volterra integral equations of the second kind, it is useful to briefly overview some aspects of these equations. A nonlinear

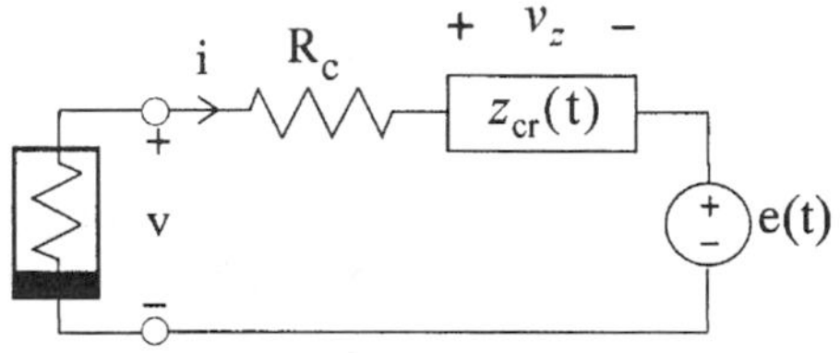

Figure 9.9. Left-end (right-end) equivalent circuit of an imperfect two-conductor line connected to nonlinear resistors.

Volterra integral equation of the second kind is an equation of the form

$$y(t) - \int_a^t K(t, \tau; y(\tau))d\tau = q(t), \tag{9.10}$$

where $y(t)$ is the unknown (e.g., Linz, 1985; Brunner and van der Houwen, 1986). The right-hand side $q(t)$ and the *kernel* $K(t, \tau; y)$ are known functions. If $K(t, \tau; y)$ depends linearly on y, that is,

$$K(t, \tau; y) = k(t, \tau)y, \tag{9.11}$$

the equation is said to be linear. When the kernel $k(t, \tau)$ is of the form (difference kernels)

$$k(t, \tau) = k(t - \tau), \tag{9.12}$$

the corresponding integral equations are referred to as a *linear convolution equation of the second kind.*

Many theorems may be found in the literature on the existence and the uniqueness of these kinds of integral equations (e.g., Linz, 1985; Brunner and van der Houwen, 1986). These theorems typically consider Volterra integral equations in the so-called *normal form*, that is, the form given by Eq. (9.10), where the kernel $K(t, \tau; y)$ must be single-valued and defined for any value of the unknown y.

If the kernel $K(t, \tau; y)$ is continuous and locally Lipschitz with respect to y (see §9.3.2) for any bounded y, and the solution remains bounded for any, t, then Eq. (9.10) has one, and only one, continuous differentiable solution. As with the ordinary differential equation (9.8), the local Lipschitz property guarantees only the existence and the uniqueness within a time interval of nonzero length, whereas the boundedness of the solution assures its existence and uniqueness in the whole interval $(0, \infty)$, even if the kernel $K(t, \tau; y)$ is not *Lipschitz* with respect to y in the entire interval $(-\infty, +\infty)$. The same results hold if the kernel has isolated singularities that are absolutely integrable (e.g., Linz, 1985).

9.4.2. Uniqueness Condition for Imperfect Lines

To reduce the system of equations (9.9) to a Volterra integral equation in normal form, we have to eliminate from it all the unknown variables except for v_z. Therefore, we must solve the algebraic equations

$$\begin{cases} i + g(v) = 0, \\ v - R_c i - (v_z + e) = 0, \end{cases} \tag{9.13}$$

which amounts to looking for the functions F_v and F_i such that

$$v = F_v(u), \tag{9.14}$$

$$i = F_i(u), \tag{9.15}$$

for all the solutions of system of equations (9.9), where, instead of v_z, we have introduced the new unknown u defined as

$$u \equiv v_z + e. \tag{9.16}$$

For the function $F_i(u)$ we have:

$$F_i(u) = \frac{F_v(u) - u}{R_c} \quad \text{or } F_i(u) = -g[F_v(u)]. \tag{9.17}$$

By substituting the expressions (9.15) and (9.16) in the third equation of the system of equations (9.9) we obtain the nonlinear convolution equation of the second kind

$$u(t) = e(t) + \int_0^t z_{cr}(t - \tau)F_i[u(\tau)]d\tau. \tag{9.18}$$

For this equation to be in normal form, the nonlinear function $F_i(u)$ must be single valued, and defined for any value of the unknown u. This is only possible if the system of equations (9.13) admits one, and only one, solution for any value of $u = v_z + e$.

The system of equations (9.13) describes the resistive part of the circuit represented in Fig. 9.9. We can augment it to a complete equation system of a resistive circuit. Indeed, we simply add the equation of the "substitution source"

$$v_z = E_z, \tag{9.19}$$

for the branch that carries the dynamic one-port with impulse response $z_c(t)$. The resulting system of equations describes a resistive circuit, which bears a resemblance to the concept of *associated resistive circuit* introduced by Hasler and Neirynck (1986) in their study on the global state equations of nonlinear lumped circuits.

It is evident that the associated resistive circuit of the circuit of Fig. 9.9 coincides with the resistive circuit of Fig. 9.4, provided that in the latter circuit e is replaced by u. In consequence, the conditions assuring the existence of the normal form of the integral equation (9.18) are the same as those that assure the existence and the uniqueness of the solution in the ideal line case. Therefore, the existence of the normal form of the integral equation (9.18) is guaranteed if condition (9.6) is satisfied. If the characteristic of the nonlinear resistor is differentiable, the function $F_i(u)$ also is differentiable. Furthermore, if the characteristic of the nonlinear resistor is piecewise linear, the function $F_i(u)$ also is piecewise linear.

Let us assume that the condition (9.6) is satisfied. Then, Eq. (9.18) is actually in normal form. As seen in the previous paragraph, once having put the integral equation in the normal form, we should verify that the kernel

$$K(t-\tau;u) = z_{cr}(t-\tau)F_i(u) \tag{9.20}$$

satisfies the conditions required for both the existence and uniqueness of the solution.

The function $z_{cr}(t)$ is the remainder of the impulse response $z_c(t)$. For transmission lines without skin effect, the function $z_{cr}(t)$ is always bounded; furthermore, it is always continuous, except at $t = 0$, where it has a discontinuity of the first kind. Therefore, if $g(v)$ is differentiable, the kernel is locally *Lipschitz*. The kernel is globally *Lipschitz* when $g(v)$ is piecewise linear. As with Eq. (9.8), because the nonlinear resistor is weakly active, any solution of Eq. (9.18) must remain bounded for any t (see Appendix C). In conclusion, the conditions that allow us to recast Eq. (9.9) as a Volterra integral equation of the second kind in normal form are also sufficient to guarantee the existence and uniqueness of the solution.

For transmission lines with skin effect, the remainder $z_{cr}(t)$, although no longer bounded, is absolutely integrable, and all of the foregoing results still hold (Maffucci and Miano, 1999c).

9.4.3. A Circuit with an Additional Parasitic Reactance

As in the ideal case, the system of algebraic-integral equations (9.9) leads to indeterminacy when the condition (9.6) is not satisfied. In fact, as we have previously shown, the solution of the associated resistive circuit in general is not unique, and hence, the kernel $z_{cr}(t-\tau)F_i[u(\tau)]$ of the integral equation (9.18) may be multivalued. As a consequence, the circuit equations (9.9) are an inadequate model of the corresponding physical system.

As in the ideal case, the uniqueness of the solution may be assured by adding a capacitor in parallel to the voltage controlled

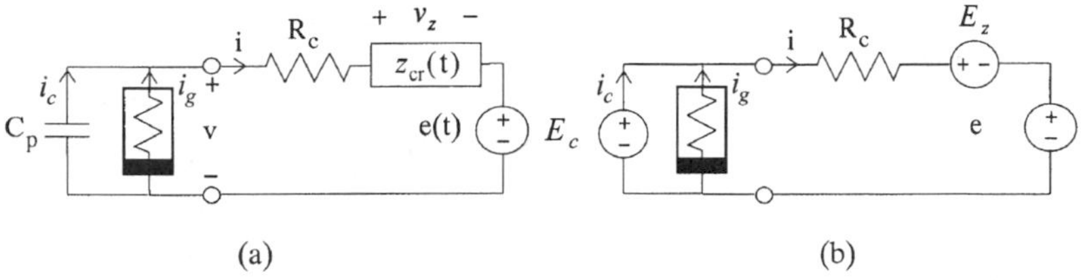

Figure 9.10. The circuit of Fig. 9.9 with an additional parasitic capacitor (a); its associated resistive circuit (b).

resistor (see Fig. 9.10a). The equations of this circuit are

$$\begin{cases} i(t) - i_g(t) - i_c(t) = 0, \\ i_g(t) + g[v(t)] = 0, \\ v(t) - R_c i(t) - v_z(t) - e(t) = 0, \\ v_z(t) = \displaystyle\int_0^t z_{cr}(t-\tau) i(\tau) d\tau, \\ v(t) = -\dfrac{1}{C_p} \displaystyle\int_0^t i_c(\tau) d\tau + v(0). \end{cases} \tag{9.21}$$

This system may be recast as a system of two Volterra integral equations in normal form by eliminating all the unknown variables, except for v_z and v. To do this, we must solve the algebraic equations

$$\begin{cases} i - i_g - i_c = 0, \\ i_g + g(v) = 0, \\ v - R_c i - v_z - e = 0. \end{cases} \tag{9.22}$$

These equations describe the resistive part of the circuit represented in Fig. 9.10a. We can introduce an associated resistive circuit as in the previous paragraph by adding the equations

$$v_z = E_z \quad \text{and } v = E_c \tag{9.23}$$

for the branches that carry, respectively, the dynamic one-port with impulse response $z_{cr}(t)$ and the capacitor. The resulting associated resistive circuit is shown in Fig. 9.10b. It is obvious that this circuit admits one, and only one, solution for any value of E_z and E_c, whatever they may be, because the substitution voltage source E_c is in parallel with the voltage-controlled nonlinear resistor.

For current-controlled resistors, when condition (9.7) is not satisfied, the uniqueness can be assured by adding a parasitic inductance in series to the current controlled resistor. In this case, the associated resistive circuit is obtained by substituting the inductor with an independent current source.

9.5. A GLIMPSE AT THE NUMERICAL SOLUTION FOR IMPERFECT TWO-CONDUCTOR TRANSMISSION LINES: ASSOCIATED DISCRETE CIRCUIT

The reduction of the circuit equations to a system of Volterra integral equations of the second kind in normal form is not an efficient method to evaluate the solution. Since the problem can be solved only numerically, it is more convenient to solve directly the network equations, rather than to solve numerically the Volterra integral equations to which the problem can be reduced. We shall tackle this problem by referring to the circuit of Fig. 9.9.

For ideal lines $z_{cr}(t) = 0$, the system of nonlinear algebraic equations (9.9) can be solved effectively by using, for instance, a numerical algorithm based on the Newton-Raphson approximation method. For imperfect lines, $z_{cr}(t) \neq 0$, the solution of the circuit equation is more difficult because we have to solve a system of nonlinear algebraic convolution equations. Now we shall deal with this problem.

An approximate solution of the system of equations (9.9) may be computed by the following iterative procedure. Let us introduce a uniform discretization of the time axis, with the time step size Δt. Let us suppose we know the solution at the time instants $t_h = h\Delta t$ for $h = 0, 1, \ldots, m-1$. Since the kernel $z_{cr}(t-\tau)$ is absolutely integrable, an approximation of the solution at the time instant t_m can be obtained by approximating the integrals on the right-hand side of the linear convolution equation by a numerical integration rule, and solving the resulting system of nonlinear algebraic equations. Several techniques are available to approximate the integral. Here we shall use the very simple numerical scheme based on the trapezoidal rule. The results that we shall show later hold for multistep integration rules as well (e.g., Linz, 1985).

Let us consider the case in which z_{cr} is everywhere bounded, that is, transmission lines without skin effect. By using the trapezoidal rule the convolution integral may be approximated as

$$\int_0^{t_m} z_{cr}(t_m - \tau)i(\tau)d\tau \cong \frac{\Delta t}{2} z_{cr}^{(0)} I^{(m)} + S^{(m)}, \tag{9.24}$$

where

$$S^{(m)} = \Delta t \sum_{r=1}^{m-1} z_{cr}^{(m-r)} I^{(r)} + \frac{\Delta t}{2} z_{cr}^{(m)} I^{(0)}, \tag{9.25}$$

and

$$I^{(h)} = i(t_h), \tag{9.26}$$

$$z_{cr}^{(h)} = z_{cr}(t_h) \quad \text{for } h \geqslant 0. \tag{9.27}$$

By definition we shall assume $\Sigma_{i=m}^{n} f_i = 0$ for $n < m$. Note that $S^{(m)}$ depends only on the time sequence $I^{(0)}, I^{(1)}, \ldots, I^{(m-1)}$.

The trapezoidal integration rule is *consistent*, that is, in Eq. (9.24) the sum converges in an appropriate sense to the definite integral as $\Delta t \to 0$, because the kernel $z_{cr}(t - \tau)$ is bounded and continuous (e.g., Linz, 1985). The error due to the approximation is proportional to Δt^2 for $\Delta t \to 0$; this dependence of the error is a reflection of the second-order convergence of the trapezoidal rule. To obtain an effective approximation, the time step size Δt must be smaller than the smallest characteristic time of the problem.

By using Eq. (9.24) to replace the convolution integral in the system of equations (9.9), we obtain the "approximate" circuital equations

$$\begin{cases} I^{(m)} + g(V^{(m)}) = 0, \\ V^{(m)} - R_c I^{(m)} \quad V_z^{(m)} - E^{(m)}, \\ V_z^{(m)} - \dfrac{z_{cr}^{(0)} \Delta t}{2} I^{(m)} = S^{(m)}, \end{cases} \tag{9.28}$$

where

$$V^{(m)} = v(t_m), \quad V_z^{(m)} = v_z(t_m), \quad E^{(m)} = e(t_m). \tag{9.29}$$

Equations (9.28) define a system of recurrence equations in the unknowns $V^{(m)}$, $V_z^{(m)}$, and $I^{(m)}$. Note that, at generic step m, $S^{(m)}$ is known because it depends only on $I^{(0)}, I^{(1)}, \ldots, I^{(m-1)}$.

The system of equations (9.28) has a useful circuit interpretation. It is easy to verify that the Eqs. (9.28) are the equations of the resistive circuit shown in Fig. 9.11. By comparing this circuit with the equivalent circuit shown in Fig. 9.9, we note that the nonlinear resistor, the linear resistor (with resistance R_c), and the voltage source e are left intact but the dynamic one-port with impulse response z_{cr} is

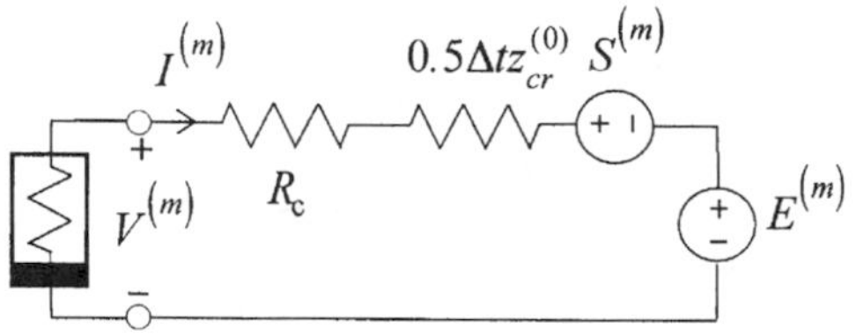

Figure 9.11. Associated discrete circuit corresponding to the dynamic circuit of Fig. 9.9.

replaced by a linear resistor of resistance $0.5\Delta t z_{cr}^{(0)}$ and an independent voltage source $S^{(m)}$. This resistive circuit is the *associated discrete circuit* of the dynamic circuit shown in Fig. 9.9. We are extending to networks composed of transmission lines and lumped circuits a well-known concept of the computer-aided analysis of lumped circuits (e.g., Chua and Lin, 1975).

Remark

When we solve numerically a system of Volterra integral equations of the second kind or a system of algebraic integral equations that can be reduced to a system of Volterra integral equations of the second kind, we have to distinguish between two kinds of stability.

A method is called *numerically stable* if the dominant error satisfies an integral equation having the same kernel as the original equation: The error due to the time discretization can not grow faster than the solution itself, and eventually it saturates if the solution is bounded (e.g., Linz, 1985). The trapezoidal method is numerically stable for a sufficiently small Δt (e.g., Linz, 1985), but it is not easy to say exactly what is meant by "sufficiently small."

A numerical approximation is said to be *A-stable* (absolutely stable) if the error stays bounded for any choice of the time step size Δt. It is well known that the trapezoidal method applied to an ordinary differential equation is *A*-stable like the backward Euler method. Unfortunately, the trapezoidal method, as well as the Euler backward method, when applied to Volterra integral equations of the second kind is not, in general, *A*-stable (e.g., Linz, 1985). In consequence, a preliminary study of the value of Δt assuring the numerical stability is necessary every time we numerically solve Volterra integral equations of the second kind. ◇

It is most remarkable that the proof of the convergence of the numerical solution of these equations can be obtained from one unified principle (e.g., Linz, 1985), which reads roughly as follows:

> *consistency and numerical stability imply the convergence of the numerical solution if:*
>
> *(a) the original equations have one and only one solution;*
> *(b) the approximate equations have exactly one solution; and*
> *(c) the solutions depend continuously on the data in a certain way.*

Therefore, four important questions arise. Does the equation system (9.28) have a solution for any m? If a solution exists, is it unique? If the solution of the actual model is unique, does the numerical model admit a unique solution? Do the solutions depend continuously on the data of the problem?

The existence and uniqueness of the solution of the resistive circuit of Fig. 9.11 bear a resemblance to the existence and uniqueness of the solution of the corresponding associated resistive circuit, provided that resistance R_c is replaced with

$$R_{\text{eff}} = R_c(1 + 0.5\Delta t z_{cr}^{(0)}/R_c). \tag{9.30}$$

Therefore, all the results we have found in the study of the existence and uniqueness for the ideal line case are still applicable. In particular, the solution of Eqs. (9.28) is unique and depends continuously on the data of the problem if

$$\frac{dg}{dv} > -\frac{1}{R_{\text{eff}}}. \tag{9.31}$$

This inequality is always satisfied if the characteristics of the resistors are monotonically increasing, that is, the resistors are both voltage and current controlled and they are locally passive. Instead, it may not be satisfied if the characteristic of the resistor has tracts with negative slopes.

The numerical solution of the system of equations (9.28) is meaningful only if the original Eqs. (9.9) have a unique solution. If inequality (9.6) is satisfied, there exists a sufficiently small time step Δt to satisfy condition (9.31) also, and vice versa, and hence the system of equations (9.28) admits one, and only one solution, depending continuously on the data of the problem. In this case the numerical solution converges to the actual one as $\Delta t \to 0$.

When the condition (9.6) is not satisfied, the original equations may admit several solutions, and hence the condition (9.31) also is not satisfied even if Δt is arbitrarily small. As a consequence, the numerical model (9.28) admits several solutions, and the discrete time sequence approximating the solution of the actual circuit is no longer unique.

Let us consider again the case in which the nonlinear resistor has a characteristic curve like that of a tunnel diode, (see Fig. 9.5a). When the condition (9.31) is not satisfied, the qualitative behavior of the function $V^{(m)} = F_n(U^{(m)})$, defined by the solution of the system of equations (9.28), where $U^{(m)} = S^{(m)} + E^{(m)}$, is similar to that of function $F(e)$ shown in Fig. 9.5b. In this case F_n is not single valued and the sequence defined by the numerical model (9.28) is not unique: $F_n(U)$ consists of three monotone branches and the solution may "jump" uninterruptedly between them for $U^{\downarrow} \leqslant U \leqslant U^{\uparrow}$, where $U^{\downarrow}$ and $U^{\uparrow}$ are the two branch points.

Remark

It is easy to show that for current-controlled resistors the condition equivalent to inequality (9.31) is

$$\frac{dr}{di'} > -R_{\text{eff}}. \quad \diamond \tag{9.32}$$

9.5.1. Effects of an Additional Parasitic Reactance

Let us now investigate the numerical solution of the circuit obtained by inserting a parasitic capacitor in parallel with the nonlinear resistor that is supposed to be voltage controlled (see Fig. 9.12a). By applying again the trapezoidal integration rule to approximate the integral operators present in the system of equations (9.21) we obtain the "approximate" circuital equations

$$\begin{cases} I^{(m)} - I_g^{(m)} - I_c^{(m)} = 0, \\ I_g^{(m)} + g(V^{(m)}) = 0, \\ V^{(m)} - R_c I^{(m)} - V_z^{(m)} = E^{(m)}, \\ V_z^{(m)} - \dfrac{\Delta t z_{cr}^{(0)}}{2} I^{(m)} = S^{(m)}, \\ V^{(m)} + \dfrac{\Delta t}{2C_p} I_c^{(m)} = Q^{(m)}, \end{cases} \tag{9.33}$$

where the term $Q^{(m)}$ is given by

$$Q^{(m)} = v(0) - \frac{\Delta t}{C_p} \sum_{r=1}^{m-1} I_c^{(r)} - \frac{\Delta t}{2C_p} I_c^{(0)}. \tag{9.34}$$

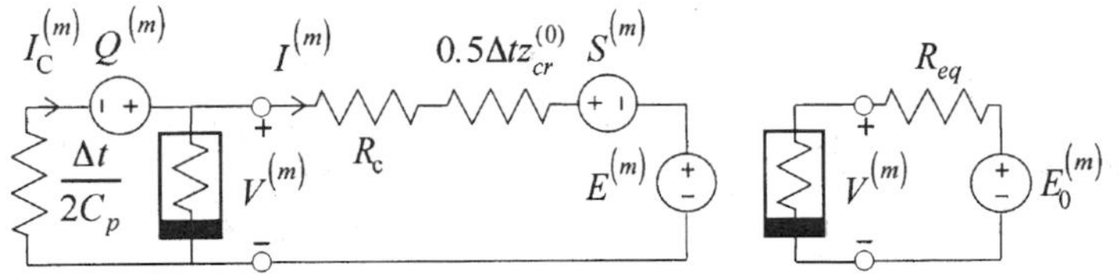

(a) (b)

Figure 9.12. Associated discrete circuit corresponding to the dynamic circuit of Fig. 9.10a (a), and its equivalent circuit (b).

Equations (9.33) have a circuit interpretation too. They are the equations of the resistive circuit shown in Fig. 9.12a. The parasitic capacitor is replaced by a linear resistor with resistance $\Delta t/(2C_p)$ and an independent voltage source $Q^{(m)}$. Figure 9.12b shows the equivalent circuit obtained representing the linear part through the Thévenin theorem. The equivalent resistance R_{eq} is given by

$$R_{eq} = \frac{R_{\text{eff}}}{1 + \dfrac{2\tau_p}{\Delta t} + \tau_p \dfrac{z_{cr}^{(0)}}{R_c}} = R_c \frac{1 + \dfrac{\Delta t z_{cr}^{(0)}}{2R_c}}{1 + \dfrac{2\tau_p}{\Delta t} + \tau_p \dfrac{z_{cr}^{(0)}}{R_c}}, \tag{9.35}$$

where $\tau_p = C_p R_c$, and the equivalent voltage source $E_0^{(m)}$ is given by

$$E_0^{(m)} = \frac{R_{eq}}{R_{\text{eff}}} (S^{(m)} + E^{(m)}) + \frac{2\tau_p}{\Delta t} \frac{R_{eq}}{R_c} Q^{(m)}. \tag{9.36}$$

Thus the resistive circuit shown in Fig. 9.12a has one, and only one, solution for any values of the "sources" if the following inequality is satisfied

$$\frac{dg}{dv} > -\frac{1}{R_{eq}}. \tag{9.37}$$

For any $C_p \neq 0$, $R_{eq} \to 0$ when $\Delta t \to 0$, and thus there exists a sufficiently small Δt such that the inequality (9.37) is satisfied, even if the characteristic of the terminal resistor does not obey (9.6) and hence the condition (9.31) is not satisfied. Therefore, the system of equations (9.33) always has a unique solution that converges to the actual solution for $\Delta t \to 0$.

Figure 9.13 shows the qualitative behavior of the function $V^{(m)} = F_{np}(E_0^{(m)})$ implicitly defined by the system of equations (9.33), for

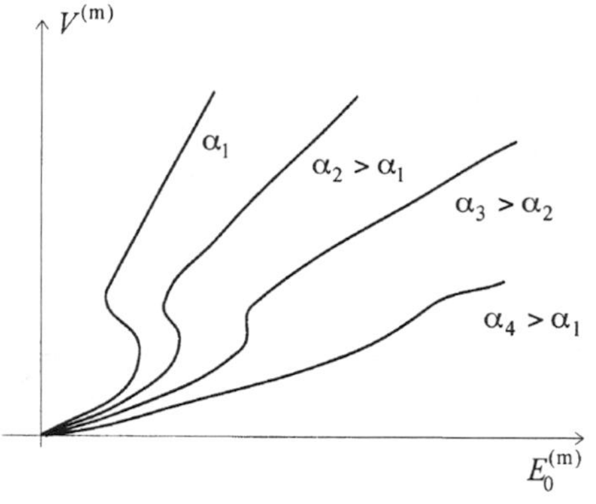

Figure 9.13. Function $V^{(m)}=F_{np}(E_0^{(m)})$ for increasing values of $\alpha=\tau_p/\Delta t$.

increasing values of $\alpha = \tau_p/\Delta t$. For low values of α the function F_{np} has three branches, whereas for high values of α, F_{np} becomes single valued.

Proceeding as in the foregoing, it is possible to show that when the characteristic of a current-controlled resistor does not obey Eq. (9.7), the condition equation (9.32) is not satisfied and the uniqueness of the numerical solution is guaranteed by adding a parasitic inductor in series with it.

9.5.2. Numerical Solution of an Ill-Posed Circuit by Artificially Enforcing the Continuity of the Terminal Voltage

As shown here, by adding parasitics it is possible to guarantee the uniqueness of both the actual and the numerical solution. Among all the possible approximate solutions, the one (it is unique) that assures the time continuity of the voltage and the current at the line ends is captured. Unfortunately, the resulting network will be characterized by a system of *stiff differential equations* due to the very high "characteristic frequencies" introduced by the parasitics.

Another approach, which circumvents these difficulties, is to ignore the parasitics and allow the solution of Eqs. (9.28) to be multivalued. The ambiguities arising from such multivalued functions may then be solved by using the inertia and jump postulates as in the ideal case discussed in §9.3.2. Nevertheless, hereafter we shall show that it is possible to continue using the ill-posed model, provided that a suitable algorithm is adopted to solve the nonlinear algebraic equations (9.28) (Maffucci and Miano, 1999c).

The parasitic capacitor that we should add in parallel with the nonlinear resistor guarantees the uniqueness of the solution because it imposes the continuity of the voltage $v(t)$. In this way, if the solution is initially on a stable branch of $V^{(m)} = F_n(U^{(m)})$, where $U^{(m)} = S^{(m)} + E^{(m)}$, it will remain there until it reaches a branch point, at which time it takes a jump to the other stable branch, in a time interval that goes to zero as $C_p \downarrow 0$.

The same result can be obtained by solving the Eqs. (9.28) by the Newton-Raphson method with a suitable constraint: To initiate the Newton-Raphson iteration at the mth time step we use as an initial guess the solution at the previous time step.

By eliminating $V_z^{(m)}$ and $I^{(m)}$ in the system of equations (9.28) we obtain the nonlinear equation

$$H(V^{(m)}; U^{(m)}) \equiv V^{(m)} + R_{\text{eff}} g(V^{(m)}) - U^{(m)} = 0. \tag{9.38}$$

By applying the Newton-Raphson method, the solution of Eq. (9.38) can be obtained through the iterative algorithm

$$V^{(m)}(0) = \hat{V}, \tag{9.39}$$

$$V^{(m)}(k) = V^{(m)}(k-1) - \left.\frac{H}{dH/dV}\right|_{V^{(m)}(k-1)} \quad \text{for } k = 1, 2, \ldots \tag{9.40}$$

where $V^{(m)}(k)$ stands for the solution at the kth iteration of the Newton-Raphson algorithm and $\hat{V}$ is a suitable initial guess. If $V^{(m-1)}$ is on a stable branch of F_n, by imposing as the initial guess

$$V^{(m)}(0) = V^{(m-1)}, \tag{9.41}$$

we enforce the Newton-Raphson algorithm to capture the solution $V^{(m)}(k)$ that remains on that stable branch, until a branch point is reached. When the solution approaches a branch point, for example, the branch point $U^{\uparrow}$, and U varies in such a way as to exceed this value, the Newton-Raphson algorithm may be opportunely modified to overcome the "impasse" point $dH/dV = 0$ by means, for instance, of the secant methods and norm-reducing check (e.g., Ortega and Rheinboldt, 1970). In this way, the solution is naturally forced to jump on the other stable branch of F_n. Note that $dH/dV = 0$ at the branch points.

In Fig. 9.14 the function $H(V; U)$ is plotted vs V for different values of U by considering a characteristic curve of the same type as that of a tunnel diode. The zeros of $H(V; U)$ are given by the intersections of the curves and the abscissa axis (they are indicated with circles and squares).

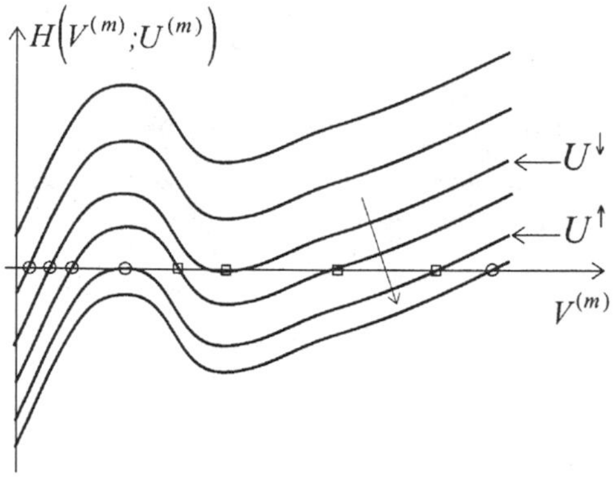

Figure 9.14. Behavior of the function $H(V; U)$ for different values of U (the arrow indicates the direction of increasing values of U).

Let us consider a voltage waveform $e(t)$ of the type shown in Fig. 9.15a. The qualitative behavior of the voltage $v(t)$ obtained by imposing the condition (9.41) is shown in Fig. 9.15b; when $U^{(m)}$ increases, the continuity of the solution, enforced by Eq. (9.41), forces the numerical iterations to converge towards the zeros indicated with circles in Fig. 9.14, thereby avoiding the other possible solutions represented with squares. The "jump" happens only when U has reached and exceeded the value $U^{\downarrow}$. Instead, when $U^{(m)}$ decreases, the continuity of the solution, enforced by Eq. (9.41), forces the numerical iterations to converge towards the zeros indicated with squares, which then avoids the other possible solutions represented with circles.

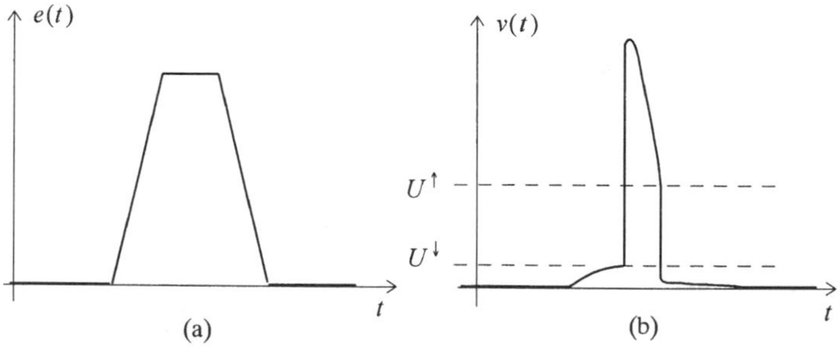

Figure 9.15. (a) Voltage source waveform $e(t)$; (b) terminal voltage $v(t)$.

Remark

The forementioned results can be easily extended to lines with skin effect because for them the kernel is absolute integrable, although unbounded at $t = 0$. For example, for a line with normal skin effect (see Eq. (5.98)) by using the trapezoidal rule, the convolution integral appearing in the system of equations (9.21) may be approximated as

$$\int_0^{t_m} z_{cr}(t_m - \tau)i(\tau)d\tau \cong \left(\frac{4}{3}\theta R_c\sqrt{\frac{\Delta t}{\pi}} + \frac{\Delta t}{2}z_{cb}^{(0)}\right)I^{(m)} + S^{(m)}, \qquad (9.42)$$

where $S^{(m)}$ is yet given by Eq. (9.25) with $(t_m > 0)$

$$z_{cr}^{(m)} = \frac{\theta}{\sqrt{\pi t_m}} + z_{cb}^{(m)}. \qquad (9.43)$$

Therefore, all the results obtained for lines without skin effect still hold, provided that in all of the preceding expressions $z_{cr}^{(m)}$ is defined as in Eq. (9.43) and that $\Delta t z_{cr}^{(0)}/2$ is replaced by

$$\frac{\Delta t z_{eq}^{(0)}}{2} \equiv \frac{\Delta t z_{cb}^{(0)}}{2} + \frac{4}{3}\theta R_c\sqrt{\frac{\Delta t}{\pi}}. \qquad (9.44)$$

Similar results hold for transmission lines with anomalous skin effect and only the expressions of $z_{eq}^{(0)}$ and $z_{cr}^{(m)}$, with $m \geqslant 1$, change. For more details see Maffucci and Miano (1999c).

9.6. WELL-POSEDNESS OF THE NETWORK EQUATIONS

In §9.3 and 9.4 we have shown that the existence and uniqueness of the solution are guaranteed if:

(i) it is possible to reduce the network equations to a system of Volterra integral equations in the normal form;
(ii) the kernels of the integral equations are locally Lipschitz; and
(iii) the solutions are bounded.

The existence and uniqueness of the solution of the Volterra integral equations in normal form are assured if the characteristic curves of the nonlinear elements are regular and every resistive element can produce at most the electrical power $I|v|$, if the resistor is voltage controlled, or $V|i|$, if the resistor is current controlled, that is, they are weakly active. The regularity of the characteristic curve implies the local Lipschitz property, and hence, the local existence and uniqueness. Instead, the weakly active property implies the boundedness of the solution, and hence the global existence and uniqueness. Therefore, the solution of the circuit exists and is unique if the circuit equations can be transformed into a system of Volterra integral equations in normal form. This result holds in the most general case of transmission lines connecting generic lumped circuits.

In this section we shall study the transition from the equations of the circuit shown in Fig. 9.2, which are algebraic, differential and integral equations, to a system of Volterra integral equations in normal form, as in §9.4.

Only for the sake of clarity shall we refer to the network shown in Fig. 9.16, where there are only two dynamic elements, a linear time-invariant capacitor and a linear time-invariant inductor. The results that we shall obtain still hold if more than two dynamic elements, even nonlinear and time varying, are considered. The three-port $\mathcal{N}$ represents an arbitrary interconnection of two-terminal, multiterminal, and multiport resistive elements.

It is useful to formulate the characteristic equations of the dynamic one-ports in the integral form. In this way, all the equations describing the behavior of the dynamic elements, either those with lumped parameters or those with distributed parameters, have the same form.

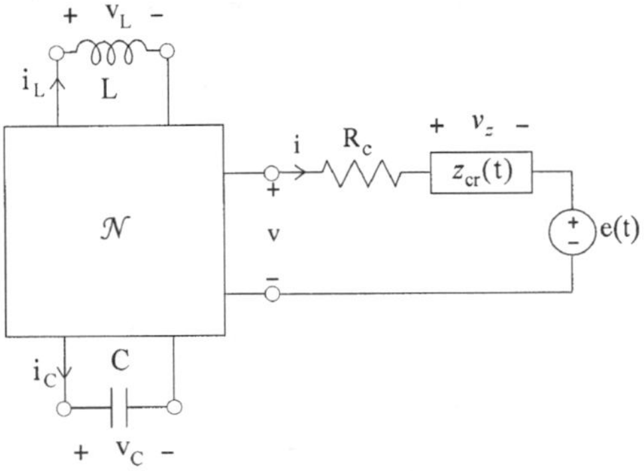

Figure 9.16. Left-end (right-end) equivalent circuit of a two-conductor line connecting a generic lumped resistive network $\mathcal{N}$ connected to two dynamic one-ports.

This circuit is described by a set of algebraic and linear integral convolution equations. The equations of the dynamic one-ports are

$$v_C(t) = \frac{1}{C}\int_0^t i_C(\tau)d\tau + v_C(0), \tag{9.45}$$

$$i_L(t) = \frac{1}{L}\int_0^t v_L(\tau)d\tau + i_L(0), \tag{9.46}$$

$$v_z(t) = \int_0^t z_{cr}(t-\tau)i(\tau)d\tau, \tag{9.47}$$

whereas the equations describing the resistive part are

$$v(t) - R_c i(t) - v_z(t) - e(t) = 0, \tag{9.48}$$

$$\mathbf{N}(\mathbf{v}, \mathbf{i}, v_C, i_C, v_L, i_L, v, i) = \mathbf{0}; \tag{9.49}$$

$\mathbf{v} = |v_1, \ldots, v_n|^{\mathrm{T}}$, $\mathbf{i} = |i_1, \ldots, i_n|^{\mathrm{T}}$ are the vectors that represent, respectively, the voltages and the currents of the resistive elements that are inside $\mathcal{N}$. The system of equations (9.49) is composed by $2n + 3$ algebraic equations consisting of $n + 3$ linearly independent Kirchhoff equations and n characteristic equations of the resistive elements of $\mathcal{N}$. Therefore, the overall system is constituted by $2n + 7$ equations and $2n + 7$ unknowns.

To reduce the system of equations (9.45)–(9.49) to a system of Volterra integral equations, we have to eliminate from it all the unknown variables except for v_C, i_L, and v_z. It is clear, then, that in order to reduce the circuit equations to a system of Volterra integral equations in normal form, we need to be able to express the "nonstate circuit variables," and in particular, the capacitor current i_C, the inductor voltage v_L, and the current i at the line end, as functions of the "state" and source variables, by means of single valued functions. Therefore, we must solve the algebraic equations (9.48) and (9.49), which amounts to looking for a vector function $\mathbf{q}$, such that

$$\begin{vmatrix} i_C \\ v_L \\ i \end{vmatrix} = \mathbf{q}(v_c, i_L, v_z; t), \tag{9.50}$$

for all the solutions of the system of equations (9.48) and (9.49).

The system of equations (9.48) and (9.49) describes the resistive part of the circuit represented in Fig. 9.16. We can augment it to a

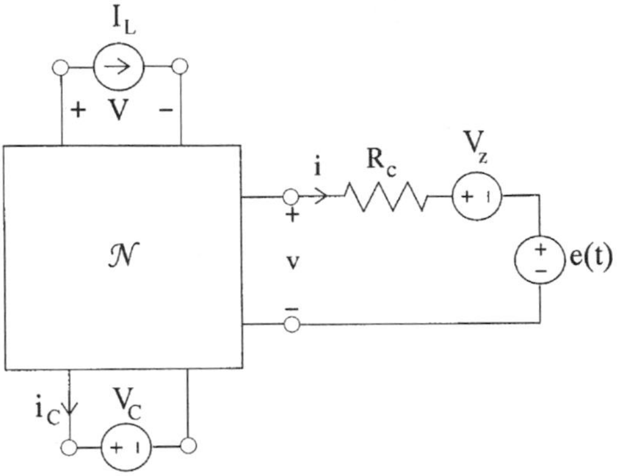

Figure 9.17. Associated resistive circuit of the dynamic circuit of Fig. 9.16.

complete equation system of a resistive circuit as we have done in §9.4. Indeed, we simply add the equation

$$v_C = V_C, \tag{9.51}$$

for the branch which carries the capacitor, the equation

$$i_L = I_L, \tag{9.52}$$

for the branch which carries the inductor, and the equation

$$v_z = V_z, \tag{9.53}$$

for the branch that carries the dynamic one-port with impulse response $z_{cr}(t)$. The resulting system of equations describes the *associated resistive circuit* (see Fig. 9.17) of the dynamic circuit shown in Fig. 9.16.

Remark

The associated resistive circuit of the dynamic circuit represented in Fig. 9.16 is obtained by replacing the one-port, described by the impulse response $z_{cr}(t)$, with a constant voltage source, the capacitor with a constant voltage source, and the inductor with a constant current source. In order to distinguish these from the other source, we call them *substitution sources*. This definition extends that given by Hasler (Hasler and Neirynck, 1986). ◇

The result obtained is very significant: the "nonstate" variables, i_C, v_L, and i can be expressed at each instant as functions of the sole "state variables" v_C, i_L, and v_z and of the independent source, through the associated resistive circuit. Thus it is evident that the condition necessary and sufficient to express the nonstate circuit variables of a dynamic circuit as functions of the state variables, by means of single valued functions, is that the associated resistive circuit admits one and only one solution for every admissible state value. In this way, the possibility of reducing the circuit equations to a system of Volterra integral equations in normal form is brought back to the study of the existence and the uniqueness of the solution for a resistive circuit.

The problems connected to the existence and the uniqueness of the solution of nonlinear resistive circuits are beyond the scope of this book. They are dealt with in detail in Hasler and Neirynck (1986). It is, however, useful here to recall some major results.

In the presence of nonlinearities, a resistive circuit may have no solutions, one solution, or several solutions. The pathological situations, wherein a resistive network, even linear, may not admit solutions, are those where there is incongruency or dependence between the constitutive relations of some circuit elements and the Kirchhoff equations regulating the interaction. Obviously, the incongruency or the dependence is entirely in the model used to represent the real circuit; a more realistic model, which includes the disregarded effects, would overcome every problem of incongruency or dependence.

If the circuit $\mathcal{N}$ is linear the associated resistive circuit also is linear. In general the existence and the uniqueness of the solution of a linear resistive network are guaranteed, if there are no loops consisting of voltage sources only, no cut-sets consisting of current sources only, and the resistances of the circuit resistors are strictly positive. Therefore, in these cases it is possible to reduce the circuit equations to a system of Volterra integral equations in normal form if:

(i) the dynamic elements do not have loops consisting of only capacitors and independent voltage sources and cut-sets consisting of only inductors and independent current sources; and
(ii) all the circuit resistances are strictly positive.

Even if the nonlinear case does not lend itself to a systematic general treatment, the previous criterion continues to be valid if the characteristics of the nonlinear resistive lumped elements are defined for all voltage and current values and are strictly increasing. Instead, if the circuit also contains resistors with nonmonotone characteristics,

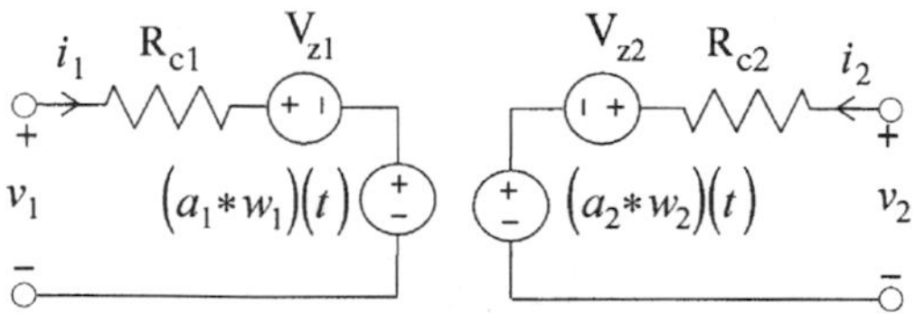

Figure 9.18. Associated resistive circuit of a generic two-conductor line.

the associated resistive circuit can have more than one solution, as we have demonstrated in §9.4, and so a normal form Volterra integral equation system may not exist and the solution may not be unique. From the physical point of view this is another very interesting example of an incomplete model. The incongruency can be resolved by resorting to a higher-order model that takes into account the effects of disregarded parasitic reactances. In general these reactances impose the continuity of the "state variable," namely the control variable of the nonlinear resistors, as shown in the previous sections.

The analysis developed in this paragraph is easily extended to a generic network composed of two-conductor transmission lines and lumped nonlinear circuits. The *associated resistive circuit* of the overall network is obtained by replacing the generic transmission line with the resistive two-port shown in Fig. 9.18. Note that, at the actual time instant t, the two ports behave as if they did not interact between them: What happens at one end at the actual time t does not depend on what happens at the same time at the other end, due to the delay introduced by the finite value of the propagation velocity.

9.7. NUMERICAL SOLUTION OF THE NETWORK EQUATIONS

The reduction of the circuit equations to a system of Volterra integral equations in normal form is not an efficient method to evaluate the solution. Since the problem can be only solved numerically, the direct solution of the network is more convenient than is the solution of the system of Volterra integral equations. Again, for the sake of clarity, we shall refer to the circuit shown in Fig. 9.16.

By approximating the integrals in the integral equations (9.45)–(9.47) by the trapezoidal rule already described in §9.5 we obtain the approximated equations of algebraic type

$$\begin{cases} V_C^{(m)} - \dfrac{\Delta t}{2C} I_C^{(m)} = Q_V^{(m)}, \\ I_L^{(m)} - \dfrac{\Delta t}{2L} V_L^{(m)} = Q_I^{(m)}, \\ V_z^{(m)} - \dfrac{\Delta t z_{cr}^{(0)}}{2} I^{(m)} = S^{(m)}, \\ V^{(m)} - R_c I^{(m)} - V_z^{(m)} = E^{(m)}, \\ \mathbf{N}(\mathbf{V}^{(m)}, \mathbf{I}^{(m)}, V_C^{(m)}, I_C^{(m)}, V_L^{(m)}, I_L^{(m)}, V^{(m)}, I^{(m)}) = \mathbf{0}. \end{cases} \tag{9.54}$$

where $S^{(m)}$ is given by Eq. (9.25) and

$$Q_V^{(m)} = v_C(0) + \frac{\Delta t}{C} \sum_{r=1}^{m-1} I_C^{(r)} + \frac{\Delta t}{2C} I_C^{(0)}, \tag{9.55}$$

$$Q_L^{(m)} = i_L(0) + \frac{\Delta t}{L} \sum_{r=1}^{m-1} V_L^{(r)} + \frac{\Delta t}{2L} V_L^{(0)}. \tag{9.56}$$

The generic variable $A^{(m)}$ indicates the value of the variable $a(t)$ at the time instant $t_m = m\Delta t$. Note that, at the mth time step, $S^{(m)}$, $Q_V^{(m)}$, and $Q_I^{(m)}$ can be considered as if they were known because they depend solely on the values of the solution at the previous time steps.

The system of equations (9.54) has the following circuit interpretation. It is easy to verify that Eqs. (9.54) are the equations of the resistive circuit shown in Fig. 9.19. By comparing this circuit with the associated resistive circuit shown in Fig. 9.17 we note that the three-port $\mathcal{N}$, the linear resistor with resistance R_c, and the voltage source e are left intact, but the dynamic one-port with current based impulse response $z_{cr}(t)$ is replaced by a linear resistor of resistance $0.5\Delta t z_{cr}^{(0)}$ in series to an independent voltage source $S^{(m)}$, the capacitor by a resistor of resistance $0.5\Delta t/C$ in series to an independent voltage source $Q_V^{(m)}$, and the inductor by a resistor of resistance $2L/\Delta t$ in parallel with an independent current source $Q_I^{(m)}$. This is the *associated discrete circuit* of the dynamic circuit shown in Fig. 9.16, see §9.5.

The adjective "discrete" is used to emphasize that the parameters in the model are discrete in nature, that is, they differ from one time step to another. The adjective "associated" is used to emphasize that the models associated with different integration algorithms are different. In this book we always refer to the integration trapezoidal rule, which is an implicit integration rule, because it is at the same time simple and accurate. Other associated discrete circuit models may be easily derived by any explicit or implicit numerical integration rule.

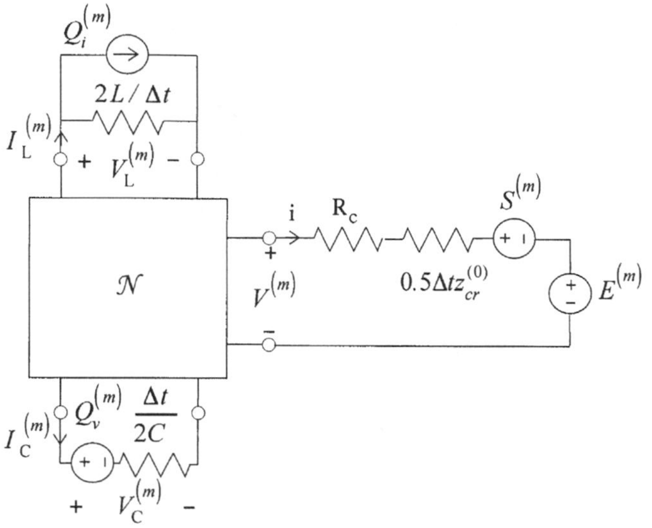

Figure 9.19. Associated discrete circuit of the network shown in Fig. 9.16.

However, by changing the integration rule, only the circuit parameters vary, whereas the topology of the associated discrete circuit remains unchanged. The reader is referred to the existing literature for more details on this subject (e.g., Chua and Lin, 1975).

We immediately observe that the resistive circuit of Fig. 9.19 "tends" to the associated resistive circuit of Fig. 9.17 for $\Delta t \to 0$. This implies that, if the associated resistive circuit has one, and only one solution, and hence the actual circuit has one, and only one, solution, the numerical approximated equation (9.54) has one, and only one, solution that converges to the actual solution for $\Delta t \to 0$.

However, the importance of the resistive circuit shown in Fig. 9.19 arises from another motive too. The use of the associated discrete circuit reduces the transient analysis of the dynamic circuit of Fig. 9.16 to the dc analysis of the resistive circuit of Fig. 9.19. By assigning the values of $v_C(0)$ and $i_L(0)$, we know the values of the voltage and the current source associated, respectively, with the capacitor and inductor. Then we can solve the associated discrete circuit at the first time step by any efficient method, such as modified nodal analysis combined with the Newton-Raphson method (e.g., Chua and Lin, 1975). The associated discrete model is then "updated." In this way, the values of $Q_V^{(m)}$, $Q_I^{(m)}$, and $S^{(m)}$ are changed according to their

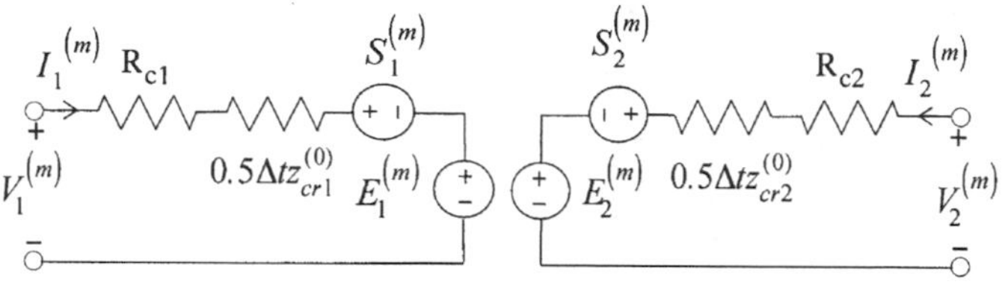

Figure 9.20. Associated discrete circuit of a generic two-conductor transmission line.

definitions. This updated resistive network is then solved to yield the solution at the time instant t_2 and so on. Clearly, this procedure can be iterated as many times as necessary. The most widely used circuit simulator, SPICE, is based on the concept of associated discrete circuit.

The analysis developed in this paragraph is easily extended to a generic network composed of two-conductor transmission lines and lumped nonlinear circuits. The *associated discrete circuit* corresponding to the overall network is obtained by replacing each transmission line by the resistive two-port shown in Fig. 9.20. Note that at the actual time step, the two ports behave as if they did not interact between them.

Remark

The associated discrete circuit of a transmission line can also be obtained by stepwise integration of line equations in characteristic form (see §8.6). ◇

9.8. LUMPED CIRCUITS CONNECTED THROUGH MULTICONDUCTOR TRANSMISSION LINES

The previous sections have dealt with the problem of the existence and the uniqueness of the solution of networks consisting of lumped circuits and two-conductor transmission lines. As we shall briefly outline in this section, the study of the existence and the uniqueness of the solution of networks composed of multiconductor transmission lines and lumped circuits may be performed in the same way. Furthermore, even the numerical solution of this general case may be achieved by following the same procedure as already shown in the previous section.

9.8.1. Associated Resistive Circuit

The equivalent representation of Thévenin type of each end of a generic multiconductor line is shown in Fig. 9.3.

Let us indicate with $v_{zk}(t)$, for $k = 1, 2, \ldots, n$, the contribution to voltage $v_k(t)$ given by $(z_{cr} * i)(t)$,

$$v_{zk}(t) = \sum_{h=1}^{n} \int_0^t z_{cr,kh}(t - \tau) i_h(\tau) d\tau. \tag{9.57}$$

The entries of the impulse response $z_{cr}(t)$ are absolute integrable functions. In the absence of skin effect, these functions are bounded and continuous everywhere for $t \geqslant 0^+$, whereas when there is the skin effect they have singularities like those of two-conductor lines. Then, voltage v_k is given by

$$v_k = e_k + v_{zk} + \sum_{h=1}^{n} R_{c,kh} i_h, \tag{9.58}$$

where $R_{c,kh}$ is the kh entry of matrix R_c. Voltage e_k may be considered as if it were known, because it depends only on the history on the time interval $(0, t - T_1)$, where T_1 is the one-way transit time relevant to the fastest propagation mode.

We may now proceed as done for the two-conductor line case, reasoning in the same way. The associated resistive circuit of the dynamic multiport representing the line behavior at a generic end is shown in Fig. 9.21: it is composed of a linear resistive multiport with resistance matrix R_c, the voltage sources $e_1, e_2, \ldots, e_n$, and the substitution voltages $v_{z1} = V_{z1}, v_{z2} = V_{z2}, \ldots, v_{zn} = V_{zn}$.

9.8.2. Lines Connecting Resistive Multiports

It is instructive to consider the case in which a multiconductor line connects resistive lumped circuits. Let us assume that each lumped termination may be modeled as a voltage controlled multi-port,

$$\mathbf{i} = -\mathbf{g}(\mathbf{v}), \tag{9.59}$$

where $\mathbf{g}(\mathbf{v})$ is a regular vector function. Furthermore, let us assume that the nonlinear resistive multiport is weakly active, that is, there exists a positive constant h such that $\mathbf{v}^T\mathbf{i} \leqslant h\|\mathbf{v}\|$; $\mathbf{v}^T\mathbf{i}$ represents the electrical power supplied by the multiport. If the Jacobian matrix $\partial\mathbf{g}/\partial\mathbf{v}$ of the nonlinear function $\mathbf{g}(\mathbf{v})$ satisfies the condition

$$\frac{\mathbf{v}^T(\partial\mathbf{g}/\partial\mathbf{v})\mathbf{v}}{\mathbf{v}^T \mathrm{R}_c^{-1}\mathbf{v}} > -1 \quad \text{for any } \mathbf{v} \neq \mathbf{0}, \tag{9.60}$$

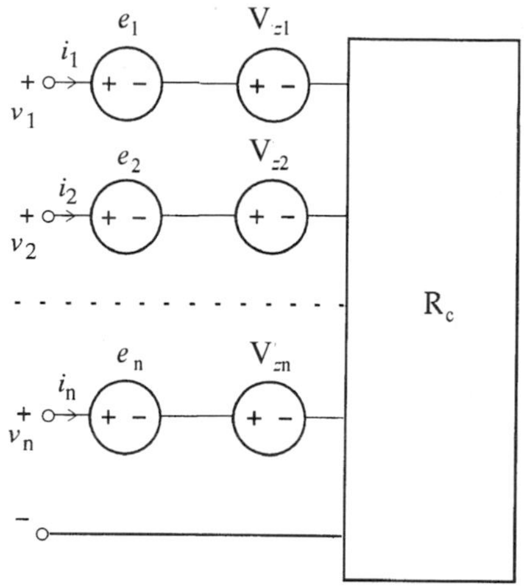

Figure 9.21. Associated resistive circuit of the dynamic multiport representing the behavior of a generic end of a multiconductor transmission line.

then the corresponding associated resistive circuit has one, and only one, solution. This statement can be proved by following the same reasoning as that made in the two-conductor case, §9.3 and 9.4. For details, we advise the reader to consult Maffucci and Miano (1998), where the uniqueness problem for multiconductor lines connecting nonlinear resistive multiports is dealt with in greater depth.

Instead, if all the ports of the terminal element are current controlled, then inequality (9.60) becomes

$$\frac{\mathbf{i}^T(\partial\mathbf{r}/\partial\mathbf{i}')\mathbf{i}}{\mathbf{i}^T\mathrm{R}_c\mathbf{i}} > -1 \quad \text{for any } \mathbf{i} \neq \mathbf{0}, \tag{9.61}$$

where $\mathbf{i}' = -\mathbf{i}$ and $\mathbf{r}(\mathbf{i}')$ is the characteristic of the terminal resistive multiport according to the normal convention for the reference directions of its currents and voltages. The conditions (9.60) and (9.61) are always satisfied if the characteristic curves of the resistors are strictly locally passive, that is if the incremental conductance matrix $\partial\mathbf{g}/\partial\mathbf{v}$ and the incremental resistance matrix $\partial\mathbf{r}/\partial\mathbf{i}'$ are positive definite.

Let us consider, for example, a three-conductor transmission line terminated with two weakly active voltage controlled resistors

$$i_{21} = -g_1(v_{21}), \quad i_{22} = -g_2(v_{22}), \tag{9.62}$$

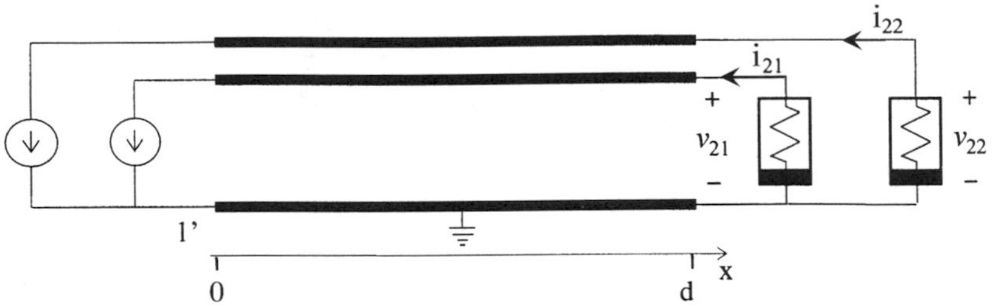

Figure 9.22. A three-conductor line connected to two nonlinear resistors.

as shown in Fig. 9.22, where g_1 and g_2 are continuous functions.

The inequality (9.60) is satisfied if

$$\frac{\mathbf{v}^T \operatorname{diag}(\xi_1, \xi_2)\mathbf{v}}{\mathbf{v}^T \mathrm{G}_c \mathbf{v}} > -1, \tag{9.63}$$

where $\mathrm{G}_c = \mathrm{R}_c^{-1}$, and $\xi_1 = dg_1/dv_1$, $\xi_2 = dg_2/dv_2$ are the differential conductances of the two nonlinear resistors.

The condition (9.63) may be represented in the plane (ξ_1, ξ_2), as shown in Fig. 9.23. This condition requires that the working point always has to lie in the region Σ, limited by the equilateral hyperbola

$$(\xi_1 + G_{c11})(\xi_2 + G_{c22}) - G_{c12}^2 = 0, \tag{9.64}$$

where G_{cij} is the ijth entry of the characteristic conductance matrix

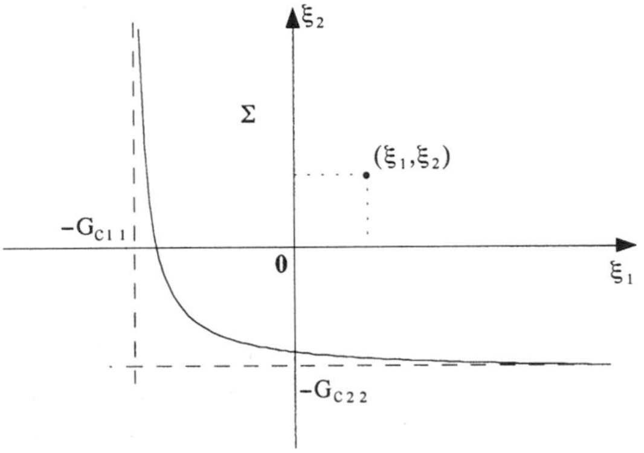

Figure 9.23 Graphic representation of the condition equation (9.63).

G_c. When the coupling between the conductors is disregarded, $G_{c12} = 0$, and hence the equilateral hyperbola coincides with the asymptotes $\xi_1 = -G_{c11}$ and $\xi_2 = -G_{c22}$. Therefore, the coupling effects reduce the region of the plane (ξ_1, ξ_2) in which the condition (9.63) holds.

9.8.3. Associated Discrete Circuit

By using the trapezoidal rule, the convolutions (9.57) are approximated through the algebraic relations

$$V_{zk}^{(m)} = \frac{\Delta t}{2} \sum_{h=1}^{n} z_{cr,kh}^{(0)} I_h^{(m)} + S_k^{(m)}, \tag{9.65}$$

where

$$S_k^{(m)} = \sum_{h=1}^{n} \left[\Delta t \sum_{r=1}^{m-1} z_{cr;kh}^{(m-r)} I_h^{(r)} + \frac{\Delta t}{2} z_{cr;kh}^{(m)} I_h^{(0)} \right]. \tag{9.66}$$

Let us introduce the matrix

$$\Delta \mathrm{R} = \frac{\Delta t}{2} z_{cr}(0^+), \tag{9.67}$$

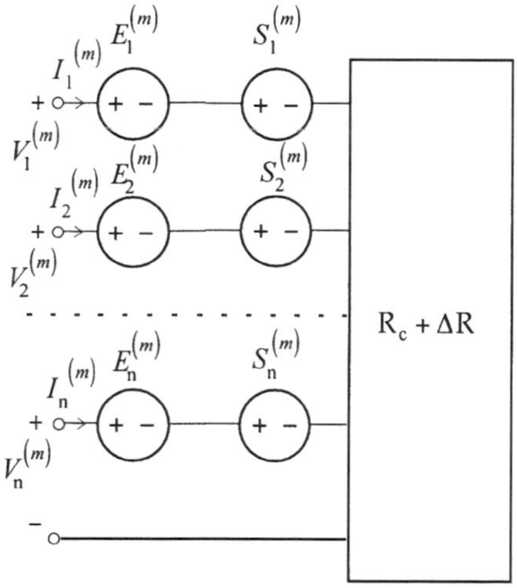

Figure 9.24. Associated discrete circuit of the linear dynamic multiport representing the terminal behavior of a generic multiconductor transmission line.

and the vector

$$\mathbf{S}^{(m)} = |S_1^{(m)}, S_2^{(m)}, \ldots, S_n^{(m)}|. \tag{9.68}$$

Then the associated discrete model, obtained with the trapezoidal rule, of the linear dynamic multiport representing the terminal behavior of a generic multiconductor transmission line is given by

$$\mathbf{V}^{(m)} = (\mathrm{R}_c + \Delta\mathrm{R})\mathbf{I}^{(m)} + \mathbf{S}^{(m)} + \mathbf{E}^{(m)}. \tag{9.69}$$

The corresponding associated discrete circuit is shown in Fig. 9.24: It is composed of a linear resistive multiport with resistance matrix $(\mathrm{R}_c + \Delta\mathrm{R})$ and the voltage sources $(E_1^{(m)} + S_1^{(m)})$, $(E_2^{(m)} + \mathrm{S}_2^{(m)}), \ldots, (E_n^{(m)} + S_n^{(m)})$.

CHAPTER 10

Qualitative Analysis of an Ideal Two-Conductor Line Connected to Nonlinear Resistors: Periodic Solutions, Bifurcations, and Chaos

10.1. INTRODUCTION

The main goal of network analysis is to determine the voltages and the currents in networks. In the previous chapter, we first discussed the conditions that ensure the existence and uniqueness of the solution of networks composed of linear transmission lines and nonlinear lumped elements, and then described a simple and effective numerical method to solve them.

The advent of powerful computer and network analysis software greatly simplifies network analysis. However, if we rely entirely on the numerical simulation, we never get a global description of the network behavior, unless we carry out a huge number of numerical computations. Lacking any prior general information about the qualitative behavior of the network solutions, simulations could be nothing but blind groping. "Thus, the numerical analysis should be complemented by a qualitative analysis, one that concentrates on the general properties of the circuit, properties that do not depend on the particular set of circuit parameters" (Hasler, 1995).

The qualitative analysis of nonlinear lumped circuits has been greatly developed in the last 30 years. Many general results have been found: the existence of multiple steady-state solutions, their stability, the bifurcations, and the presence of chaotic dynamics. The contribution of Martin Hasler has been one of the most important.

There exist phenomena that can be observed only when nonlinear lumped circuit elements are considered. The state of an unstable linear lumped circuit diverges when the time tends to infinity; instead, this may happen in a finite time if nonlinear elements are present.

An autonomous linear circuit possesses a sole equilibrium state, to which the state of the circuit, if stable, tends, irrespective of the initial conditions. Instead, several equilibria may exist in autonomous nonlinear circuits. The state of the circuit reaches one of them, solely depending on the initial conditions. An autonomous linear circuit may oscillate, provided that it has at least a pair of pure imaginary natural frequencies. This is an ideal case, unreachable, in practice, because of the losses. However, even if it were possible to deal with the ideal case, the amplitude of the oscillations would depend on the initial conditions. Instead, an autonomous nonlinear circuit may show stable periodic oscillations — limit cycles — having period and amplitude independent of the initial conditions. There are autonomous nonlinear circuits that have both stable equilibria states and stable limit cycles. The state of the circuit will tend to one of these possible trajectories, depending on the initial conditions.

In general, the steady-state solutions of an autonomous circuit are continuously deformed as a circuit parameter is varied, although their nature remains unchanged. The steady-state solution is what remains after the transient has decayed to zero, and can be either a constant or a periodic solution. Conversely, at very specific values of the parameter the solutions change in a qualitative way. These values are called bifurcation points. A stable equilibrium state may bifurcate in three equilibrium states, two of which are stable and one unstable, or in an unstable equilibrium state and a stable periodic solution. A stable periodic solution with period T may bifurcate in an unstable periodic solution with period T, and a stable periodic solution with period $2T$. This is the so-called *period-doubling bifurcation*. Almost periodic solutions may possibly arise. These types of qualitative behavior are characterized by solutions whose asymptotic behavior, for $t \to \infty$, has a discrete frequency spectrum.

Nonlinear circuits may have steady-state solutions that are extremely sensitive to initial conditions, follow an erratic movement, and have a broad continuous spectrum. They are bounded, locally unstable, and nonperiodic. Such behaviors are the telltale manifestation of *chaos*.

The reader is referred to the excellent book written by Hasler and Neirynck (1986) and to the excellent *Special Issue of the IEEE Proceedings* edited by Chua (1987) for a complete and comprehensive treatment of this subject.

We can not end this book without trying to extend the idea of the qualitative analysis developed by Hasler to circuits composed by lumped and distributed elements. We realized immediately that this would be a very difficult task. The main difficulty is due to the fact that the characteristic relations describing the terminal behavior of the lines are difference-convolution relations with delays. However, we feel that we have succeeded in this task, at least in the very simple case of an ideal two-conductor line connecting two nonlinear resistors. In this chapter we shall outline the problem and illustrate some general results.

The state equations relevant to this circuit are nonlinear difference equations with one delay. These equations are studied by recasting them as a nonlinear one-dimensional (1D) map, in which the time t is no longer a continuous variable but a sequence of discrete values. By studying the main properties of these maps, we shall show that bifurcations, periodic oscillations, and chaos may arise when at least one terminal resistor is active and the other is nonlinear (Sharkovsky *et al.*, 1993; De Menna and Miano, 1994; Corti *et al.*, 1994).

Self-oscillations have been observed in a transmission line connected at one end to a tunnel diode in series with a dc bias voltage and at the other end to a short circuit (Nagumo and Shimura, 1961). The asymptotic stability of a system of nonlinear networks interconnected by two-conductor lossless transmission lines has been studied in de Figueiredo and Ho (1970).

Many examples represented by a similar mathematical model are seen in mechanical systems, in particular in stringed and wind instruments (Nagumo and Shimura, 1961). Witt has considered the self-oscillations occurring in a string of a violin and developed the method of analysis on which our approach is based (Witt, 1937).

We hope that the few new elements outlined in this chapter can be a stimulus to extend the qualitative analysis to a generic network composed of nonlinear lumped and linear distributed circuits. In the last section we shall briefly outline the main difficulty arising when two-conductor lines with losses are considered.

10.2. STATE EQUATIONS IN NORMAL FORM FOR AN IDEAL TWO-CONDUCTOR LINE CONNECTED TO NONLINEAR RESISTORS: FORMULATION IN TERMS OF A SCALAR MAP $u_{n+1} = f(u_n)$

Let us consider an ideal transmission line connecting two weakly active nonlinear resistors (see §9.3) that we assume, for example, to be voltage controlled. The transmission line may be represented through the equivalent circuit of Thévenin type introduced in Chapter 2. Therefore, the study of the dynamics of such a circuit may be performed by referring to the equivalent circuit represented in Fig. 10.1.

The circuit equations are (for $t \geqslant 0$)

$$i_1(t) + g_1[v_1(t); t] = 0, \tag{10.1}$$

$$v_1(t) - R_c i_1(t) = w_1(t), \tag{10.2}$$

$$i_2(t) + g_2[v_2(t); t] = 0, \tag{10.3}$$

$$v_2(t) - R_c i_2(t) = w_2(t), \tag{10.4}$$

$$w_1(t+T) = 2v_2(t) - w_2(t), \tag{10.5}$$

$$w_2(t+T) = 2v_1(t) - w_1(t). \tag{10.6}$$

The functions $g_1(v_1; t)$ and $g_2(v_2; t)$, which we assume to be continuous, represent the characteristics of the nonlinear resistors, in accordance with the normal convention. Both the resistors are assumed to be, in general, time varying; R_c is the characteristic resistance and T is the one-way transit time of the line.

Equations (10.1)–(10.4) describe the behavior of the resistive part of the circuit, whereas the pair of linear difference equations with one delay (10.5) and (10.6) govern the dynamics of the state variables of the circuit, w_1 and w_2. Equations (10.5) and (10.6) have to be solved

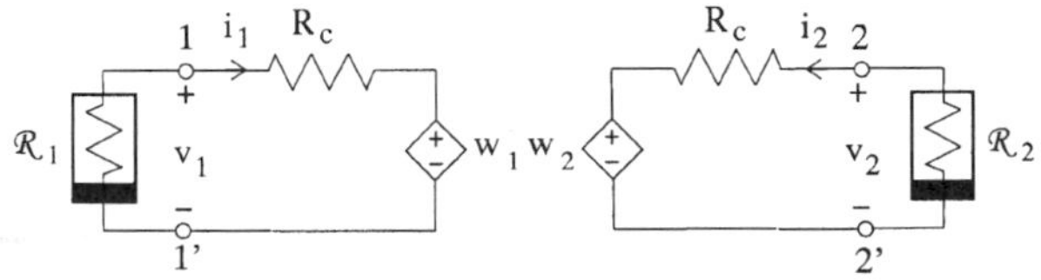

Figure 10.1. Equivalent circuit of an ideal line connecting two nonlinear one-ports.

with the initial conditions

$$w_1(t) = w_{10}(t) \quad \text{for } 0 \leqslant t \leqslant T, \tag{10.7}$$

$$w_2(t) = w_{20}(t) \quad \text{for } 0 \leqslant t \leqslant T, \tag{10.8}$$

where the functions w_{10} and w_{20}, defined on the time interval $(0, T)$, are related to the initial distribution of the voltage and current along the line according to Eqs. (2.58), (2.59), (2.86), and (2.87).

According to the uniqueness criteria discussed in the previous chapter (see §9.3), to assure the uniqueness of the solution of the circuit of Fig. 10.1 we have to impose, for any t

$$\frac{dg_1}{dv_1} > -\frac{1}{R_c}, \tag{10.9}$$

$$\frac{dg_2}{dv_2} > -\frac{1}{R_c}. \tag{10.10}$$

To obtain the global state equations in normal form of the circuit, we have to eliminate all the variables, except for w_1 and w_2 from the circuit equations (10.1)–(10.6). Therefore, we must solve the nonlinear algebraic equations (10.1) and (10.2) in order to eliminate the variables v_1 and i_1, which amounts to looking for the function F_1, such that

$$v_1(t) = F_1[w_1(t); t], \tag{10.11}$$

for all voltages v_1, the solution of the pair of equations (10.1) and (10.2). In the same manner, we must solve the nonlinear algebraic equations (10.3) and (10.4), in order to eliminate the variables v_2 and i_2, which amounts to looking for the function F_2, such that

$$v_2(t) = F_2[w_2(t); t], \tag{10.12}$$

for all voltages v_2, the solution of the pair of equations (10.3) and (10.4). Using Eqs. (10.11) and (10.12), from Eqs. (10.5) and (10.6) we obtain

$$w_1(t+T) = N_1\{w_2(t); t\} \quad \text{for } t > 0, \tag{10.13}$$

$$w_2(t+T) = N_2\{w_1(t); t\} \quad \text{for } t > 0, \tag{10.14}$$

where the nonlinear algebraic functions N_1 and N_2 are defined as follows:

$$N_1\{w_2(t); t\} = 2F_2[w_2(t); t] - w_2(t), \tag{10.15}$$

$$N_2\{w_1(t); t\} = 2F_1[w_1(t); t] - w_1(t). \tag{10.16}$$

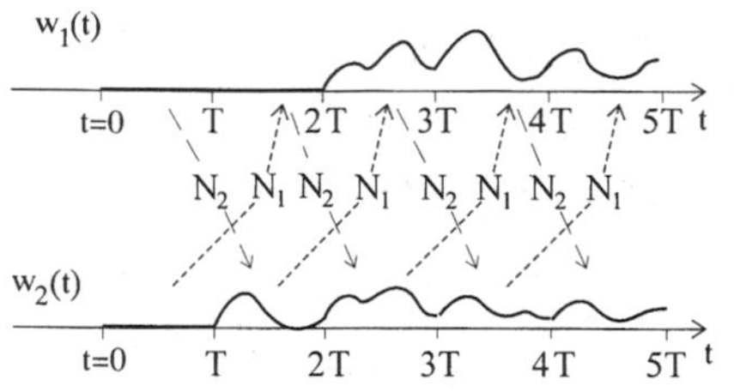

Figure 10.2. Sketch of the recurrence defined by the nonlinear difference equations (10.13) and (10.14), for a line initially at rest.

Equations (10.13) and (10.14) are the *state equations in normal form* of the circuit of Fig. 10.1. They are a system of two nonlinear difference equations with one delay that have to be solved with the initial conditions (10.7) and (10.8).

Equations (10.13) and (10.14) allow the solution to be determined recursively. The initial conditions (10.6) and (10.7) fix $w_1(t)$ and $w_2(t)$ for $0 \leqslant t \leqslant T$. By using Eqs. (10.13) and (10.14) we determine, respectively, $w_1(t)$ and $w_2(t)$, for $T \leqslant t \leqslant 2T$. Once $w_1(t)$ and $w_2(t)$ are known for $T \leqslant t \leqslant 2T$, we determine $w_1(t)$ and $w_2(t)$ for $2T \leqslant t \leqslant 3T$, and so forth (see Fig. 10.2).

The equation system (10.13) and (10.14) may be decoupled and rewritten as follows

$$w_1(t + 2T) = \mathcal{M}_1\{w_1(t); t\}, \tag{10.17}$$

$$w_2(t + 2T) = \mathcal{M}_2\{w_2(t); t\}, \tag{10.18}$$

where the 1D continuous and single-valued functions $\mathcal{M}_1$ and $\mathcal{M}_2$ are given by

$$\mathcal{M}_1 = N_1\{N_2[w_1(t); t]; t\}, \tag{10.19}$$

$$\mathcal{M}_2 = N_2\{N_1[w_2(t); t]; t\}. \tag{10.20}$$

The time difference equations (10.19) and (10.20) have to be solved, respectively, with the initial conditions

$$w_1(t) = w_1^{(0)}(t) \quad \text{for } 0 \leqslant t \leqslant 2T, \tag{10.21}$$

$$w_2(t) = w_2^{(0)}(t) \quad \text{for } 0 \leqslant t \leqslant 2T, \tag{10.22}$$

where

$$w_1^{(0)}(t) = \begin{cases} w_{10}(t) & \text{for } 0 \leqslant t \leqslant T, \\ N_1[w_{20}(t-T); t-T] & \text{for } T \leqslant t \leqslant 2T, \end{cases} \tag{10.23}$$

$$w_2^{(0)}(t) = \begin{cases} w_{20}(t) & \text{for } 0 \leqslant t \leqslant T, \\ N_2[w_{10}(t-T); t-T] & \text{for } T \leqslant t \leqslant 2T, \end{cases} \tag{10.24}$$

Indeed, we need to solve only Eq. (10.19) or (10.20). For example, once we have evaluated $w_1(t)$ by solving Eq. (10.19), $w_2(t)$ may be determined immediately through Eq. (10.14).

For time-invariant resistors, the functions $\mathscr{M}_1$ and $\mathscr{M}_2$, and, hence, the difference equations (10.19) and (10.20), do not depend explicitly on the time. In this case we say that the difference equations are *autonomous*, as for differential equations.

It is evident that according to Eqs. (10.17) the value of w_1 at time instant t, will only depend on its value at the previous time instant $t - 2T$. Therefore, to the generic value $w_1^{(0)}(t_0)$ of the initial condition, where $0 \leqslant t_0 \leqslant 2T$, there corresponds a discrete sequence depending only on that value. We may think of the circuit as if it were composed of an infinite number of *nonlinear uncoupled oscillators*. These remarks suggest formulating the difference equation (10.17) as a recurrence equation, in which the time is no longer a continuous variable but a discrete sequence defined by

$$t_n = 2nT + t_0 \quad \text{for } n = 0, 1, 2, \ldots \quad \text{and} \quad 0 \leqslant t_0 \leqslant 2T. \tag{10.25}$$

The equation governing the dynamics of the samples

$$u_n = w_1(t_n) \quad \text{for } n = 0, 1, 2, \ldots, \tag{10.26}$$

is

$$u_{n+1} = f(u_n; n; t_0), \tag{10.27}$$

where

$$f(u; n, t_0) = \mathscr{M}_1(u; nT + t_0). \tag{10.28}$$

Note that the generic nth sample, u_n, also depends on t_0; for each t_0 we have a different discrete time sequence, and hence a different sequence $u_1, u_2, \ldots, u_n$. The function f is always continuous with respect to u.

This chapter is concerned mainly with the study of the qualitative behavior of the dynamics when the terminal resistors are time invariant. In this case, the nonlinear function f defined by Eq. (10.28) does not explicitly depend on n and t_0, and Eq. (10.28) reduces to the "time-invariant" equation

$$u_{n+1} = f(u_n), \tag{10.29}$$

where $f(u) = \mathscr{M}_1(u)$.

Remark

The state equations of lossless multiconductor transmission lines terminated with nonlinear resistors are a system of nonlinear difference equations with multidelays. For lines embedded in transversally homogeneous dielectrics the dynamics are described by multidimensional difference equations with one delay. The reader is referred to Thompson and Stewart (1986) and Mira (1987) for a thorough and comprehensive treatment of the two-dimensional (2D) nonlinear difference equations with one delay. ◇

10.3. A GLIMPSE AT THE SCALAR MAPS

In the literature the discrete difference equation (10.29) (or Eq. (10.27)) is often called scalar *map f*. In this paragraph, we shall recall some basic concepts and the qualitative dynamics relevant to a scalar map. In particular, we shall deal with the stability and the asymptotic behavior of the sequence of iterates of u_0 under the function f

$$u_0, f(u_0), f(f(u_0)), f(f(f(u_0))), \ldots. \tag{10.30}$$

The reader is referred to the many excellent books existing in the literature for a more thorough and comprehensive treatment of the subject (e.g., Thompson and Stewart, 1986; Schuster, 1988; Devaney, 1989; Hale and Koçak, 1992; Ott, 1993).

As typical when embarking on a journey through an unknown territory, maps and dictionaries are essential for traversing the wilderness of nonlinear dynamics. Consequently, in the following the basic parlance and jargon necessary to become familiar with the dynamics described by nonlinear maps are introduced.

A sequence of the type (10.30) is called *orbit* of u_0. Note that the orbit (10.30) is a set of discrete points, not an interval. In fact, this is the main reason for the rich dynamics of maps, even in one dimension. The orbits of a scalar map may be described through an efficient geometric method, the *phase portrait*. This is a picture on the real line of all orbits. The terminology scalar map actually connotes the geometric process of taking one point to another.

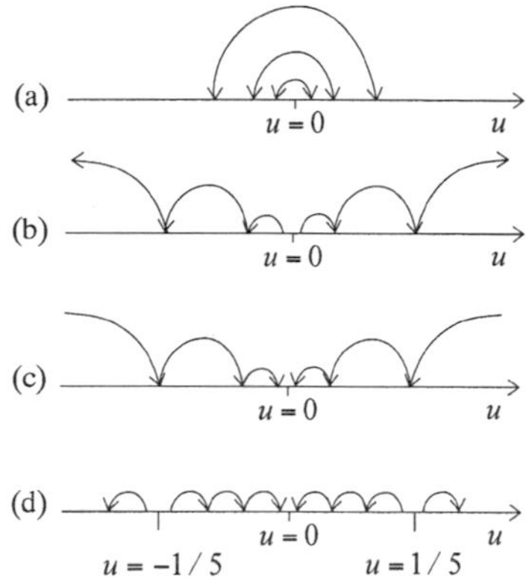

Figure 10.3. Some phase portraits.

Figure 10.3 depicts the phase portrait of some simple maps (e.g., Devaney, 1989). All the orbits generated by $f(u) = -u$ are periodic with period 2, as sketched in Fig. 10.3a. Figure 10.3b shows that the map $f(u) = 3u$ generates diverging orbits; for $u_0 > 0$ all the orbits tend to $+\infty$ as $n \to \infty$, whereas for $u_0 < 0$ they tend to $-\infty$. All the orbits generated by $f(u) = u/3$ tend to zero as $n \to \infty$, whatever the value of the initial condition u_0; see Fig. 10.3c. The orbits of the map $f(u) = u^5$ have a very interesting property — some orbits tend to zero, others diverge depending on the value of the initial condition u_0; see Fig. 10.3d.

There are some orbits that, although particularly simple, play a central role in the qualitative study of the dynamics of the map f; these are the *fixed points* and the *periodic orbits*.

Fixed Points

The point $\bar{u}$ is called *fixed point* of f if $f(\bar{u}) = \bar{u}$. ◇

Note that the fixed points of f remain fixed under iterations of the map, as in the case of equilibrium points of differential equations. However, in computing fixed points, one has to find the solutions of the algebraic equation

$$u - f(u) = 0. \tag{10.31}$$

Periodic Orbits

A point u^* is called a *periodic point* if $f^m(u^*) = u^*$ where m is an integer. The least positive m for which $f^m(u^*) = u^*$ is called period of the orbit. The set of all the iterates of a periodic point is called the *periodic orbit*. Here f^m indicates m times the composition of the map with itself,

$$f^m(u) = f(f^{m-1}(u)) = \underbrace{f(f(f(\cdots f(u)\cdots)))}_{m \text{ times}} \tag{10.32}$$

$$f^1(u) = f(u). \quad \diamond$$

Note that the periodic point u^* is the fixed point of the map f^m defined by Eq. (10.32). Furthermore, a fixed point of f is also a fixed point of f^m.

10.3.1. Stair-Step Diagrams

In general, the solution of the scalar map (10.29) may be followed by using a simple geometric method, called *stair-step diagrams*. We first plot the graph of the function f, as well as the diagonal, that is the 45° line (see Fig. 10.4). Since $u_{n+1} = f(u_n)$, we report the horizontal axis as u_n, and the vertical axis as u_{n+1}. The vertical line starting from u_0 meets the graph of f at the point $(u_0, f(u_0)) = (u_0, u_1)$. The horizontal line starting from this point intersects the diagonal at the point (u_1, u_1). The vertical line starting from this point intersects the horizontal axis at the point u_1. By repeating the same steps we can obtain u_2, u_3, etc. This is a powerful tool to evaluate a generic orbit u_0 for any type of mapping.

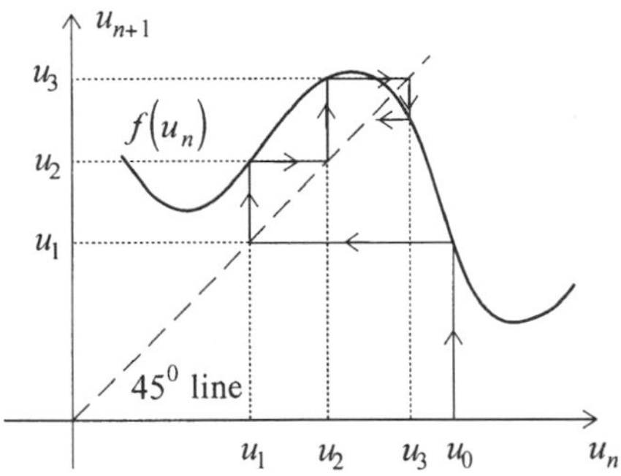

Figure 10.4. An example of stair-step diagram.

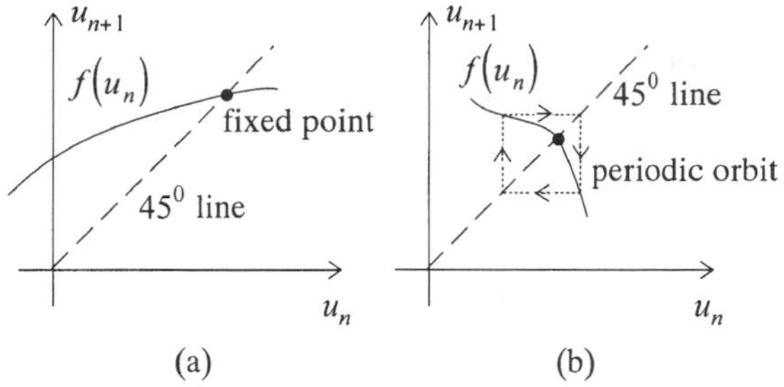

Figure 10.5. Fixed points (a) and periodic orbits (b) in the plane (u_n, u_{n+1}).

It is important to observe that the fixed points of f correspond to the points of intersection of the graph of f with the 45° line (Fig. 10.5a), and the periodic orbits to the closed stair-step paths (Fig. 10.5b).

10.3.2. Linear Map

A fundamental concept in the analysis of the qualitative behavior is that of *stable* and *unstable* linear circuit. A linear circuit is (asymptotically) stable if all its natural frequencies lie within the left complex half-plane (without the imaginary axis), whereas the circuit is unstable if there is at least one natural frequency on the right of the imaginary axis (see §2.7.2). The solutions of an unstable circuit with close initial conditions diverge more and more from one another, as $t \to \infty$, and each of them eventually diverges. Obviously these remarks also apply to the linear maps describing the sequence of the samples $u_1, u_2, \ldots, u_n, \ldots$ of the solution $w_1(t)$.

Let us consider the case in which the line connects two linear resistors. It is easy to show that (see §2.7),

$$N_1(w_2) = \Gamma_2 w_2, \tag{10.33}$$

$$N_2(w_1) = \Gamma_1 w_1, \tag{10.34}$$

where Γ_1 and Γ_2 are the reflection coefficients associated to the two linear resistors, defined by Eqs. (2.101) and (2.102). When the resistance of a linear resistor is positive, the absolute value of the reflection coefficient is always <1, whereas it is >1 when the resistance is negative. For short and open circuits the absolute value of the reflection coefficient is equal to 1, and it is equal to -1 for short circuits and $+1$ for open circuits.

The map of an ideal two-conductor line connecting two linear resistors is

$$u_{n+1} = \gamma u_n, \tag{10.35}$$

where the constant parameter γ is given by

$$\gamma = \Gamma_1 \Gamma_2. \tag{10.36}$$

Therefore, the orbit of u_0 is the set of points

$$u_n = \gamma^n u_0 \quad n = 0, 1, 2, \ldots. \tag{10.37}$$

The fixed point of the map equation (10.37) is $u = 0$. When $|\gamma| < 1$, $|u_n|$ monotonically decreases toward zero for $n \to \infty$, and hence it tends to the sole fixed point $u = 0$, whatever the value of the initial condition. In these cases the fixed point is *asymptotically stable*. Therefore, the state variables w_1 and w_2 tend to zero for $t \to \infty$. For $|\gamma| > 1$, $|u_n|$ monotonically increases toward infinity for $n \to \infty$, thus the fixed point is *unstable*. The state variables w_1 and w_2 diverge for $t \to \infty$. Furthermore, for $\gamma > 0$ the orbit monotonically increases or decreases on one side of the fixed point, much like an orbit of a linear first order ordinary differential equation. Instead, for $\gamma < 0$ the orbit changes sign alternatively and is no longer monotone, a behavior with no counterpart in first order differential equations.

Finally, for $|\gamma| = 1$, we have $|u_n| = |u_0|$ for any n, and hence, the fixed point is *stable* — for $\gamma = 1$ the orbit is the constant sequence of u_0, whereas for $\gamma = -1$ the orbit is an alternating sequence of two values, $+u_0$ and $-u_0$. This is a periodic orbit of minimal period 2. Therefore, for $\gamma = 1$, the state variables w_1 and w_2 oscillate periodically with the minimal period $2T$, whereas they oscillate with the minimal period $4T$ for $\gamma = -1$. Note that the amplitudes and shapes of these oscillations depend only on the initial conditions.

These considerations agree with those outlined in Chapter 2, where we have studied a line connected to two linear resistive circuits. In fact expression (10.37) may be rewritten as

$$u_n = e^{\lambda(t_n - t_0)} u_0 \quad n = 0, 1, 2, \ldots, \tag{10.38}$$

where λ are the natural frequencies given by,

$$\lambda = \frac{1}{2T} ln\, \gamma. \tag{10.39}$$

This result agrees with the expression of the natural frequencies of the system of difference equations (2.99) and (2.100) (see Chapter 2); remember that the logarithmic function has infinite branches in the complex plane and thus the expression on the right-hand side of (10.39) assumes infinite values.

10.3.3. Affine Linear Map

The dynamics of a line connecting two linear resistive lumped circuits with independent sources (see §2.7), is described by a map of the kind

$$u_{n+1} = \gamma u_n + b_n, \tag{10.40}$$

where b_n is a known sequence taking into account the effects of the sources.

For constant sources or periodic sources with period T (in this case the sources are in resonance with the line), b_n is independent of n, $b_n = b_0$ — for stationary sources, b_0 does not even depend on t_0, whereas for periodic sources it does. In these cases, for $\gamma \neq 1$ the map has the fixed point given by

$$\bar{u} = \frac{b_0}{1-\gamma}, \tag{10.41}$$

and the solution of Eq. (10.40) is

$$u_n = \gamma^n u_0 + (1-\gamma^n)\bar{u} = \gamma^n u_0 + \frac{1-\gamma^n}{1-\gamma} b_0. \tag{10.42}$$

When $|\gamma| < 1$, the orbit of u_0 tends eventually to the fixed point $\bar{u}$, whatever the value of u_0, whereas, for $|\gamma| > 1$, the solution diverges. For $\gamma = -1$ the orbit is a sequence of two values, $2\bar{u} - u_0$ and u_0, and thus depends on the initial condition.

Finally, for $\gamma = 1$ the solution is given by

$$u_n = u_o + nb_0; \tag{10.43}$$

the term nb_0 increasing linearly with n, results from the resonance between the sources and the line and the absence of dissipation. The weakly active property only implies that the solution has to be bounded at finite time instants. In fact, the orbit equation (10.43) is bounded at finite, and eventually diverges. In addition, when there is dissipation, the solution is eventually bounded, too.

10.3.4. Bounded Solutions and Stability

In the domain of nonlinear circuits, the range of all possible behaviors is much wider then the stability-instability alternative relevant to linear ones.

For linear circuits, the stability and the instability have both global and local aspects; the most spectacular property of an unstable linear circuit is that its solution diverges for $t \to \infty$ and, conversely, only stable solutions are bounded.

For nonlinear circuits, the stability is both a local and a global concept. Instead, the *local instability* means a divergence of solutions with close initial conditions, and *global instability* means an unlimited increase of the solution. Solutions that are locally unstable may remain always bounded.

Eventually Uniformly Bounded Orbits

The orbits of f are *eventually uniformly bounded* if there is a constant K so that, for any initial condition u_0, there is a time t_M such that

$$|u_n| \leqslant K \quad \text{for } n > M. \quad \diamond \tag{10.44}$$

In this definition the adverb *uniformly* indicates that K does not depend on the particular solution, that is, on the initial condition u_0, whereas the adverb *eventually* indicates that the solution is bounded by a constant K independent of u_0 after a sufficiently long time.

If we require that two different orbits generated by f have to be separated only by a distance of $\varepsilon > 0$ at the time t_n, it is sufficient to choose their initial conditions at a sufficiently small distance $\delta > 0$, because the solution depends continuously on the initial conditions. However, in general, there may not exist a minimum δ (positive) different from zero that guarantees a fixed maximum distance ε for every time instant t_n; it may happen that δ must be chosen smaller and smaller as t_n increases. In these cases the orbits diverge from one another, although they start from very close initial conditions. We say that they are *locally unstable*.

An adequate definition of a stable solution is the one proposed by Hasler and Neirynck (1986): solutions with sufficiently close initial conditions that remain close throughout the future are stable.

Stable, Unstable and Asymptotically Stable Orbits

An orbit $u_0, u_1, u_2, \ldots$ of the map f is stable, if, for every $\varepsilon > 0$, there exists a $\delta > 0$ such that any orbit $v_0, v_1, v_2, \ldots$ with

$$|u_0 - v_0| < \delta, \tag{10.45}$$

satisfies the inequality

$$|u_n - v_n| < \varepsilon, \tag{10.46}$$

for any $n > 0$. If this is not the case, the orbit is unstable.

The orbit of u_0 is *asymptotically stable* if it is stable and

$$\lim_{n\to\infty} |u_n - v_n| = 0. \quad \diamond \tag{10.47}$$

The qualitative behavior of the dynamics of the map f depends on the stability of its fixed points and of the fixed points of its iterates.

Let f be a continuous and differentiable map. A fixed point $\bar{u}$ of f is asymptotically stable if $|f'(\bar{u})| < 1$, and it is locally unstable if $|f'(\bar{u})| > 1$.

The proof of this property may be found in many books (e.g., Hale and Koçak, 1992). However, we can convince the reader through the following considerations. In analogy with the linearization about an equilibrium point of an ordinary differential equation, we expect, under certain conditions, that the local behavior of the map f in the neighborhood of the fixed point $\bar{u}$ is the same as that of the affine map

$$u_{n+1} = f'(\bar{u})(u_n - \bar{u}) + \bar{u}. \tag{10.48}$$

This map has the fixed point $\bar{u}$. Its graph is the straight line τ tangent to the graph of f at the fixed point $\bar{u}$ (see Fig. 10.6). Therefore we can infer the stability of the fixed point $\bar{u}$ of the map f to be the same as the stability of the fixed point of the affine map (10.48). Then, it is evident from the stair-step diagrams in Fig. 10.6 that the fixed point $\bar{u}$ is asymptotically stable if $|f'(\bar{u})| < 1$ and unstable if $|f'(\bar{u})| > 1$.

Note that the behavior of the orbit near the fixed point is completely different, depending on whether $f'(\bar{u}) > 0$ or $f'(\bar{u}) < 0$. In fact, like linear maps, if $f'(\bar{u}) > 0$, then in a neighborhood of $\bar{u}$, the orbits remain on one side of $\bar{u}$ (Figs. 10.6c and 10.6d), whereas, for $f'(\bar{u}) < 0$, the orbits oscillate around $\bar{u}$ (Figs. 10.6a and 10.6b).

Hyperbolic Fixed Point

A fixed point $\bar{u}$ of f is said to be hyperbolic if $|f'(\bar{u})| \neq 1$. $\diamond$

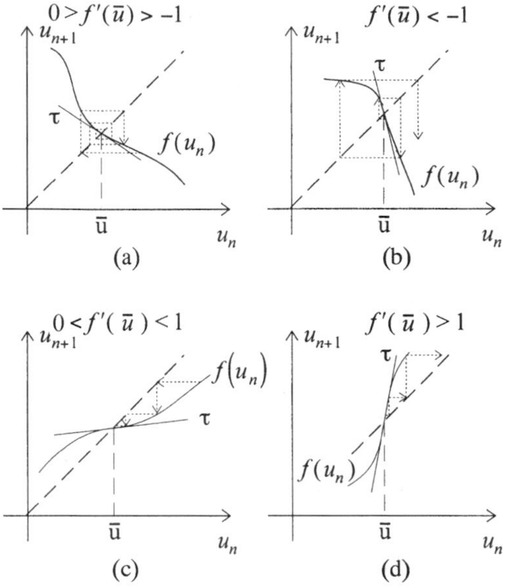

Figure 10.6. Orbits near an asymptotically stable fixed point, (a) and (c), and near an unstable fixed point, (b) and (d).

If a fixed point is hyperbolic, it is either asymptotically stable or unstable, and the stability type is determined by the first-order derivative of f. Instead, the stability of a nonhyperbolic fixed point is not determined by the first derivative of the map. Consider the quadratic map $f(u) = u + u^2$. The point $u = 0$ is a nonhyperbolic fixed point because $f'(0) = 1$. From the stair-step diagram shown in Fig. 10.7 it is evident that $u = 0$ is unstable, because it attracts on the left and repels on the right.

In the same manner we may analyze the stability of periodic orbits, by studying the stability of the fixed points of the mth iterates f^m.

10.3.5. Steady-State Solution

One of the most important characteristics of a circuit is its *steady-state solution*, namely, its asymptotic behavior for $t \to \infty$ corresponding to a given initial condition.

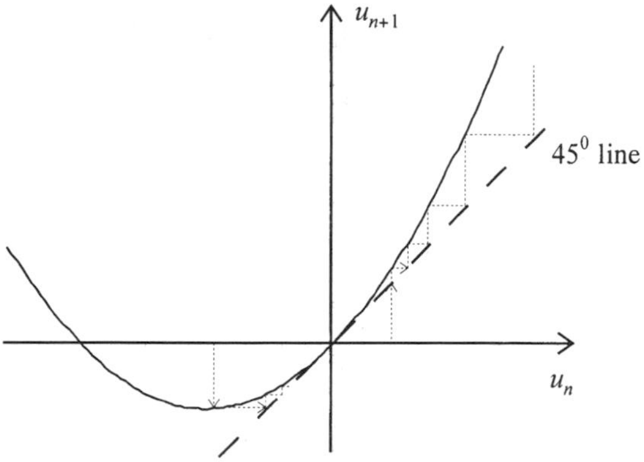

Figure 10.7. An example of nonhyperbolic fixed point.

Limit Set

A point u^∞ is a limit point of the orbit of u_0 if there is a time sequence $u_{m_1}, u_{m_2}, \ldots, u_{m_q}, \ldots$ such that

$$\lim_{m_q \to \infty} |u_{m_q} - u^\infty| = 0. \tag{10.49}$$

The *limit set* of the orbit of u_0 is the set of the limit points of the orbit of u_0. ◇

Invariant Set

A set of points S is invariant under the action of the map f, if for any $u_0 \in S$, we have $u_n \in S$ for any value of n. ◇

It is evident that the limit set of each orbit of a time-invariant map is invariant.

Steady-State Solution

Let the orbit of u_0 be a stable bounded sequence generated from f. If there is a stable orbit $u_0^\infty, u_1^\infty, u_2^\infty, \ldots$ located in the limit set of the orbit of u_0, and such that

$$\lim_{m \to \infty} |u_m - u_m^\infty| = 0, \tag{10.50}$$

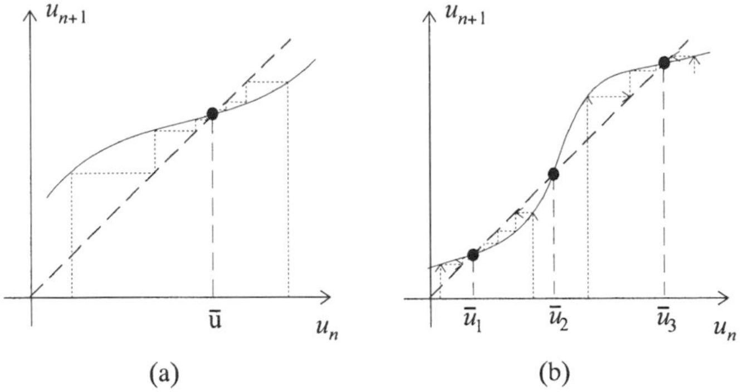

Figure 10.8. Examples of maps with one (a) and more (b) fixed points.

then the orbit u_0^∞, u_1^∞, $u_2^\infty, \ldots$ is the *steady-state solution* associated with the orbit of u_0. ◇

It is evident that the steady-state solution associated with a stable bounded orbit of an autonomous circuit is unique. It is either a fixed point, a periodic orbit, or an almost periodic orbit.

When a map has only one fixed point $\bar{u}$ that is asymptotically stable, all the orbits converge toward it, whatever the initial condition. In this case the map has only one steady-state (see Fig. 10.8a). There exist maps that have more than one fixed point, see for example, the map shown in Fig. 10.8b. In this case two fixed points are asymptotically stable, $\bar{u}_1$ and $\bar{u}_3$, whereas the other $\bar{u}_2$, is unstable. Any orbit starting to the left (or right) of $\bar{u}_2$ converges toward $\bar{u}_1$ (or $\bar{u}_3$). In this case we have two distinct steady-state solutions, $\bar{u}_1$ and $\bar{u}_3$; which of them is reached depends only on the initial condition. Analogous considerations may be made for periodic orbits.

Basins of Attraction and Globally Asymptotically Stable Solutions

Let u_0, u_1, $u_2, \ldots$ be a sequence generated from the map f that is asymptotically stable. The *basin of attraction* of the orbit of u_0 is the set of initial conditions $\hat{u}_0$ such that

$$\lim_{n \to \infty} |u_n - \hat{u}_n| = 0, \tag{10.51}$$

where $\hat{u}_n$ is the orbit of $\hat{u}_0$. If the basin of attraction of the orbit $u_0, u_1, u_2, \ldots$ is the whole real axis, the orbit of u_0 is said to be *globally asymptotically stable.* ◇

The fixed point of the map shown in Fig. 10.8a is globally asymptotically stable. The basin of attraction of the fixed point $\bar{u}_1$ of the map shown in Fig. 10.8b is the half-straight line $u < \bar{u}_2$, whereas the basin of attraction of the fixed point $\bar{u}_3$ is the half-straight line $u > \bar{u}_2$. The boundary between the two different basins is the unstable fixed point $\bar{u}_2$.

Neighborhood of a Set

A neighborhood of a set S is an open set that contains S. ◇

Attractive Set

The set S is attractive if all the orbits that start from an initial condition $u_0 \in U_0$, where U_0 is a neighborhood of S, end up in any neighborhood of S after a finite lapse of time. ◇

Attractor

An *attractor* is an attractive set that is minimum in some sense. ◇

Asymptotically stable fixed points and asymptotically stable periodic orbits are attractors. As we shall see in §10.7 more complex attractors exist, known as *strange attractors* characterizing the chaotic solutions.

10.4. PASSIVITY, EVENTUAL PASSIVITY, AND LOCAL PASSIVITY

The actual expressions for N_1 and N_2, and hence, for the nonlinear map f given by Eq. (10.29), depend only on the characteristic relations of the terminal resistors. However, as we have briefly outlined in the previous section, the main qualitative properties of the dynamics generated by a map are strictly related to its local behavior in the neighborhood of the fixed points and periodic orbits, and in particular on its first-order derivative with respect to u. As we shall show later, the local behavior of the map $f(u)$ is strictly related to the local behavior of the characteristic curves of the terminal resistors, in particular to the incremental electrical power absorbed by them. Furthermore, as we shall see, the boundedness of the solution is strictly related to the concept of eventual passivity.

Here we shall recall some basic concepts such as *passivity*, *eventual passivity* and *local passivity*. For more details see, for instance, Hasler and Neirynck (1986).

The *passive resistors* are resistive elements that can absorb only electrical power, whereas the *active resistors* can also supply electrical power. The *eventually passive resistors* are, in general, active resistors that always absorb electrical power, as soon as the current or the voltage exceeds a certain value. It is evident that all the active physical devices are eventually passive. If the circuit of Fig. 10.1 is composed only of eventually passive resistors the solution is always bounded, even for $t \to \infty$. It might even be possible that the incremental power absorbed by an element will always be positive, irrespective of the two solutions to which it is related. This is roughly the case of the *locally passive* resistors. This property, as we shall show in the next section, assures the asymptotic stability of the solution of the circuit of Fig. 10.1.

Passive and Active Resistive Elements

Let v and i be the voltage and the current of a resistive element whose reference directions are in accordance with the normal convention. The resistor is said to be *passive* if, for any pair (v, i) of current and voltage which satisfies its constitutive relation, we have

$$vi \geqslant 0. \tag{10.52}$$

If this is not the case, the resistor is said to be *active*. If the inequality (10.52) is strict, except when $v = 0$ and $i = 0$, then the resistor is said to be *strictly passive*. ◇

The characteristic curve of a passive resistor crosses the point (0, 0) of the (v, i) plane if it is continuous. Linear resistors with positive resistance, *pn*-junction diodes and tunnel diodes are examples of strictly passive resistors, whereas a voltage source, a current source, and a linear resistor with negative resistance are active elements. A short circuit and an open circuit are simply passive. The main difference between a passive and a strictly passive resistor is that in the latter the dissipation mechanism acts until the rest condition is reached. In general, any one-port composed only of strictly passive resistive elements is strictly passive. This is a simple consequence of the Tellegen's theorem.

A *weakly active* resistive element (see §9.3), is a one-port that can produce at most the electrical power $I|v|$ if it is voltage controlled, or the electrical power $V|i|$ if it is current controlled, where I and V are two constants. All the passive elements are weakly active; an independent source also is weakly active, whereas an active linear resistor is not weakly active.

Eventually Passive Resistive Elements

An active resistive element is said to be *eventually passive* if there is a constant $K > 0$, such that

$$vi \geqslant 0 \quad \text{for} \quad \frac{v^2}{R^*} + R^* i^2 > K^2 \tag{10.53}$$

where R^* is a normalization resistance; for example, we may assume $R^* = 1\Omega$. If the first inequality of Eq. (10.53) is strict, then the resistor is said to be *strictly eventually passive.* ◇

The instantaneous electrical power that can be supplied by an eventually passive resistive element is bounded. A generic one-port, composed only of independent sources and passive resistors, is eventually passive, provided that any voltage source is connected in series with an eventually passive resistor and any current source is connected in parallel with an eventually passive resistor.

Locally Passive Resistor

A resistor is said to be *locally passive* at a point (v_0, i_0) of its characteristic if, for any point of the characteristic in a neighborhood of (v_0, i_0), the inequality

$$\Delta v \Delta i = (v - v_0)(i - i_0) \geqslant 0 \tag{10.54}$$

is satisfied. It is said to be strictly locally passive at (v_0, i_0) if the inequality (10.54) is strictly satisfied, unless for $v = v_0$ and $i = i_0$. If the resistor is (strictly) locally passive at every point of the characteristic, we simply call it (strictly) locally passive. ◇

Linear resistors with positive resistance, *pn*-junction diodes and independent voltage sources connected in series with a linear resistor with positive resistance or with a *pn*-junction diode, are examples of locally passive resistors. A generic one-port, composed only of independent sources and passive resistors with monotonically increasing characteristic curves, is locally passive.

Locally Active Resistor

A resistor is said to be *locally active* in a region D of its characteristic curve if for any point $(v_0, i_0) \in D$, there is a neighborhood of (v_0, i_0) where

$$\Delta v \Delta i = (v - v_0)(i - i_0) < 0. \quad \diamond \tag{10.55}$$

The tunnel diodes or the thyristors with disconnected gate (the four-layer diodes) are examples of locally active resistors. One-ports composed of operational amplifiers, transistors, controlled sources, and independent sources may be locally active.

10.5. SOME GENERAL PROPERTIES OF THE DYNAMICS

The functions $v_1 = F_1(w_1)$ and $v_2 = F_2(w_2)$ (see (10.11) and (10.12)) are implicitly defined by the nonlinear equations (see Eqs. (10.1) to (10.4))

$$v_1 + R_c g_1(v_1) = w_1, \tag{10.56}$$

$$v_2 + R_c g_2(v_2) = w_2. \tag{10.57}$$

Because both resistors are assumed to be weakly active and the characteristic curves are assumed to be continuous, we have

$$v_1 \to \pm\infty \Leftrightarrow w_1 \to \pm\infty, \tag{10.58}$$

$$v_2 \to \pm\infty \Leftrightarrow w_2 \to \pm\infty. \tag{10.59}$$

Under the assumptions (10.9) and (10.10), functions F_1 and F_2, and, hence, functions N_1 and N_2, are single valued and continuous according to the implicit function theorem (e.g., Korn and Korn, 1961). As both the terminal resistors are assumed to be weakly active, both functions F_1 and F_2 , and, hence, functions N_1 and N_2, are bounded for any bounded value of w_1 and w_2 (see §9.3). In consequence, the function f defined by Eq. (10.28) is single valued and continuous too. Therefore, there exists one, and only one, sequence with the initial condition u_0. The solution depends continuously on t and on the initial condition. In particular, the function f is bounded for any bounded value of u, and hence, any sequence generated by it has to be bounded at any finite time. As we have shown in the previous section (§10.3.2), the solution may diverge for $t \to \infty$ if there are resistors with negative resistance. The solution also diverges for $t \to \infty$ when there is no dissipation and the sources are in resonance with the line.

As we shall see later, the solution of the circuit of Fig. 10.1 is eventually uniformly bounded if both the terminal resistors are eventually passive. The circuit tends to a rest state for $t \to \infty$ when both resistors are strictly passive.

10.5.1. Study of the Boundedness of the Solution Through the Lyapunov Function Method

A very simple and elegant method is available to prove that a solution is bounded, or that a solution is asymptotically stable; it is the *Lyapunov function method* (e.g., Hasler and Neirynck, 1986; Hale and Koçak, 1992). In this section the approach applied by Hasler and Neirynck to nonlinear lumped circuits is extended to the circuit of Fig. 10.1, which is composed of lumped and distributed elements.

The electrical power absorbed by the line at the ends may be expressed as (see §2.6)

$$\mathscr{E}(t+T) - \mathscr{E}(t) = i_1(t)v_1(t) + i_2(t)v_2(t), \tag{10.60}$$

where

$$\mathscr{E}(t) = \frac{1}{4R_c}[w_1^2(t) + w_2^2(t)]. \tag{10.61}$$

The function $\mathscr{E}$ is strictly positive definite. The set of pairs (w_1, w_2) such that the variable $\mathscr{E}$ is bounded, is also bounded; furthermore, there is a one-to-one correspondence between v_1 and w_1, and between v_2 and w_2. Let us indicate with $P(t)$ the electrical power globally absorbed by both the terminal resistors,

$$P(t) = v_1(t)g_1[v_1(t);t] + v_2(t)g_2[v_2(t);t]. \tag{10.62}$$

It is immediate that

$$\mathscr{E}(t+T) - \mathscr{E}(t) = -P(t). \tag{10.63}$$

When both terminal resistors are strictly passive, we have

$$P(t) > 0 \tag{10.64}$$

for $w_1(t)$ and $w_2(t)$ different from zero, and $P(t) = 0$ for $w_1(t) = w_2(t) = 0$. Therefore, for any value of $\mathscr{E}$ different from zero, the condition (10.64) holds, whereas $P(t) = 0$ only for $\mathscr{E} = 0$. In these cases the function $\mathscr{E}$ is a *strictly Lyapunov function* of the circuit shown in Fig. 10.1.

It follows that the function $\mathscr{E}(t)$ decreases along any solution of the circuit of Fig. 10.1 when both terminal resistors are strictly passive. In particular, any solution located inside an interval $(0, \mathscr{E}^*)$ at a given instant is no longer able to escape to the outside. Consequently, the solution is bounded. Furthermore, as the decrease of $\mathscr{E}$ along the solution is strict, apart from, $\mathscr{E} = 0$, $\mathscr{E}(t)$ tends eventually to its minimum value and, consequently, the solution tends to a rest state.

When both terminal resistors are passive and not strictly passive, it may happen that $P(t) = 0$ for $\mathscr{E} \neq 0$. In these cases the solution is always bounded, but the circuit may eventually tend to a state different from rest. The most simple situation in which this occurs is, for example, when both resistors are short circuits.

Note that to apply the Lyapunov function method it is not necessary to know the state equations of the overall network. This is a great feature.

Let us now consider the more general situation in which at least one terminal resistor is eventually passive. It is evident that in this case there may be a bounded set of pairs (w_1, w_2) in which the sign of the power P is negative: they correspond to the operating conditions of the eventually passive terminal resistor belonging to the region where it is active. Therefore, there exists a value of $\mathscr{E}$, which we indicate with $\mathscr{E}_c$, such that for $\mathscr{E} \geqslant \mathscr{E}_c$ the power P globally absorbed by the resistors is strictly positive, whereas for $\mathscr{E} < \mathscr{E}_c$ it may be negative. In these cases the function $\mathscr{E}$ is a *strictly Lyapunov function* of the circuit for $\mathscr{E}_c \leqslant \mathscr{E} < \infty$. Therefore, when the line is connected to two eventually passive resistors, the solution eventually enters the interval $(0, \mathscr{E}_c)$, whatever the initial condition, and hence, it is eventually uniformly bounded.

In conclusion, if both the resistors are eventually passive, the solutions of the circuit shown in Fig. 10.1 are always bounded, even if they may be locally unstable. Furthermore, the circuit tends eventually to rest if both the resistors are strictly passive.

10.5.2. Local Behavior of the Map $u_{n+1} = f(u_n)$

As we have briefly outlined in §10.3, the main qualitative properties of the dynamics generated by a map are strictly related to its local behavior in the neighborhood of the fixed points and periodic orbits, and in particular on its first-order derivative with respect to u.

The general expression of the derivative of the map f with respect to u is obtained by differentiating both members of Eq. (10.19) with respect to w_1, and by placing $w_1 = u$. We obtain

$$\frac{df}{du} = \left.\frac{dN_1}{dw}\right|_{w=N_2(u)} \frac{dN_2}{du}. \tag{10.65}$$

The expressions of dN_1/dw_2 and dN_2/dw_1 are obtained from Eqs. (10.15) and (10.16), respectively,

$$\frac{dN_1}{dw_2} = 2\frac{dF_2}{dw_2} - 1, \tag{10.66}$$

$$\frac{dN_2}{dw_1} = 2\frac{dF_1}{dw_1} - 1. \tag{10.67}$$

By differentiating both members of Eqs. (10.56) and (10.57) we obtain, respectively,

$$\frac{dF_1}{dw_1} = \left.\frac{1}{1 + R_c dg_1/dv_1}\right|_{v_1=F_1(w_1)}, \tag{10.68}$$

$$\frac{dF_2}{dw_2} = \left.\frac{1}{1 + R_c dg_2/dv_2}\right|_{v_2=F_2(w_2)}. \tag{10.69}$$

Finally, by substituting the expressions (10.68) and (10.69) in the expressions (10.66) and (10.67), we obtain

$$\frac{dN_1}{dw_2} = \left.\frac{1 - R_c dg_2/dv_2}{1 + R_c dg_2/dv_2}\right|_{v_2=F_2(w_2)}, \tag{10.70}$$

$$\frac{dN_2}{dw_1} = \left.\frac{1 - R_c dg_1/dv_1}{1 + R_c dg_1/dv_1}\right|_{v_1=F_1(w_1)}. \tag{10.71}$$

Observe that the signs of dN_1/dw_2 and dN_2/dw_1 depend only on the signs of the respective numerators, because of the conditions (10.9) and (10.10), which we have to impose in order for the problem to be well-posed.

In the following we shall study the main properties of the nonlinear functions N_1 and N_2, and hence of the map $f(u)$, as well as the qualitative behavior of the sequence generated by it, by referring to the two main classes of nonlinear resistors, locally passive and locally active resistors.

10.5.3. Qualitative Behavior of the Solution for Locally Passive Resistors

Let us assume that both the characteristic curves of the terminal resistors are continuous and strictly increasing, that is

$$\frac{dg_1}{dv_1} > 0, \tag{10.72}$$

$$\frac{dg_2}{dv_2} > 0. \tag{10.73}$$

If a voltage controlled resistor $i = g(v)$, is locally passive, then $dg/dv \geqslant 0$. Conversely, if $dg/dv > 0$, then the resistor is strictly locally passive. We have already shown that the circuit of Fig. 10.1 tends eventually to the rest state if both resistors are strictly passive. Now we shall analyze the main features of the map f, and hence, the qualitative behavior of its solutions when both terminal resistors are locally passive. Remember that a locally passive resistor may be eventually passive.

The conditions (10.72) and (10.73) imply, respectively:

$$\left|\frac{dN_1}{dw_2}\right| < 1, \tag{10.74}$$

$$\left|\frac{dN_2}{dw_1}\right| < 1, \tag{10.75}$$

as evident from their expressions (10.70) and (10.71). Therefore, for strictly locally passive resistors the map f is everywhere contracting, that is,

$$\left|\frac{df}{du}\right| < 1. \tag{10.76}$$

A map of this type can not have more than one fixed point. Indeed, if f had, for example, two fixed points, there would be an interval of value of u in which Eq. (10.76) would not be satisfied. It is also evident that there exists always a fixed point. Only a map of the kind shown in Fig. 10.9 has no fixed points, but in our case we exclude this possibility because if this were so, the solution would be unbounded.

In conclusion, when the line connects two strictly locally passive resistors the orbit of u_0 tends asymptotically to the unique fixed point of the map f, $\bar{u}$ (Fig. 10.10), whatever the values of u_0 and the time instant t_0. Therefore, the voltage wave w_1 tends asymptotically to the

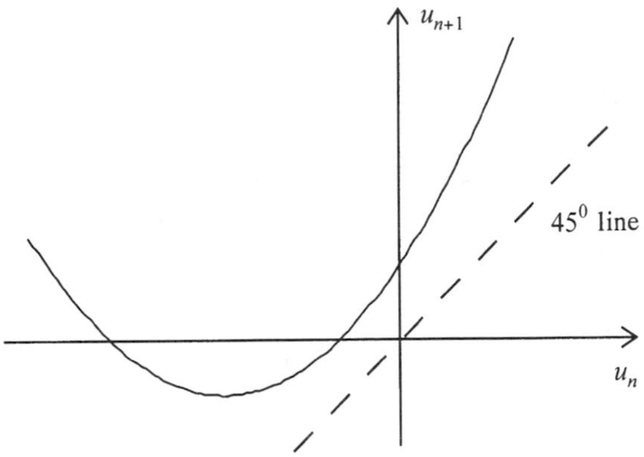

Figure 10.9. An example of a map without fixed points.

constant value $\bar{u}$, and w_2 tends asymptotically to the constant value $N_2(\bar{u})$. When both terminal resistors are strictly passive, the point $\bar{u} = 0$ is the fixed point of f. If at least one terminal resistor is active then $\bar{u} \neq 0$.

The asymptotic stability of the fixed point $\bar{u}$ could be easily proved by using the Lyapunov function method and noting that both of the terminal resistors behave as if they were passive with respect to the

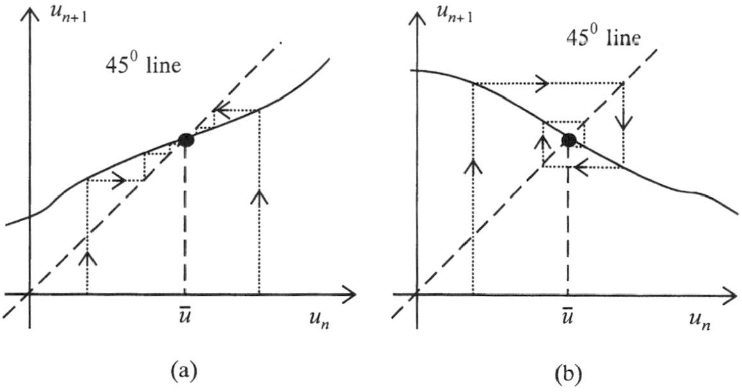

Figure 10.10. Maps with one asymptotically stable fixed point:

(a) $0 < \dfrac{dN_1}{dw_2} \cdot \dfrac{dN_2}{dw_1} < 1$ and (b) $-1 < \dfrac{dN_1}{dw_2} \cdot \dfrac{dN_2}{dw_1} < 0$.

stationary operating conditions corresponding to the fixed point $\bar{u}$. In the steady state, the voltage and the current distributions along the line are uniform.

10.6. QUALITATIVE BEHAVIOR OF THE SOLUTION FOR LOCALLY ACTIVE RESISTORS: A GLIMPSE AT THE BIFURCATIONS

What happens when at least one of the two resistors of the circuit of Fig. 10.1 is locally active? In this case we may have

$$|f'(u)| > 1 \tag{10.77}$$

in some intervals.

Let us consider the case in which the resistor $\mathscr{R}_1$ is a linear passive resistor with resistance R_1, whereas the resistor $\mathscr{R}_2$ is locally active and has the characteristic curve shown in Fig. 10.11. With this choice, the map f may be expressed analytically through a simple piecewise-linear function. Furthermore, this map, as we shall see, shows very rich dynamics. A nonlinear resistor with the characteristic shown in Fig. 10.11 may be realized by using operational amplifiers, diodes, and linear resistors (e.g., Chua and Ayrom, 1985).

The analytical expression of $g_2(v_2)$ is

$$g_2(v_2) = \begin{cases} G_+(v_2 - E_2) - G_-E_2 & \text{for } v_2 \geqslant E_2, \\ -G_-v_2 & \text{for } -E_2 \leqslant v_2 \leqslant E_2, \\ G_+(v_2 + E_2) + G_-E_2 & \text{for } v_2 \leqslant -E_2, \end{cases} \tag{10.78}$$

where G_+, G_-, and E_2 are positive parameters. It is a continuous piecewise-linear function. The central piece has negative slope, whereas the slopes of the other two pieces are positive. This is the simplest

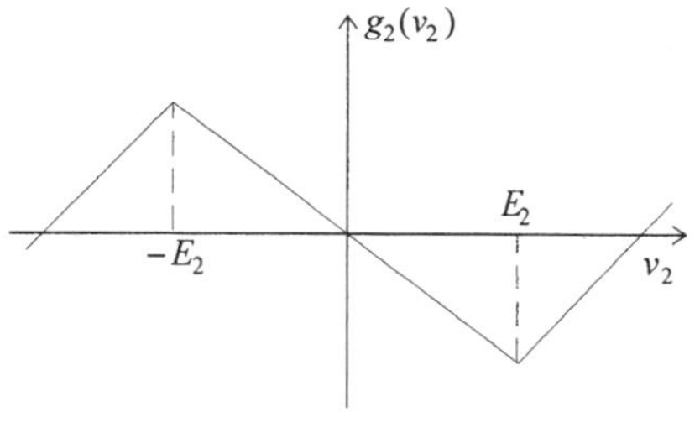

Figure 10.11. Characteristic of a piecewise-linear locally active resistor.

way to make a nonlinear resistor that is both active, locally active and eventually passive.

In this case the function $N_2(w_1)$ is given by

$$N_2(w_1) = \Gamma_1 w_1, \tag{10.79}$$

where Γ_1 is the reflection coefficient of the linear resistor $\mathscr{R}_1$:

$$\Gamma_1 = \frac{R_1 - R_c}{R_1 + R_c}. \tag{10.80}$$

Because $R_1 > 0$, it is $-1 \leqslant \Gamma_1 \leqslant 1$.

Unlike for locally passive resistors, the condition (10.10) is not naturally satisfied for locally active resistors, but it has to be enforced explicitly. Thus, some constraints have to be imposed to render the problem well posed. In this case, the condition (10.10) yields

$$1 - R_c G_- > 0. \tag{10.81}$$

If Eq. (10.81) were not satisfied, we could not neglect the effect of the parasitic capacitances characterizing the "physical" component whose static behavior is described by Eq. (10.78).

The function $v_2 = F_2(w_2)$ can be expressed analytically. Under the assumption (10.81) we obtain

$$F_2(w_2) = \begin{cases} \dfrac{w_2 - W_2}{1 + R_c G_+} + E_2 & \text{for } w_2 \geqslant W_2, \\ \dfrac{w_2}{1 - R_c G_-} & \text{for } -W_2 \leqslant w_2 \leqslant W_2, \\ \dfrac{w_2 + W_2}{1 + R_c G_+} - E_2 & \text{for } w_2 \leqslant -W_2 \end{cases} \tag{10.82}$$

where

$$W_2 = (1 - R_c G_-)E_2. \tag{10.83}$$

Note that for $E_2 > 0$, it is always $W_2 > 0$ because of the constraint (10.81). By substituting the expression (10.82) in Eq. (10.15) for $N_1(w_2)$ we obtain

$$N_1(w_2) = \begin{cases} \Gamma_+(w_2 - W_2) + \Gamma_- W_2 & \text{for } w_2 \geqslant W_2, \\ \Gamma_- w_2 & \text{for } -W_2 \leqslant w_2 \leqslant W_2, \\ \Gamma_+(w_2 + W_2) - \Gamma_- W_2 & \text{for } w_2 \leqslant -W_2, \end{cases} \tag{10.84}$$

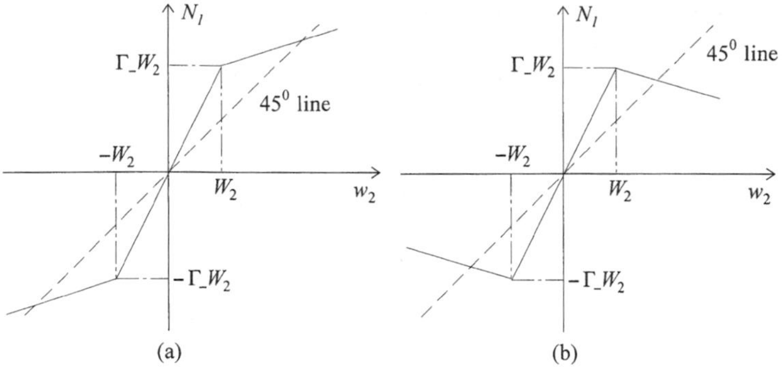

Figure 10.12. Graphs of the function $N_1(w_2)$ for $\Gamma_- > 1$: (a) $0 < \Gamma_+ < 1$, and (b) $-1 < \Gamma_+ < 0$.

where

$$\Gamma_+ = \frac{1 - G_+ R_c}{1 + G_+ R_c}, \quad \Gamma_- = \frac{1 + G_- R_c}{1 - G_- R_c}. \tag{10.85}$$

As $G_- > 0$, the constraint (10.81) implies that Γ_- is always greater than unity, that is, $\Gamma_- > 1$, whereas $|\Gamma_+| < 1$ because $G_+ > 0$. Therefore, the possible shapes of $N_1(w_2)$ are those shown in Fig. 10.12.

The nonlinear map $f(u)$ is given by

$$f(u) = \Gamma_1 \begin{cases} \Gamma_+(u - W_2) + \Gamma_- W_2 & \text{for } u \geqslant W_2, \\ \Gamma_- u & \text{for } -W_2 \leqslant u \leqslant W_2, \\ \Gamma_+(u + W_2) - \Gamma_- W_2 & \text{for } u \leqslant -W_2. \end{cases} \tag{10.86}$$

It is evident that by choosing opportunely $\Gamma_1 \Gamma_-$, it is possible to satisfy the condition (10.77) for $-W_2 \leqslant u \leqslant W_2$; for $|u| > W_2$, the absolute value of df/du is always less than unity. As we shall now see, some or all fixed points of f may be unstable, and multiple steady-state stationary solutions or periodic solutions will arise.

The point $\bar{u} = 0$ is always a fixed point of the map f. It is asymptotically and globally stable for $|\Gamma_1| < 1/|\Gamma_-|$: In this case the dissipative effects due to the passive linear resistor prevail on the amplification action of the locally active resistor and any orbit converges toward $\bar{u} = 0$ whatever the initial condition. What happens for $|\Gamma_1| > 1/|\Gamma_-|$? The fixed point $\bar{u} = 0$ is unstable for $|\Gamma_1 \Gamma_-| > 1$. As we shall see in the following, the actual behavior of the map will depend on the sign of Γ_1. The fixed point $\bar{u} = 0$ becomes nonhyperbolic for $|\Gamma_1 \Gamma_-| = 1$ (see §10.3.4).

The study of changes in the qualitative structures of the dynamics of a map and, in general, of a differential equation as the parameters vary, is called *bifurcation theory*. At a given parameter value, a map is said to have *stable orbit structure* if the qualitative structure of the orbit does not change for sufficiently small variations of the parameter. A parameter value for which the map does not have a stable orbit structure is called a *bifurcation value*, and the map is said to be at a *bifurcation point*. It is evident that hyperbolic fixed points are insensitive to small variations of the parameters, whereas nonhyperbolic fixed points may be sensitive.

10.6.1. Pitchfork Bifurcation of a Nonhyperbolic Fixed Point

Let us consider the case in which Γ_1 is positive, and hence it varies between zero and unity. Figures 10.13a and 10.13b show, respectively, the graph of f for $\Gamma_1 < 1/\Gamma_-$ and $\Gamma_1 > 1/\Gamma_-$. In this case it is $f'(0) > 0$, therefore the orbit of the map is monotonic.

Now we shall investigate the bifurcation that the nonhyperbolic fixed point $\bar{u} = 0$ with $\Gamma_1 = 1/\Gamma_-$ is likely to undergo when Γ_1 is subjected to a perturbation. As the parameter Γ_1 passes through the value $1/\Gamma_-$, the fixed point $u = 0$ loses its stability and two new asymptotically stable fixed points bifurcate from it. The basin of attraction of the fixed point $+\bar{U}$ is the half-straight line $u > 0$, and the basin of attraction of the fixed point $-\bar{U}$ is the half-straight line $u < 0$.

There is a very useful graphical method for depicting the important dynamic features of such a map as the parameter Γ_1 varies. This

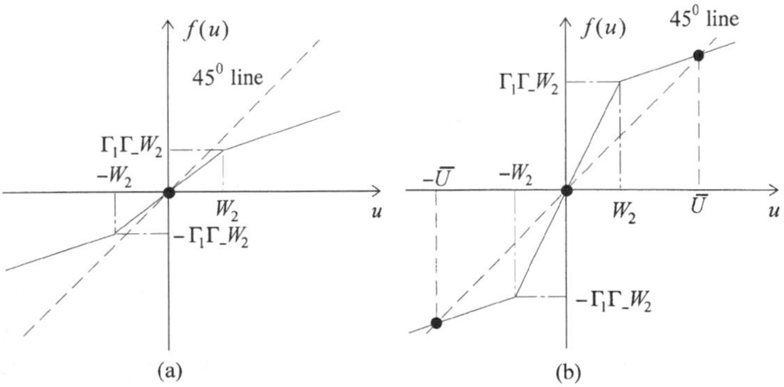

Figure 10.13. Graphs of the map $f(u)$ given by Eq. (10.86) for $0 < \Gamma_1 < 1$: (a) $0 < \Gamma_1 < 1/\Gamma_-$ and (b) $1/\Gamma_- < \Gamma_1 < 1$.

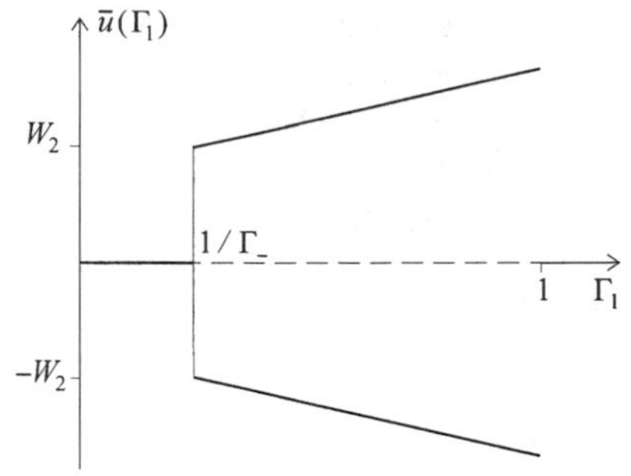

Figure 10.14. The fixed points of f versus Γ_1: pitchfork bifurcation.

method consists in drawing the curve representing the fixed points as the parameter Γ_1 varies between 0 and 1 (Fig. 10.14). To represent the stability types of these fixed points, we label stable fixed points with solid curves and unstable fixed points with dashed curves. The resulting picture is called *bifurcation diagram*. A bifurcation diagram of the type shown in Fig. 10.14 is known as *pitchfork bifurcation* because of its shape. For $\Gamma_1 > 1/\Gamma_-$ the circuit of Fig. 10.1 has two asymptotically stable stationary solutions, and it behaves as a bistable multivibrator.

10.6.2. Period-Doubling Bifurcation of a Nonhyperbolic Fixed Point

Let us consider the case in which Γ_1 is negative, and hence it varies between -1 and 0. Figures 10.15a and 10.15b show, respectively, the graph of f for $\Gamma_1 > -1/\Gamma_-$ and $\Gamma_1 < -1/\Gamma_-$. In this case the map has always only one fixed point $\bar{u} = 0$. Furthermore, it is $f'(0) < 0$, therefore the orbit of the map is not monotonic, and it flips a point at the right of $\bar{u} = 0$ to the left of $\bar{u} = 0$, and vice versa.

For $\Gamma_1 > -1/\Gamma_-$ the fixed point $\bar{u} = 0$ is globally asymptotically stable. Now we shall investigate the bifurcation undergone by the nonhyperbolic fixed point $\bar{u} = 0$ with $\Gamma_1\Gamma_- = -1$ when Γ_1 is subjected to a perturbation. Because the only fixed point becomes unstable when the parameter Γ_1 passes through the value $-1/\Gamma_-$ and the orbit has to remain bounded, it is plausible that the odd iterates converge to a limit point, say, u^* and the even iterates converge to $f(u^*)$. If this is the case, then $f^2(u^*) = u^*$ and $f(u^*) \neq u^*$, that is, u^* is a periodic point of period 2.

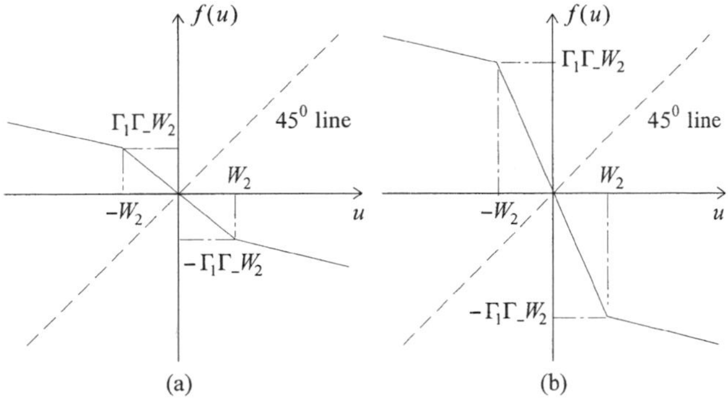

Figure 10.15. Graphs of the map $f(u)$ given by Eq. (10.86) for $-1 < \Gamma_1 < 0$: (a) $-1/\Gamma_- < \Gamma_1 < 0$ and (b) $-1 < \Gamma_1 < -1/\Gamma_-$.

In fact, the second iterate of f, shown in Fig. 10.16, has three fixed points; that is, the point $u = 0$, which is unstable, and the two asymptotically stable points $u = \pm u^*$. Therefore, the map f for $-1 < \Gamma_1 < -1/\Gamma_-$ has two asymptotically stable periodic orbits with period 2; namely, the periodic orbit u^*, $f(u^*)$, and the periodic orbit $-u^*$, $-f(u^*)$. The basin of attraction of the periodic orbit u^*, $f(u^*)$ is the half-straight line $u > 0$, and the basin of attraction of the periodic orbit $-u^*$, $-f(u^*)$ is the half-straight line $u < 0$. The circuit of Fig. 10.1 behaves as an oscillator.

This kind of bifurcation is called *period-doubling* or *flip bifurcation*. The bifurcation diagram is shown in Fig. 10.17. The solid line shows the two periodic points of the periodic orbit u^*, $f(u^*)$, whereas

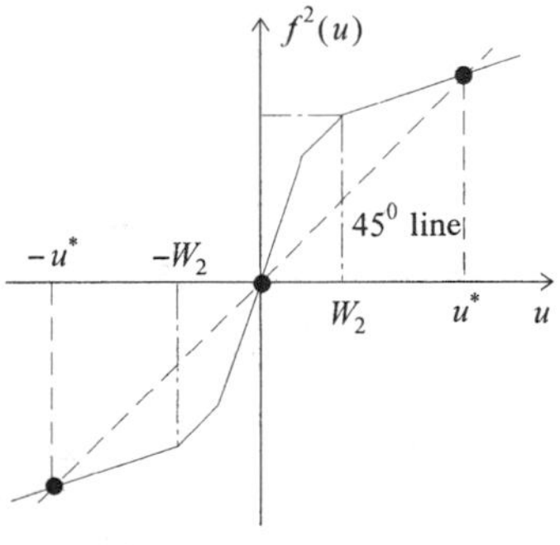

Figure 10.16. Second iterate of the map f represented in Fig. 10.15b.

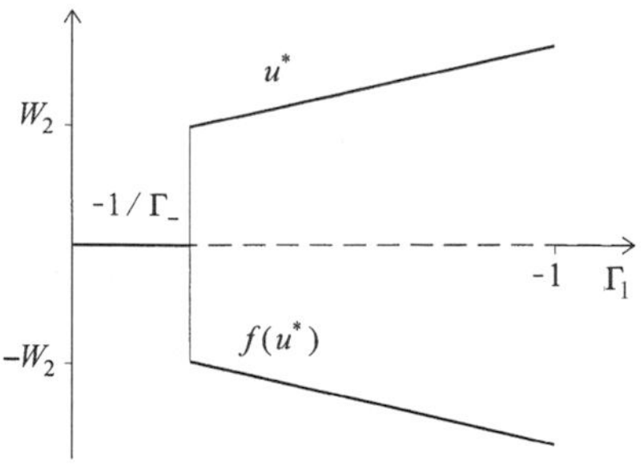

Figure 10.17. Period-doubling bifurcation for the map f represented in Fig. 10.16.

the dashed line indicates the unstable fixed point. Despite its resemblance, this diagram should not be confused with that of pitchfork bifurcations of equilibrium points shown in Fig. 10.14.

10.7. A GLIMPSE AT THE BEHAVIOR OF NONINVERTIBLE MAPS: CHAOTIC DYNAMICS

More complex dynamics may arise if the resistor $\mathscr{R}_1$, although passive and locally passive, is nonlinear. In this section and in the following one we shall deal with these questions by referring to two particular cases that are at the same time very simple and extremely interesting for the richness of their dynamics.

10.7.1. A Tent Map

Let us consider the case in which the resistor $\mathscr{R}_1$ has the continuous piecewise-characteristic curve shown in Fig. 10.18. The analytical expression of $g_1(v_1)$ is given by

$$g_1(v_1) = \begin{cases} G_u(v_1 - E_1) + G_d E_1 & \text{for } v_1 \geqslant E_1, \\ G_d v_1 & \text{for } v_1 \leqslant E_1 \end{cases}, \tag{10.87}$$

where G_u, G_d, and E_1 are positive parameters with $G_u > G_d$. A nonlinear resistor with the characteristic shown in Fig. 10.18 may be realized by using diodes and linear resistors (e.g., Chua, Desoer, and Kuh, 1987).

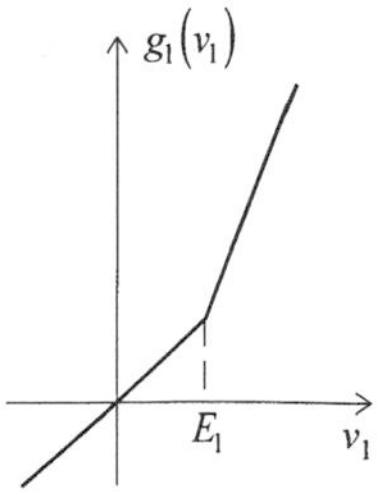

Figure 10.18. Characteristic of a piecewise-linear locally passive resistor.

It is easy to verify that in this case the function $v_1 = F_1(w_1)$ is given by

$$F_1(w_1) = \begin{cases} \dfrac{w_1 - W_1}{1 + R_c G_u} + E_1 & \text{for } w_1 \geqslant W_1, \\ \dfrac{w_1}{1 + R_c G_d} & \text{for } w_1 \leqslant W_1, \end{cases} \tag{10.88}$$

where

$$W_1 = (1 + R_c G_d) E_1. \tag{10.89}$$

The analytical expression of the function $N_2(w_1)$ is

$$N_2(w_1) = \begin{cases} \Gamma_u(w_1 - W_1) + \Gamma_d W_1 & \text{for } w_1 \geqslant W_1, \\ \Gamma_d w_1 & \text{for } w_1 \leqslant W_1 \end{cases}, \tag{10.90}$$

where

$$\Gamma_u = \frac{1 - G_u R_c}{1 + G_u R_c}, \quad \Gamma_d = \frac{1 - G_d R_c}{1 + G_d R_c}. \tag{10.91}$$

Two of the possible shapes of the function $N_2(w_1)$ are illustrated in Fig. 10.19. Note that $N_2(w_1)$ may be nonmonotonic as shown in Fig. 10.19b.

The map $f(u)$ is obtained by the composition of the function $N_1(z)$ given by (10.84), with the function $z = N_2(u)$ given by Eq. (10.90) (see §10.2):

$$f(u) = N_1(N_2(u)). \tag{10.92}$$

Several shapes are possible for the map f, depending on the actual values of the parameters of N_1 and N_2. In the following, only one case, the most significant one, will be analyzed because the dynamics

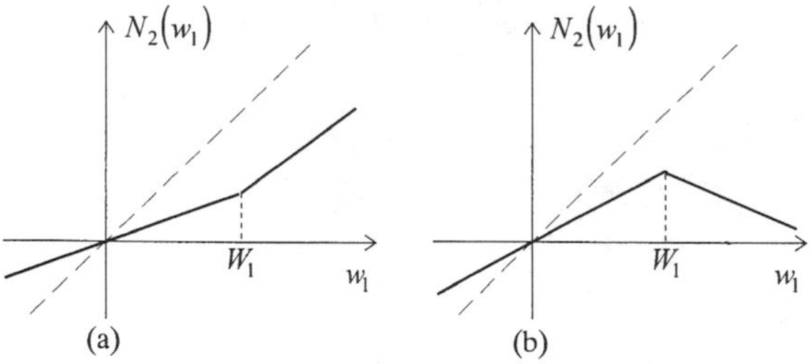

Figure 10.19. Possible shapes of the function $N_2(w_1)$ for $0 < \Gamma_d < 1$ and (a) $0 < \Gamma_u < 1$, (b) $-1 < \Gamma_u < 0$.

arising from the other cases are similar to those obtained in the previous section.

Let us consider the map f obtained by assuming

$$0 < \Gamma_d < 1, \quad \Gamma_u = -\Gamma_d, \quad W_2 > 2W_1 \quad \text{and} \quad 0 < \Gamma_+ < 1. \tag{10.93}$$

Furthermore, let us introduce the parameter

$$\beta = \Gamma_- \Gamma_d. \tag{10.94}$$

In this case, f is *nonmonotonic*, that is, it is equal to zero at $u = 0$ and $u = 2W_1$, and it has a maximum at $u = W_1$:

$$f_{\max} = \beta W_1. \tag{10.95}$$

For $\beta < 1$ the map has only one fixed point $\bar{u} = 0$ that is globally asymptotically stable. The resulting map f is of the type shown in Fig. 10.20. For $\beta = 1$ all the points belonging to the interval $(-W_2, W_1)$ are nonhyperbolic fixed points of the map and are stable.

What happens for $\beta > 1$?

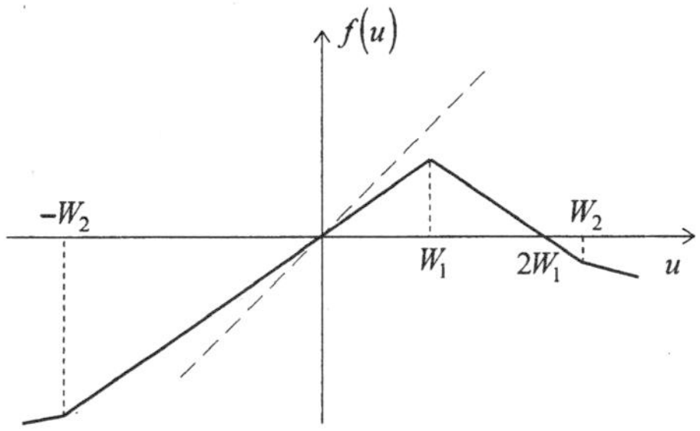

Figure 10.20. The map f given by Eq. (10.92) for $\beta < 1$.

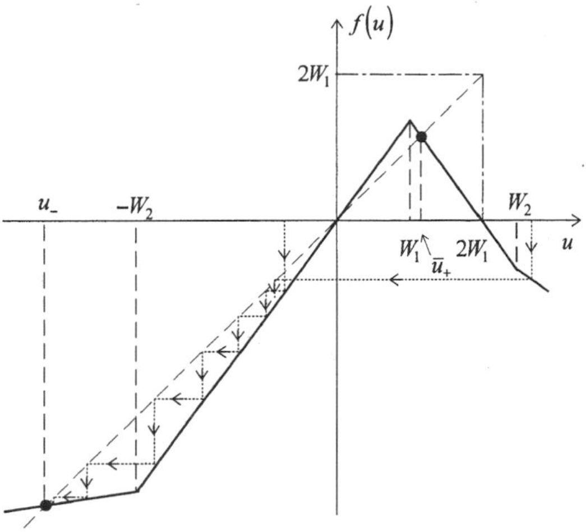

Figure 10.21. The map f given by Eq. (10.92) for $1 < \beta < 2$.

Let us first investigate the situations in which $0 \leqslant \beta \leqslant 2$ (see Fig. 10.21). The fixed point $\bar{u} = 0$ becomes unstable and two other fixed points arise — $\bar{u}_-$ and $\bar{u}_+$. Unlike the pitchfork bifurcation described in the previous section, in this case the fixed point $\bar{u}_+$ is always unstable. The fixed point $\bar{u}_-$ is always asymptotically stable. All the orbits with $u_0 < 0$ and $u_0 > 2W_1$ converge toward the fixed point $\bar{u}_-$ for $n \to \infty$, as the stair-step diagram of Fig. 10.21 shows.

What can we say for the orbits with $0 < u_0 < 2W_1$?

Because we have $f_{\max} \leqslant 2W_1$ for $\beta \leqslant 2$, the orbits starting from $(0, 2W_1)$ remain confined to $(0, 2W_1)$; we say that the function f maps the interval $(0, 2W_1)$ onto itself. Therefore, these orbits can not eventually move off to the stable fixed point u_-. This is a direct consequence of the noninvertibility of the map in the interval $(0, 2W_1)$.

The dynamics of the orbits starting from $(0, 2W_1)$ may be studied by referring to the *tent map* (e.g., Schuster, 1988; Ott, 1993)

$$u_{n+1} = \beta(W_1 - |W_1 - u_n|). \tag{10.96}$$

This is the simplest *one-dimensional noninvertible map* exhibiting chaotic dynamics. For $0 \leqslant \beta < 1$ the tent map has only one fixed point that is asymptotically stable, whereas for $1 < \beta \leqslant 2$ it has two fixed points, both unstable. Even if the two fixed points are unstable, all the orbits starting from $(0, 2W_1)$ remain confined to the interval $(0, 2W_1)$ for $\beta \leqslant 2$.

What can we say about the dynamics if there are no stable fixed points and the orbits are constrained to stay in a bounded interval? In

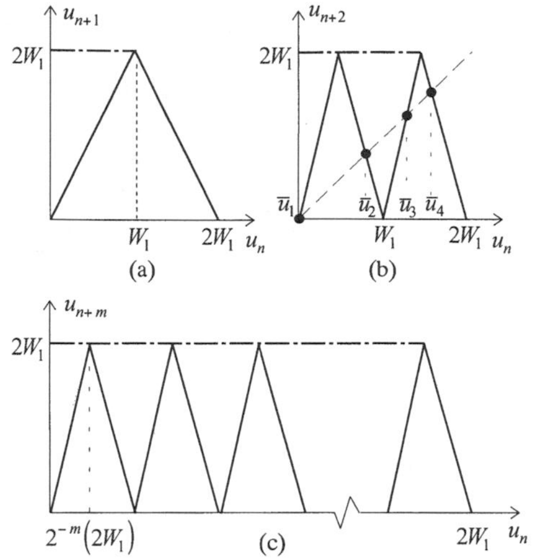

Figure 10.22. The tent map (a), its second (b) and *m*th iterates (c), for $\beta = 2$.

such cases, it may happen that the attracting sets of the orbits are periodic orbits, which can be determined by composing the map with itself.

Let us first consider the case $\beta = 2$. Figure 10.22a shows the tent map for $\beta = 2$, and Fig. 10.22b shows its second iterate. We note that if u_n is equal to 0, W_1 or $2W_1$, then two applications of the tent map yield $u_{n+2} = 0$, while if u_n is either $W_1/2$ or $3W_1/2$, then two applications of the tent map yield $u_{n+2} = 2W_1$. Furthermore, the variation of u_{n+2} with u_n between these points must be linear. The second iterate of the tent map has four fixed points; $\bar{u}_1 = 0$ and $\bar{u}_4 = \bar{u}_+$ are the two fixed points of the tent map, whereas the fixed points $\bar{u}_2$ and $\bar{u}_3$ are two periodic points. The orbit of the periodic solution is $\bar{u}_2, \bar{u}_3$.

The fixed points of the second iterate are all unstable because for $\beta = 2$ the absolute value of the slope of the tent map is everywhere greater than unity. Therefore, the tent map has an unstable periodic solution with period 2 besides the two fixed points $\bar{u} = 0$ and $\bar{u}_+$. It is easy to verify that for $\beta = 2$, the tent map has periodic orbits with arbitrary period m. Figure 10.22c shows a generic mth iterate. These periodic orbits also are unstable.

The set of all periodic points of the tent map with $\beta = 2$ is *dense* in the interval $(0, 2W_1)$ (e.g., Ott, 1993), that is, for any $0 < u < 2W_1$ and ε, no matter how small ε is, there is at least one periodic point in

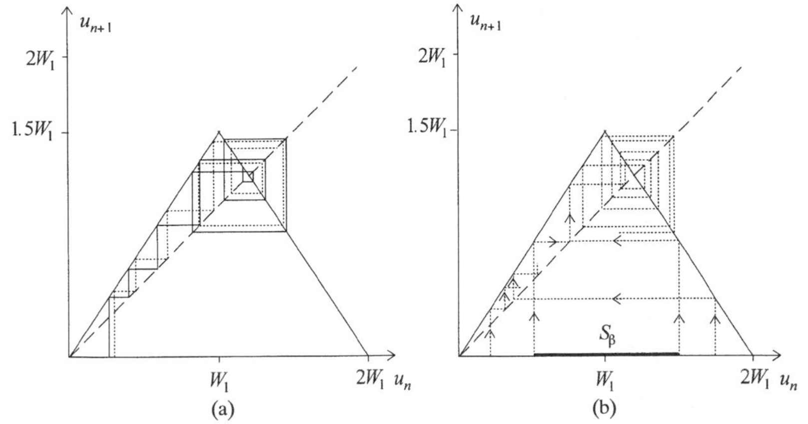

Figure 10.23. Stair-step diagrams for the tent map with $\beta = 3/2$: (a) sensitivity on the initial conditions; (b) attracting set S_β.

$(u - \varepsilon, u + \varepsilon)$. However, the set of all periodic points is a *countable* infinite set, while the set of points of $(0, 2W_1)$ is uncountable. Therefore, randomly chosen initial conditions do not produce periodic orbits.

For $1 < \beta < 2$ there are still periodic orbits. They are all unstable for the same reasons as seen for the case $\beta = 2$.

The tent map for $1 < \beta \leqslant 2$ has no periodic attracting sets because all the periodic orbits are unstable. If the attracting set is neither a fixed point nor a periodic orbit, what is the motion of the trajectory on the interval $(0, 2W_1)$?

You should convince yourself, by using either the stair-step diagram or a computer simulator, that the orbits starting in the interval $(0, 2W_1)$ remain bounded in the interval $(0, 2W_1)$ and, more importantly, that two orbits starting close to each other, say, on the left of $u = W_1$, diverge until they are "folded back" by mapping from the right-hand side of the map (Fig. 10.23a). Because we have $|df/du| > 1$ for $1 < \beta \leqslant 2$ on the whole interval $(0, 2W_1)$, the orbits with close initial conditions diverge exponentially.

The motion of the trajectory is mixing because f maps the interval $(0, 2W_1)$ onto itself. Any orbit starting from the interval $(0, 2W_1)$ eventually moves toward the interval $S_\beta = [(2 - \beta)W_1, \beta W_1]$ and all the orbits starting inside the interval S_β remain there because the tent map maps the interval S_β onto itself (Fig. 10.23b). Therefore, the interval S_β is an attractive set for all the orbits starting from the interval $(0, 2W_1)$. Finally, any orbit that enters the interval S_β is

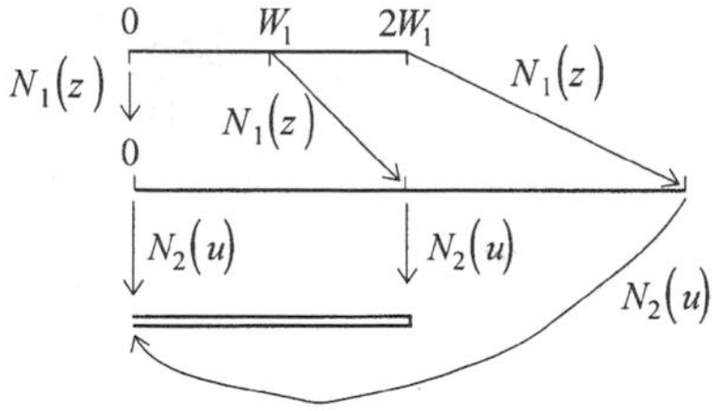

Figure 10.24. Stretching and back-folding mechanism of the tent map for $\beta = 2$.

locally unstable, follows an erratic oscillation that always stays inside S_β, and never repeats itself. Such solutions and the attractor are called *chaotic*.

Let us refer, for example, to the case $\beta = 2$ for which S_β is the entire interval $(0, 2W_1)$, and let us assume that $\Gamma_- = 2$ and $\Gamma_d = 1$. Figure 10.24 shows the action of the map $f(u) = N_1(N_2(u))$ on the interval S_β as consisting of two steps. In the first step the interval $(0, 2W_1)$ is uniformly stretched twice its original length. In the second step, the stretched interval is folded on itself so that the segment is now contained in the original interval. The *stretching* leads to exponential divergence of nearby trajectories. The *folding* process, which keeps the orbit inside S_β, is a direct consequence of the noninvertibility of the tent map.

Remark

The stretching and back-folding mechanisms are not only a way to describe the action of the tent map, but they have an evident physical meaning (see Fig. 10.25). The stretching is given by the resistor $\mathscr{R}_2$ operating in the locally active region, and the back folding is given by the nonlinear passive and locally passive resistor $\mathscr{R}_1$. ◇

Other examples of piecewise-linear maps may be found in De Menna and Miano (1994); and Corti, Miano, and Verolino (1996).

10.7.2. The Lyapunov Exponent and Chaotic Transients

We have already noted that adjacent points become separated under the action of the map for $\beta > 1$. The *Lyapunov exponent* (e.g., Schuster, 1988; Ott, 1993) measures this exponential separation. Let

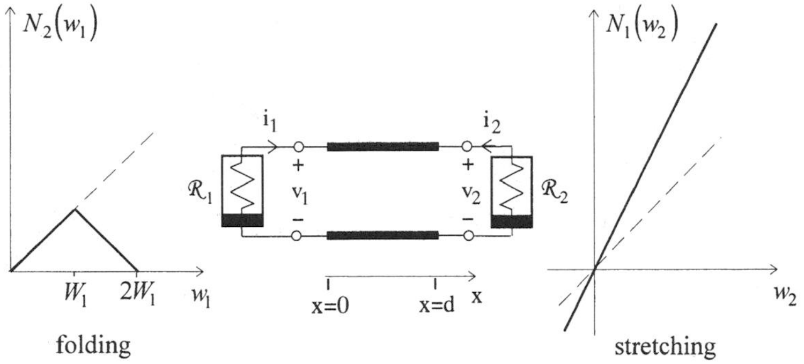

Figure 10.25. Physical interpretation of the action of the tent map.

us consider two orbits with nearby initial conditions, as shown in Fig. 10.26, and let us introduce the parameter $\lambda(u_0)$ such that

$$\varepsilon \exp[n\lambda(u_0)] = |f^n(u_0 + \varepsilon) - f^n(u_0)|. \tag{10.97}$$

In the limits $\varepsilon \to 0$ and $n \to \infty$ the definition (10.97) leads to

$$\lambda(u_0) = \lim_{n\to\infty} \frac{1}{n} ln \left|\frac{df^n}{du}\right|_{u=u_0} = \lim_{n\to\infty} \frac{1}{n} \sum_{i=0}^{n-1} ln|f'(u_i)|. \tag{10.98}$$

The quantity $e^{\lambda(u_0)}$ represents the average factor by which the distance between closely adjacent orbits becomes stretched after one iteration. For the tent map, the Lyapunov exponent is independent of the initial condition u_0 and its expression is (Schuster, 1988)

$$\lambda = ln(\beta). \tag{10.99}$$

In this case the average stretching factor $e^{\lambda(u_0)}$ is equal to β. The Lyapunov exponent changes its sign at $\beta = 1$; for $\beta < 1$ it is negative and all the orbits are asymptotically stable, whereas for $\beta > 1$ it is positive and nearby orbits locally diverge.

Any orbit of the tent map returns to a given neighborhood of any point of S_β an infinite number of times for $n \to \infty$. The irregular

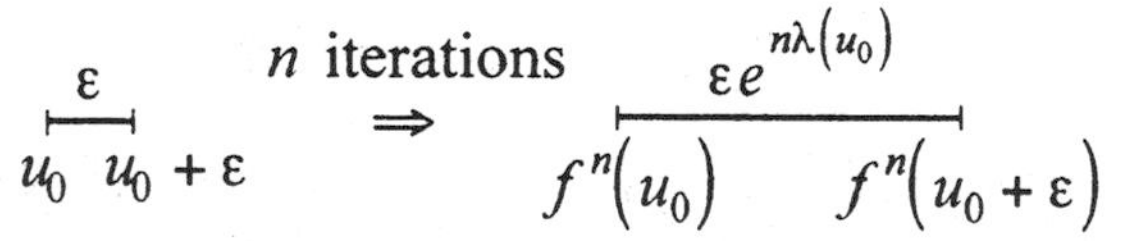

Figure 10.26. The local exponential separation of two nearby orbits.

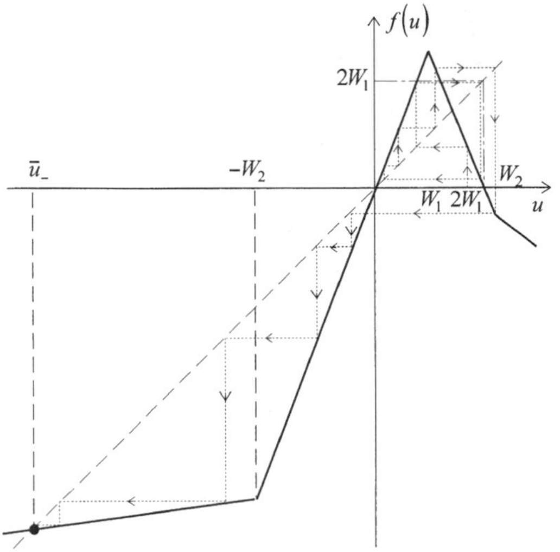

Figure 10.27. Stair-step diagram of the map *f* given by Eq. (10.92) for $\beta > 2$.

sequence of iterates $u_0, u_1, u_2, \ldots$, uniformly covers the interval S_β for $n \to \infty$ and the correlation function, measuring the correlation between iterates that are m steps apart, is a delta function (e.g., Schuster, 1988). Thus the chaotic attractor of the tent map is the whole interval S_β.

What happens for $\beta > 2$? When $\beta > 2$ the orbits starting from $(0, 2W_1)$ eventually escape from it, and hence move off the asymptotically stable fixed point $\bar{u}_-$ (Fig. 10.27). In such cases we may have chaotic transients (e.g., Ott, 1993). Let us consider an orbit starting inside the interval $(0, 2W_1)$ when $\beta = 2 + \varepsilon$, where ε is an arbitrarily small positive number. The term "chaotic transient" refers to the fact that this orbit may spend a long time inside $(0, 2W_1)$ before it leaves and moves off to the asymptotically stable fixed point $\bar{u}_-$ because the interval $(0, 2W_1)$ is the chaotic attractor of all the orbits starting from $(0, 2W_1)$ when $\beta = 2$.

10.7.3. A Unimodal Map

Now we shall deal with another interesting example of a chaotic circuit. Let us consider the case in which the resistor $\mathscr{R}_1$ is a *p-n* (exponential) diode, connected in parallel with a linear resistor with resistance R_1:

$$g_1(v_1) = I_s[\exp(v_1/(\eta V_T)) - 1] + v_1/R_1; \tag{10.100}$$

here, I_s is the inverse saturation current, V_T is the volt-equivalent temperature, and η is the emission coefficient. The other port is connected to the locally active piecewise resistor whose characteristic is given by Eq. (10.78).

As far as the function $v_1 = F_1(w_1)$ is concerned, we observe that $F_1(0) = 0$ and that its second-order derivative is given

$$\frac{d^2F_1}{dw_1^2} = -\frac{R_c}{(1 + R_c dg_1/dv_1)^3}\frac{d^2g_1}{dv_1^2}. \tag{10.101}$$

Thus, in this case we have both $dF_1/dw_1 > 0$ and $d^2F_1/dw_1^2 < 0$. Finally, the asymptotic behavior of F_1 is given by

$$v_1 = F_1(w_1) \approx \frac{1}{1 + R_c/R_1}(w_1 + R_cI_s) \quad \text{for } w_1 \to -\infty, \tag{10.102}$$

$$v_1 = F_1(w_1) \approx \eta V_T \ln[w_1/(R_cI_s)] \quad \text{for } w_1 \to +\infty. \tag{10.103}$$

Figure 10.28 shows a qualitative behavior of F_1.

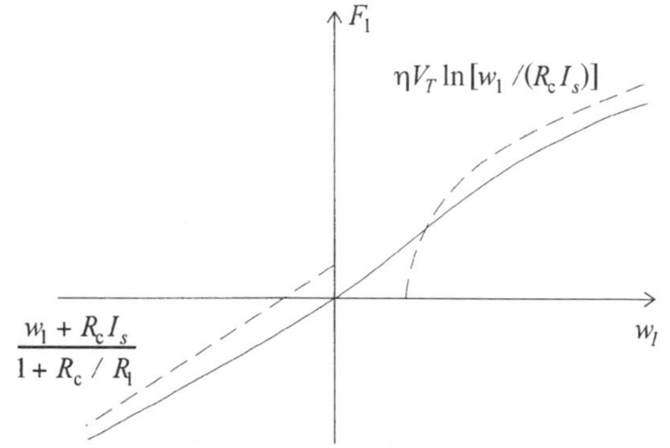

Figure 10.28. A qualitative behavior of the function $F_1(w_1)$ when the line is connected to the nonlinear resistor with characteristic equation (10.78) and the p-n diode.

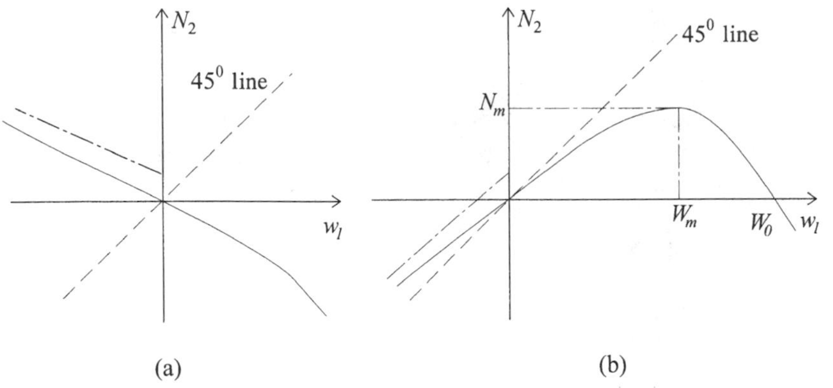

Figure 10.29. Qualitative behavior of the function $N_2(w_1)$: (a) $R_c > R_1$ and (b) $R_c < R_1$.

The first-order derivative of $N_2(w_1)$ is given by (see Eq. (10.71))

$$\frac{dN_2}{dw_1} = \left. \frac{1 - R_c\left(\dfrac{1}{R_1} + \dfrac{e^{v_1/\eta V_T}}{R_s}\right)}{1 + R_c\left(\dfrac{1}{R_1} + \dfrac{e^{v_1/\eta V_T}}{R_s}\right)} \right|_{v_1 = F_1(w_1)}, \tag{10.104}$$

where

$$R_s = \frac{\eta V_T}{I_s} \tag{10.105}$$

is the incremental resistance of the *pn*-junction diode at $v_1 = 0$. It is evident that for $R_c > R_1$, the first-order derivative of N_2 with respect to w_1 is always negative, and the function $N_2(w_1)$ is strictly decreasing (Fig. 10.29a).

By using the main properties of the function $F_1(w_1)$ we can determine the qualitative behavior of the function $N_2(w_1)$ for $R_c < R_1$. We have the following properties:

(a) $N_2(0) = 0, \quad N_2(W_0) = 0,$ (10.106)

(b) $\left.\dfrac{dN_2}{dw_1}\right|_{w_1 = W_m} = 0,$ (10.107)

$$\text{(c)}\ N_2(w_1) \approx \Gamma_1 w_1 + \frac{2R_c}{1 + R_c/R_1} I_s \quad \text{for } w_1 \to -\infty, \tag{10.108}$$

$$\text{(d)}\ N_2(w_1) \approx 2\eta V_T ln(w_1/R_c I_s) - w_1 \quad \text{for } w_1 \to +\infty, \tag{10.109}$$

where

$$W_0 = 2\eta V_T z^*(\beta), \quad W_m = \eta V_T \left(1 + \frac{R_c}{R_1}\right) ln\, \beta + R_c I_s (\beta - 1), \tag{10.110}$$

$$\beta = \frac{R_s}{R_c}\left(1 - \frac{R_c}{R_1}\right), \quad \Gamma_1 = \frac{R_1 - R_c}{R_1 + R_c}, \tag{10.111}$$

and $z^*(\beta)$ is the nontrivial solution of the nonlinear equation

$$e^z - \beta z = 1. \tag{10.112}$$

For $R_c < R_1$, it is $\beta > 0$ and $\Gamma_1 > 0$. Therefore, Eq. (10.112) admits two solutions, namely, $z = 0$ and $z^* > 0$. The first-order derivative of N_2 is zero at $w_1 = W_m$, where the function N_2 attains its maximum value

$$N_m = R_c I_s (\beta\, ln\, \beta - \beta + 1). \tag{10.113}$$

The function N_2 is strictly increasing for $w_1 < W_m$, and strictly decreasing for $w_1 > W_m$. Furthermore, the Schwarzian derivative (Singer, 1978)

$$S(N_2) = \frac{N_2'''(w_1)}{N_2'(w_1)} - \frac{3}{2}\left(\frac{N_2''(w_1)}{N_2'(w_1)}\right)^2, \tag{10.114}$$

is negative. This condition ensures that the function $\sqrt{|N_2'(w_1)|}$ is *convex*, and while this condition might seem to be quite unnatural, it has a very important role in the general theory of the scalar maps, as Singer has discovered (Singer, 1978).

We have obtained a very important result. By varying the resistance R_1 the shape of the function N_2 can be changed continuously, passing from an invertible function for $R_c > R_1$ (see Fig. 10.29a) to a function that is noninvertible for $R_c < R_1$ (see Fig. 10.29b). As we have already seen in the previous example, the noninvertibility of the map produces many interesting phenomena.

Let us now analyze in detail the map f that we obtain by composing the function N_2, when $R_c < R_1$, with a function $N_1(w_2)$ of the kind shown in Fig. 10.12a. Furthermore, let us assume that $W_2 > N_m$. In this manner, we can vary arbitrarily the slope of the

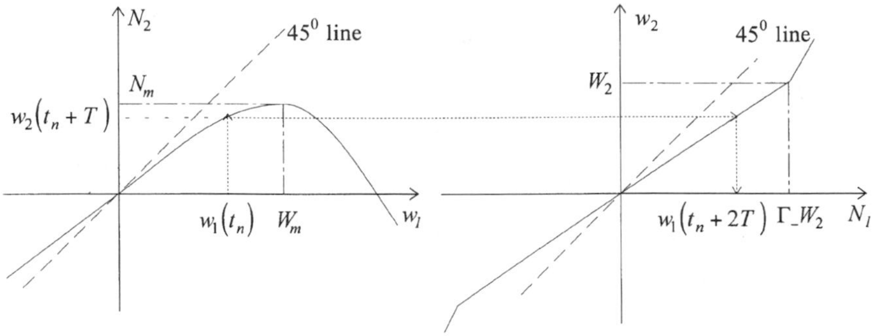

Figure 10.30. Composition of the functions N_1 and N_2.

noninvertible part of the map $f(u) = N_1(N_2(u))$. The "composition" that gives this map may be discussed graphically as in Fig. 10.30, where the graphs of the functions $N_1(w_2)$ and $N_2(w_1)$ are reported. Starting from $w_1(t_n)$ through the map N_2 we evaluate $w_2(t_n + T)$. By drawing the horizontal line starting from the point $[w_1(t_n), w_2(t_n + T)]$, we intersect the graph of the function $N_1(\cdot)$; this intersection gives $w_1(t + 2T)$.

The qualitative behavior of the map f obtained in this way is illustrated through the graphs shown in Figs. 10.31 to 10.33. The map f always has a fixed point at $u = 0$, a maximum at $u = W_m$, where its value is given by

$$F_m = \Gamma_- N_m, \tag{10.115}$$

and vanishes at $u = W_0$. Furthermore, for

$$\Gamma_- \leqslant \Gamma_c \equiv \frac{W_0}{N_m}, \tag{10.116}$$

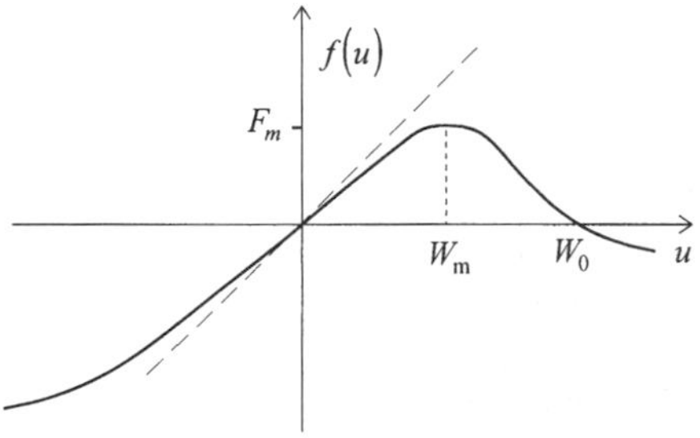

Figure 10.31. Plot of the map f, for $\Gamma_- < 1/\Gamma_0$.

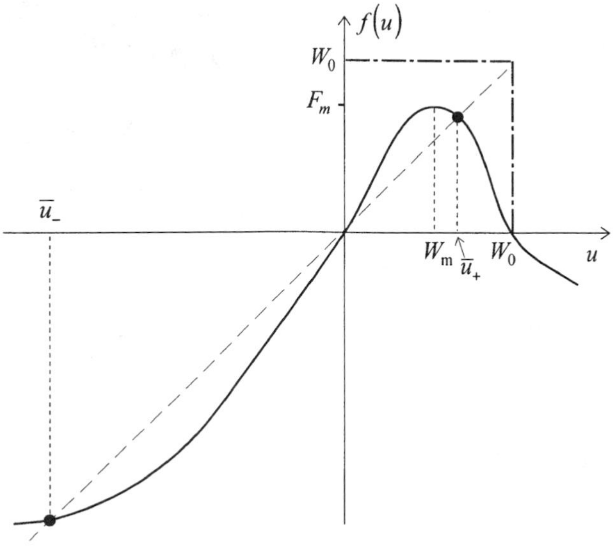

Figure 10.32. Plot of the map f for $1/\Gamma_0 < \Gamma_- < \Gamma_c$.

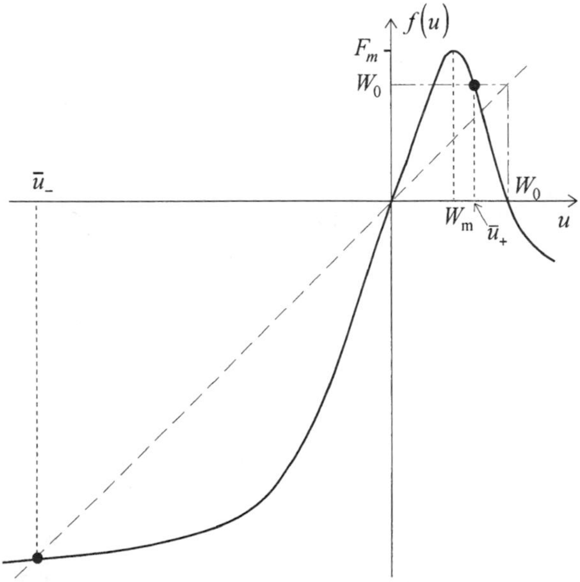

Figure 10.33. Plot of the map f for $\Gamma_- > \Gamma_c$.

f maps the interval $(0, W_0)$ onto itself, and hence any orbit starting inside $(0, W_0)$ remains there. The first derivative of $N_2(u)$ at $u = 0$ is given by

$$N_2'(0) = \Gamma_0 = \frac{1 - R_c\left(\dfrac{1}{R_1} + \dfrac{1}{R_s}\right)}{1 + R_c\left(\dfrac{1}{R_1} + \dfrac{1}{R_s}\right)}. \tag{10.117}$$

Note that $\Gamma_c > 1/\Gamma_0$.

For $\Gamma_- < 1/\Gamma_0$, the map f has only one fixed point $\bar{u} = 0$, which is globally asymptotically stable (Fig. 10.31). At $\Gamma_- = 1/\Gamma_0$ the fixed point $\bar{u} = 0$ undergoes a pitchfork bifurcation. For $\Gamma_- > 1/\Gamma_0$ the map f has three fixed points $\bar{u} = 0, \bar{u}_-$, and $\bar{u}_+$ (Fig. 10.32); the fixed point $\bar{u} = 0$ is always unstable, and the fixed point $\bar{u}_-$ is always asymptotically stable, whereas the fixed point $\bar{u}_+$ is asymptotically stable in a certain interval $(1/\Gamma_0, \Gamma')$: for $\Gamma_- = \Gamma'$. For $\Gamma_- > \Gamma'$, period-doubling bifurcations, chaotic dynamics and other phenomena occur, as we shall see soon.

For $\Gamma_- > 1/\Gamma_0$, all the orbits with $u_0 < 0$ or $u_0 > W_0$ converge asymptotically toward $\bar{u}_-$ (see Figs. 10.32 and 10.33). The behavior of the dynamics for $0 < u_0 < W_0$ may be quite different depending on the value of Γ_-. If $\Gamma_- \leqslant \Gamma_c$ any orbit starting inside $(0, W_0)$ remains there, and it may converge to the fixed point $\bar{u}_+$, to a periodic attractor or to a chaotic attractor depending on the value of Γ_-, as we shall see soon. By contrast, for $\Gamma_- > \Gamma_c$, orbits starting inside the interval $(0, W_0)$ may eventually escape from it and hence move off to the fixed point $\bar{u}_-$ (Fig. 10.33). As with the tent map, chaotic transient phenomena may appear for $\Gamma_- = \Gamma_c + \varepsilon$.

10.7.4. The Logistic Map

Let us now study in detail the qualitative behavior of the dynamics of f when $\Gamma_- \leqslant \Gamma_c$ and $0 < u < W_0$. The map f has only one maximum at $u = W_m$, is monotonically decreasing for $u < W_m$, has a negative Schwarzian derivative for $0 < u < W_0$, and maps $(0, W_0)$ onto itself. A map that has these properties is called a *unimodal map*. Consequently, the dynamics of f for $0 < u < W_0$ and $\Gamma_- \leqslant \Gamma_c$ exhibit the same qualitative behavior as that of the *logistic map* (e.g., Schuster, 1988; Thompson and Stewart, 1986)

$$z_{n+1} = F(z_n) = rz_n(1 - z_n), \tag{10.118}$$

when the parameter r varies in the interval $(0, 4)$ and $0 \leqslant z_0 \leqslant 1$.

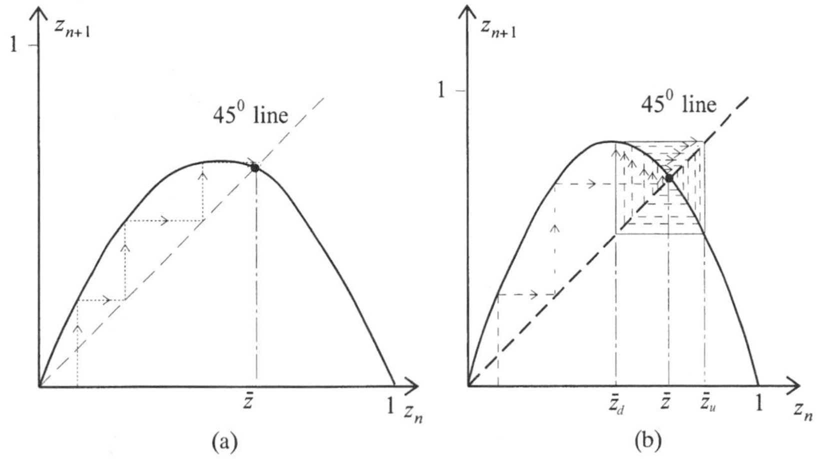

Figure 10.34. Stair-step diagrams of the logistic map for (a) $1 < r < 3$ and for (b) $3 < r < r_1$.

This map was introduced by Verhulst in 1845 to describe the growth of a population in a closed area (e.g., Peitgen and Richter, 1986). In 1976, May found that the orbits of the map (10.118) display, as a function of the parameter r, a rather complicated behavior that becomes chaotic at a critical value of r (May, 1976). Peitegen and Richter proposed using the logistic map to simulate saving accounts with a self-limiting rate of interest (e.g., Schuster, 1988).

The logistic map maps the unit interval $(0, 1)$ onto itself. For $0 < r < 1$ it has only one fixed point $\bar{z} = 0$, which is always asymptotically stable. The interval $(0, 1)$ is the basin of attraction of $\bar{z} = 0$. At $r = 1$, when the slope of $F(z)$ at the fixed point $\bar{z} = 0$ is equal to -1, the equilibrium bifurcates by pitchfork bifurcation, and a new asymptotically stable fixed point $\bar{z} = 1 - 1/r$ arises (Fig. 10.34a). This fixed point is asymptotically stable for $1 < r < r_0 = 3$. All the orbits starting in the interval $(0, 1)$ converge towards this fixed point $\bar{z}$. At the critical value r_0, the slope of $F(z)$ at the fixed point $\bar{z}$ equals -1, the equilibrium bifurcates by flip or period-doubling bifurcation, and a unique and stable periodic orbit with period 2 arises (Fig. 10.34b). Figure 10.35 shows the second iterate of the logistic map for (a) $r < r_0$ and (b) $r > r_0$.

As r increases further at a certain value $r_1 > r_0$, the orbit with period 2 becomes unstable (all the fixed points of the second iterate become unstable). As r crosses the value r_1 the period-2 orbit

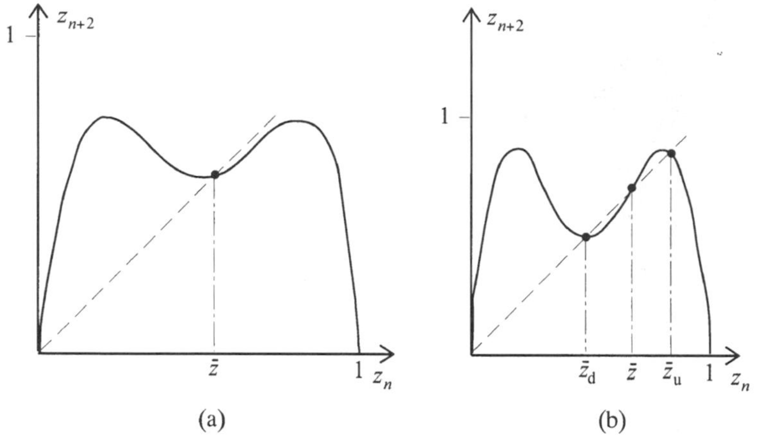

Figure 10.35. The second iterate of the logistic map for (a) $r < 3$ and (b) $3 < r < r_1$.

bifurcates to a period-4 orbit; each period point of the second iterate map undergoes a flip bifurcation.

Increasing r from r_1 to a certain value r_2 the period-4 orbit becomes unstable, and a stable period-8 orbit appears and remains stable for a range $r_2 < r \leqslant r_3$. This process of period doubling continues, successively producing an infinite cascade of periodic orbits. In general, let us indicate with $r_{m-1} < r \leqslant r_m$ the range in which the periodic orbit with period 2^m is stable. There is an accumulation point of an infinite number of period-doubling bifurcations at finite r value, denoted by r_∞:

$$r_\infty \equiv \lim_{m \to \infty} r_m = 3.5699456\ldots. \tag{10.119}$$

This phenomenon is called the *Feigenbaum cascade of period doubling*, and is a universal property of the unimodal maps in the sense that it occurs not only in the logistic map, but in any unimodal map. The values r_m, where the number of periodic points changes from 2^{m-1} to 2^m, scale as

$$r_m = r_\infty - \text{constant}\,\delta^{-m} \quad \text{for } m \gg 1. \tag{10.120}$$

The distances d_m between the points in a 2^m-cycle that are closest to

$z = 1/2$ have constant ratios:

$$\frac{d_m}{d_{m+1}} = -\alpha \quad \text{for } m \gg 1. \tag{10.121}$$

The Feigenbaum constants α and δ do not depend on the particular unimodal map and have values

$$\delta = 4.6692016091\ldots, \quad \alpha = 2.5029078750\ldots. \tag{10.122}$$

What happens beyond $r = r_\infty$, for $r_\infty < r \leqslant 4$?

Chaotic dynamics arise, as in the tent map. Any orbit starting from the interval $(0, 1)$ eventually moves toward the interval $S_r = [1 - r/4, r/4]$ and all the orbits starting inside the interval S_r remain there because the logistic map maps the interval S_r onto itself. Therefore, the interval S_r is an attractive set for all the orbits starting from the interval $(0, 1)$. Any orbit that enters the interval S_r is locally unstable, follows an erratic oscillation that always stays inside S_r, and never repeats itself. The "chaotic regime" is interrupted by r-windows where the asymptotic behavior of the orbits is again periodic.

Figure 10.36a shows the accumulation of 100 points of the iterates for $n > 400$ as a function of r together with the Lyapunov exponent λ,

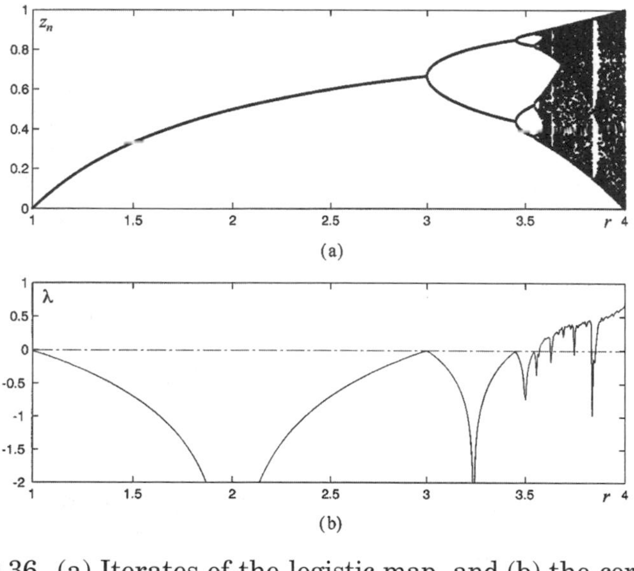

Figure 10.36. (a) Iterates of the logistic map, and (b) the corresponding Lyapunov exponent.

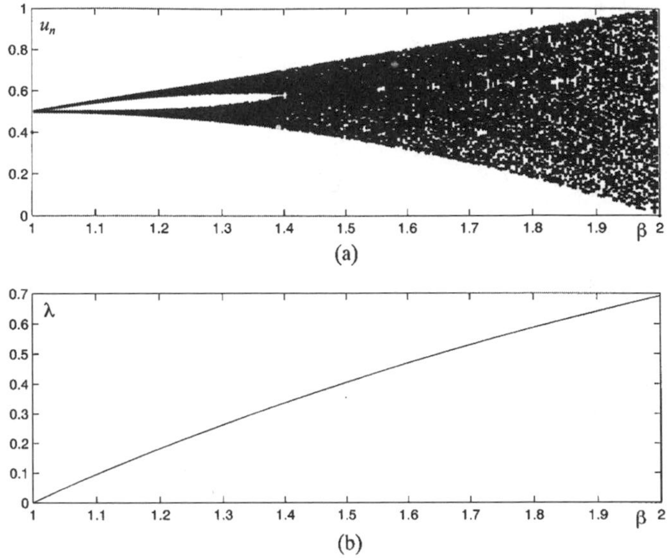

Figure 10.37. (a) Iterates of the tent map, and (b) the corresponding Lyapunov exponent.

Fig. 10.36b, obtained via Eq. (10.98). For $0 < r < r_\infty$, where only periodic attractors exist, the Lyapunov exponent is always negative: It becomes zero at the flip bifurcation points. However, for $r_\infty < r < 4$, λ is mostly positive, indicating chaotic behavior. The chaotic behavior is interrupted by narrow "windows" where the solution again converges to periodic attractors. Note that an asymptotically stable periodic solution with period 3 arises for $r_\infty < r < 4$. These solutions originate from another type of bifurcation mechanism called *tangent bifurcation* (for details, see Ott, 1993).

Unlike the tent map, the density of iterates of the logistic map for $n \to \infty$ is not constant on the attracting set S_r; it is higher at the ends of the interval S_r than in the middle (e.g., Schuster, 1988).

Remark

Unlike the unimodal maps, the tent map is characterized by an abrupt transition to the chaos (Fig. 10.37a); in Fig. 10.37b the behavior of the corresponding Lyapunov exponent is shown (see Eq. (10.99)). The abrupt transition is a consequence of the fact that the tent map has no stable periodic orbits. In general, the behavior of noninvertible piecewise-linear maps is strongly influenced by the number of segments; with two segments the transition to the chaos is abrupt and period doubling bifurcation points are absent, while with more segments, period doubling bifurcation points appear and their number rises with the number of segments (Corti *et al.*, 1994). ◇

More details on the unimodal maps relevant to transmission lines terminated with *pn*-junction diodes may be found in (Corti *et al.*, 1994; Corti, Miano, and Verolino, 1996).

10.7.5. The Spatio-Temporal Chaos

Once we have evaluated the orbit of $u_0(t_0)$, namely, $u_1(t_0)$, $u_2(t_0), \ldots, u_n(t_0), \ldots,$ for any value of t_0 belonging to the interval $(0, 2T)$, we can determine the spatial profiles of the backward and forward voltage waves $v^-(t + x/c)$ and $v^+(t + T - x/c)$ at any time instant. For instance, we have for the backward voltage wave

$$v^-[2nT + x/c] = \tfrac{1}{2}u_n(x/c). \tag{10.123}$$

The forward voltage wave may be obtained by using Eq. (10.14).

When the map has only one fixed point that is globally asymptotically stable, the spatial profile of the backward voltage wave tends to be uniform for $t \to \infty$. However, if the map has two asymptotically stable fixed points u_+ and u_-, then the spatial profile tends to a piecewise-constant spatial distribution for $t \to \infty$ that assumes the values u_+ and u_-. When the map has an asymptotically stable periodic-$2m$ orbit, the spatial profile again tends to a piecewise constant spatial distribution for $t \to \infty$ characterized by $2m$ stable values. When the map has a chaotic attractor, the backward voltage wave tends to a spatial profile that is highly irregular; transition to chaotic behavior in the spatial profile is observed. All these behaviors are shown in Fig. 10.38 for the tent map and in Fig. 10.39 for the logistic map. This is an example of the so-called *spatio-temporal chaos* (Moon, 1992; De Menna, and Miano, 1994; Corti *et al.*, 1994).

10.8. LOSSY TRANSMISSION LINES

In the previous sections it has been shown that bifurcations, self-oscillations, period doubling and chaos can arise in ideal two-conductor transmission lines terminated with nonlinear resistors.

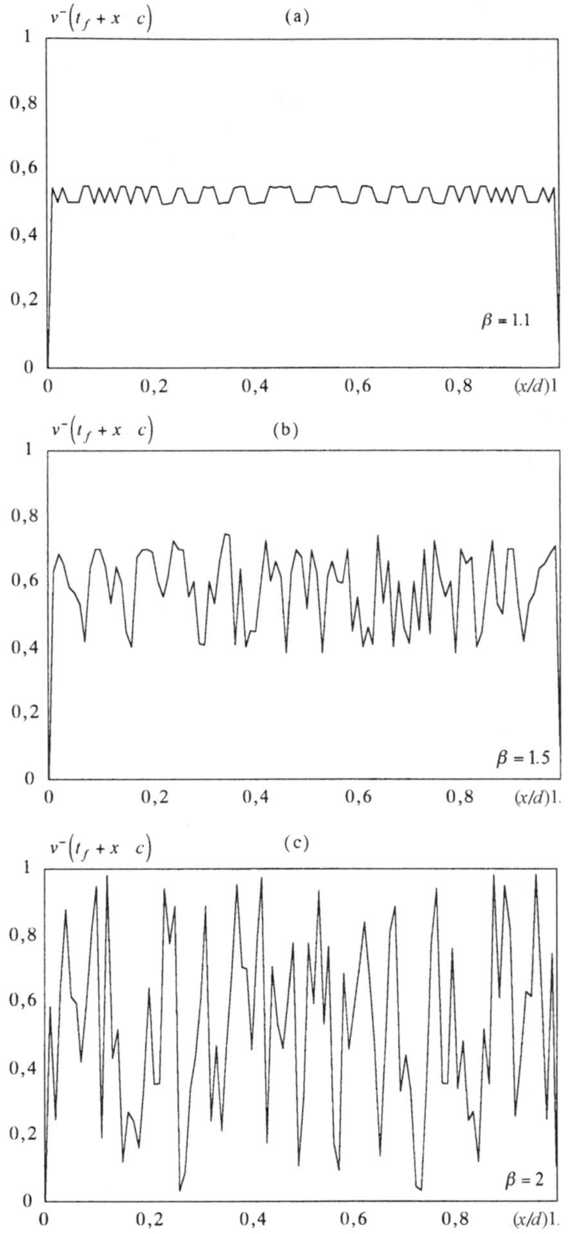

Figure 10.38. Spatial profile of the backward wave at $t_f = 1001T$ generated by the tent map for different values of β; the initial profile is a ramp.

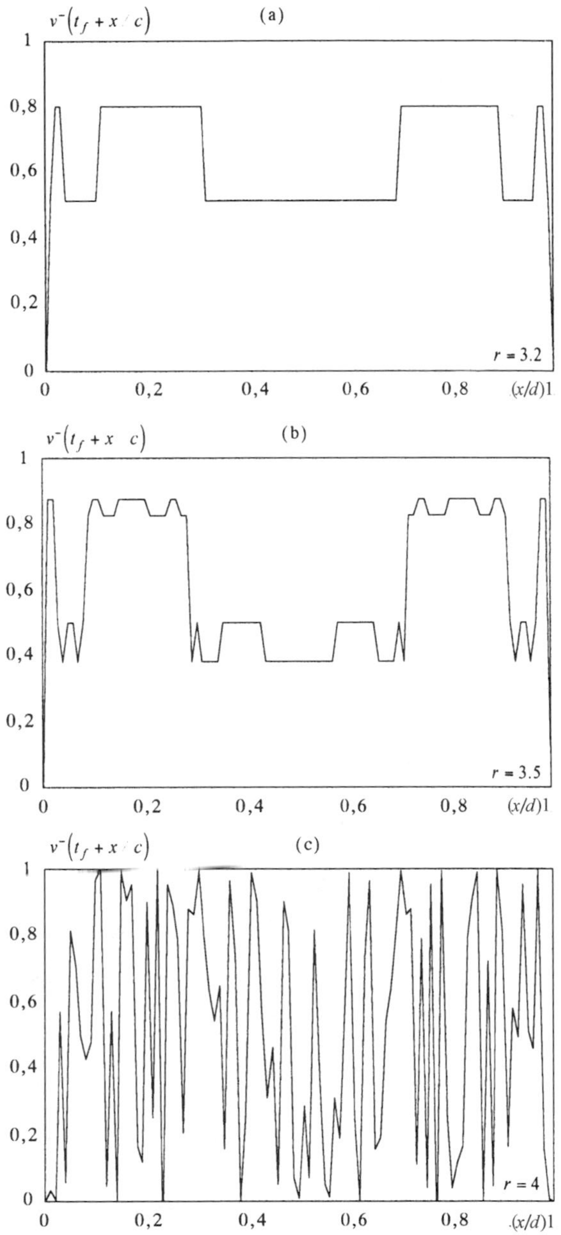

Figure 10.39. Spatial profile of the backward wave at $t_f = 1001T$ generated by the logistic map for different values of r; the initial profile is a ramp.

What happens when we consider the effects of the losses along the line?

The state equations of the circuit of Fig. 10.1 for lossy lines with frequency-independent parameters are (see Chapter 4)

$$i_1(t) + g_1[v_1(t); t] = 0, \tag{10.124}$$

$$v_1(t) - R_c i_1(t) = \int_0^t z_{cr}(t-\tau) i_1(\tau) d\tau + w_1(t), \tag{10.125}$$

$$i_2(t) + g_2[v_2(t); t] = 0, \tag{10.126}$$

$$v_2(t) - R_c i_2(t) = \int_0^t z_{cr}(t-\tau) i_2(\tau) d\tau + w_2(t), \tag{10.127}$$

$$w_1(t+T) = e^{-\mu T}[2v_2(t) - w_2(t)] + \int_0^t p_r(t-T-\tau)[2v_2(\tau) - w_2(\tau)] d\tau, \tag{10.128}$$

$$w_2(t+T) = e^{-\mu T}[2v_1(t) - w_1(t)] + \int_0^t p_r(t-T-\tau)[2v_1(\tau) - w_1(\tau)] d\tau, \tag{10.129}$$

where the parameter μ is given by Eq. (4.6), $z_{cr}(t)$ is given by Eq. (4.58) and $p_r(t)$ is given by Eq. (4.95). The losses along the line give rise to two different phenomena, namely, a simple damping of the signals and a "wake" (see Chapter 4).

When the Heaviside condition is satisfied $\nu = 0$, both $z_{cr}(t)$ and $p_r(t)$ are equal to zero (see Chapter 4), and the system of equations (10.124) to (10.129) reduces to the system of equations (10.1) to (10.6) except for the damping factor $\exp(-\mu T)$. Therefore, all the results we have described in the previous two sections hold provided that the functions N_1 and N_2, defined in Eqs. (10.15) and (10.16), respectively, are multiplied by the damping factor $\exp(-\mu T)$.

In the more general case of $\nu \neq 0$ the state equations (10.124) to (10.129) are a system of nonlinear difference-convolution equations with one delay. The state at the generic time instant t depends both on its value at the time instant $t - T$ and on its whole time history in the interval $(0, t - T)$ because of the wake generated by the dispersion in time due to the losses.

Equations (10.124) to (10.129) for $\nu \neq 0$ can not be solved analytically. Preliminary results obtained by solving them numerically show that bifurcations, self-oscillations, period-doubling and chaos survive also in transmission lines with pronounced losses (Corti *et al.*, 1998). This seems to be very promising when we are looking for these phenomena in actual transmission lines. Nevertheless, an accurate study of the qualitative behavior of the solutions of the equations (10.124) to (10.129) is needed to understand clearly the action of the wake introduced by the dispersion on the steady-state behavior of the circuit.

Appendix A

Some Useful Notes on the Matrix Operators

In this Appendix, we shall recall some basic concepts and properties of the matrix operators that we have used in this book. The reader is referred to the many excellent books existing in the literature for a thorough and comprehensive treatment of this subject (e.g., Korn and Korn, 1961; Horn and Johnson, 1985).

A.1. PRELIMINARY DEFINITIONS

Let **R** and **C** denote, respectively, the set of the real and complex numbers. Let A_{ij} denote the (i,j) entry of a square $n \times n$ matrix A supposed to be real. The matrix A defines a linear transformation of the vector space $\mathbf{C}^n$ into itself:

$$\mathbf{y} = \mathrm{A}\mathbf{x}, \quad \mathbf{x}, \mathbf{y} \in \mathbf{C}^n. \tag{A.1}$$

The matrix A is said to be *positive definite* if

$$\mathbf{x}^{*\mathrm{T}}\mathrm{A}\mathbf{x} > 0 \quad \text{for any } \mathbf{x} \neq \mathbf{0} \in \mathbf{C}^n, \tag{A.2}$$

where $\mathbf{x}^*$ indicates the conjugates of $\mathbf{x}$.

The *transpose* of the matrix A is the square matrix A^T whose generic entry is given by

$$\mathrm{A}^\mathrm{T}_{ij} = \mathrm{A}_{ji}, \quad i, j = 1, \ldots, n. \tag{A.3}$$

The matrix A is *symmetric* if $\mathrm{A} = \mathrm{A}^\mathrm{T}$.

The *inverse* of a square $n \times n$ matrix A is the matrix A^{-1} such that

$$AA^{-1} = I, \tag{A.4}$$

where I is the identity $n \times n$ matrix. The matrix A^{-1} exists iff $\det(A) \neq 0$. A symmetric and positive definite matrix is invertible.

Let us indicate with an apex one of the two preceding matrix operations. The following reverse-order law holds:

$$(AB)' = B'A'. \tag{A.5}$$

Note that, in general, the matrix product does not commutate, namely

$$AB \neq BA, \tag{A.6}$$

even if A and B are symmetric. In general, the product of two symmetric matrices gives rise to a matrix that is not symmetric.

A.2. THE EIGENVALUE PROBLEM $\mathbf{Au} = \lambda\mathbf{u}$

Let us consider the equation

$$A\mathbf{u} = \lambda\mathbf{u}, \quad \lambda \in \mathbf{C}. \tag{A.7}$$

If a scalar λ and nonzero $\mathbf{u}$ satisfy this equation, they are called, respectively, eigenvalue of A and eigenvector of A associated with λ. The eigenvalues are the roots of the characteristic polynomial

$$p_A(\lambda) = \det(\lambda I - A). \tag{A.8}$$

As the polynomial $p_A(\lambda)$ has degree n, the matrix A has n eigenvalues, counting multiplicities.

Let us indicate with $\lambda_1, \lambda_2, \ldots, \lambda_n$ the n eigenvalues of A and let us assume that there exist n linearly independent eigenvectors $\mathbf{u}_1, \mathbf{u}_2, \ldots, \mathbf{u}_n$ ($\mathbf{u}_h$ is the eigenvector corresponding to the eigenvalue λ_h). Thus, the matrix

$$T = |\mathbf{u}_1 \vdots \mathbf{u}_2 \vdots \cdots \vdots \mathbf{u}_n| \tag{A.9}$$

is invertible and is the similarity transformation that diagonalizes A,

$$T^{-1}AT = \operatorname{diag}(\lambda_1, \lambda_2, \ldots, \lambda_n). \tag{A.10}$$

The matrix T is called *modal* matrix of A.

It is well known that, if the eigenvalues of a matrix are all distinct, the corresponding eigenvectors are all linearly independent. In reality, this is a sufficient but not necessary condition for the eigenvectors to be independent. The algebraic multiplicity m_j of the

one, whatever the algebraic multiplicity of each of them is.

Let us consider the generic eigenvalue λ_h of A and the corresponding eigenvector $\mathbf{u}_h$. It is immediate that

$$\lambda_h = \frac{\mathbf{u}_h^{\mathrm{T}} \mathrm{A} \mathbf{u}_h}{\mathbf{u}_h^{\mathrm{T}} \mathbf{u}_h}. \tag{A.24}$$

Therefore, all the eigenvalues are positive if the matrix A is positive definite. □

A.3. THE GENERALIZED EIGENVALUE PROBLEM Au = λBu

Let us consider the *generalized eigenvalue problem*

$$\mathrm{A}\mathbf{u} = \lambda \mathrm{B}\mathbf{u}, \tag{A.25}$$

where A and B are symmetric real matrices and B is positive definite; the matrix B is invertible because of its properties. It is evident that all the eigenvalues and eigenvectors of the matrix D defined as

$$\mathrm{D} = \mathrm{B}^{-1}\mathrm{A} \tag{A.26}$$

are solutions of the generalized eigenvalue problem (A.25) and vice versa. Unfortunately, in general, the matrix D is not symmetric, even if it is given by the product of two symmetric matrices, and hence, we cannot apply the results obtained in the previous section.

The matrix B is symmetric and positive definite, therefore the bilinear form

$$\langle \mathbf{a}^*, \mathbf{b} \rangle = \mathbf{a}^{*\mathrm{T}} \mathrm{B} \mathbf{b} \tag{A.27}$$

defines a scalar product in $\mathbf{C}^n$.

The following properties hold for the solutions of the generalized eigenvalue problem (A.25) (Korn and Korn, 1961):

(i) all the eigenvalues λ and eigenvectors $\mathbf{u}$ are real;
(ii) the eigenvectors $\mathbf{u}_i$ and $\mathbf{u}_j$ corresponding to two different eigenvalues $\lambda_i \neq \lambda_j$, are mutually orthogonal with respect to the scalar product $\langle \mathbf{a}, \mathbf{b} \rangle$ defined by Eq. (A.27), namely

$$\langle \mathbf{u}_i, \mathbf{u}_j \rangle = \mathbf{u}_i^{\mathrm{T}} \mathrm{B} \mathbf{u}_j = 0; \tag{A.28}$$

and

(iii) for each eigenvalue λ_i with algebraic multiplicity m_i there are exactly m_i linearly independent eigenvectors.

If, in addition, A is positive definite, then:

(iv) the eigenvalues are positive.

PROOF. The properties (i), (ii), and (iv) can be demonstrated by reasoning in the same way as was done for thc properties (i), (ii), and (iv) of the eigenvalue problem $A\mathbf{u} = \lambda\mathbf{u}$ dealt with in the previous section. To demonstrate property (iii) we have to proceed in a slightly different way.

Let us again assume that $\lambda_1, \lambda_2, \ldots, \lambda_{n-3}$ are distinct and λ_{n-2}, λ_{n-1}, λ_n are coincident, and let us indicate with $\mathbf{u}_{n-2}$ an eigenvector corresponding to the eigenvalue $\lambda_{n-2} = \lambda_{n-1} = \lambda_n$. The eigenvectors $\mathbf{u}_1$, $\mathbf{u}_2, \ldots, \mathbf{u}_{n-3}$, $\mathbf{u}_{n-2}$ are orthogonal between them with respect to the scalar product $\langle \mathbf{a}, \mathbf{b} \rangle$.

Let us indicate with S_{n-2} the subspace of $\mathbf{R}^n$ spanned by the set of eigenvectors $\mathbf{u}_1, \mathbf{u}_2, \ldots, \mathbf{u}_{n-2}$ and with S_2 the subspace of dimension 2 of $\mathbf{R}^n$ orthogonal S_{n-2} with respect to the scalar product $\langle \mathbf{a}, \mathbf{b} \rangle$. Let $\mathbf{v}$ be a generic element of S_2 and $\mathbf{w}$ a generic element of S_{n-2},

$$\mathbf{w} = \sum_{i=1}^{n-2} a_i \mathbf{u}_i. \tag{A.29}$$

Since the matrix A is symmetric, the matrix D given by Eq. (A.26) maps any element of S_2 into itself. In fact, due to the symmetry of A we have

$$\langle \mathbf{w}, D\mathbf{v} \rangle = \mathbf{w}^T BD\mathbf{v} = \mathbf{w}^T A\mathbf{v} = \mathbf{v}^T A\mathbf{w} = \sum_{i=1}^{n-2} a_i \lambda_i \langle \mathbf{v}, \mathbf{u}_i \rangle = 0. \tag{A.30}$$

At this point we can proceed in the same manner as was done for the eigenvalue problem $A\mathbf{u} = \lambda\mathbf{u}$ dealt with in the previous section. □

A.4. FUNCTIONS OF A MATRIX OPERATOR

The most general definition of the function $f(M)$ of a square $n \times n$ matrix operator M is based on the spectral decomposition of M (e.g., Lancaster and Tismenetsky, 1985). Provided that the matrix M has n linearly independent eigenvectors $\mathbf{M}_1, \mathbf{M}_2, \ldots, \mathbf{M}_n$, it may be expressed as

$$M = T\,\mathrm{diag}(m_1, m_2, \ldots, m_n)T^{-1} \tag{A.31}$$

where $m_1, m_2, \ldots, m_n$ are the eigenvalues of M, and

$$T = |\mathbf{M}_1 \vdots \mathbf{M}_2 \vdots \cdots \vdots \mathbf{M}_n|$$

is the corresponding modal matrix. Then the function $f(\mathrm{M})$ is defined as

$$f(\mathrm{M}) = \mathrm{T}\,\mathrm{diag}(f(m_1)\quad f(m_2), \ldots, f(m_n))\mathrm{T}^{-1}. \tag{A.32}$$

The matrix $f(\mathrm{M})$ so defined has the same eigenvectors as those of the matrix M, and its eigenvalues are $f(m_1)$, $f(m_2), \ldots, f(m_n)$. It is diagonalizable by the same similarity transformation that diagonalizes M. For example, for the exponential function we have

$$\exp(\mathrm{M}) = \mathrm{T}\,\mathrm{diag}(e^{m_1}, e^{m_2}, \ldots, e^{m_n})\mathrm{T}^{-1}, \tag{A.33}$$

and for the square root function we have

$$\sqrt{\mathrm{M}} = \mathrm{T}\,\mathrm{diag}(\sqrt{m_1}, \sqrt{m_2}, \ldots, \sqrt{m_n})\mathrm{T}^{-1}. \tag{A.34}$$

The properties

$$\sqrt{\mathrm{M}}\sqrt{\mathrm{M}} = \mathrm{M}, \tag{A.35}$$

$$(\sqrt{\mathrm{M}})^{-1} = \sqrt{\mathrm{M}^{-1}}, \tag{A.36}$$

$$\frac{d}{dx}\exp(\mathrm{Mx}) = \mathrm{M}\exp(\mathrm{M}x) = \exp(\mathrm{M}x)\mathrm{M} \tag{A.37}$$

are immediate. Note that, in general, $\sqrt{\mathrm{AB}} \neq \sqrt{\mathrm{BA}}$ and $\exp(\mathrm{A}+\mathrm{B}) \neq \exp(\mathrm{A})\exp(\mathrm{B}) \neq \exp(\mathrm{B})\exp(\mathrm{A})$. The equality holds only if the matrices A and B have the same modal matrix.

A.5. THE PERTURBATION OF A MATRIX OPERATOR: ASYMPTOTIC BEHAVIOR OF THE EIGENVALUES

In this section we consider the problem of determining the asymptotic behavior, as $\varepsilon \to 0$ (ε is real), of the eigenvalues of the $n \times n$ matrix

$$\Lambda = \Lambda^{(0)} + \varepsilon\Lambda^{(1)} + \varepsilon^2\Lambda^{(2)} + \varepsilon^3\Lambda^{(3)} + O(\varepsilon^4) \quad \text{as } \varepsilon \to 0 \tag{A.38}$$

under the assumption that the eigenvalues of Λ and $\Lambda^{(0)}$ are real.

In particular we shall show that the ith eigenvalue of this matrix, β_i, can always be put in the following form

$$\beta_i = \beta_i^{(0)} + \beta_i^{(1)}\varepsilon + \beta_i^{(2)}\varepsilon^2 + O(\varepsilon^3) \quad \text{as } \varepsilon \to 0 \tag{A.39}$$

where $\beta_i^{(0)}$ is the ith eigenvalue of the matrix $\Lambda^{(0)}$.

The eigenvalue $\beta_i(\varepsilon)$ is a solution of the equation

$$D(\beta_i) = \det(\beta_i I - \Lambda) = \det(\beta_i I - \Lambda^{(0)} - \varepsilon\Lambda^{(1)} - \ldots) = 0. \tag{A.40}$$

Now, let us rewrite $D(\beta_i)$ as

$$D(\beta_i) = D_0(\beta_i) + \varepsilon D_1(\beta_i) + \varepsilon^2 D_2(\beta_i) + \dots \tag{A.41}$$

where

$$D_0(\beta_i) = \det(\beta_i I - \Lambda_0). \tag{A.42}$$

The function $D(\beta_i)$ can be expanded in Taylor series in a suitable neighborhood of $\beta_i^{(0)}$, so that

$$\begin{aligned} D(\beta_i) = {} & D_0(\beta_i^{(0)}) + \varepsilon D_1(\beta_i^{(0)}) + \dots \\ & + (\beta_i - \beta_i^{(0)})[D_0'(\beta_i^{(0)}) + \varepsilon D_1'(\beta_i^{(0)}) + \dots] + \\ & + \frac{(\beta_i - \beta_i^{(0)})^2}{2}[D_0''(\beta_i^{(0)}) + \varepsilon D_1''(\beta_i^{(0)}) + \dots] + \dots . \end{aligned} \tag{A.43}$$

Now, let us suppose that $\beta_i^{(0)}$ is a simple eigenvalue of Λ_0: as a consequence $D_0'(\beta_i^{(0)}) \neq 0$, and from Eq. (A.43) it is possible to write

$$\beta_i(\varepsilon) - \beta_i^{(0)} = -\frac{1}{D_0'(\beta_i^{(0)})} \frac{D_0(\beta_i^{(0)}) + \varepsilon D_1(\beta_i^{(0)}) + \dots}{1 + \varepsilon D_1'(\beta_i^{(0)})/D_0'(\beta_i^{(0)}) + \varepsilon^2 D_2'(\beta_i^{(0)})/D_0'(\beta_i^{(0)}) + \dots} \tag{A.44}$$

and hence

$$\beta_i(\varepsilon) = \beta_i^{(0)} + \varepsilon\beta_i^{(1)} + \varepsilon^2\beta_i^{(2)} + \dots, \tag{A.45}$$

which demonstrates Eq. (A.39).

Now, in order to face the most general case, we shall deal with a multiple eigenvalue $\bar{\beta}_0$ of the matrix Λ_0, whose algebraic multiplicity is $m < n$. Without loss of generality, we may assume the remaining n-m eigenvalues to be simple, and hence we can put them in the form (A.39). Consequently we have

$$D_0'(\bar{\beta}_0) = D_0''(\bar{\beta}_0) = \cdots = D_0^{(m-1)}(\bar{\beta}_0) = 0, \quad D_0^{(m)}(\bar{\beta}_0) \neq 0. \tag{A.46}$$

Now, by using Eqs. (A.43) and (A.46), it is not difficult to obtain, for $\varepsilon \to 0$ (Deif, 1982),

$$\prod_{i=1}^{m} [\beta_i(\varepsilon) - \bar{\beta}_0) \approx (-1)^m m! \frac{D_1(\bar{\beta}_0)}{D_0^{(m)}(\bar{\beta}_0)} \varepsilon, \tag{A.47}$$

$$\sum_{i=1}^{m} (\beta_i(\varepsilon) - \bar{\beta}_0) \approx -\left[\sum_{i=m+1}^{n} \lambda_i^{(1)} + \frac{D_1^{(n-1)}(\bar{\beta}_0)}{(n-1)!}\right]\varepsilon, \tag{A.48}$$

from which we conclude that $\beta_i(\varepsilon)$ can be expanded into a series of $\sqrt[m]{\varepsilon}$ [*Puiseux* expansion (Deif, 1982)], as follows

$$\beta_i(\varepsilon) = \bar{\beta}_0 + \eta_i^{(1)} \sqrt[m]{\varepsilon} + \eta_i^{(2)} \sqrt[m]{\varepsilon^2} + \cdots + \eta_i^{(m)} \varepsilon + \eta_i^{(m+1)} \sqrt[m]{\varepsilon^{m+1}} + \cdots \quad \text{(A.49)}$$

for $i = 1, \ldots, m$. Note that $\eta_i^{(k)}$ is always a complex number, unless k is an integer multiple of m; in this case $\eta_i^{(k)}$ becomes a real number (Deif, 1982). Under the assumption that $\beta_i(\varepsilon)$ and $\bar{\beta}_0$ are real, it must be $\eta_i^{(k)} = 0$, except for $\eta_i^{(m)}, \eta_i^{(2m)}, \ldots$; therefore, Eq. (A.49) again reduces to Eq. (A.39).

Appendix B

Some Useful Notes on the Laplace Transform

In this Appendix, we shall recall the basic concepts and properties of the Laplace transform and of the analytic functions. In particular, in Table B.I we shall resume the properties of the Laplace transforms and in Table B.2 we shall give the analytical expressions of some Laplace transforms used in this book. The reader is referred to the many excellent books for a thorough and comprehensive treatment of these arguments (e.g., Korn and Korn, 1961; Smirnov, 1964b; Doetsch, 1970; Guest, 1991).

B.1. GENERAL CONSIDERATIONS

Let $f(t)$ be a generic single-valued function that is equal to zero for $t < 0$. The Laplace transform

$$F(s) = \int_{0^-}^{\infty} f(t)e^{-st}\,dt, \tag{B.1}$$

associates a unique image function $F(s)$ of the complex variable $s = \sigma + i\omega$ with every object function $f(t)$ such that the improper integral (B.1) exists (e.g., Korn and Korn, 1961; Doetsch, 1970; Guest, 1991). The greatest lower bound σ_a of the real part of s for which the improper integral (B.1) converges *absolutely* and *uniformly* is called *abscissa of absolute convergence* of the Laplace transform. The half plane $Re\{s\} > \sigma_a$ is called *region of convergence* of the Laplace trans-

Table B.1. Some Properties of the Laplace Transform (Korn and Korn, 1961)

Operation	Object function	Image function
Differentiation of the object function ... if $f'(t)$ exists for all $t > 0$	$f'(t)$	$sF(s) - f(0^-)$
Integration of the object function ... if $f'(t)$ exists for all $t > 0$	$\int_0^t f(t)dt + C$	$\frac{F(s)}{s} + \frac{C}{s}$
Convolution of the object functions	$(f_1 * f_2)(t) \equiv \int_{0^-}^{\infty} f_1(\tau)f_2(t-\tau)d\tau$	$F_1(s)F_2(s)$
Translation (shift) of the object function ... if $f(t) = 0$ for $t < 0$	$f(t - b)$	$e^{-bs}F(s)$
Translation of the image function	$e^{at}f(t)$	$F(s - a)$
Initial value theorem ... if the limit on the right exists	$f(0^+) = \lim_{s\to\infty} sF(s)$	
Final value theorem ... if both the limits exist	$\lim_{t\to\infty} f(t) = \lim_{s\to 0} sF(s)$	

form. The function $F(s)$ is *analytic* for $Re\{s\} > \sigma_a$. Conversely, provided that certain hypotheses are satisfied, the integral

$$\frac{1}{2\pi i}\int_{\sigma - i\infty}^{\sigma + i\infty} F(s)e^{st}\,ds = \begin{cases} 0 & \text{for } t < 0 \\ f(t) & \text{for } t > 0 \end{cases} \quad \text{for } \sigma > \sigma_c, \tag{B.2}$$

reproduces $f(t)$ using $F(s)$. The transform (B.2) is the inverse of (B.1). When $\sigma_a < 0$, the integral in the inversion formula (B.2) may be evaluated along the imaginary axis, $\sigma = 0$, and the inversion reduces to the inverse Fourier transform.

The region of definition of the analytic function $F(s)$ defined by (B.1) for $Re\{s\} > \sigma_a$ can usually be extended by analytic continuation (e.g., Smirnov, 1964b), so as to include the entire s plane with the exception of singular points situated to the left of the abscissa of absolute convergence. We thus say that the Laplace transform will include not only the function $F(s)$ as given in its region of convergence, but also its *analytical continuation* whether or not the results are unique.

The singularity of an analytic function $F(s)$ is any point where it is not analytic. The point $s = a$ is an *isolated singularity* of the function $F(s)$ if it is analytic for $0 < |s - a| < \delta$, where δ is an arbitrarily small positive number, but not for $s = a$. The function $F(s)$ is analytic at infinity iff $F(1/s)$ is analytic at the origin.

If the function $F(s)$ is single valued, it can be expanded into a Laurent series of integer powers of $(s - a)$ in the neighborhood of an isolated singularity $s = a$ (e.g., Smirnov, 1964b). In this case there are three possibilities:

(i) the series does not contain terms with negative powers of $(s - a)$, *removable singularity*;
(ii) the series contains a finite number of terms with negative powers of $(s - a)$, *pole* of order m; and
(iii) the series contains an infinite number of terms with negative powers of $(s - a)$, *essential singularity*.

Let us now consider the case in which the analytic continuation gives rise to a multivalued function (e.g., Smirnov, 1964b). The analytic continuation along two different routes surrounding the singular point will lead to nonunique results. This type of singularity is called *branch point*, and the various possible distinct sets of values generated by the continuation are called *branches* of the function $F(s)$. All the possible branches of the analytical continuation of the function $F(s)$ may be obtained by the process in which one winds about the branch point as many times as may be necessary. The branch point $s = a$ is said to be a *branch point of the* $(m - 1)$th *order* if the analytical continuation of $F(s)$ in the neighborhood of the branch point gives rise to m different branches. For example, $\sqrt{s}$ has two branch points of the first order, one at $s = 0$ and the other at $s = \infty$. For the function $\sqrt{(s - a)(s - b)}$ the points a and b are branch points of the first order. The function $ln(s)$ has two branch points of the infinite order at $s = 0$ and $s = \infty$.

Table B.2. Some Laplace Transforms

Nr.	$F(s)$	$f(t)$	References
1	$\exp[-a\sqrt{(s+\beta)^2-\alpha^2}]$	$e^{-a\beta}\delta(t-a) + a\alpha e^{-\beta t}\frac{I_1(\alpha\sqrt{t^2-a^2})}{\sqrt{t^2-a^2}}u(t-a)$	Ghizzetti and Ossicini, 1971
2	$s\frac{e^{-a\sqrt{(s+\beta)^2-\alpha^2}}}{\sqrt{(s+\beta)^2-\alpha^2}}$	$e^{-a\beta}\delta(t-a) + \frac{d}{dt}[e^{-\beta t}I_0(\alpha\sqrt{t^2-a^2})]u(t-a)$	Ghizzetti and Ossicini, 1971
3	$\frac{1}{s}\sqrt{\frac{s+2b}{s+2a}}$	$e^{-(a+b)t}I_0[(a-b)t]u(t) + 2b\int_0^t e^{-(a+b)\tau}I_0[(a-b)\tau]d\tau$	Ghizzetti and Ossicini, 1971

4	$e^{-\alpha\sqrt{s}}$	$\dfrac{\alpha}{2\sqrt{\pi}t^{3/2}}e^{-\alpha^2/4t}u(t)$	Doetsch, 1970; Ghizzetti and Ossicini, 1971
5	$\dfrac{1}{s^{\alpha}}$ (α arbitrarily real)	$\begin{cases} \dfrac{t^{\alpha-1}}{\Gamma(\alpha)}u(t) & \text{for } \alpha > 0 \\ \dfrac{1}{\Gamma(\alpha)}\mathrm{Pf}[t^{\alpha-1}u(t)] & \text{for } \alpha < 0, \alpha \neq -1, -2, \ldots \\ \delta^{(n)}(t) & \text{for } \alpha = -n = 0, -1, -2, \ldots \end{cases}$	Doetsch, 1970
6	$\sqrt{s^2+\alpha^2}-s$	$\alpha\dfrac{J_1(\alpha t)}{t}u(t)$	Doetsch, 1970; Ghizzetti and Ossicini, 1971
7	$\exp(-\tau\sqrt{s^2+\alpha^2})$	$\delta(t-\tau)-x\tau\dfrac{J_1(\alpha\sqrt{t^2-\tau^2})}{\sqrt{t^2-\tau^2}}u(t-\tau)$	Ghizzetti and Ossicini, 1971

With the symbol $\mathrm{Pf}[t^{-\lambda}u(t)]$ we indicate the *pseudofunction* (Doetsch, 1970).

$$\mathrm{Pf}[t^{-\lambda}u(t)] = D^n\left[\frac{(-1)^n}{(\lambda-1)\cdots(\lambda-n)}t^{-\lambda+n}u(t)\right], \quad (\lambda > 1, \text{ not an integer}, -\lambda+n > -1, n \text{ an integer})$$

where with $D^n f(t)$ we indicate the nth distribution derivative. Note that $t^{-\lambda}$, for $\lambda > 0$, is continuous at all points t except $t = 0$.

In the neighborhood of a branch point $s = a$ of the $(m-1)$th order, the function $F(s)$ can be expanded into a Laurent series of integer powers of the argument $s' = \sqrt[m]{s-a}$ (e.g., Smirnov, 1964b). If the expansion contains only terms with negative powers of s', the branch point is of *regular type*. If the expansion contains a finite number of terms with negative powers of s', the branch point is of the *polar type*, whereas if the expansion contains an infinite number of terms with negative powers of s', the branch point is of the *essential singularity type*.

All these definitions can be extended to the point at infinity by using the inversion formula $\hat{s} = 1/s$.

Remark

It is a characteristic of the Laplace transform that it tends towards zero when the variable s tends toward ∞ in its region of convergence. Instead, the Laplace transforms of distributions do not possess this property. Actually, when $s \to \infty$ these Laplace transforms tend towards a constant value different from zero or toward infinity although never more strongly than a power of s (e.g., Doetsch, 1970). If the function $F(s)$ vanishes at least as $1/s$ for $s \to \infty$ in its region of convergence, then the corresponding time domain object function $f(t)$ is bounded everywhere (e.g., Korn and Korn, 1961). ◇

B.2. THE ASYMPTOTIC BEHAVIOR OF THE OBJECT FUNCTION FOR $t \to \infty$

The behavior of $f(t)$ for $t \to \infty$ depends upon the singularities of $F(s)$, which are situated to the left of its region of convergence. These singularities could be poles, essential singularity points, or branch points (we are implicitly assuming that there are only isolated singularities).

The influence of the poles and essential singular points on the behavior of the object function for $t \to \infty$ is completely different from the influence of the branch points. Therefore, we have to distinguish between them.

The prediction of the asymptotic behavior of $f(t)$ for $t \to \infty$ is strictly related to the behavior of $F(s)$ near its singular points. Let us first consider the case in which the analytical continuation of the function $F(s)$ is single valued. Suppose that $F(s)$ is analytic for $Re\{s\} \leqslant \sigma_a$, with the possible exception of the poles λ_n $(n = 0, 1, 2, \ldots$ and

$\sigma_a = Re\{\lambda_0\} \geqslant Re\{\lambda_1\} \geqslant, \ldots)$, where $F(s)$ has the respective principal parts

$$\frac{a_1^{(n)}}{s - \lambda_n} + \frac{a_2^{(n)}}{(s - \lambda_n)^2} + \cdots + \frac{a_{m_n}^{(n)}}{(s - \lambda_n)^{m_n}}. \tag{B.3}$$

In general, $F(s)$ may have an infinite number of poles. Furthermore, let us assume that $F(s)$ vanishes for $s \to \infty$ at least as $1/s^\alpha$ with $\alpha > 1$. Then, the asymptotic behavior of $f(t)$ for $t \to \infty$ is given by (Doetsch, 1970),

$$f(t) \approx \sum_{n=0}^{\infty} \left[a_1^{(n)} + a_2^{(n)} t + \cdots + \frac{a_{m_n}^{(n)}}{(m_n - 1)!} t^{m_n - 1} \right] e^{\lambda_n t}. \tag{B.4}$$

In particular, the leading term of $f(t)$ for $t \to \infty$ is strictly related to the singularities with the largest real part, if the function $F(s)$ has a finite number of poles.

If $\hat{\lambda}$ is an essential singular point, the principal part of $F(s)$ in a neighborhood of $s = \hat{\lambda}$ would be represented not by a finite sum but by an infinite series.

Let us now consider the case in which the singularities of $F(s)$ are only branch points. The function $F(s)$, which is supposed to be analytic for $Re\{s\} \leqslant \sigma_a$, with the possible exception of the branch point λ_b, has the following asymptotic expansion:

$$F(s) \approx \sum_{m=0}^{\infty} c_m (s - \lambda_b)^{\gamma_m} \quad \text{as } s \to \lambda_b (Re\{\gamma_0\} \leqslant Re\{\gamma_1\} \leqslant, \ldots), \tag{B.5}$$

and tends toward zero for $s \to \infty$. Then, the asymptotic behavior of $f(t)$ for $t \to \infty$ is given by (Doetsch, 1970),

$$f(t) \approx e^{\lambda_b t} \sum_{m=0}^{\infty} \frac{c_m}{\Gamma(-\gamma_m)} \frac{1}{t^{\gamma_m + 1}} \left(\frac{1}{\Gamma(-\gamma_m)} = 0 \quad \text{for } \gamma_m = 0, 1, 2, \ldots \right), \tag{B.6}$$

where $\Gamma(u)$ is the gamma function (e.g., Abramowitz and Stegun, 1972). For instance, the asymptotic behavior for $t \to \infty$ of the inverse Laplace transform of $\sqrt{s - \lambda_b}$ is given by $e^{\lambda_b t}/[\Gamma(-1/2)t^{3/2}]$, whereas the asymptotic behavior of the inverse Laplace transform of $1/\sqrt{s - \lambda_b}$ is given by $e^{\lambda_b t}/[\Gamma(1/2)t^{1/2}]$.

When there are many branch points, the asymptotic behavior of $f(t)$ is completely determined by the local behavior of $F(s)$ near the branch point with the largest real part. If there are many branch points with the same largest real part, the asymptotic behavior of $f(t)$ is given by the superposition of the asymptotic behavior corresponding to each of them.

Finally, if the function $F(s)$ has both poles and branch points, the asymptotic behavior of $f(t)$ is given by the superposition of the asymptotic behavior corresponding to the branch points with the largest real part and the asymptotic behavior corresponding to the poles that are to the right of the branch points.

The reader is referred to Doetsch (1970) for a thorough and comprehensive treatment of this argument.

Appendix C

Some *a priori* Estimates

C.1. *a priori* ESTIMATES FOR THE SOLUTION OF EQ. (9.8)

We now show that the solution of the equation

$$R_c C_p \frac{dv}{dt} + v + R_c g(v) = e, \tag{C.1}$$

with the initial condition

$$v(t = t_0) = v_0, \tag{C.2}$$

satisfies the inequality

$$v^2(t) \leqslant v_0^2 + \frac{1}{2R_c C_p} \int_{t_0}^{t} [R_c I + 2|e(\tau)|]^2 \, d\tau. \tag{C.3}$$

Multiplying Eq. (C.1) by v we obtain

$$\frac{R_c C_p}{2} \frac{dv^2}{dt} + v^2 + R_c g(v) v = e(t) v. \tag{C.4}$$

Now integration yields

$$v^2(t) = v_0^2 - \frac{2}{R_c C_p} \int_{t_0}^{t} \{v^2(\tau) + R_c g[v(\tau)] v(\tau) - e(\tau) v(\tau)\} d\tau. \tag{C.5}$$

As the terminal resistor is weakly active (see §9.2), we have

$$-g(v)v \leqslant |v|I. \tag{C.6}$$

where I is a positive constant. This and the classical inequalities (e.g., Zeidler, 1990)

$$2|xy| \leqslant \frac{x^2}{c} + cy^2 \quad c > 0, \qquad \int_{t_0}^{t} v(\tau)e(\tau)d\tau \leqslant \int_{t_0}^{t} |v(\tau)e(\tau)|d\tau, \tag{C.7}$$

imply (with $c = 1/2$) the inequality (C.3).

If the "forcing term" $e(t)$ is bounded, the inequality (C.3) assures us that the solutions of Eq. (C.1) remain bounded in its interval of definition, that is, there exists a bounded constant B such that

$$|v(t)| \leqslant B \tag{C.8}$$

for any time interval in which the solution exists.

C.2. *a priori* ESTIMATES FOR THE SOLUTION OF EQ. (9.18)

We now show that the solution of the Volterra integral equation

$$u(t) = e(t) + \int_{0}^{t} K[t - \tau; u(\tau)]d\tau, \tag{C.9}$$

where

$$K(t - \tau; u) = \frac{z_{cr}(t - \tau)}{R_c}[F_v(u) - u], \tag{C.10}$$

is bounded for any t because the terminal resistor is weakly active. The function $F_v(u)$ expresses the voltage v as a function of the auxiliary voltage u defined by Eq. (9.16).

Let us assume that there exist two positive continuous functions $a(t)$ and $b(t)$ such that

$$|K(t - \tau, u)| \leqslant a(t - \tau)|u| + b(t - \tau). \tag{C.11}$$

Let $y(t)$ be the solution of the linear Volterra integral equation of the second kind

$$y(t) = |e(t)| + \int_{0}^{t} [a(t - \tau)y(\tau) + b(t - \tau)]d\tau. \tag{C.12}$$

This equation admits a bounded solution $y(t)$, because its kernel and the known function $e(t)$ are continuous (e.g., Linz, 1985). Then, the solution of Eq. (C.9) satisfies the inequality (e.g., Lakshmikantham and Leela, 1969)

$$|u(t)| \leqslant y(t), \tag{C.13}$$

that is, the solutions of Eq. (C.9) are bounded for any time instant in which they exist.

Now we show that there exist two positive continuous functions $a(t)$ and $b(t)$, such that the inequality (C.11) holds because the nonlinear resistor is weakly active (see §9.2). The kernel $K(t-\tau;u)$ by definition satisfies the following inequality:

$$K(t-\tau;u) \leqslant \left|\frac{z_{cr}(t-\tau)}{R_c}\right| (|F_v(u)| + |u|). \tag{C.14}$$

From property (9.2) we obtain

$$\begin{aligned} u &= v + R_c g(v) \geqslant v - R_c I \quad \text{for } v \geqslant 0, \\ u &= v + R_c g(v) \leqslant v + R_c I \quad \text{for } v \leqslant 0, \end{aligned} \tag{C.15}$$

therefore

$$\begin{aligned} F_v(u) &= v \leqslant u + R_c I \quad \text{for } v \geqslant 0, \\ F_v(u) &= v \geqslant u - R_c I \quad \text{for } v \leqslant 0. \end{aligned} \tag{C.16}$$

Thus, the following inequality holds:

$$|F_v(u)| \leqslant |u| + R_c I. \tag{C.17}$$

By combining inequalities (C.14) and (C.17) we have the following inequality:

$$K(t-\tau;u) \leqslant \left|\frac{z_{cr}(t-\tau)}{R_c}\right| (2|u| + R_c I) \tag{C.18}$$

that implies inequality (C.11).

Appendix D

Tables of Equivalent Representations of Transmission Lines

In this appendix the Thévenin and Norton equivalent circuits describing the terminal behavior of transmission lines are reported, both in the Laplace and in the time domain. The lines are assumed to be initially at rest and no distributed independent sources are considered.

The reference conventions for the currents and the voltages at the line ends are those shown in Fig. 2.10 and in Fig. 3.3.

Table D.1. Two-Conductor Lines, Laplace Domain

Representation	Output equations	State equations
Thévenin	$V_1(s)=Z_c(s)I_1(s)+W_1(s)$ $V_2(s)=Z_c(s)I_2(s)+W_2(s)$	$W_1(s)=P(s)[2V_2(s)-W_2(s)]$ $W_2(s)=P(s)[2V_1(s)-W_1(s)]$
Norton	$I_1(s)=Y_c(s)V_1(s)+J_1(s)$ $I_2(s)=Y_c(s)V_2(s)+J_2(s)$	$J_1(s)=P(s)[-2I_2(s)+J_2(s)]$ $J_2(s)=P(s)[-2I_1(s)+J_1(s)]$
Nonuniform lines (Thévenin)	$V_1(s)=Z_{c1}^{+}(s)I_1(s)+A_1(s)W_1(s)$ $V_2(s)=Z_{c2}^{-}(s)I_2(s)+A_2(s)W_2(s)$	$W_1(s)=P^{-}(s)[2V_2(s)-W_2(s)]$ $W_2(s)=P^{+}(s)[2V_1(s)-W_1(s)]$

Line	Characteristic impedance	Global propagation operator
Lossless	$Z_c(s)=R_c=G_c^{-1}=\sqrt{L/C}$	$P(s)=\exp(-sT)$ $\quad T=d/c$
Lossy (RLGC)	$Z_c(s)=Y_c^{-1}(s)=\sqrt{\dfrac{R+sL}{G+sC}}$	$P(s)=\exp(-d\sqrt{(R+sL)(G+sC)})$
Lossy (general case)	$Z_c(s)=Y_c^{-1}(s)=\sqrt{\dfrac{Z(s)}{Y(s)}}$	$P(s)=\exp(-d\sqrt{Z(s)Y(s)})$
Asymptotic expansion	$Z_c(s)=R_c+Z_{cr}(s)$	$P(s)=e^{-sT}[P_p(s)+P_r(s)]$

Note: for nonuniform lines see Chapter 7.

Table D.2. Two-Conductor Lines, Time Domain

Representation	Output equations	State equations
	Lossless lines	
Thévenin	$v_1(t)=R_c i_1(t)+w_1(t)$	$w_1(t)=2v_2(t-T)-w_2(t-T)$
	$v_2(t)=R_c i_2(t)+w_2(t)$	$w_2(t)=2v_1(t-T)-w_1(t-T)$
Norton	$i_1(t)=G_c v_1(t)+j_1(t)$	$j_1(t)=-2i_2(t-T)+j_2(t-T)$
	$i_2(t)=G_c v_2(t)+j_2(t)$	$j_2(t)=-2i_1(t-T)+j_1(t-T)$
	Lossy lines (RLGC)	
Thévenin	$v_1(t)=R_c i_1(t)+(z_{cr}*i_1)(t)+w_1(t)$	$w_1(t)=e^{-\mu T}[2v_2(t-T)-w_2(t-T)]$ $+(p_r*[2v_2-w_2])(t-T)$
	$v_2(t)=R_c i_2(t)+(z_{cr}*i_2)(t)+w_2(t)$	$w_2(t)=e^{-\mu T}[2v_1(t-T)-w_1(t-T)]$ $+(p_r*[2v_1-w_1])(t-T)$
Norton	$i_1(t)=G_c v_1(t)+(y_{cr}*v_1)(t)+j_1(t)$	$j_1(t)=e^{-\mu T}[-2i_2(t-T)+j_2(t-T)]$ $+(p_r*[-2v_2+j_2])(t-T)$
	$i_2(t)=G_c v_2(t)+(y_{cr}*v_2)(t)+j_2(t)$	$j_2(t)=e^{-\mu T}[-2i_1(t-T)+j_1(t-T)]$ $+(p_r*[-2i_1+j_1])(t-T)$
	Lossy lines (general case)	
Thévenin	$v_1(t)=R_c i_1(t)+(z_{cr}*i_1)(t)+w_1(t)$	$w_1(t)=(p*[2v_2-w_2])(t)$
	$v_2(t)=R_c i_2(t)+(z_{cr}*i_2)(t)+w_2(t)$	$w_2(t)=(p*[2v_1-w_1])(t)$
Norton	$i_1(t)=G_c v_1(t)+(y_{cr}*v_1)(t)+j_1(t)$	$j_1(t)=(p*[-2v_2+j_2])(t)$
	$i_2(t)=G_c v_2(t)+(y_{cr}*v_2)(t)+j_2(t)$	$j_2(t)=(p*[-2i_1+j_1])(t)$
	Nonuniform lossless lines	
Thévenin	$v_1(t)=R_{c1} i_1(t)+(z_{c1r}^{+}*i_1)(t)$ $+(a_{1r}*w_1)(t)+w_1(t)$	$w_1(t)=(p^{-}*[2v_2-w_2])(t)$
	$v_2(t)=R_{c2} i_2(t)+(z_{c2r}^{-}*i_2)(t)$ $+(a_{2r}*w_2)(t)+w_2(t)$	$w_2(t)=(p^{+}*[2v_1-w_1])(t)$

Table D.3. Two-Conductor Lines, Impulse Responses

	Current-based impulse response of a perfectly matched line
General expression	$z_c(t)=R_c\delta(t)+z_{cr}(t)$
Lossless lines	$z_{cr}(t)=0$
Lossy lines (RLGC)	$z_{cr}(t)=\nu R_c e^{-\mu t}[I_0(\nu t)+I_1(\nu t)]u(t)$ μ, ν (see Eqs. (4.6), (4.8))
Lossy lines (skin-effect) $Z(s)=R+K\sqrt{s}+sL$	$z_{cr}(t)=\frac{R_c K}{2L}\frac{1}{\sqrt{\pi t}}u(t)+z_{cb}(t)$
	$z_{cb}(t)=L^{-1}\left[Z_c(s)-R_c-\frac{R_c K}{2L}\frac{1}{\sqrt{s}}\right]$
	Voltage-based impulse response of a perfectly matched line
General expression	$y_c(t)=G_c\delta(t)+y_{cr}(t)$
Lossless lines	$y_{cr}(t)=0$
Lossy lines (RLGC)	$y_{cr}(t)=\nu G_c e^{-\mu t}[I_1(\nu t)-I_0(\nu t)]u(t)$ μ, ν (see Eqs. (4.6), (4.8))
Lossy lines (skin-effect) $Z(s)=R+K\sqrt{s}+sL$	$y_{cr}(t)=\frac{-K}{2R_c L}\frac{1}{\sqrt{\pi t}}u(t)+y_{cb}(t)$
	$y_{cb}(t)=L^{-1}\left[Y_c(s)-\frac{1}{R_c}+\frac{K}{2R_c L}\frac{1}{\sqrt{s}}\right]$
	Global propagation impulse response
General expression	$p(t)=p_p(t-T)+p_r(t-T)$
Lossless lines	$p_p(t)=\delta(t)$ $p_r(t)=0$
Lossy lines (RLGC)	$p_p(t)=e^{-\mu T}\delta(t)$ μ (see Eq. (4.6)
	$p_r(t)=T\nu e^{-\mu(t+T)}\frac{I_1[\nu\sqrt{(t+T)^2-T^2}]}{\sqrt{(t+T)^2-T^2}}u(t)$ ν (see Eq. (4.8))
Lossy lines (skin-effect) $Z(s)=R+K\sqrt{s}+sL$	$p_p(t)=e^{-\mu T}\delta(t;\alpha)$ μ, α (see Eq. (5.78))
	$\delta(t;\alpha)=\frac{\alpha}{2t^{3/2}\sqrt{\pi}}e^{-\alpha^2/4t}u(t)$
	$p_r(t)=L^{-1}[e^{sT}P(s)-e^{-(\mu T+\alpha\sqrt{s})}]$

Note: for nonuniform lines see Chapter 7.

Table D.4. Multiconductor Lines, Laplace Domain

Representation	Output equations	State equations
Thévenin	$\mathbf{V}_1(s)=Z_c(s)\mathbf{I}_1(s)+\mathbf{W}_1(s)$ $\mathbf{V}_2(s)=Z_c(s)\mathbf{I}_2(s)+\mathbf{W}_2(s)$	$\mathbf{W}_1(s)=P(s)[2\mathbf{V}_2(s)-\mathbf{W}_2(s)]$ $\mathbf{W}_2(s)=P(s)[2\mathbf{V}_1(s)-\mathbf{W}_1(s)]$
Norton	$\mathbf{I}_1(s)=Y_c(s)\mathbf{V}_1(s)+\mathbf{J}_1(s)$ $\mathbf{I}_2(s)=Y_c(s)\mathbf{V}_2(s)+\mathbf{J}_2(s)$	$\mathbf{J}_1(s)=P_I(s)[-2\mathbf{I}_2(s)+\mathbf{J}_2(s)]$ $\mathbf{J}_2(s)=P_I(s)[-2\mathbf{I}_1(s)+\mathbf{J}_1(s)]$ $P_I(s) = Z^{-1}(s)P(s)Z(s)$

Line	Characteristic impedance	Global propagation operator
Lossless	$R_c=G_c^{-1}=C^{-1}\sqrt{CL}$ $=\sqrt{(LC)^{-1}}\,L$	$P(s)=T_v\,\text{diag}[e^{-sT_1},\ldots,e^{-sT_n}]T_v^{-1}$ $=\sum_{h=1}^{n}\mathbf{e}_h\mathbf{s}_h^T e^{-sT_h}\quad T_h=d/c_h$ $LC=T_v\,\text{diag}\left(\frac{1}{c_1^2},\ldots,\frac{1}{c_n^2}\right)T_v^{-1}$
Lossless, homogeneous dielectric	$R_c=G_c^{-1}=C^{-1}/c=cL$	$P(s)=I\exp(-sT)$
Lossy (RLGC)	$Z_c(s)$ $=\sqrt{[(R+sL)(G+sC)]^{-1}}(R+sL)$ $Y_c(s)=Z_c^{-1}(s)$	$P(s)=\exp(-d\sqrt{(R+sL)(G+sC)})$
Lossy (general case)	$Z_c(s)=Y_c^{-1}(s)=\sqrt{(Z(s)Y(s))^{-1}}Z(s)$	$P(s)=\exp(-d\sqrt{Z(s)Y(s)})$
Asymptotic expansion	$Z_c(s) = R_c+Z_{cr}(s)$	$P(s)=P_p(s)+P_r(s)$

Table D.5. Multiconductor Lines, Time Domain

Representation	Output equations	State equations
Thévenin	$\mathbf{v}_1(t)=R_c\mathbf{i}_1(t)+(z_{cr}*\mathbf{i}_1)(t)+\mathbf{w}_1(t)$ $\mathbf{v}_2(t)=R_c\mathbf{i}_2(t)+(z_{cr}*\mathbf{i}_2)(t)+\mathbf{w}_2(t)$	$\mathbf{w}_1(t)=(p*[2\mathbf{v}_2-\mathbf{w}_2])(t)$ $\mathbf{w}_2(t)=(p*[2\mathbf{v}_1-\mathbf{w}_1])(t)$
Norton	$\mathbf{i}_1(t)=G_c\mathbf{v}_1(t)+(y_{cr}*\mathbf{v}_1)(t)+\mathbf{j}_1(t)$ $\mathbf{i}_2(t)=G_c\mathbf{v}_2(t)+(y_{cr}*\mathbf{v}_2)(t)+\mathbf{j}_2(t)$	$\mathbf{j}_1(t)=(p_I*[-2\mathbf{i}_2+\mathbf{j}_2])(t)$ $\mathbf{j}_2(t)=(p_I*[-2\mathbf{i}_1+\mathbf{j}_1])(t)$

Table D.6. Multiconductor Lines, Impulse Responses

	Current-based impulse response of a perfectly matched line
General expression	$z_c(t) = \mathrm{R}_c\delta(t) + z_{cr}(t)$
Lossless lines	$z_{cr}(t) = 0$
Lossy lines (RLGC)	$z_{cr}(t) = L^{-1}[\mathrm{Z}_c(s) - \mathrm{R}_c]$
Lossy lines (skin-effect) $\mathrm{Z}(s) = \mathrm{R} + \mathrm{K}\sqrt{s} + s\mathrm{L}$	$z_{cr}(t) = \dfrac{\Psi^{(1)}}{\sqrt{\pi t}} u(t) + z_{cb}(t)$ $\Psi^{(1)}$ (see Eq. (6.238))
	$z_{cb}(t) = L^{-1}\left[\mathrm{Z}_c(s) - \mathrm{R}_c - \dfrac{\Sigma^{(1)}}{\sqrt{s}}\right]$
	Voltage-based impulse response of a perfectly matched line
General expression	$y_c(t) = \mathrm{G}_c\delta(t) + y_{cr}(t)$
Lossless lines	$y_{cr}(t) = 0$
Lossy lines (RLGC)	$y_{cr}(t) = L^{-1}[\mathrm{Y}_c(s) - \mathrm{G}_c]$
Lossy lines (skin-effect) $\mathrm{Z}(s) = \mathrm{R} + \mathrm{K}\sqrt{s} + s\mathrm{L}$	$y_{cr}(t) = \dfrac{\Sigma^{(1)}}{\sqrt{\pi t}} u(t) + y_{cb}(t)$ $\Sigma^{(1)}$ (see Eq. (6.231))
	$y_{cb}(t) = L^{-1}\left[\mathrm{Y}_c(s) - \mathrm{G}_c - \dfrac{\Psi^{(1)}}{\sqrt{s}}\right]$
	Global propagation impulse response
General expression	$p(t) = p_p(t) + p_r(t)$
Lossless lines	$p_p(t) = \sum_{h=1}^{n} \mathbf{e}_h \mathbf{s}_h^T \delta(t - T_h)$ $p_r(t) = 0$
Lossless lines, homogeneous dielectric	$p_p(t) = \mathrm{I}\delta(t - T)$ $p_r(t) = 0$
Lossy lines (RLGC)	$p_p(t) = \sum_{h=1}^{n} \mathbf{e}_h \mathbf{s}_h^T e^{-\mu_h T_h} \delta(t - T_h)$ μ_h, T_h (see Eq. (6.188))
	$p_r(t) = L^{-1}\left[\mathrm{P}(s) - \sum_{h=1}^{n} \mathbf{e}_h \mathbf{s}_h^T e^{-(s+\mu_h)T_h}\right]$
Lossy lines (skin-effect) $\mathrm{Z}(s) = \mathrm{R} + \mathrm{K}\sqrt{s} + s\mathrm{L}$	$p_p(t) = \sum_{h=1}^{n} \mathbf{e}_h \mathbf{s}_h^T e^{-\mu_h T_h} \delta(t - T_h; \alpha_h)$ μ_h, T_h, α_h (see Eq. (6.224))
	$\delta(t; \alpha) = \dfrac{\alpha}{2t^{3/2}\sqrt{\pi}} e^{-\alpha^2/4t} u(t)$
	$p_r(t) = L^{-1}\left[\mathrm{P}(s) - \sum_{h=1}^{n} \mathbf{e}_h \mathbf{s}_h^T e^{-(s+\mu_h)T_h - \alpha_h\sqrt{s}}\right]$

References

Abramowitz, M. and Stegun, I.A. (1972). *Handbook of Mathematical Functions*, New York: Dover Publications.

Amari, S. (1991). Comments on An exact solution for the nonuniform transmission line problem. *IEEE Trans. Microwave and Techniques* **39**(3):611–612.

Ames, W.F. (1977). *Numerical Methods for Partial Differential Equations*, New York: Academic Press.

Baum, C. E. (1988). High-frequency propagation on nonuniform multiconductor transmission lines in uniform media. *Intern. Journal of Numerical Modelling*. **1**:175 – 188.

Bender, C.M. and Orszag, S.A. (1978). *Advanced Mathematical Methods for Scientists and Engineers*, Auckland: McGraw-Hill.

Bobbio, S., Di Bello, C., and Gatti, E. (1980). Analysis of a time-varying line with constant characteristic resistance. *Alta Frequenza*, **XLIX**(3):247–250.

Braiton, R.K. and Moser, J.K. (1964). A theory of nonlinear networks. *Quart. Appl. Math.* **12**:1–33, 81–104.

Branin, F.H., Jr. (1967). Transient analysis of lossless transmission lines. *Proc. IEEE* **55**:2012–2013.

Braunisch, H. and Grabinski, H. (1998). Time-domain simulation of large lossy interconnect systems on conducting substrates. *IEEE Trans. Circuits and Systems-I: Fundamental Theory and Applications*. **45**(9):909–918.

Brews, J.R. (1986). Transmission line models for lossy waveguide interconnections in VLSI. *IEEE Trans. Electron Devices* **ED-33**:1356 – 1365.

Bronshtein, I.N. and Semendyayev, K.A., 1985. *Handbook of Mathematics*, Berlin: Springer.

Brunner, H. and van der Houwen, P.J. (1986). *The Numerical Solution of Volterra Equations*, Amsterdam: North-Holland.

Burkhart, S.C. and Wilcox, R.B. (1990). Arbitrary pulse shape synthesis via nonuniform transmission lines. *IEEE Trans. Microwave Theory Tech.* **38**:1514–1518.

Canavero, F.G., Daniele, V., and Graglia, R.D. (1988). Electromagnetic pulse interaction with multiconductor transmission lines. *Electromagnetic* **8**:293–310.

Chang, E.C. and Kang, S.M. (1996). Transient simulation of lossy coupled transmission lines using iterative linear least square fitting and piecewise recursive convolution. *IEEE Trans. Circ. and Syst.-I* **CAS-43**:923–932.

Chua, L.O. and Rohrer, R.A. (1965). On the dynamic equations of a class of nonlinear RLC networks, *IEEE Trans. Circuit Theory* **CT-12**:475–489.

Chua, L.O. and Alexander, G.R. (1971). The effects of parasitic reactances on nonlinear networks. *IEEE Trans. Circuit Theory* **CT-18**:520–532.

Chua, L.O., Desoer, C.A., and Kuh, E.S. (1987). *Linear and Nonlinear Circuits*, New York: McGraw-Hill.

Chua, L.O. and Lin, P.M. (1975). *Computer Aided Analysis of Electronic Circuits*, Englewood Cliffs: Prentice-Hall.

Chua, L.O. and Ayrom, F. (1985). Designing nonlinear single op-amp circuits: a cook-book approach. *International Journal on Circuit Theory and Applications* **13**:235–268.

Chua, L.O. (Ed.). (1987). *Special Issue on Chaotic Systems. IEEE Proc.* **75**(8).

Collin, R.E. (1992). *Foundation of Microwave Engineering*, New York: McGraw-Hill.

Corti, L., De Menna, L., Miano, G., and Verolino, L. (1994). Chaotic dynamics in an infinite-dimensional electromagnetic system. *IEEE Trans. Circuit and Systems* **41**(11):730–736.

Corti, L., Miano, G., and Verolino, L. (1996). Bifurcation and chaos in transmission lines. *Electrical Engineering (Archiv für Elektrotechnik)* **79**(3):165–171.

Corti, L., Maffucci, A., Miano, G., and Verolino, L. (1997). Time-domain two-port representations of a lossy line. *Electrical Engineering (Archiv für Elektrotechnik)* **80**(4):235–240.

Corti, L., De Magistris, M., De Menna, L., and Miano, G. (1998). Chaotic dynamics in lossy lines. *Proceedings EMC'98* Roma, September 14–18.

Courant, R. and Hilbert, D. (1989). *Methods of Mathematical Physics*, New York: John Wiley & Sons.

Dai, W.W.M. (Guest ed.). (1992). *Special Issue on Simulation, Modeling, and Electrical Design of High-Speed and High-Density Interconnects. IEEE Trans. Circ. and Syst.-I* **CAS-39**.

D'Amore, M., Sarto, S. (1996a). Simulation models of a dissipative transmission line above a lossy ground for a wide-frequency range: Part I: Single conductor configuration. *IEEE Trans. Electromag. Compat.*, **38**:127–138.

D'Amore, M., Sarto, S. (1996b). Simulation models of a dissipative transmission line above a lossy ground for a wide-frequency range: Part II: Multiconductor configuration. *IEEE Trans. Electromag. Compat.*, **38**:139–149.

Deif, A.S. (1982). *Advanced Matrix Theory*, New York: John Wiley.

De Menna, L. and Miano, G. (1994). Chaotic dynamics in a simple electromagnetic system. *Il Nuovo Cimento* — B **109B**:911–916.

Deutsch, A. *et al.* (1990). High-speed signal propagation on lossy transmission lines. *IBM Journal Res. Devel.* **34**:601–616.

Deutsch, A. *et al.* (1995). Modeling and characterization of long on-chip interconnections for high-performance microprocessors. *IBM Journal Res. Devel.* **39**:547–567.

Deutsch, A. *et al.* (1997). When are transmission line effects important for on-chip interconnections? *IEEE Trans. Microwave Theory and Techniques* **45**(10):1836–1846.

Devaney, R.L. (1989). *An Introduction to Chaotic Dynamical Systems*, Reading, MA: Addison-Wesley.

Djordjevic, A.R., Sarkar, T.K., and Harrington, R.F. (1986). Analysis of lossy transmission lines with arbitrary nonlinear terminal networks. *IEEE Trans. Microwave Theory and Techniques* **MTT-34**(6):660–666.

Djordjevic, A.R., Sarkar, T.K., and Harrington, R.F. (1987). Time-domain response of multiconductor transmission lines. *Proc. IEEE* **75**(6):743–764.

Doetsch, G. (1970). *Introduction to the Theory and Application of the Laplace Transformation*, Berlin: Springer-Verlag.

El-Zein, A., Haque, M., Chowdhury, S. (1994). Simulating nonuniform lossy lines with frequency dependent parameters by the method of characteristics. *IEEE International Symposium on Circuits and Systems, ISCAS'94*, **4**:327–330.

Erdélyi, A. (1956). *Asymptotic Expansions*, New York: Dover Publications, Inc.

Fano, R.M., Chu, L.J., and Adler, R.B. (1960). *Electromagnetic Energy Transmission and Radiation*, New York: Wiley, pp. 535–548.

de Figueiredo, R.J.P. and Ho, C.Y. (1970). Absolute stability of a system of nonlinear networks interconnected by lossless transmission lines. *IEEE Trans. Circuit Theory* **CT-17**(4):575–584.

Franceschetti, G. (1997). *Electromagnetics*, New York: Plenum Press.

Ghizzetti, A. and Ossicini, A. (1971). *Trasformate di Laplace e Calcolo Simbolico*, Torino: UTET.

Gordon, C., Blazeck, T., and Mittra, R. (1992). Time-domain simulation of multiconductor transmission lines with frequency-dependent losses. *IEEE Trans. Computer Aided Design* **CAD-11**:1372–1387.

Green, M.M. and Wilson, A.N., Jr. (1992). How to identify unstable dc operating points. *IEEE Trans. Circuits and Systems-I* **CAS-39**:820 – 832.

Grotelüschen, E., Dutta, L. S., and Zaage, S. (1994). Quasi-analytical analysis of Broadband properties of multiconductor transmission lines on semiconducting substrates. *IEEE Trans. Components, Packaging, Manufact. Tech.* **17**:376–382.

Guest, P.B. (1991). *Laplace Transforms and an Introduction to Distributions*, New York: Ellis Horwood.

Hale, J.K. and Koçak, H. (1992). *Dynamics and Bifurcations*, New York: Springer Verlag.

Hale, J.K. and Verduyn Lunel, S.M. (1993). *Introduction to Functional Differential Equations*, New York: Springer Verlag.

Hasler, M. and Neirynck, J. (1986). *Nonlinear Circuits*, Norwood: Artech House.

Hasler, M. (1995). Qualitative analysis, in *The Circuits and Filters Handbook*, Wai-Kai Chen, Editor in Chief, New York: CRC and IEEE Press.

Haus, H.A. and Melcher, J.R. (1989). *Electromagnetic Fields and Energy*, Englewood Cliffs. Prentice-Hall.

Hayden, L.A. and Tripathi, V.K. (1991). Nonuniformly coupled microstrip transversal filters for analog signal processing. *IEEE Trans. Microwave Theory Tech.* **39**:47–53.

Heaviside, O. (1951). *Electromagnetic Theory*, Volume I (first edition 1893; second edition 1951), London: E.&F.N. SPON LTD.

Horn, J. (1899). Ueber eine lineare differentialgleichung zweiter ordnung mit einem willkürlichen parameter. *Math. Ann.* **52**:271 – 292.

Horn, R.A. and Johnson, C.R. (1985). *Matrix Analysis*, New York: Cambridge University Press.

Kittel, C. (1966). *Introduction to Solid State Physics*, New York: John Wiley and Sons.

Korn, G.A. and Korn, T.M. (1961). *Mathematical Handbook for Scientists and Engineers*, New York: McGraw-Hill.

Kuznetsov, D.B. and Schutt-Ainé, J.E. (1996). Optimal transient simulation of transmission lines. *IEEE Trans. Circ. and Syst.-I* **CAS-43**:110–121.

Kuznetsov, D.B. (1998). Efficient circuit simulation of nonuniform transmission lines. *IEEE Trans. Micr. Theory and Techn.* **46**:546–550.

Lakshmikantham, V. and Leela, S. (1969). *Differential and Integral Inequalities*, New York: Academic Press.

Lancaster, P. and Tismenetsky, M. (1985). *The Theory of Matrices*, New York: Academic Press.

Landau, L.D. and Lifshitz, E.M. (1958). *Quantum Mechanics*, Oxford: Pergamon Press.

Lefschetz, S. (1977). *Differential Equations: Geometric Theory*, New York: Dover Publications.

Lighthill, M.J. (1958). *Introduction to Fourier Analysis and Generalised Funcions*, Cambridge: Cambridge University Press.

Lin, S. and Kuh, E.S. (1992). Transient simulation of lossy interconnects based on the recursive convolution formulation. *IEEE Trans. Circ. and Syst.-I* **39**(11):879–892.

Lindell, I.V. and Gu, Q. (1987). Theory of time-domain quasi-TEM modes in inhomogeneous multiconductor lines. *IEEE Trans. Microwave Theory and Techniques* **MTT-35**:893–897.

Linz, P. (1985). Analytical and numerical methods for Volterra equations, *SIAM Studies in Applied Mathematics*, Philadelphia.

Lu, K. (1997). An efficient method for analysis of arbitrary nonuniform transmission lines. *IEEE Trans. Micr. Theory and Techn.* **45**:9–14.

Maffucci, A. and Miano, G. (1998). On the dynamic equations of linear multiconductor transmission lines with terminal nonlinear multiport resistors. *IEEE Trans. Circ. and Syst.-I* **CAS-45**:812–829.

Maffucci, A. and Miano, G. (1999a). A new method to evaluate the impulse responses of multiconductor lossy transmission lines, *13th International Symposium on Electromagnetic Compatibility*, Zurich, pp. 121–126, February.

Maffucci, A. and Miano, G. (1999b). Irregular terms in the impulse response of a multiconductor lossy transmission line. *IEEE Trans. Circ. and Syst.-I* **46**:788–805.

Maffucci, A. and Miano, G. (1999c). On the uniqueness of the numerical solution of nonlinearly loaded lossy transmission lines. *International Journal of Circuit Theory and Applications* **27**:455–472.

Maffucci, A. and Miano, G. (1999d). Representation in time domain of nonuniform two-conductor transmission lines as two-ports. Two-port characterization of nonuniform transmission lines through WKB method, *Internal Reports*, Department of Electrical Engineering, University of Naples Federico II, October.

Maffucci, A. and Miano, G. (2000a). An accurate time-domain model of transmission lines with frequency-dependent parameters. *International Journal on Circuit Theory and Applications*, **28**:263–280.

Maffucci, A. and Miano, G. (2000b). Circuital representation of nonuniform transmission lines. *Proc. the 4th IEEE Workshop on Signal Propagation on Interconnects,* Magdeburg, Germany, 17–19 May.

Maio, I., Pignari, S., and Canavero, F. (1994). Influence of the line characterization on the transient analysis of nonlinearly loaded lossy transmission lines. *IEEE Trans. Circ. and Syst.-I* **CAS-41**:197 – 209.

Marcuwitz, N. and Schwinger, J. (1951). On the representation of the electric and magnetic fields produced by currents and discontinuities in waveguides. *Jour. Applied Physics* **22**:806–819.

Mao, J.F. and Li, Z.F. (1992). Analysis of the time response of multiconductor transmission lines with frequency-dependent losses by the method of convolution-characteristics. *IEEE Trans. Microwave Theory and Techniques*. **40**(4):637–644.

Mao, J.F. and Kuh, E.S. (1997). Fast simulation and sensitivity analysis of lossy transmission lines by the method of characteristics. *IEEE Trans. Circuits and Systems-I* **44**(5):391–401.

Marx, K.D. (1973). Propagation modes, equivalent circuits, and characteristic terminations for multiconductor transmission lines with inhomogeneous dielectrics. *IEEE Trans. Microwave Theory and Tech.* **MTT-21**:450–457.

Matick, R.E. (1995). *Transmission Lines for Digital and Communication Networks*, New York: IEEE Press.

May, R.M. (1976). Simple mathematical models with very complicated dynamics. *Nature* **261**:459–467.

Miano, G. (1997). Uniqueness of solution for linear transmission lines with nonlinear terminal resistors. *IEEE Trans. Circuit and Systems-I* **44**(7):569–582.

Mira, C. (1987). *Chaotic Dynamics*, Singapore: World Scientific.

Montrose, M.I. (1998). *EMC and the Printed Circuit Board*, New York: IEEE Press.

Moon, F.C. (1992). *Chaotic and Fractal Dynamics*, New York: John Wiley & Sons.

Motorola, Inc. (1989). Transmission line effects in PCB applications, AN1051/D.

Nagumo, J. and Shimura, M. (1961). Self-oscillation in a transmission line with a tunnel diode. *Proc. IRE* 1281–1291.

Nahman, N. S. and Holt, D. R. (1972). Transient analysis of coaxial cables using the skin effect approximation $A + B\sqrt{s}$. *IEEE Trans. Circ. Theory* **19**(5):443–451.

Nayfeh, A.H. (1973). *Perturbation Methods*, New York: John Wiley & Sons.

Olyslager, F., De Zutter, D., and de Hoop, A.T. (1994). New reciprocal circuit model for lossy waveguide structures based on the orthogonality of the eigenmodes. *IEEE Trans. Microwave Theory and Tech.* **42**:2261–2269.

Orhanovic, N., Tripathi, V.K., and Wang, P. (1990). Time domain simulation of uniform and nonuniform multiconductor lossy lines by the method of characteristics. *IEEE MTT-S Intern. Microwave Symp. Digest*, 1191–1194.

Ortega, J.M. and Rheinboldt, W.C. (1970). *Iterative Solution of Nonlinear Equations in Several Variables*, New York: Academic Press.

Ott, E. (1993). *Chaos in Dynamical Systems*, New York: Cambridge University Press.

Paul, C.R. (1992). *Introduction to Electromagnetic Compatibility*, New York: John Wiley & Sons.

Paul, C.R. (1994). *Analysis of Multiconductor Transmission Lines*, New York: John Wiley & Sons.

Peitgen, H.O. and Richter, P.H. (1986). *The Beauty of Fractals*, Berlin: Springer.

Romeo, F. and Santomauro, M. (1987). Time domain-simulation of n-coupled transmission lines. *IEEE Trans. Microwave Theory Tech.* **35**:131–137.

Roychowdhury, J.S., Newton, A.R., and Pederson, D.O. (1994). Algorithms for the transient simulation of lossy interconnect. *IEEE Trans. Computer Aided Design* **CAD-13**:96–104.

Schelkunoff, S.A. (1955). Conversion of Maxwell's equations into generalized telegraphists's equations. *Bell Syst. Tech. J.* **34**:995 – 1043.

Schutt-Ainé, J.E. and Mittra, R. (1988). Scattering parameter transient analysis of transmission lines loaded with nonlinear terminations. *IEEE Trans. Microwave Theory Tech.* **MTT-36**:529–536.

Schutt-Ainé, J.E. (1992). Transient analysis of nonuniform transmission lines. *IEEE Trans. Circ. Syst.-I* **39**:378–385.

Schuster, H.G. (1988). *Deterministic Chaos*, Veinheim: VCH.

Semlyen, A. and Dabuleanu, A. (1975). Fast and accurate switching transient calculations on transmission lines with ground return using recursive convolutions. *IEEE Trans. Power Apparatus Systems* **PAS-94**:561–571.

Sharkovsky, A.N., Mastrenko, Y., Deregel, P., and Chua, L.O. (1993). Dry turbulence from a time-delayed Chua's circuit. *J. Circuits Syst. Comp.* **3**:645–668.

Silvester, P.P. and Ferrari, R.L. (1990). *Finite Elements for Electrical Engineers*, New York: Cambridge University Press.

Singer, D. (1978). Stable orbits and bifurcations of maps of the interval. *SIAM Jour. Appl. Math.* **35**:260–267.

Smirnov, V.I. (1964a). *A Course Of Higher Mathematics*, Volume II, Oxford: Pergamon Press.

Smirnov, V.I. (1964b). *A Course Of Higher Mathematics*, Volume III, Part 2, Oxford: Pergamon Press.

Stratton, J.A. (1941). *Electromagnetic Theory*, New York: McGraw-Hill.

Sugai, I. (1960). D'Alembert's method for nonuniform transmission lines, *Proc. IRE* **48**:823–824.

Sugai, I. (1961). A new exact method of nonuniform transmission lines. *Proc. IRE* **49**:627–628.

Sunde, E.D. (1968). *Earth Conduction Effects in Transmission Systems*, New York: Dover.

Thompson, J.M.T. and Stewart, H.B. (1986). *Nonlinear Dynamics and Chaos*, New York: John Wiley and Sons.

Tripathi, V.K. and Sturdivant, R. (Guest Ed.). (1997). *Special Issue on Interconnects and Packaging. IEEE Trans. Microwave Theory and Techniques* **45**(10).

Weinberg, S. (1995). *The Quantum Theory of Fields: Foundations*, New York: Cambridge University Press.

Whitham, G.B. (1974). *Linear and Nonlinear Waves*, New York: John Wiley & Sons.

Witt, A. (1937). Sur la théorie de la corde de violon, *Tech. Phys. USSR*, **4**:261–288.

Yu, Q. and Kuh, E.S. (1996). An accurate time domain interconnect model of transmission line networks. *IEEE Trans. Circ. and Syst.-I* **43**:200–208.

Zeidler, E. (1986). *Nonlinear Functional Analysis and its Applications*, vol. I, New York: Springer-Verlag.

Index